Learning and Analytics in Intelligent Systems

Volume 1

Series Editors

George A. Tsihrintzis, University of Piraeus, Piraeus, Greece
Maria Virvou, University of Piraeus, Piraeus, Greece
Lakhmi C. Jain, Faculty of Engineering and Information Technology, Centre for
Artificial Intelligence, University of Technology Sydney, NSW, Australia;
University of Canberra, Canberra, ACT, Australia; KES International,
Shoreham-by-Sea, UK; Liverpool Hope University, Liverpool, UK

The main aim of the series is to make available a publication of books in hard copy form and soft copy form on all aspects of learning, analytics and advanced intelligent systems and related technologies. The mentioned disciplines are strongly related and complement one another significantly. Thus, the series encourages cross-fertilization highlighting research and knowledge of common interest. The series allows a unified/integrated approach to themes and topics in these scientific disciplines which will result in significant cross-fertilization and research dissemination. To maximize dissemination of research results and knowledge in these disciplines, the series publishes edited books, monographs, handbooks, textbooks and conference proceedings.

More information about this series at http://www.springer.com/series/16172

George A. Tsihrintzis · Maria Virvou ·
Evangelos Sakkopoulos ·
Lakhmi C. Jain
Editors

Machine Learning Paradigms

Applications of Learning and Analytics in Intelligent Systems

 Springer

Editors
George A. Tsihrintzis
University of Piraeus
Piraeus, Greece

Evangelos Sakkopoulos
University of Piraeus
Piraeus, Greece

Maria Virvou
Department of Informatics
University of Piraeus
Piraeus, Greece

Lakhmi C. Jain
Faculty of Engineering and Information
Technology
Centre for Artificial Intelligence
University of Technology Sydney
NSW, Australia

University of Canberra
Canberra, ACT, Australia

KES International
Shoreham-by-Sea, UK

Liverpool Hope University
Liverpool, UK

ISSN 2662-3447   ISSN 2662-3455   (electronic)
Learning and Analytics in Intelligent Systems
ISBN 978-3-030-15630-5   ISBN 978-3-030-15628-2   (eBook)
https://doi.org/10.1007/978-3-030-15628-2

Library of Congress Control Number: 2019935546

This Springer imprint is published by the registered company Springer Nature Switzerland AG
The registered company address is: Gewerbestrasse 11, 6330 Cham, Switzerland

To our beloved daughters, Evina,
Konstantina and Andreani
George A. Tsihrintzis and Maria Virvou

To my beloved family
Evangelos Sakkopoulos

To my beloved family
Lakhmi C. Jain

Foreword

Systems operating in the real world are most often faced with modeling inaccuracies due to nonlinearities, chaotic phenomena, presence of a plethora of degrees of freedom, inherent uncertainties, time variability of the context, interference, and noise. In order for systems to perform the tasks for which they are designed with accuracy and efficiency, intelligence needs to be incorporated in them in the form of modules which continuously collect and process data from the environment. At the same time, *Learning* and *Analytics* have emerged as two active technological research areas which may allow *Intelligent Systems* to track change, to self-adapt, and to improve their task execution performance.

Intelligent Systems, Learning, and Analytics are strongly related scientific disciplines and complement one another significantly. Thus, a unified/integrated approach to themes and topics in these disciplines will result in significant cross-fertilization, research advancement, and new knowledge creation.

The book at hand constitutes the inaugurating volume in the new Springer series on **Learning and Analytics in Intelligent Systems**. Specifically, the Editors have included an editorial note (Chapter "Applications of Learning and Analytics in Intelligent Systems"), as well as an additional 15 chapters covering *Applications of Learning and Analytics in Intelligent Systems* in seven areas, namely:

- *Learning and Analytics in Intelligent Medical Systems.*
- *Learning and Analytics in Intelligent Power Systems.*
- *Learning and Analytics in Performance Assessment.*
- *Learning and Analytics in Intelligent Safety and Emergency Response Systems.*
- *Learning and Analytics in Intelligent Social Media.*
- *Learning and Analytics in Intelligent Imagery and Video.*
- *Learning and Analytics in Integrated Circuits.*

The book audience includes professors, graduate students, practitioners, and researchers in *Applications of Learning and Analytics in Intelligent Systems* and other related areas. As such, it is self-contained and its chapters are appropriately grouped into seven parts. An extensive list of references at the end of each chapter guides readers to probe further into application areas of interest to them.

I believe that the Editors have done an outstanding job in addressing the pertinent topics and associated problems. I consider the book to be a great addition to the area of *Applications of Learning and Analytics in Intelligent Systems*. I am confident that it will help professors, graduate students, researchers, and practitioners to understand and explore further Learning and Analytics methods and apply them in real-world Intelligent Systems.

Patras, Greece   Dr. Vassilios Verykios
Professor in the School of Science and Technology
Director of the Big Data Analytics and Anonymization Lab
Hellenic Open University

Preface

Learning and *Analytics* are two technological areas which have been attracting very significant research interest over the recent years.

On one hand, in particular, there is significant research effort worldwide to incorporate *learning* abilities into machines, i.e., develop mechanisms, methodologies, procedures, and algorithms that allow machines to become better and more efficient at performing specific tasks, either on their own or with the help of a supervisor/instructor. The umbrella term used for all the various relevant approaches is *Machine Learning*.

On the other hand, *(Data) Analytics* is the term devised to describe specialized processing techniques, software, and systems aiming at extracting information from extensive data sets and enabling their users to draw conclusions, to make informed decisions, to support scientific theories, and to manage hypotheses.

Intelligent Systems are systems which collect data from their environment, process them, and, in response, self-adapt so as to perform the tasks for which they are designed with ever-increasing accuracy and efficiency. Needless to say, intelligent systems can benefit significantly from incorporation of learning and analytics modules into them.

The volume at hand constitutes the inaugurating volume in the new Springer series on *Learning and Analytics in Intelligent Systems*. The new series aims mainly at making available a publication of books in hard copy form and soft copy form on all aspects of learning, analytics, and advanced intelligent systems and related technologies. The mentioned disciplines are strongly related and complement one another significantly. Thus, the series encourages cross-fertilization highlighting research and knowledge of common interest. The series allows a unified/integrated approach to themes and topics in these scientific disciplines which will result in significant cross-fertilization and research dissemination. To maximize dissemination of research results and knowledge in these disciplines, the series will publish edited books, monographs, handbooks, textbooks, and conference proceedings.

The research book at hand is directed towards professors, researchers, scientists, engineers, and students of all disciplines. An extensive list of references at the end of each chapter guides readers to probe further into the application areas of interest to them. We hope that they all find it useful in their works and researches.

We are grateful to the authors and the reviewers for their excellent contributions and visionary ideas. We are also thankful to Springer for agreeing to publish this book. Last, but not least, we are grateful to the Springer staff for their excellent work in producing this book.

Piraeus, Greece   George A. Tsihrintzis
Piraeus, Greece   Maria Virvou
Piraeus, Greece   Evangelos Sakkopoulos
Canberra, Australia   Lakhmi C. Jain

Contents

Chapter 1
Applications of Learning and Analytics in Intelligent Systems

George A. Tsihrintzis, Maria Virvou, Evangelos Sakkopoulos
and Lakhmi C. Jain

Abstract The book at hand constitutes the inaugural volume in the new Springer series on *Learning and Analytics in Intelligent Systems*. The series aims at making available a publication of books in hardcopy and soft-copy form on all aspects of learning, analytics and advanced intelligent systems and related technologies. These disciplines are strongly related and complement one another significantly. Thus, the new series encourages a unified/integrated approach to themes and topics in these disciplines which will result in significant cross-fertilization, research advancement and new knowledge creation. To maximize dissemination of research results and knowledge, the series will publish edited books, monographs, handbooks, textbooks and conference proceedings. The book at hand is directed towards professors, researchers, scientists, engineers and students. An extensive list of references at the end of each chapter guides readers to probe further into application areas of interest to them.

Learning is defined in the Cambridge Dictionary of English [1] as "*The activity of obtaining knowledge.*" A similar and perhaps a bit more general definition is given in the Oxford Dictionary of English [2] as "*The acquisition of knowledge or*

G. A. Tsihrintzis (✉) · M. Virvou · E. Sakkopoulos
University of Piraeus, Piraeus, Greece
e-mail: geoatsi@unipi.gr

M. Virvou
e-mail: mvirvou@unipi.gr

E. Sakkopoulos

e-mail: sakkopul@gmail.com

L. C. Jain
University of Technology Sydney, NSW, Australia
e-mail: jainlakhmi@gmail.com

Liverpool Hope University, Liverpool, UK
e-mail: jainlc2002@yahoo.co.uk

University of Canberra, Canberra, Australia

© Springer Nature Switzerland AG 2019
G. A. Tsihrintzis et al. (eds.), *Machine Learning Paradigms*,
Learning and Analytics in Intelligent Systems 1,
https://doi.org/10.1007/978-3-030-15628-2_1

skills through study, experience, or being taught." That is, a *learner* is involved in a *learning paradigm*, who seeks to expand his/her knowledge and/or improve his/her skills and/or become better and more efficient at performing specific tasks. There are various learning paradigms, in which the learner works either on his/her own or with the help of a supervisor/instructor.

In most circumstances, a learning paradigm refers to a process oriented towards a *human* learner. Such learning paradigms are quite common and include self-study or instructor-based frameworks. Since the late 1950s, however, learning paradigms have appeared for the training of machines in performing simpler or more complicated tasks. Indeed, Arthur Samuel seems to be the first to coin the term *Machine Learning* to describe learning paradigms in which "*computers [attain] the ability to learn without being explicitly programmed*" [3].

Since the appearance of the term *Machine Learning*, extensive research has been conducted towards achieving its goal. Various paradigms have appeared, as a one-fits-all solution (a "panacea") has not yet been discovered and most probably does not exist. It is not surprising that not a single machine learning paradigm is most suitable for all classes of learning tasks. In biological and naturally-occurring mechanisms, a similar learning variety is observed. For example, humans possess biological neural networks to learn and process acoustic, visual and other signals [4], while at the same time they have an immune system to recognize antigens and produce a proper response to them [5]. Similarly, naturally-occurring genetic (evolutionary) processes are very common [6, 7], while swarm intelligence optimization procedures are used by ant and termite colonies and bird and fish swarms [8]. In Machine Learning, we often mimic these and other biological and naturally-occurring processes in such advanced learning frameworks as Deep Learning [9], Artificial Immune Systems [10], Artificial Chemistry [11] or Artificial Life [12].

At the same time, the 4th Industrial Revolution seems to be unfolding, characterized by synergies between physical, digital, biological and energy sciences and technologies, which are sewn together by the availability of extensive and diverse amounts of *data* [13–15].

Data Analytics is the term devised to describe specialized processing techniques, software and systems aiming at extracting information from extensive data sets and enabling their users to draw conclusions, to make informed decisions, to support scientific theories and to manage hypotheses [16–18]. The need for Data Analytics arises in most modern scientific disciplines, including engineering, natural, computer and information sciences, education, economics, business, commerce, environment, healthcare, and life sciences.

Ultimately, learning and analytics need to be embedded in *Intelligent Systems*, i.e. systems which collect data from their environment, process them and, in response, self-adapt so as to perform the tasks for which they are designed with ever increasing accuracy and efficiency [19].

The book at hand explores many of the scientific and technological areas in which Learning and Analytics are incorporated in Intelligent Systems and, thus, may play a significant role in the years to come. The book comes as the fifth volume under the general title MACHINE LEARNING PARADIGMS and follows two related monographs [10, 20] and two edited collections of chapters [21, 22].

More specifically, the book at hand consists of an editorial chapter (Chap. 1) and an additional fifteen (15) chapters. All chapters in the book were invited from authors who work in the corresponding areas of Learning and Analytics as applied on Intelligent Systems and are recognized for their research contributions. In more detail, the chapters in the book are organized into seven parts, as follows:

The first part of the book consists of three chapters devoted to *Learning and Analytics in Intelligent Medical Systems*.

Specifically, Chap. 2, by Anna Karen Garate Escamilla, Amir Hajjam El Hassani and Emmanuel Andres, is on *"A Comparison of Machine Learning Techniques to Predict the Risk of Heart Failure."* The authors explore supervised machine learning techniques used in different investigations to predict whether a patient will have a heart failure or not.

Chapter 3, by Jose Liñares Blanco, Marcos Gestal, Julián Dorado and Carlos Fernandez-Lozano, is on *"Differential Gene Expression Analysis of RNA-seq Data using Machine Learning for Cancer Research."* The authors address Machine Learning-based approaches for gene expression analysis using RNA-seq data for cancer research and compare it with a classical gene expression analysis approach.

Chapter 4, by Evangelos Karampotsis, George Dounias and Jan Jantzen, is on *"Machine Learning Approaches for Pap-Smear Diagnosis: An Overview."* The authors apply Machine Learning techniques in medical data analysis problems, with the goal to optimize the Pap-Smear or Pap-Test diagnosis.

The second part of the book consists of two chapters devoted to *Learning and Analytics in Intelligent Power Systems*.

Specifically, Chap. 5, by Miltiadis Alamaniotis, is on a *"Multi-Kernel Analysis Paradigm Implementing the Learning from Loads Approach for Smart Power Systems."* The author presents a new machine learning paradigm, focusing on the analysis of recorded electricity load data. The presented paradigm utilizes a set of multiple kernel functions to analyze a load signal into a set of components.

Chapter 6, by John A. Paravantis, Athanasios Ballis, Nikoletta Kontoulis, and Vasilios Dourmas, is on *"Conceptualizing and Measuring Energy Security: Geopolitical Dimensions, Data Availability, Quantitative and Qualitative Methods."* The authors discuss how empirical geographical, energy, socioeconomic, environmental and geopolitical data could be collected and a novel geopolitical energy security index could be developed with reference to appropriate statistical techniques.

The third part of the book consists of three chapters devoted to *Learning and Analytics in Performance Assessment*.

Specifically, Chap. 7, by Apostolos Meliones and George Makrides, is on *"Automated Stock Price Motion Prediction Using Technical Analysis Datasets and Machine Learning."* The authors study, develop and evaluate a prototype stock motion prediction system using Technical Analysis theory and indicators, in which the process of forecasting is based on Machine Learning.

Chapter 8, by Athanasios T. Ballis and John A. Paravantis, is on *"Airport Data Analysis Using Common Statistical Methods and Knowledge-Based Techniques."*

The authors outline an analysis performed for 28 Greek island airports aiming at revealing key aspects of demand (seasonality and connectivity) and of supply (practical capacity of airports, characteristics of airplanes and air services).

Chapter 9, by Gregory Koronakos, is on "*A Taxonomy and Review of the Network Data Envelopment Analysis Literature.*" The author describes the underlying notions of network DEA methods and their advantages over the classical DEA ones. He also conducts a critical review of the state-of-the art methods in the field and provides a thorough categorization of a great volume of network DEA literature in a unified manner.

The fourth part of the book contains two chapters devoted to **Learning and Analytics in Intelligent Safety and Emergency Response Systems**.

Specifically, Chap. 10, authored by Joao Rico, Jose Barateiro, Juan Mata, Antonio Antunes and Elsa Cardoso, is on "*Applying Advanced Data Analytics and Machine Learning to Enhance the Safety Control of Dams.*" The authors present the full data lifecycle in the safety control of large-scale civil engineering infrastructures (focused on dams), from the data acquisition process, data processing and storage, data quality and outlier detection, and data analysis. A strong focus is made on the use of machine learning techniques for data analysis, where the common multiple linear regression analysis is compared with deep learning strategies, namely recurrent neural networks.

Chapter 11, by Minsung Hong and Rajendra Akerkar, is on "*Analytics and Evolving Landscape of Machine Learning for Emergency Response.*" The authors review the use of machine learning techniques to support the decision-making processes for the emergency management and discuss their challenges.

The fifth part of the book contains two chapters devoted to **Learning and Analytics in Intelligent Social Media**.

Specifically, Chap. 12, authored by Paraskevas Koukaras and Christos Tjortjis, is on "*Social Media Analytics, Types and Methodology.*" The authors discuss, elaborate and categorize various relevant concepts as they arise in mining tasks (supervised and unsupervised), while presenting the required process and its steps to analyze data retrieved from Social Media ecosystems.

Chapter 13, by Athanasia Kolovou, is on "*Machine Learning Methods for Opinion Mining in Text: The Past and the Future.*" The author presents a comprehensive overview of the various methods used for sentiment analysis and how they have evolved in the age of big data.

The sixth part of the book contains two chapters devoted to **Learning and Analytics in Intelligent Imagery and Video**.

Specifically, Chap. 14, by Jeffrey W. Tweedale, is on "*Ship Detection using Machine Learning and Optical Imagery in the Maritime Environment.*" The author assesses five Convolutional Neural Network models for their ability to identify ship types in images from remote sensors.

Chapter 15, by Iyiola E. Olatunji and Chun-Hung Cheng, is on "*Video Analytics for Visual Surveillance and Applications: An Overview and Survey.*" The authors present a detailed survey of advances in video analytics research and its subdomains such as behavior analysis, moving object classification, video summarization, object detection, object tracking, congestion analysis, abnormality detection and information fusion from multiple cameras.

Finally, the seventh part of the book contains one chapter devoted to *Learning and Analytics in Integrated Circuits*.

Specifically, Chap. 16, by John Liaperdos, Angela Arapoyanni and Yiorgos Tsiatouhas, is on *"Machine Learning in Alternate Testing of Integrated Circuits."* The authors present the Alternate Test paradigm is presented as it is probably the most well-established machine learning application in the field of integrated circuit testing.

In this book, we have presented some of the emerging scientific and technological areas in which Learning and Analytics arise as tools for building better Intelligent Systems. The book comes as the fifth volume under the general title MACHINE LEARNING PARADIGMS, following two related monographs [17, 18] and two edited collections of chapters [19, 20]. Societal demand continues to pose challenging problems, which require ever more efficient tools, new methodologies, and intelligent systems to de devised to address them. Machine Learning is a relatively new technological area, which has been receiving a lot of interest worldwide in terms both of research and applications. Thus, the reader may expect that additional volumes on other aspects of Machine Learning Paradigms and their application areas will appear in the near future.

References

1. https://dictionary.cambridge.org/dictionary/english/learning?q=LEARNING
2. https://en.oxforddictionaries.com/definition/learning
3. A. Samuel, Some studies in machine learning using the game of checkers. IBM J. **3**(3), 210–229 (1959)
4. https://en.wikipedia.org/wiki/Neural_network
5. https://en.wikipedia.org/wiki/Immune_system
6. https://en.wikipedia.org/wiki/Evolutionary_computation
7. D.B. Fogel, What is evolutionary computation?, IEEE Spectr. 26–32 (2000)
8. https://en.wikipedia.org/wiki/Swarm_intelligence
9. J. Patterson, A. Gibson, *Deep Learning: A Practicioner's Approach* (O' Reilly Media, California, USA, 2017)
10. D.N. Sotiropoulos, G.A. Tsihrintzis, Machine learning paradigms—artificial immune systems and their applications in software personalization, volume 118, in *Intelligent Systems Reference Library Book Series* (Springer, 2017)
11. W. Banzhaf, L. Yamamoto, *Artificial Chemistries* (MIT Press, 2015)
12. C. Adami, *Introduction to Artificial Life* (Springer, New York, 1998)
13. K. Schwabd, The fourth industrial revolution—what it means and how to respond, in *Foreign Affairs*, December 12, 2015, (https://www.foreignaffairs.com/articles/2015-12-12/fourth-industrial-revolution)
14. J. Toonders, Data is the new oil of the digital economy, *Wired* (https://www.wired.com/insights/2014/07/data-new-oil-digital-economy/)
15. https://www.economist.com/news/leaders/21721656-data-economy-demands-new-approach-antitrust-rules-worlds-most-valuable-resource
16. https://en.wikipedia.org/wiki/Data
17. https://en.wikipedia.org/wiki/Data_analysis

18. https://searchdatamanagement.techtarget.com/definition/data-analytics
19. G.A. Tsihrintzis, M. Virvou, L.C. Jain (eds.), Intelligent computing systems: emerging application areas, vol. 627 in *Studies in Computational Intelligence Book Series* (Springer, 2016)
20. A.S. Lampropoulos, G.A. Tsihrintzis, Machine learning paradigms—applications in recommender systems, volume 92, in *Intelligent Systems Reference Library Book Series* (Springer, 2015)
21. G.A. Tsihrintzis, D.N. Sotiropoulos, L.C. Jain (eds.), Machine learning paradigms—advances in data analytics, volume 149, in *Intelligent Systems Reference Library Book Series* (Springer, 2019)
22. M. Virvou, E. Alepis, G.A. Tsihrintzis, L.C. Jain (eds.), Machine learning paradigms—advances in learning analytics, volume 158, in *Intelligent Systems Reference Library Book Series* (Springer, 2019)

Part I
Learning and Analytics in Intelligent
Medical Systems

Chapter 2
A Comparison of Machine Learning Techniques to Predict the Risk of Heart Failure

Anna Karen Garate Escamilla, Amir Hajjam El Hassani
and Emmanuel Andres

Abstract The chapter explores the supervised machine learning techniques used in different investigations to predict whether a patient will have a heart failure or not. The research focuses on the articles that used the "Cleveland Heart Disease Data Set" from the UCI Machine Learning Repository and performs a comparison of the different techniques used to find the best performance. A conclusion of the best technique is also provided in this chapter. Some examples of the techniques are C4.5 tree, Naïve Bayes, Bayesian Neural Networks (BNN), Support Vector Machine (SVM), Artificial Neural Network (ANN), Nearest Neighbor (KNN).

Keywords Machine learning · Heart failure · Classification · Hybrid algorithm

2.1 Introduction

The WHO (World Health Organization) lists cardiovascular diseases as the leading cause of death worldwide [1] which represents 31% of people dying every year. Heart Failure (HF) is a known condition of cardiovascular disease, it occurs when the heart cannot pump enough blood to support the body. Learning the symptoms and characteristics of a heart disease is vital for experts in the medical field to acknowledge the risk of the patients. Among the most correlated factors affecting the heart diseases are the alcohol consumption, the use of tobacco, the lack of exercise, an unhealthy diet, a high blood pressure, diabetes, a high total of cholesterol, a family history of heart disease, an overweight and obesity [1]. Also, the American Heart

A. K. Garate Escamilla · A. Hajjam El Hassani (✉)
Nanomedicine Lab, Univ. Bourgogne Franche-Comte, UTBM, 90010 Belfort, France
e-mail: amir.hajjam@utbm.fr

E. Andres
Service de Médecine Interne, Diabète et Maladies métaboliques de la Clinique Médicale B,
CHRU de Strasbourg, Strasbourg, France

Faculté de Médecine de Strasbourg, Centre de Recherche Pédagogique en Sciences de la Santé,
Université de Strasbourg (UdS), Strasbourg, France

© Springer Nature Switzerland AG 2019
G. A. Tsihrintzis et al. (eds.), *Machine Learning Paradigms*,
Learning and Analytics in Intelligent Systems 1,
https://doi.org/10.1007/978-3-030-15628-2_2

Association lists symptoms of HF, such as shortness of breath, irregular heartbeats, weight gain (1 or 2 kg per day), weakness, difficulty sleeping, swelling of the legs, nausea, sweating, chronic cough and high heart rate [2]. The symptoms of HF are not easy to be seen by practitioners for their nature of being common or confused with the signs of aging.

The HF condition becomes more sever in the last stages; patients need a constant monitoring to do not risk their own life. Therefore, early diagnosis of patients is vital. In recent years, the availability of new medical data has become a new opportunity for health specialists to improve diagnosis, especially cardiovascular diseases. In hospitals, practitioners have increased the use of computer technologies and IT tools to have a better decision-making and avoid errors caused by poor data processing.

Machine Learning has positioned itself as an important solution to assist the diagnosis of patients in the medical industry. The term machine learning is applied when the task to be performed is complex and long to be programmed, such as medical records, analysis of electronic health information, predictions of pandemics, analysis of operational costs and human genomics [3]. Machine learning is an analytical tool which learns from the experience and help users identifies patterns without being programmed.

Currently there are different studies intended to predict and diagnose heart diseases using machine learning techniques. This study [4] proposed a classification of electrocardiogram (ECG) signals through a deep neural network (DNN), with an improved in accuracy and a faster computation. Other research [5] classified cardiac arrhythmias using computer-aided diagnostic (CAD) systems to diagnose cardiovascular disorders. The CAD system classified five type beats: normal (N), premature ventricular contraction (PVC), premature atrial contraction (APC), left bundle branch block (LBBB) and right bundle branch block (RBBB). The prediction is with the use of support vector machine (SVM), obtaining an accuracy of 98.60% $\pm$ 0.20% with raw data, 96.30% $\pm$ 0.44% with PCA and 97.60% $\pm$ 0.19% with Fisher Score (FS). The authors [6] proposed a clinical decision support system (CDSS) for the analysis of HF, which compares the performance of neural network (NN), the support vector machine (SVM), a system with fuzzy rules, classification and regression tree (CART) and random forest. The best accuracy was 87.6% using the CART model. This article [7] used a SVM technique to diagnose heart disease in patients with diabetes, obtaining an accuracy of 94.60% and predicting features such as age, sex, blood pressure and blood sugar.

The present work focused on machine learning classification techniques that are able to predict whether a patient has a HF. To achieve this, a comparison of the different techniques is made using the same data set. The Cleveland data set from the UCI Machine Learning Repository [8] is a data set often referenced in the literature when the accuracy performed is needed. Twenty-five studies were selected to compare which classification method had the best accuracy. Some of the techniques are: decision tree, Naïve Bayes, neuronal network, random forest, k-nearest neighbors, logistic regression and hybrid algorithm.

Remaining of this paper is organized as follows. A short summary of machine learning is explained in Sect. 2. Detail descriptions of the methodology are presented in Sect. 3. An overview of the proposed systems is reported in Sect. 4. Finally, Sect. 5 concludes the work.

2.2   Theoretical Background

Machine Learning (ML) is a method employed in data science to learn from past experiences after being programmed. The information is constantly growing, that nowadays the traditional methods cannot be used to solve the problems of the world. The analysis of the data aims to predict real models and needs a system that can be reliable, solid and trustful, such as ML [9].

Machine Learning works with an input of training data set after learning the structure patterns from the data. In this learning phase, the output is a model that will be used in the resting phase. The testing phase uses another data set and test it in the model built in the training phase, the output of this phase is the prediction.

Machine Learning algorithms perform with two types of data: labeled data and unlabeled data. Supervised learning works with labeled data, when the features are given. The labeled data can be numerical or categorical. Regression algorithms are used to make predictions with numerical data and classification algorithms with categorical data. Supervised learning entails a learning map between a set of input X variables and quantitative output Y or qualitative output G (referred for group), this mapping predicts the output for unknown data. Unsupervised learning used unlabeled data to identify the patterns present in the data set, mostly clustering algorithms. Unsupervised learning does not have an explicit output or an evaluation associated with each input, rather it brings the aspects of the structure of the input that can be captured in the output. Classification models are widely used in healthcare area for the prediction of medical diseases.

2.3   Data and Methods

2.3.1   Dataset

The data set used in the research is the Cleveland data set from the UCI Machine Learning Repository. The data set contains 76 features, but most of the existing articles used only the subset of 14 features described in Table 1. The categorical feature is Num and contains if a patient has a presence or absence of a heart disease. The values 1–4 of the original data set were transformed in one value that is the presence (1) of a heart disease and absence continues as a value 0.

The data set most used in other studies is the Cleveland data set with 303 patients. The distribution of class is the 54% for heart disease absence and the 46% for heart

Table 1 Features of the heart disease data set

Number	Code	Feature	Description	Domain	Data type
1	Age	Age	Age in years	28–77	Real
2	Sex	Sex	1 = male; 0 = female	0, 1	Binary
3	Cp	Chest pain type	1 = typical angina; 2 = atypical angina; 3 = non-angina pain; 4 = asymptomatic	0, 1, 2, 3, 4	Categorical
4	Trestbps	Resting blood pressure (mg)	At the time of admission in hospital	80–200	Real
5	Chol	Serum cholesterol (mg)		86–603	Real
6	Fbs	Fasting blood sugar >120 mg/dl	1 = yes; 0 = no	0, 1	Binary
7	Restecg	Resting electrocardiographic results	0 = normal; 1 = ST-T wave abnormal; 2 = left ventricular hypertrophy	0, 1, 2	Categorical
8	Thalach	Maximum heart rate achieved		60–202	Real
9	Exang	Exercise induced angina	1 = yes; 0 = no	0–1	Binary
10	Oldpeak	ST depression induced by exercise relative to rest		0–6.2	Real
11	Slope	The slope of the peak exercise ST segment	1 = upsloping; 2 = flat; 3 = downsloping	1, 2, 3	Categorical
12	Ca	Number of major vessels (0–3) colored by fluoroscopy		0, 1, 2, 3	Real
13	Thal	Thal	3 = normal; 6 = fixed defect; 7 = reversible defect	3, 6, 7	Categorical
14	Num	The predicted attribute	0 = no presence; 1 = presence	0, 1	Binary

disease present. The data in the data set is mostly complete; only six instances are partially incomplete.

2.3.2   Evaluation Process

There is no homogeneous cleansing process, but there are two methods mostly used in the investigations: (1) eliminate incomplete rows and (2) fill incomplete rows with the average result.

Heart failure is a disease that causes millions of deaths. In medicine, the importance of an early diagnosis is vital for the patient. It is expected that classification algorithms will have good results and will not return to the doctors a false result. Therefore, the confusion matrix helps healthcare practitioners to have a clear idea if the system has a high performance.

The elements that contain the confusion matrix are: (1) true positive (TP), which are the patients who had a heart disease and were diagnosed correctly; (2) true negative (TN), which are the patients who did not have a heart disease and were diagnosed correctly; (3) false negative (FN), which are the patients who had heart disease and were diagnosed incorrectly; and (4) false positive (FP), which are the patients who did not have heart disease and were diagnosed incorrectly. In medical field, the FN are the most dangerous predictions.

From the confusion matrix, different performance can be calculated. For the comparison, only the accuracy is taken account since is the one most used in the investigations. Accuracy measures the instance classified correctly [10]. The formula to calculate the accuracy is given by

$$Accuracy = \frac{TP + TN}{TP + FP + FN + TN} \tag{1}$$

2.3.3   Weka

Some studies that worked with the Cleveland data set used Weka [11] for prediction. WEKA is a computer tool specialized in machine learning and developed by the University of Waikato in New Zealand. The algorithm solves practical problem of the world using Java. Weka has models for pre-processing, classification, regression, clustering, association rules and visualization.

2.4   Overview of Proposed Systems

Fifty-six models in twenty-five studies had been selected to make the comparison. It was taken into account the classification models which used the Cleveland data set with two classes. The author, name of the method, accuracy, tool, number of features and percentage of data trained and tested are used. The accuracy of the models is the most important metric for the comparison.

2.4.1 Logistic Regression

Logistic regression is a binary classification response used to describe the information and explain the relationship between the dependent variable and the independent variable. It is a linear method described by the two-class equation [12]

$$f(w) := \sum_{i=1}^{n} y_i \log p(x_i; \beta) + \log(1 - p(x_i; \beta)). \tag{2}$$

Table 2 shows the logistic regression study that used the Cleveland data set. The work of [13] computed an accuracy of 84.80% using all the features.

2.4.2 Decision Tree

The use of decision tree is widely used in ML for being easy to comprehend (explaining to someone a tree and the leaves are easier to understand block than the black-box of neural network) [9]. Decision tree break the classification into partitions (leaves) from the starting point (root). It is the most used technique in the data set selected.

The most common decision tree algorithm is ID3 (Iterative Dichotomiser 3) by Quinlan [14], which consists in generate all the possible decision trees that are correct and select the simplest. The metrics used by ID3 is called Entropy and was presented by Shannon [15], which is the amount of impurity in the features given by the equation [9]

$$Entropy(p) = -\sum_{i} p_i \log_2 p_i, \tag{3}$$

The idea is to choose the training set with which the entropy will decrease considering the features of each partitions. This is known as information gain and is defined by the equation

Table 2 Logistic regression accuracy

Author	Method	Accuracy	Tool	Features	Data set
Khanna et al. 2015 [13]	Logistic regression	84.80%	–	13	50% training, 50% testing

$$Gain(S, F) = Entrophy(S) - \sum_{f \,\epsilon\, values(F)} \frac{|S_f|}{|S|} Entropy(S_f) \qquad (4)$$

where S is the set of examples, F is a possible features, and $|S_f|$ is the number of S that have value f for F.

Previous studies used an advanced extension of ID3 called C4.5 [16]. C4.5 uses a method called post-pruning, which consists on reducing the tree by removing the preconditions and improving the accuracy on training data.

Table 3 shows the decision tree studies that used the Cleveland data set. The work of [17] computed the best result with an accuracy of 89.14% using all the features. Some of the results had accuracy between the 77 and 79%.

2.4.3   Random Forest

Random forest [18] is a collection of decision trees predictors where each tree depends on the value of an independent random vector. It gives a remarkable performance concerning practical problems. The idea is to make each tree with a random subset of data and train them. The output is the majority vote for classification to predict the class [9]. Random forest uses the technique of bagging to reduce the variance of a function.

For a classification problem, let $\hat{C}_b(x)$ be the output (predictor) of the bth random forest tree by

Table 3 Decision tree accuracy

Author	Method	Accuracy (%)	Tool	Features	Data set
Chaki et al. 2015 [39]	Decision tree C4.5	77.56	WEKA	13	–
El-Bialy et al. 2015 [36]	Decision tree C4.5	78.54	WEKA	13	–
Sen 2017 [40]	Decision tree C4.5	77.56	WEKA	13	–
Khan et al. 2016 [17]	Decision tree C4.5	89.14	WEKA	13	70% training, 30% testing
Kumar et al. 2018 [20]	Decision tree C4.5	83.40	WEKA	13	–
Olsson et al. 2015 [41]	Decision tree C4.5	78.37	WEKA	13	–
Shouman et al. 2011 [42]	Decision tree C4.5	84.10	WEKA	13	–
Srinivas et al. 2010 [24]	Decision tree C4.5	82.50	WEKA	13	–

$$\hat{C}_{rf}^{B}(x) = majority\ vote\{\hat{C}_b(x)\}_1^B \qquad (5)$$

Table 4 shows the random forest studies that used the Cleveland data set. The work of [17] computed the best result with an accuracy of 89.25% using all the features. The other results were similar with an accuracy of 81% [19] and 80% [20].

2.4.4 K-Nearest Neighbor

K-nearest neighbor (KNN) is simple classifier algorithm that finds the closet trained pointed when is asked for an unknown point and predicts the output according to the distance between them. The distance is measure by Euclidean distance method

$$d(a, b) = \sqrt{\sum_{i=1}^{n}(a_i - b_i)^2} \qquad (6)$$

where the point a and point b are two point in the space. The k-nearest neighbor formula can be compute by [9]

$$E((y - \hat{f}(x))^2) = \sigma^2 + ''\left[f(x) - \frac{1}{k}\sum_{i=0}^{k}f(x_i)\right]^2 + \frac{\sigma^2}{k} \qquad (7)$$

If k is small, there would not be a lot of neighbors that leads a high variance for the small amount of information. If k is big, the variance will decrease.

Table 5 shows the KNN studies that used the Cleveland data set. The work of [21] computed the best result with an accuracy of 97.30% using all the features. Most of the results had accuracy between the 75 and 76%.

Table 4 Random forest accuracy

Author	Method	Accuracy (%)	Tool	Features	Data set
Kodati et al. 2018 [19]	Random forest	81	WEKA	13	–
Khan et al. 2016 [17]	Random forest	89.25	WEKA	13	70% training, 30% testing
Kumar et al. 2018 [20]	Random forest	80	WEKA	13	–

Table 5 KNN accuracy

Author	Method	Accuracy (%)	Tool	Features	Data set
Acharya 2017 [36]	KNN	82	Scikit-learn	13	70% training, 30% testing
Sen 2017 [40]	K-nearest neighbour (1Bk)	76.24	WEKA	13	–
Kodati et al. 2018 [19]	K-nearest neighbour (1Bk)	75.30	WEKA	13	–
Shouman et al. 2012 [21]	KNN	97.40	–	13	–
Kumar et al. 2018 [20]	K-nearest neighbour (1Bk)	76.24	WEKA	13	–

2.4.5 Artificial Neuronal Network

Artificial neural network (ANN) is a method that emulates the way brain works. It consists of a number of nodes (neurons) that are connected through links. Just like the brain, ANN is designed to adapt and learn from past experiences. To model a neuron [9] is needed: (1) a set of weighted inputs w_i that made the synapses work; (2) an adder that sum the inputs, similar to the cell membrane to collect electrical charge; and (3) the activation function that decides if the neuron will have a peak for the inputs. The equation is computed by

$$h = \sum_{i=1}^{m} w_i x_i \qquad (8)$$

where x_i represents the set of input nodes multiplying the weight of the synapse x_i and be sum. With this information, the neurons must decide if they are going to work and produce the output or a 0.

Based on this work, to initialize an ANN is needed to set the weights w_{ij} to random numbers. The algorithm is divided in two phases. The training phase looks for T interactions or until there is not a mistake. For each input, there is required the activation of each neuron j using the activation function g given by

$$y_j = g\left(\sum_{i=0}^{m} w_{ij} x_i\right) = \begin{cases} 1 \, if \, \sum_{i=0}^{m} w_{ij} x_i > 0 \\ 0 \, if \, \sum_{i=0}^{m} w_{ij} x_i \leq 0 \end{cases} \qquad (9)$$

For the recall phase, the activation of each neuron j is given by

$$y_j = g\left(\sum_{i=0}^{m} w_{ij}x_i\right) = \begin{cases} 1 \, if \, w_{ij}x_i > 0 \\ 0 \, if \, w_{ij}x_i \leq 0 \end{cases} \tag{10}$$

Table 6 shows the ANN studies that used the Cleveland data set. The work of [22] computed the best result with an accuracy of 98.16% using all the features. Most of the results had accuracy between the 85 and 90%.

2.4.6 SVM

Support Vector Machine (SVM) was presented [23] as an impressive linear and non-linear classification algorithm. The SVM find linear boundaries in a higher space to generate a better output separation and improve the prediction. The equation that computes SVM is given by

$$min_x \frac{1}{2}x^T t_i t_j K x + q^T x \tag{11}$$

with $Gx \leq h$, $Ax = b$.

Table 7 shows the SVM studies that used the Cleveland data set. The work of [24] computed the best result with an accuracy of 88% using all the features. Most of the results had accuracy between the 84 and 88%.

Table 6 ANN accuracy

Author	Method	Accuracy (%)	Tool	Features	Data set
Khanna et al. 2015 [13]	A generalized regression neural network (GRNN)	89	–	13	50% training, 50% testing
Wadhonkar et al. 2015 [22]	Neural network (MLP)	98.16	–	13	90% training, 10% testing
Khan et al. 2016 [17]	Neural network (MLP)	92.73	WEKA	13	70% training, 30% testing
Vivekanandan et al. 2017 [22]	Neuronal network	85	–	13	–
Srinivas et al. 2010 [24]	Neural network (MLP)	89.70	WEKA	13	–
Das et al. 2009 [43]	Neural network	89.01	–	13	70% training, 30% testing
Caliskan et al. 2017 [44]	Deep neural network	85.20	–	13	70% training, 30% testing
Abushariah et al. 2014 [45]	ANN	87.04	MATLAB	13	80% training, 20% testing

2.4.7   Naïve Bayes

Naïve Bayes is a classifier based on the theorem of Bayes with strong independence assumptions between the features. It works with the assumption of using observation of the problem to make a prediction. The theorem of Bayes equation is expressed by

$$P(A|B) = \frac{P(A|B)P(A)}{P(B)} \tag{12}$$

where A and B are independent of each other, A is conditional of B, and B is conditional of A.

Table 8 shows Naïve Bayes studies that used the Cleveland data set. The work of [24] computed the best result with an accuracy of 83.7% using all the features. Most of the results had accuracy between the 80 and 84%.

2.4.8   OneR

OneR is a classifier that generates one rule for each predictor; the rule consist in select the feature with the smallest total error [25].

Table 9 shows the OneR study that used the Cleveland data set. The work of [17] computed an accuracy of 91.21% using all the features.

Table 7 SVM accuracy

Author	Method	Accuracy (%)	Tool	Features	Data set
Chaki et al. 2015 [39]	SPV (SMO)	84.12	WEKA	13	–
Khanna et al. 2015 [13]	SVM (linear)	87.60	–	13	50% training, 50% testing
Sen 2017 [40]	SPV (SMO)	84.16	WEKA	13	–
Kodati et al. 2018 [19]	SPV (SMO)	84	WEKA	13	–
Kumar et al. 2018 [20]	SPV (SMO)	84.16	WEKA	13	–
Srinivas et al. 2010 [24]	SPV (SMO)	88	WEKA	13	–

Table 8 Naïve Bayes accuracy

Author	Method	Accuracy (%)	Tool	Features	Data set
Chaki et al. 2015 [39]	Naive Bayes	83.50	WEKA	13	–
Acharya 2017 [36]	Naive Bayes	83	Adaboost	9	70% training, 30% testing
Sen 2017 [40]	Naive Bayes	83.50	WEKA	13	–
Kodati et al. 2018 [19]	Naive Bayes	83.70	WEKA	13	–
Kumar et al. 2018 [20]	Naive Bayes	83.50	WEKA	13	–
Olsson et al. 2015 [41]	Naive Bayes	79.90	WEKA	13	–
Srinivas et al. 2010 [24]	Naive Bayes	82	WEKA	13	–

2.4.9   ZeroR

ZeroR is a simple classification method that relies in the output and ignores the features [26]. ZeroR as a predictor is not strong as the other models, but its strength relay on benchmarking for other classification methods.

Table 10 shows the ZeroR study that used the Cleveland data set. The work of [17] computed an accuracy of 79.56% using all the features.

2.4.10   Hybrid

A hybrid algorithm combines two or more algorithms to make the prediction, resulting in a better algorithm. Table 11 shows the hybrid algorithms used in different research.

Table 9 OneR accuracy

Author	Method	Accuracy	Tool	Features	Data set
Khan et al. 2016 [17]	OneR	91.21%	WEKA	13	70% training, 30% testing

Table 10 ZeroR accuracy

Author	Method	Accuracy	Tool	Features	Data set
Khan et al. 2016 [17]	ZeroR	79.56%	WEKA	13	70% training, 30% testing

Table 11 Hybrid accuracy

Author	Method	Accuracy (%)	Tool	Features	Data set
Kumar-Sen et al. 2013 [27]	Neuro-Fuzzy in 2 layers	96	MATLAB	9	80% training, 20% testing
Ziasabounchi et al. 2014 [29]	ANFIS in 5 layers	85	MATLAB	7	80% training, 20% testing
Acharya 2017 [36]	SBS+SVC	82	Adaboost	9	70% training, 30% testing
Acharya 2017 [36]	SBS+Random forest	80	Adaboost	9	70% training, 30% testing
Uyar et al. 2017 [34]	GA based on RFNN	97.78	–	13	85% training, 15% testing
Bhuvaneswari et al. 2012 [33]	GA+NN	94.17	–	13	70% training, 30% testing
Santhanam et al. 2013 [37]	PCA+Regression	92	–	7 ± 1	–
Santhanam et al. 2013 [37]	PCA1+NN	95.20	–	8 ± 1	–
Parthiban et al. 2008 [35]	GA+CANFIS	82	Neuro Solution	13	–
Vivekanandan et al. 2017 [31]	Fuzzy AHP and a feed-forward neural network	83	–	9	–
Mokeddem et al. 2013 [28]	GA+Bayes Naive (BN)	85.50	–	7	–
Mokeddem et al. 2013 [28]	GA+SVM	83.82	–	7	–
Mokeddem et al. 2013 [28]	BFS+MLP	80.53	–	6	–
Mokeddem et al. 2013 [28]	BFS+C4.5	78.55	–	6	–
Abushariah et al. 2014 [28]	Neuro-Fuzzy	75.93	MATLAB	13	80% training, 20% testing
Soni et al. 2011 [38]	WAC	81.51	–	13	–

Neuro-Fuzzy (artificial neuronal networks and fuzzy logic) is one of the most used hybridization techniques in computer science. It combines the fuzzy human reasoning with the connections of neuronal networks. The studies [27] and [28] used this model to compute the accuracy of the data set with results of 96% and 75.93% respective. The study [29] used an adaptive Neuro-Fuzzy interface system called ANFIS [30] having an accuracy of 85%. Another investigation [31] used a Fuzzy AHP and a feed-forward neural network having an accuracy of 83%.

Genetic algorithm (GA) is a method that wants to move one population to a new population using natural selection [32]. The work of [28] proposed different solutions using GA and other models; this is using GA+BN with an accuracy of 85.50% and GA+SVM with an accuracy of 83.82%. In another study [33], the author used GA+NN with a result of 94.17%.

There are some studies based on Neuro-Fuzzy and GA. The paper of [34] used a GA based on a RFNN (recurrent fuzzy neural networks) computing an accuracy of 97.78%. Another hybrid based on Neuro-Fuzzy is CANFIS (Coactive Neuro-Fuzzy interface system) combined with the use of genetic algorithm (GA) [35] with an accuracy of 82%.

The study of [36] proposed the use of SBS (Sequential Backward Elimination) combined with SVC (support vector classifier) and Random Forest (RF). The SBS begins with the original features and creates a subset with the most significant ones, reducing the less significant features. The hybrid of SBS+SVC computed and accuracy of 82%, while SBS+RF gave an accuracy of 80%.

The paper of [37] used PCA (Principal Component Analysis) hybrids. PCA is a statistical procedure that converts the correlated features into a new set of uncorrelated features with the objective of loss the less amount of information. PCA1 is a method that takes the component values of the feature, the one that gets the highest gets the rank1 and consecutively. The use of PCA and regression computed an accuracy of 92%, while the use of PCA1+NN computed the 95.20% accuracy.

BFS (Best First Search) is a search algorithm that explores a graph and finds the node with the best score. The study of [28] computed an accuracy of 80.53% using the hybrid of BFS+MLP and for BFS+C4.5 an accuracy of 78.55%.

WAC (Weighted Associative Classifier) uses weighted support and confidence framework to create the association rule from the data set. The study of [38] used WAC (Weighted Associative Classifier) to predict heart failure and computed an accuracy of 81.51%.

2.5 Comparison Results

According to the comparison between the models, when the feature selection was used, the better is the prediction of the heart failure, this can be observed in the results of the hybrid algorithms. It should be considered that many of the studies do not clarify the division used for training and testing nether is mentioned the cross-validation.

Table 12 Best models accuracy

Author	Method	Accuracy (%)	Tool	Features	Data set
Khanna et al. 2015 [13]	Logistic regression	84.80	–	13	50% training, 50% testing
Khan et al. 2016 [17]	Decision tree C4.5	89.10	WEKA	13	70% training, 30% testing
Khan et al. 2016 [17]	Random forest	89.25	WEKA	13	70% training, 30% testing
Shouman et al. 2012 [21]	KNN	97.40	–	13	–
Wadhonkar et al. 2015 [22]	Neural network (MLP)	98.16	–	13	90% training, 10% testing
Khanna et al. 2015 [13]	SVM (linear)	87.60	–	13	50% training, 50% testing
Khan et al. 2016 [17]	OneR	91.21	WEKA	13	70% training, 30% testing
Khan et al. 2016 [17]	ZeroR	79.56	WEKA	13	70% training, 30% testing
Kodati et al. 2018 [19]	Naive Bayes	83.70	WEKA	13	–
Uyar et al. 2017 [34]	GA based on RFNN	97.78	–	13	85% training, 15% testing
Santhanam et al. 2013 [37]	PCA1+NN	95.20	–	8 ± 1	–

Therefore, Table 12 contains the best model presented in their respective methods. The studies presented by logistic regression, OneR and ZeroR only had one model each. Logistic regression computed an accuracy of 84.80%, this is a low result when is compared with other models. The model of OneR had a high accuracy of 91.21%. ZeroR had the smallest accuracy of all with 79.56%.

The highest decision tree result was of 89.10% using Weka C4.5. This value is not consistent with the other decision tree models, which computed values at least 5% lower. In the case of random forest, the value computed is also higher than the other models in previous investigations. The best accuracy was of 89.25% and was computed by the Weka tool. The case of KNN is similar than decision tree and random forest. The accuracy presented of 97.40% is extremely higher when is compared to the other methods, which had, at highest, a value of 82%.

Artificial neuronal network computed the best results of the algorithms that are not hybrid. The lowest value obtained was of 85%, but most of the models overpass the 89%. The best accuracy was of 98.16%, with a training data set of 90%, which is higher than the recommendation. It also has to be taken into account that 90% of the data was used for training and only 10% of the data for testing; this distribution is greater than what is expected.

The results presented by SVM and Naïve Bayes are consistent among their models. With the small oscillation of accuracy of 4%, in both cases, all the models presented similar accuracy, with the best model of 87.60% for SVM and 83.70% in the case of Naïve Bayes, which is the second lower accuracy.

The hybrid algorithm computed the best results throughout the study. The genetic algorithm combined with artificial neuronal network had the highest accuracy result with 97.78%; the other studies also had remarkable results. The 85% of the data set is used for training, which is higher than recommended and could influence the high performance. The use of PCA with neural network presented a high accuracy with 95.20%, which is better than all the not-hybrid algorithms.

Based on the above results and comparisons, the hybrid models accuracy are greater than the non-hybrid models in the prediction of heart failure using the Cleveland data set. In addition, the use of feature selection computed high performance, such as the case of Neuro-Fuzzy with 96% of accuracy and the use of nine features, the use of ANFIS with 85% of accuracy and the selection of seven features, the PCA models with an accuracy of 92% with 7 ± 1 (PCA+regression) and 95.20% with 8 ± 1 (PCA1+NN) and the use of GA and Naïve Bayes with 85.50% of accuracy and seven features.

2.6   Conclusions

Machine learning models help healthcare systems to find new ways to improve the quality of patient's life. Classification is a well-known solution for predicting heart failure diseases and labels the condition of the patient according with some features. A comparative analysis regarding the different methods is presented in this chapter. The main objective is to present to the reader the different machine learning techniques used in the past. At the end of the comparison, certain value information was founded in the accuracy. According to the results, the highest classification techniques were obtained by the hybrid algorithm of Neuro-Fuzzy.

The most important limitation of this study is the lack of information presented from some investigations, so it is difficult to make a recommendation in the systems when the number of the features, some techniques and the distribution of the data set are not proportionated.

References

1. WHO Homepage (2018), http://www.who.int/cardiovascular_diseases/en/. Last Accessed 19 June 2018
2. HEART Homepage (2018), http://www.heart.org/HEARTORG/Conditions/HeartFailure/Heart-Failure_UCM_002019_SubHomePage.jsp. Last Accessed 19 June 2018
3. S. Shalev-Shwartz, S. Ben-David, *Understanding Machine Learning: From Theory to Algorithms* (Cambridge University Press, New York, 2016)

4. M.M. Al Rahhal et al., Deep learning approach for active classification of electrocardiogram signals. Inf. Sci. **345**, 340–354 (2016)
5. A.F. Khalaf, M.I. Owis, I.A. Yassine, A novel technique for cardiac arrhythmia classification using spectral correlation and support vector machines. Expert Syst. Appl. **42**(21), 8361–8368 (2015)
6. G. Guidi, M.C. Pettenati, P. Melillo, E. Iadanza, A machine learning system to improve heart failure patient assistance. IEEE J. Biomed. Health Inform. **18**(6), 1750–1756 (2014)
7. G. Parthiban, S.K. Srivatsa, Applying machine learning methods in diagnosing heart disease for diabetic patients. Int. J. Appl. Inf. Syst. **3**(7), 25–33 (2012)
8. UCI Heart Disease Data Set (2018), http://archive.ics.uci.edu/ml/datasets/heart+disease. Last Accessed 20 June 2018
9. S. Marsland, *Machine Learning: An Algorithmic Perspective* (Chapman and Hall/CRC, 2015)
10. M. Bramer, *Principles of Data Mining* (Springer London Ltd., 2013)
11. E. Frank, M.A. Hall, I.H. Witten, *The WEKA Workbench. Online Appendix for "Data Mining: Practical Machine Learning Tools and Techniques"*, 4th edn. (Morgan Kaufmann, 2016)
12. H. Trevor et al., *The Elements of Statistical Learning: Data Mining, Inference, and Prediction* (Springer, 2017)
13. D. Khanna et al., Comparative study of classification techniques (SVM, logistic regression and neural networks) to predict the prevalence of heart disease. Int. J. Mach. Learn. Comput. **5**(5), 414–419 (2015)
14. J.R. Quinlan, Induction of decision trees. Mach. Learn. **1**(1), 81–106 (1986)
15. C.E. Shannon, A mathematical theory of communication. Bell Syst. Tech. J. **27**, 379–423, 623–656 (1948)
16. J.R. Quinlan, *C4.5: Programs for Machine Learning* (Morgan Kaufmann Publishers, 1993)
17. S.S. Khan, Prediction of angiographic disease status using rule based data mining techniques. Biol. Forum Int. J. **8**(2), 103–107 (2016)
18. L. Breiman, *Random Forest*, vol. 45 (Kluwer Academic Publishers, 2001), pp. 5–32
19. S. Kodati, Analysis of heart disease using in data mining tools Orange and Weka. Glob. J. Comput. Sci. Technol. **18–1** (2018)
20. N. Mutyala, Prediction of heart diseases using data mining and machine learning algorithms and tools. Int. J. Sci. Res. Comput. Sci. Eng. Inf. Technol. **3–3** (2018)
21. M. Shouman et al., Applying k-nearest neighbour in diagnosing heart disease patients. Int. J. Inf. Educ. Technol., 220–223 (2012)
22. M. Wadhonkar, A data mining approach for classification of heart disease dataset using neural network. Int. J. Appl. Innov. Eng. Manag. **4**(5), 426–433 (2015)
23. V.N. Vapnik, *The Nature of Statistical Learning Theory* (Springer, New York, NY, USA, 1995)
24. K. Srinivas et al., Analysis of coronary heart disease and prediction of heart attack in coal mining regions using data mining techniques, in *2010 5th International Conference on Computer Science & Education* (2010)
25. OneR (2018), http://www.saedsayad.com/oner.htm. Last Accessed 19 June 2018
26. ZeroR (2018), http://chem-eng.utoronto.ca/~datamining/dmc/zeror.htm. Last Accessed 19 June 2018
27. A.Q. Ansari, N.K. Gupta, Automated diagnosis of coronary heart disease using neuro-fuzzy integrated system, in *2011 World Congress on Information and Communication Technologies* (2011)
28. S. Mokeddem et al., Supervised feature selection for diagnosis of coronary artery disease based on genetic algorithm, in *Computer Science & Information Technology (CS & IT)* (2013)
29. N. Ziasabounchi, I. Askerzarde, ANFIS based classification model for heart disease prediction. Int. J. Electr. Comput. Sci. **14**(2), 7–12 (2014)
30. J.-S.R. Jang, ANFIS: adaptive-network-based fuzzy inference system
31. T. Vivekanandan, N.C.S.N. Iyengar, Optimal feature selection using a modified differential evolution algorithm and its effectiveness for prediction of heart disease. Comput. Biol. Med. **90**, 125–136 (2017)
32. M. Mitchell, *An Introduction to Genetic Algorithms* (MIT Press, Cambridge, MA, 1996)

33. N.G.B. Amma, Cardiovascular disease prediction system using genetic algorithm and neural network, in *2012 International Conference on Computing, Communication and Applications* (2012)
34. K. Uyar, A. Ilhan, Diagnosis of heart disease using genetic algorithm based trained recurrent fuzzy neural networks. Procedia Comput. Sci. **120**, 588–593 (2017)
35. L. Parthiban, Intelligent heart disease prediction system using CANFIS and genetic algorithm. Int. J. Biol. Med. Sci. **3**(3), 157–160 (2008)
36. A. Acharya, *Comparative Study of Machine Learning Algorithms for Heart Disease Prediction* (Helsinki Metropolia University of Applied Sciences, 2017)
37. T. Santhanam, E.P. Ephzibah, Heart disease classification using PCA and feed forward neural networks. Min. Intell. Knowl. Explor. Lect. Notes Comput. Sci., 90–99 (2013)
38. J. Soni, Intelligent and effective heart disease prediction system using weighted associative classifiers. Int. J. Comput. Sci. Eng. **5**(6), 2385–2392 (2011)
39. D. Chaki et al., A comparison of three discrete methods for classification of heart disease data. Bangladesh J. Sci. Ind. Res. **50**(4), 293 (2015)
40. S.K. Sen, Predicting and diagnosing of heart disease using machine learning algorithms. Int. J. Eng. Comput. Sci. (2017)
41. A. Olsson, D. Nordlof, *Early Screening Diagnostic Aid for Heart Disease Using Data Mining* (2015)
42. M. Shouman, T. Turner, R. Stocker, Using decision tree for diagnosing heart disease patients, in *9th Australasian Data Mining Conference*, vol. 121, pp. 23–30
43. R. Das et al., Effective diagnosis of heart disease through neural networks ensembles. Expert. Syst. Appl. **36**(4), 7675–7680 (2009)
44. A. Caliskan, M.E. Yuksel, Classification of coronary artery disease data sets by using a deep neural network. EuroBiotech J. **1**(4), 271–277 (2017)
45. M.A.M. Abushariah et al., Automatic heart disease diagnosis system based on artificial neural network (ANN) and adaptive neuro-fuzzy inference systems (ANFIS) approaches. J. Softw. Eng. Appl. **07**(12), 1055–1064 (2014)

Chapter 3
Differential Gene Expression Analysis of RNA-seq Data Using Machine Learning for Cancer Research

Jose Liñares Blanco, Marcos Gestal, Julián Dorado
and Carlos Fernandez-Lozano

Abstract Transcriptome analysis, as a tool for the characterization and understanding of phenotypic alterations in molecular biology, plays an integral role in the understanding of complex, multi-factorial and heterogeneous diseases such as cancer. Profiling of transcriptome is used for searching the genes that show differences in their expression level associated with a particular response. RNA-seq data allows researchers to study millions of short reads derived from an RNA sample using next-generation sequencing (NGS) methods. In general terms, such amount of data is difficult to understand and there is no optimal analysis pipeline for every single analysis. Classical statistical approaches are provided in different R packages (i.e. DESeq or edgeR packages). In medicine, a Machine Learning algorithm can be used for the differential expression analysis of a particular response (i.e. sick versus healthy patients) selecting the genes that are more relevant for discriminating both health outcomes, considering biological pathway information, gene relations or using integrative approaches in order to include all the information available from different curated data-sources. The main aim of our proposal is to practically address Machine Learning based approach for gene expression analysis using RNA-seq data for cancer research within the R framework and to compare it with a classical gene expression analysis approach.

3.1 Introduction

The identification of genes that are differentially expressed on the basis of different health conditions is fundamental in understanding the molecular and biological mechanisms of disease etiology and pathogenesis. To carry out this identification, transcriptome analysis is one of the main tools available to researchers today. These analyses make it possible to characterize and understand the biological and molecular

J. Liñares Blanco · M. Gestal · J. Dorado · C. Fernandez-Lozano (✉)
Faculty of Computer Science, Computer Science Department, University of A Coruña,
15071 A Coruña, Spain
e-mail: carlos.fernandez@udc.es

© Springer Nature Switzerland AG 2019
G. A. Tsihrintzis et al. (eds.), *Machine Learning Paradigms*,
Learning and Analytics in Intelligent Systems 1,
https://doi.org/10.1007/978-3-030-15628-2_3

27

bases that establish the variations in the phenotype of individuals, as is the case with diseases. Over the last few decades, microarrays have been used successfully for this type of analysis, but recently the tools for massive sequencing of complementary DNA (RNA-seq) have increased exponentially [22, 27]. From a sample of RNA, millions of small readings can be obtained as a result of sequencing the next generation of DNA complementary from the sample. From here, two main options are available: to map these sequences against a reference genome (as it could be the case when conducting a study with humans) or to perform a de novo mapping, in which no reference genome is available. In general, most studies focus on the mapping of sequences against a reference genome. The number of readings that have been correctly mapped against a gene can be used as a measure of the abundance of the sample in the analysis [23].

From this point, it is easy to determine that the data obtained from RNA-seq are mainly used to identify differences of expression between genes belonging to organisms in different conditions. However, despite the efforts made by the scientific community, there is no single accepted methodology for analyzing this type of data. Instead, we encounter different implementations. Thus, for example, within the framework R (https://www.R-project.org/) and Bioconductor we have, for instance, the packages DESeq [3], edgeR [24] or tweeDESeq [8] to mention just some.

Similarly, it is also true that machine learning techniques are increasingly being used in the analysis of real-world problems [6, 9, 11, 19]. The main objective of this work is to present the different tasks carried out during a differential expression analysis when resorting to a variety of conventional approaches (using workflows implemented in R and Bioconductor framework packages) and when resorting to an approach based on automated learning techniques on a real problem using free-access RNA-seq data. More specifically, data from The Cancer Genome Atlas (TCGA) [33] will be used for the analysis. TCGA is the result of the combined effort of the National Cancer Institute (NCI) and the National Human Genome Research Institute (NHGRI) to generate a multi-dimensional map of key genomic changes in 33 different types of cancer. Indeed, this has resulted in researchers having public and free access to data volumes that can not be tackled by individual groups, thus enabling the community of cancer researchers to join forces to improve prevention, diagnosis and treatment of a disease as complex and multifactorial as cancer [2].

Of the different types of problems that can be solved using machine learning techniques, two stand out: supervised and unsupervised problems. In the case of supervised problems there is a known output, either categorical (classification) or quantitative (regression). Two algorithms of automated learning will be used for the type of data in this work—both widely used in biological problems as they have yielded very good results: Random Forest [7] and Generalized Linear Models [13]. These techniques have been selected because they are particularly sensitive to problems such as the one analyzed in this work [31], where there is a clear unbalance between the number of variables (p) and the number of cases (n) and it is fulfilled that $n \gg p$.

However, caution should be exercised when using these types of techniques. An experiment design [10] must be followed that ensures that these techniques are not returning incorrect results because of overadjustment of the data. Each technique usually has different hyperparameters that need to be adjusted for best performance. It is precisely because of this that precautions must be taken so that this adjustment does not lead to an overfitting or overtraining of the models when learning from the examples available to them. A common practice is to perform multiple repetitions ensuring that the deviation of the error remains constant and that no bias results from the distribution of the data. Also, cross validation techniques must be used in order to, on the one hand, calculate the degree of generalization of the models and, on the other hand, prevent the hyperparameters from being too adjusted to the specific data used during the training phase.

3.2 Materials and Methods

3.2.1 RNAseq

RNAseq is a genetic ultrasequencing technique for the quantification of messenger RNA molecules to determine the profile of an individual's genetic expression at a specific time.

But, how has this technique appeared? Why has it displaced other techniques of gene expression analysis? and why is it so important for genetic study? To answer all these questions and understand the impact of this technique, let's start at the beginning.

It was in the first decade after the Second World War when scientists Francis Crick and James Watson made what is undoubtedly one of the most important discoveries of the twentieth century: the structure of the DNA double helix. In 1953 they published in the journal Nature an article entitled "Molecular structure of nucleic acids" [32]. It is thanks to this article that we have our current model on the structure of the DNA molecule.

Three years later, in October 1956, Francis himself enunciated the well-known Central Dogma of Molecular Biology in a paper entitled "Ideas on Protein Synthesis (October, 1956)". This dogma has set the guidelines for the flow of genetic information, which is stored in DNA, transferred to RNA and translated into protein. Although Francis Crick described this flow as unidirectional, it is now known that RNA can be converted to DNA by specific enzymes. That is why the use word 'dogma' is so controversial as it alludes to a a fact or a belief that cannot be questioned, a premature statement to say the least at at early time in the field of genetics. Crick himself acknowledged his mistake in using the word 'dogma', and and, funnily enough, he justified it by saying that he really did not know very well what it meant when he published his manuscript.

We are therefore at a time when a direct relationship is known to exist between protein, DNA and RNA but the underlying mechanisms involved in the conversion of a gene to a specific protein are yet not known. This knowledge has been extended thanks to the discovery of the genetic code. This code establishes that the 20 existing amino acids (the structures that make up the proteins—analogous to the nucleotides in the DNA) are encoded by one or several specific trios of nucleotides, called codons, present contiguously in the sequences of messenger RNA. Marshall Nirenberg was the first scientist to sequence the bases in each codon in 1965. As a result, he learned that the UUU and UUC codons of a messenger RNA would give the amino acid Phenylalanine. Or that the AUG codon would involve the presence of the amino acid Methionine, and the same for the 20 amino acids that are known in nature and make up the whole variety of proteins.

It is at this stage that the standard patterns followed by the flow of genetic information began to be clear. How cells convert the information in nucleotides to amino acids was known. Therefore, the DNA of a cell, specifically its genes, is nothing but the information needed to build each protein in our body. The order of their nucleotides is what determines the entire amalgam of proteins. Therefore, if you know the sequence of the genes, can you know everything? It is at this moment that the scientific community realises the importance of genetic sequencing/that genetic sequencing is a pressing need. But not everything would be a bed of roses. Think back to the 70s, with little technology, a lack of optimal resources and a great challenge ahead: 'spelling' each of the nucleotides that make up the DNA sequences of cells. How can such material be accessed? More difficult still, how can it be read?

In 1973, researchers Gilbert and Maxam were able to report 24 nucleotide bases of an enzyme gene. This work took them two years, which made an average of one base per month. It was in 1977 when Frederick Sanger, twice a Nobel Prize winner, and his collaborators developed a technology fast enough to give hope and a significant qualitative leap as regards the challenge of sequencing. Sanger sequenced the first complete nucleic acid in history, the genome of the bacteriophage Phi-X174, composed of some 5,500 nucleotide bases. The method roughly consisted in four reactions performed in different tubes in parallel. Each tube contained the same DNA replication, an enzyme that replicated the template chain (DNA polymerase), a radioactively labeled primer (necessary for polymerase action), all four nucleotides (A, C, T, G). Besides, each tube had one of the four terminating nucleotides. The terminating nucleotides had a characteristic: once they were added to the chain by the polymerase, no other nucleotide could join as they did not have the necessary molecular 'anchoring', hence the name of the method. The mechanism that follows can be easily imagined. The primer, which had been designed to be complementary to an area of the template chain, joined in its specific position so that it could be recognized by the DNA polymerase, which then added nucleotides according to the complimentary of the chain (if it recognizes a T it adds an A, if it recognizes a G it adds a C and vice versa). When by randomness the polymerase adds a terminator nucleotide, the replication stops at that instant. This way, in a tube, let us say that terminal Thymine was added to Tube 2, then tube 2 will present numerous sequences, all of them terminated in T, but of different lengths (bear in mind that in the tube there are thousands of polymerases, thousands of template sequences, thousands of nucleotides, etc.). This fact having happened in each tube, when all the sequences are run through an electrophoresis gel, they will be seen sequentially in the gel (knowing

Fig. 3.1 Representation of the electrophoretic development after a DNA sequencing using the original Sanger method. The sequence would be read from bottom to top and corresponds to the sequence complementary to the mold sequence

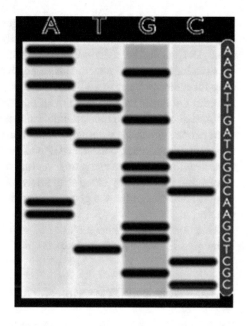

that the smaller sequences travel more distance than the larger ones). The result of electrophoresis is shown in Fig. 3.1.

The sequencing bases established by the Sanger method and its subsequent automation in the 1980s led to the sequencing the human genome. This challenge resulted in the largest biomedical research project in history: a consortium made up of hundreds of scientists from the USA, the United Kingdom, Japan, France, Germany, China and other countries whose goal was to make the first complete sequence of the genome of a human individual (3 billion DNA bases). The project lasted 13 years and its total cost was estimated at $ 3,800 million. It was mythical and so was the competition for the publication of the first draft of the human genome between the public consortium and a private company, Celera Genomics. This company, led by Craig Venter, proposed an alternative sequencing strategy: the random sequencing (shotgun), with which he had already been able to sequence the first cellular genome back in 1995. Finally, the competition ended in a tie and the publications of both sequences did not appear until February 2001 when the consortium published it in the journal Nature and Celera published it in the Science magazine. It was at that moment that a revolution in biomedical research began. Data published after the project report that economic impact of the HGP has been immense. Between 1988 and 2010, the sequencing projects associated with industries derived from the HGP generated, directly and indirectly, a total economic impact of $ 796,000 million: $ 244,000 million in personal income and 3.8 million in jobs per year. It has been estimated that the return on investment of the US federal government has been $ 141 per dollar invested in this huge project.

The reader can imagine the impact Sanger's technique had on twentieth-century biomedical research, although the needs of the researchers were not completely satisfied. Despite the fact that the automation and the parallelization of the experiments led to an increase in the sequencing speed, the sequencing times of a genome did not drop for almost a year. In addition, sequencing could not be done completely, that is, it did not cover the entire genome, but had to be done in parts by designing the appropriate primers and then assembling them all. Not to mention the fact that there were genome locations where this technique did not work very well.

It is at this moment, once the HGP and the analysis of all your data is over and the eagerness to continue investigating has vanished mainly because of the disheartening results that the project yielded that a new revolution begins. In this case it not only involves knowledge but the generation of the data: Second Generation Sequencing is born.

This term refers to new methods of sequencing that have surpassed pioneering techniques in two fundamental aspects: speed and coverage.

3.2.1.1 What is the Transcriptome?

So far, we have taken a cursory glance at the history of the sequencing of the human genome and other species. This arduous journey on which thousands of researchers have embarked over all these years has laid the foundations for future genetic research. It has, however, focused on a single level of the Central Dogma enunciated by F. Crick: DNA. But obviously, the complexity of living beings—and particularly of humans—is not confined to their DNA. It also encompasses the multiple regulatory complexes existing at the level of DNA, RNA and protein. Therefore, knowing the entire genetic sequence of an organism does not give us all the necessary information to, for example, predict a complex disease. That is why many research projects in biomedicine or basic research have focused their efforts on the study of another element of the dogma: RNA or transcriptome.

The passage of DNA to RNA is known as the transcription process. Human DNA is located in the nucleus of most of their cells, where 23 pairs of very long chains with double helical conformation can be found, which—when rolled up on themselves and with the help of specific proteins—form what we know as chromosomes. Because every living organism comes from a single cell, the genetic material of each cell that makes us up is the same. In other words, each cell of a human organism has the same sequence of about 3,000 million base pairs. Therefore, the visible difference that exists between the cells does not come from the composition of their genetic material, but from the way each cell reads it, that is to say, from the passage of DNA to RNA and subsequently to protein. For example, all human cells contain the same 20,000 genes, but not all genes are expressed in the same way in every cell. It is precisely here that lies the diversity among cells, organisms or species.

It is important to highlight the variability that exists around transcription, surely the strongest level of variability in the central dogma. Also relevant is the fact that, as we have noted above, there are some 20,000 genes in each of our cells, which very much contrasts with the expectations at the beginning of the Human Genome Project, when scientists put the figure of genes between 100,000 and 1,000,000. This relatively few number of genes further underscores the importance of genetic regulation and, at the same time, one cannot but wonder at the machinery that nature has developed so that the information necessary to create all the diversity of proteins can be compressed into so few genetic "files".

And since we are talking about numbers, there are around 20,000 genes (at the DNA level); it is estimated that there are between 105 and 106 mRNA (at the RNA level) and around 100,000 proteins. How can it be then that from 20,000 genes almost 1,000,000 molecules of messenger RNA can be formed? The molecular mechanism responsible for this event is called Alternative Splicing, which, without going into detail, is based on differently cutting an immature mRNA to obtain different mRNA molecules from the same gene.

Although many different RNA molecules are known today (lincRNA, miRNA, snRNA, sncRNA, miscRNA, etc.), it is mRNAs that contain the information to be translated into proteins. As to the other RNAs, two characteristics can be highlighted to differentiate them from mRNA: the passage of DNA to RNA is not carried out in coding sites of DNA, that is, they do not come from genes, and their function is focused on regulation of different molecular actions.

3.2.1.2   What Information Do You Offer Us and Why is It Useful?

Let's consider for a moment all the existing variety of biological functionalities. All of this variety, as far as we know, is stored in the 20,000 genes with varying degrees of complexity. Actually, the molecule that performs the function is not the nucleic acid, but the protein. But what we do know is the flow of genetic information, that dogma that F. Crick had enunciated. On the one hand, speaking crudely, we cannot count genes. They can only be read and little else. We can find some kind of mutation, but if we do not have another variable to which we can correlate it, we cannot do much. This is where RNA becomes useful for us. RNAs are molecules very similar to DNA. Therefore, if we could sequence all the DNA, identifying small sequences of mRNA should not be very difficult for us. In addition, several mRNA molecules can come out of a single gene, so we are already introducing a quantitative value. Until now, with DNA we only have had qualitative value, but if we can now know 'how much' a gene is being expressed, that will give us very useful information. For example, a gene that codes for a protein that performs structural work at the level of microtubules—remember that microtubules are part of the cytoskeleton which is analogous to the skeleton of a cell, which gives it rigidity and consistency. Let us suppose that this protein constitutes 20% of the microtubule proteins in a certain cell lineage, the expression of that gene a priori, in these cells, should have a very high expression. Now, let us suppose that we have two patients, from whom we have obtained the level of expression of said gene in that cell type and we observe that the profiles of one differ significantly with respect to the other. Furthermore, it so happens that the one presenting the lowest levels had been previously diagnosed with some type of disease, related, for example, with cognitive aspects. The data that gives us the expression of this gene is providing us with much more interpretable information than the information the genetic sequence would have provided.

The reader can readily imagine the huge possibilities that—starting from this poorly-constructed and unreal example—mRNA-level information can provide us with: from the study of different pathologies (healthy and sick) to the action of a specific drug (drug and placebo), cellular events, etc. Therefore, being able to count the specific number of mRNAs from a particular gene, from an organism at a specific time, will add one further piece to the intricate molecular puzzle. We have talked about the differential expression of a single gene, but what if we have the expression of all the genes?

3.2.1.3   Appearance of the NGS RNAseq Technique in 2008

While it is true that there were different techniques for the accounting of mRNA molecules, none achieved the success and popularity of the microarray technique. This was the most widely used technique in all biological laboratories for transcriptome analysis. Broadly speaking, expression microarrays (there are different types) are based on the principle of base complimentary. That is, knowing the sequence of a specific DNA molecule or, in this case, of mRNA, and after processing the sequence, concentrations of the complementary chain to the target were deposited in microwells, usually with some visible marker. This way, when introducing an aliquot with genetic material, it would hybridize with its previously designed complementary in the well. The greater the hybridization, the greater the intensity of the marker and therefore the greater the quantity of mRNA, which correlated with the greater expression of said gene.

The main limitation of microarrays, as the reader can well imagine, is the need for prior knowledge of the sequence. In addition, when working with related sequences, since they were partially complementary, they hybridized with each other—what was called cross-hybridization. In the case of microarrays, the output signal was not directly a count of molecules, but their concentration was inferred on the basis of the intensity of the marker, so it was difficult to infer and quantify, leading to a lack of replicability of the results.

It was from 2008 that the RNAseq technique was formally introduced, after the publication of two articles in the journal Nature Methods in which the applications of NGS techniques for the characterization of mRNA transcripts of several mice were described with an unprecedented depth and resolution. On the one hand, Sean Grimmond and his collaborators applied the SOLiD platform of Apply Biosystems for embryonic stem cells before and after differentiation, while Barbara Wold and her team applied the Illumina GA platform (Solexa) for brain, liver and muscle transcriptome mouse.

The Applied biosystems and Illumina sequencing platforms are conceptually related and offer approximately similar reading lengths and error rates. The difference between both lies in the protocols for the construction of the libraries, that is, the method by which the millions of small mRNA sequences are sequenced.

It was the second-generation sequencing techniques such as RNASeq that facilitated the identification of those genes that are highly or lowly expressed with respect to normal expression conditions in a control or, simply expressed, in some concrete biological state. The experiments of differential gene expression are carried out to detect and analyze differences in the expression of the genes of an organism under different conditions. In conclusion, these experiments are a fundamental tool for the study of the relationship between genes and biological processes or pathologies.

3.2.2   Classical Approach

In order to decide whether, for a given gene, there are statistically significant differences in the number of readings mapped for that gene under different biological conditions, a statistical test must be performed, for which, in turn, the reading count must be modelled to a certain distribution.

In this case, for the mRNA molecule counting data, RNASeq data, the dependent variable is discrete (e.g. breast cancer or lung cancer, healthy or diseased, etc.) and in a conventional way a Poisson regression model to model this type of data. The Poisson distribution is traditionally used to model the probability that a certain number of events will occur during a certain period of time, all this from an average frequency of occurrence. The probability function of the model is:

$$P[Y = y] = f(y; \mu) = e^{-\mu} \frac{\mu^y}{y!}, \qquad y \epsilon \mathbb{Z}_0^+, \qquad \mu > 0$$

and it also verifies that

$$E[Y] = var[Y] = \mu$$

The main characteristic of this distribution is that the variance of a random variable is expected to be equal to its average. This concept is called statistically equidispersion. Therefore, the use of this distribution for highly variable data such as those of RNAseq, which present large variations between samples and even between replicas of them, is not adequate. Frequently, a greater than expected variance is observed, a fact called overdispersion. Therefore, a variety of models have been developed over the years to collect the overdispersion of the data.

Specifically, overdispersion can appear for various reasons. Among the most common are:

- The data does not come from a Poisson distribution
- High variability in the data
- Lack of stability
- Errors of specification of the mean.

One solution to address this problem is to model the counting data through a Negative Binomial distribution. In general, a Negative Binomial distribution with two parameters, μ and θ, is described. The parameter μ corresponds to the mean, and θ corresponds to the overdispersion parameter, which was not taken into account in the Poisson distributions. Therefore, this distribution is equivalent to the Poisson when overdispersion is zero ($\theta = 0$). This parameter represents the coefficient of variation of the corresponding biological variables between the samples, which are those obtained from different samples that share a certain characteristic and/or condition.

Once the distribution followed by the data has been defined, and in this case, the dispersion of the data has been calculated, the differential expression of transcripts is determined from the corresponding statistical tests for the hypothesis test.

Nowadays, different implementations in various bioinformatics software platforms are used to automate the analysis of this data. In the specific case of R and Bioconductor [14], the edgeR package [20, 24], for instance, can be used. This package implements a statistical methodology of complete analysis of expression differences in RNA-seq expression data based on negative binomial distributions, empirical Bayesian estimation, exact tests, generalized linear models and quasi-likelihood tests. The DEseq package is available as well [3] and it is also based on negative binomial distributions of count data from high-throughput sequencing assays for the analysis of expression differences between genes. Finally, a more recently released package should be mentioned, i.e. tweeDEseq [8] which uses the Poisson-Tweedie family of distributions differential expression analysis.

3.2.3 Machine Learning

Broadly speaking, machine learning techniques can be defined as a type of system that performs tasks associated with artificial intelligence in which, through some type of algorithm or method, useful information is extracted from a set (usually large) of data. These tasks include pattern recognition, planning, prediction, diagnosis, reasoning, rule extraction, etc.

The techniques of machine learning are at the intersection of computer science, engineering, statistics or even biology and are applicable in all those areas that require performing some kind of interpretation or action on any type of data.

There are different approaches depending on how the process of extracting information is carried out:

- Supervised learning: this is an approach in which a set of training data formed by pairs (input, output desired) is available, so at each step the output offered by the machine learning method can be compared with the output really desired. The learning algorithms will try to minimize this error in such a way that—once the rules that link inputs and outputs are extracted—valid answers can be given to input patterns not previously seen (generalization capacity). Typical examples of this type of learning are the adjustment or approximation of functions; classification tasks such as spam labeling; detection of faces/objects in images, etc. In terms of the most common algorithms of supervised learning, the following can be mentioned: k-nearest neighbors (knn), naive bayes, support vector machines (SVM), decision trees (as random forest), Multiple Layer Perceptron (a kind of artificial neural networks (ANN)) ...

- Unsupervised learning: unlike in the case of supervised learning, the desired outputs for the set of inputs are lacking. Therefore, the algorithm will try to figure out the relationships between inputs and outputs, usually trying to establish similarity patterns between the inputs to associate them all with similar outputs (clustering). This type of system answers questions like "What can you tell me about X?", "What are the most frequent features in Y?", ... As to the typical algorithms, we could mention k-Means, DBSCAN, Expectation maximization, Bayesian networks, art neural networks ...
- Learning by reinforcement: its operation is similar to that of a game in which there are rewards and penalties through which the algorithm will try to extract the necessary information for the optimization of some type of parameter. As to the methods, the dynamic programming or the decision models of Markov should be highlighted.

But ... why should "intelligent" algorithms be designed instead of focusing on the development of specific algorithms for specific tasks? The reasons are multiple, but the following should be highlighted:

- Certain tasks are too complex to properly define them in a formal way. However, it is easy to specify them by pairs (input, output desired).
- Changing environments are too complex to be modelled using algorithms or standard methods. However, one of the advantages of machine learning algorithms is their adaptability.
- There is an ever-increasing amount of data available regarding any task with every passing day. Over-information also goes against traditional approaches because relationships of interest between the data can be lost amidst the data. Machine learning techniques such as data mining provide excellent results in situations like these.
- Scalability: Traditional algorithms usually work efficiently under certain restrictions of data volume that must be fixed at the time of their development. If this amount of data increases substantially over time (something quite usual) their performance degrades and, in the worst-case scenario, may cease to be applicable; This situation is easily remedied in Machine Learning algorithms through, for example, the inclusion of new process nodes.

In the design of an application that includes the use of Machine Learning techniques, several options may apply, but regardless of the option chosen, the following steps will always be included in one way or another:

- Data collection: from public or private databases, measurement devices or sensors... the sources of data are almost unlimited.

- Preparation and analysis of input data: adaptation of the data to the format required by the software used, detection of erroneous data, management of missing data, etc.
- Training: the central point of the learning process. The data is provided to the intelligent algorithm for the extraction of useful knowledge. There are multiple metrics to determine the optimal point at which to end this phase: absolute error, ROC curves, recall, precision, etc.
- Evaluation (Test): the final configuration of the algorithm must be tested with new data not used during the training phase to test its generalization capacity. Again, performance can be evaluated through multiple statistical measures: precision, recall, f-score, true positive rate, etc.

3.2.3.1   Random Forest

One of the most widely used techniques in the field of machine learning [7], Random Forest is based on the concept of decision trees. A decision tree is a hierarchical tree-shaped structure of nodes commonly used in classification tasks. There are three types of nodes in a decision tree, each of them with a differentiated role: the root node, the internal nodes and the leaf or terminal nodes.

A root node has no parent nodes. It usually has child or exit nodes, but it might have none. Internal nodes have a connection to their parent node and two or more connections to child nodes. Both the root node and the internal nodes incorporate a condition that makes it possible to separate the samples or examples in the different categories of the output variable. Finally, the terminal nodes or leaf nodes have a connection to the parent node but do not connect to child nodes, being terminal nodes of the tree. Each of these nodes represents a category or class of the output variable and the example is assigned to a category of the problem as output from the decision tree.

The first step to use the decision tree in a classification problem is to train the tree. This training process develops the conditions of the internal nodes and the leaf nodes to assign the examples to the different categories. Once the tree has been trained, in order to assign a concrete example to a category, start by applying the condition of the root node to the example and then descend through the nodes of the tree on the basis of whether the conditions of each node are met or not. Once you reach a leaf node, you get the category in which that example is framed.

To perform the training process, algorithms are used that build the classification conditions of the nodes. These algorithms make the tree grow by adding nodes that make a set of optical, local decisions about what is the best attribute to include in the condition of each node. The best-known algorithms are ID3, C4.5, CART and, above all, Random Forest (RF).

RF is a meta-classifier in which techniques of ensemble learning method such as the Bagging and Boosting method are used. This means that they combine several different learning algorithms. This way, they achieve better results than by using each algorithm separately. The RF algorithm is especially geared to problems of classification and regression in which the input variables have a high dimensionality. RF constructs several decision trees which are then used as classifiers of the Bagging and Random Subspace method. RF is normally used in supervised learning problems, but it can be adapted to unsupervised learning environments.

Specifically, an RF develops a set of trees that form what is called a forest or jungle. Each of these trees or classifiers contributes with a "vote" to indicate the kind of output associated with an example or vector of entry. Once all the trees issue their "vote" or exit class, the class assigned to that example will be the one with the most "votes" cast by the trees in the forest.

Next, we will see the operation of this technique in detail. For the first phase, the training phase, the bagging technique is used to generate the training data. The total of available examples is randomized with replacement, which means that only part of the initial set of data (typically about 2/3) will be part of the training set. All the examples that are not part of the training set will become part of the validation set. These data are called "out of bag" data (OOB). This process will result in some data (which are part of the initial examples) being used more than once in the training of the classifiers, so the RFs are less sensitive than other techniques to the changes in the input data (outliers, noise, etc.) [16].

Once the training and validation sets are constructed, a small subset of variables is randomly selected to form the present Random Subspace. With this step the decision tree is finished so that it performs the best division of the training set data taking into account only the subset of randomly selected variables in the previous step.

This process is repeated several times until the required number of decision trees that will form the forest is created. Each of these trees is trained with a different set of training data and a different subset of input variables. This way each variable of the initial data set will be "in the bag" for some RF tree and "out of the bag" for others.

In the case of RF, in addition to using the typical error measures (MSE, RMSE, etc.), the so-called out-the-bag error estimate is also commonly used. This measure is calculated by independently determining for each classification tree the percentage of OOB data that can be classified by that tree. The ratio between the misclassified examples of the total OOB data is an unbiased estimation of the generalization error, which in the case of RF converges as the number of trees in the forest increases.

Due to the process of selecting variables for the construction of trees, RF gives a measure of the importance of each variable of the input set for carrying out the classification process. This importance value is calculated from the OOB data. To calculate this importance, the values of each variable are changed randomly in the OOB data and, from there, the error is recalculated. If, upon replacing one variable for another, the error increases, this indicates that the first variable is an important variable in the process and its importance is given by the difference between the error values calculated for each of them.

In addition, the RF technique is relatively simple to parametrize with respect to other similar techniques since there is only one relevant parameter: the maximum depth allowed for the construction of the trees [16, 30].

3.2.3.2 Generalized Linear Models

The generalized linear model (GLM) was proposed by McCullagh and Nelder [21]. These models extend or generalize the traditional concept of linear regression for variables that do not follow a Normal distribution and unify several statistical regression models. Therefore, the output variable y_i is assumed to follow an exponential distribution of mean μ_i, being a type of function of $x_i^T \beta$.

Within the design of the GLM, the following 3 components can be distinguished:

- A random component, which will specify a conditional distribution for the output variable (Yi) given the values of the explanatory variables of the model. In the originally proposed formulation, this distribution was of the exponential type, such as Gaussian (normal), binomial, Poisson, gamma ... In later works, we have also incorporated distributions of the multivariate exponential type (such as the multinomial) or the non-exponential type (two-parameter negative-binomial) in situations in which the distribution of Yi is not specified in its entirety.
- A linear predictor, in the form of linear combination of regressors, where Xij are pre-specified functions for the explanation of variables that can also include quantitative exploratory variables, transformations of these variables, polynomial regressors, dummy regressors, etc.:

$$n_i = \alpha + \beta_1 X_{i1} + \beta_2 X_{i2} + \cdots + \beta_k X_{ik}$$

- A smooth and invertible linearizing link function g(•), which will be responsible for transforming the expectation of the response variable, to the linear predictor:

$$g(\mu_i) = \eta_i = \alpha + \beta_1 X_{i1} + \beta_2 X_{i2} + \cdots + \beta_k X_{ik}$$

GLMs have several advantages over more traditional regression systems:

- Working with transformed output variables is not required to meet the condition of working with variables with a normal distribution.
- The selection of the link function is separated from the selection of the random component so that more freedom is obtained in the application of the model.
- The model can be adjusted by means of Maximum Likelihood estimation, thus obtaining such estimator advantage as efficiency, consistency and asymptotic normality.
- A wide range of inference tools and model checking can be applied to GLMs; e.g., Wald and Likelihood ratio tests, Deviance, Residuals, Confidence intervals, Overdispersion, which is commonly used in the statistical software package.

Among the disadvantages of GLMs is the fact that the output variables have to be independent.

Specifically, in this work we propose the use of GLMNET [13], a truly optimized implementation for estimating generalized linear models with convex penalties. In particular, the penalties include the L1 (the lasso), L2 (ridge regression) and a combination of the two (the elastic net). During the learning phase, we will seek to determine which of the penalties works best. Calculated along a regularization path, the algorithm uses cyclical coordinate descent. This particular implementation is designed to work with very high dimensionality problems and in addition, advantage can be taken of data sets that are very sparse.

3.2.3.3   Differential Expression Analysis Using Machine Learning Techniques

One of the biggest problems in a differential expression analysis in RNAseq expression data is its high dimensionality. Some of these variables can be highly correlated and therefore contain irrelevant information. To address these problems character selection techniques were used. A general scheme of operation can be seen in Fig. 3.2. The performance of a supervised algorithm can be affected by the number and relevance of the input variables. The objective of these techniques is to find the subset of input variables that describes the structure of the data in the same way or better [12, 15, 29, 34] than the original data.

A possible classification [1, 11, 17, 25] of techniques for selecting variables would be the one that divides them into filter, wrapper or embedded methods.

Broadly speaking, filter methods take into account the particular relevance of each variable in the data set n by establishing a ranking in order of relevance[5], thus eliminating the variables with the worst score from the set of variables by creating a new space of input variables k for the classifier, where k < n in terms of the dimensionality of both. An operating scheme of this type of techniques can be seen in Fig. 3.3. According to [1, 26] examples of valid measures to establish the ranking of the variables are, among others, the following: Wilks lambda criterion, analysis of main components, mutual information techniques, self-organizing maps or distance measures such as: Student's T-test, Wilcoxon sum-rank test, Bhattacharyya or Relative Entropy.

The wrapper methods [17] evaluate a subset of variables to solve the classification problem and use the performance value obtained. This way, the subsets are evaluated according to the classification error, in the sense that the lower it is, the better will be

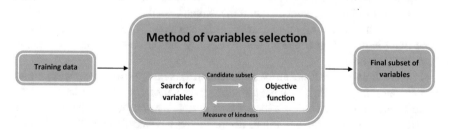

Fig. 3.2 Variable selection techniques

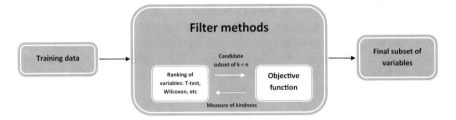

Fig. 3.3 Variable selection technique: filter

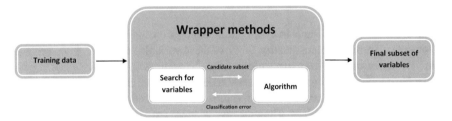

Fig. 3.4 Wrapper variable selection technique

the classifier with these variables to solve the problem. A general scheme of operation of the wrapper variable selection techniques can be seen in Fig. 3.4. There is also another set of techniques that carry out a local search in the variable space, called greedy strategies. These deterministic algorithms include forward and backward selection techniques. In forward selection techniques, for example, the process of selecting variables begins with an empty set of variables and, on the basis of the classification error value, new variables are added. Conversely, the techniques of backward selection begin the iterative process with all the variables and are eliminated in an iterative way.

The third big category are the embedded methods. These methods combine the training process with the search in the variable space. Nested methods are a particular case of this category where the search process is guided estimating the changes of the classifier performance for different subsets in the objective function [17].

The search techniques for the best subset of variables discard the variables that contribute the least to a model that improves the performance of the one that uses all the dimensions. Starting from a set of dimensionality n, there are 2n possible subsets of variables that cannot all be evaluated unless n is small. This is why heuristics are used that yield the best solution (not necessarily the optimal one) in a reasonable time (polynomial).

Filter techniques are faster and very general in the sense that they are not dependent on any classifier, although they tend to select large subsets of characteristics. On the other hand, wrapper techniques have a cost in time associated with the training of the classifier and performance values dependent on it. Overall, a good classification performance is obtained and if it is used in conjunction with mechanisms that avoid overfitting, they generalize well.

Given the high number of genes that exist in the expression RNAseq data, a filter approximation will be followed for computational cost reasons. Features were ranked using their importance measure in an unbiased Random Forest algorithm based on conditional inference [28]. In this study, the Kruskal–Wallis [18] was chosen to rank the genes because of the existence of co-expression patterns among them. This nonlinear and non-parametric method can be applied to a wide range of problems, even if they are nonlinear. It involves complex interaction effects and the number of data is much smaller than the number of predictors. This allowed us to make a first screening, thus reducing to more manageable sizes the genes most related to the pathology under study, in this case the presence or not of cancer. Subsequently, these more manageable subsets of genes will be studied with the two techniques mentioned above: RF and GLM. Each of them allows—once the training and adjustment process of hyperparameters is finished—to quantify the degree of importance (differential expression) of the gene and how it affects the final solution (prediction of the disease). In the case of RF, importance is calculated with the Gini importance index [7], which measures the total impurity increase of each variable. In the case of GLM, the importance of each variable is estimated as the addition of the regression coefficients of each predictor variable in the different fitted models.

3.2.4 Comparative Workflow

When defining the protocols for the analysis of differential expression of genes with RNAseq expression data using the two approaches mentioned in the previous sections (classical statistical analysis and Machine Learning techniques), there are some common phases, as shown in Fig. 3.5. In the sections below, the R code to perform a differential expression analysis with both approximations will be shown.

Initially, as seen in Fig. 3.5, obtaining data from TCGA, preprocessing (patient selection, quality analysis, normalization or bind databases) and biological filtering are shared tasks. From this point, the classical approach continues with a common and tagwise dispersion analysis as well as exact tests. Finally, by means of contrast test of hypothesis or null hypothesis, the genes that are differentially expressed are identified. However, Machine Learning techniques—sensitive to the size of each of the genes—need to be scaled by standardizing them with zero mean and standard deviation so that all genes affect the classification in the same way. The procedure then continues with the elimination of genes that provide the duplicate information. Then, feature selection techniques are used (in this case a filter approach will be followed, but a wrapper or even embedded could be chosen) to look for more reduced and computationally manageable subsets of genes closely related to the cancer under study. Once these subsets have been defined, it is time to apply Machine Learning techniques to predict the genetic signature more related to the characterization of the cancer under study, but to do so, as discussed above and and in agreement with [10],

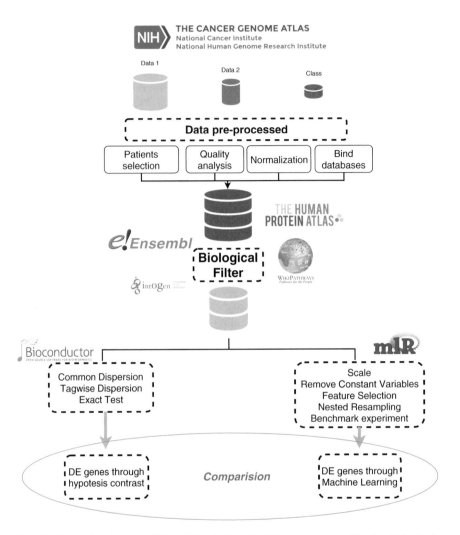

Fig. 3.5 Comparison between differential analysis methodology using conventional statistical techniques and machine learning techniques. It is observed that both methodologies share the initial part of the protocol of pre-processing of the data and they differ in the final phases

it is necessary to ensure that the models do not overadjust the data with different resampling techniques. Using the mlr package [4] the whole process can be implemented through a benchmark experiment. Finally, we can analyze which of the genes in each model have been most differentially expressed and with what importance.

Once the two approximations are completed, the similarities and differences between the two types of differential gene expression analysis in RNAseq expression data can be evaluated.

3.3 Code and Results of an Analysis with Real Data

This section shows a step-by-step example of differential expression analysis with data RNAseq of the TCGA project expression using R-language syntax. The two types of differential analysis of gene expression mentioned above and displayed in the figure of above will be shown. The problem posed is a differential analysis of gene expression between two types of patients: patients with breast cancer and lung cancer. This should serve as an example of the possibilities offered by both approaches to analyze which genes are more related to the differences between cases. A priori, there is no similarity between both types of cancer or a biological explanation that makes it necessary to perform this analysis. It is precisely this difference that it is the strength of work that will be discussed below. The use of more than 20,000 genes at the same time with machine learning algorithms during an analysis of differential gene expression for both types of tumor makes it almost impossible to be able to identify those with greater discriminatory capacity. This way, if both approaches work correctly, they should be able, at the end of the whole process discussed in the previous section, to find that smaller subset of genes capable of differentiating between both types of patients or cancers.

Next, we will briefly explain each of the phases necessary for the analysis as well as the code with R syntax executing it. A minimum knowledge on the part of the user to install the environment and the necessary packages is presumed.

3.3.1 Loading Packages

The first part of the analysis begins with the loading of the R packages necessary for correct operation.

```
require(rWikiPathways)
require(tweeDEseqCountData)
require(biomaRt)
library(vegan)
library(ggplot2)
library(ggbiplot)
library(edgeR)
library(dplyr)
library(tweeDEseq)
```

3.3.2 Loading and Searching the Data from TCGA

Subsequently, the data we will work with is loaded. In this case, it consists of two sets of data downloaded directly from the TCGA repository using the TCGA2STAT package, which requires an Internet connection. First, with the getTCGA function, the data is downloaded in the form of a list. Next, the expression array is extracted from the list as a data.frame object and its dimensions are displayed.

```
brca = getTCGA(disease = 'BRCA', data.type = 'RNASeq2')
luad = getTCGA(disease = 'LUAD', data.type = 'RNASeq2')
luad = as.data.frame(t(luad$dat))
dim(luad)
## [1]    576      20501
brca = as.data.frame(t(brca$dat))
dim(brca)
## [1]    1212      20501
```

An unbalance is observed in the starting data as we have 576 patients with lung cancer as opposed to 1212 patients with breast cancer. This could lead to problems in future analyses when using machine learning techniques. Since a sample of around 500 patients is considered statistically significant, a random resampling of patients with breast cancer is performed. For this we use the sample_n function of the dplyr package. Thus, we randomly choose the same number of breast cancer patients as of lung cancer. The different options available other than the one used to balance the data sets are beyond the scope of this chapter.

```
brca = sample_n(brca, nrow(luad))
```

3.3.3 Patient Selection

The selection of patients in this example is related to the nomenclature that the TCGA assigns to each of their patients. In the first line of code you can see the type of nomenclature. Without going into detail on what each code means, in this case we are interested in the fourth block, which in the example corresponds to '01A' in both patients. This position is the corresponding one to know whether the sample of the patient has been extracted from a solid tumor, a recurrent one, a normal sample, etc.

Keeping this in mind, and so as to make the subsequent analysis, it is necessary to coherently choose the patients to be included in the study. To this end, only those patients with code '01', corresponding to samples of solid tumors, were selected from both databases (BRCA and LUAD). Eventually, databases of 520 and 515 observations were obtained, which is an appropriate balancing of cases.

```
rownames(brca)[1:2]
## [1]      "TCGA-3C-AAAU-01A-11R-A41B-07" "TCGA-3C-AALI
    -01A-11R-A41B-07"
sample.type.brca <- sapply(rownames(brca), function(s)
    unlist(strsplit(s,"-"))[4])
sample.type.luad <- sapply(rownames(luad), function(s)
    unlist(strsplit(s,"-"))[4])
```

```
brca <- brca[grep("^01", sample.type.brca),]
luad <- luad[grep("^01", sample.type.luad),]

dim(brca)
## [1]     520     20501
dim(luad)
## [1]     515     20501
```

3.3.4 Dependent Variable Definition

Once the patients have been chosen, the dependent variable to be analyzed must be defined. In this case, the response variable will be a categorical variable that presents two classes: 'brca' or 'luad'. Thus, the differential analysis will focus on choosing the genes that have a significantly different expression between patients with breast cancer and those with lung cancer.

In addition, in the last line of code of this section the union between the two databases is made. At this stage, therefore, we have the database 'data' with all the RNAseq counts of expression and the dependent variable 'class' with the classification of all the patients. This variable must match the order with the rows of the expression matrix.

```
cb = rep('brca', length(rownames(brca)))
cl = rep('luad', length(rownames(luad)))
class = as.factor(c(cb, cl))
table(class)
## class
## brca luad
##  520  515
data = rbind(brca, luad)
dim(data)
## [1]    1035 20501
```

3.3.5 Biological Gene Filter

Due to the high dimensionality of the problem, as we have noted above, it is necessary to carry out a first common filtering between the two approximations of the genes that we want to analyze. In this case, since the problem is about two specific types of cancer, the gene filter should be consistent with the problem to be solved.

For this, three databases have been used: Ensemble, Intogen and Wikipathways. From these databases, Housekeeping genes, genes related to the cell cycle, genes related to breast cancer and genes related to lung cancer have been extracted. The code used for the extraction of these genes is shown below.

Finally, by uniting all the vectors and keeping only one copy of each gene, we obtain a total of 1103 genes.

```
#HouseKeeping genes
data("hkGenes")
ensembl = useEnsembl(biomart = "ensembl", dataset="hsapiens_gene_
    ensembl")
hkGenes<- getBM(attributes=c('ensembl_gene_id','hgnc_symbol'),
    filters ='ensembl_gene_id', values =as.vector(hkGenes), mart
    = ensembl)
hkGenes<-as.vector(unique(hkGenes$hgnc_symbol))    ## 565 HKGenes

#Cell cycle genes
wiki_cc<-getXrefList(pathway = 'WP179', systemCode = 'H')    ##
    120 wikicc_genes

#Breast Cancer genes
wiki_brca<-getXrefList(pathway = 'WP1984', systemCode = 'H')   ##
    Wikipathways

into_brca<-read.table("~/intogen-BRCA-drivers-data.tsv", sep = "\
    t", header = T)
into_brca<-as.vector(into_brca$SYMBOL)    #Intogen

#Lung Adenocarcinome  WP2512
wiki_luad<-getXrefList(pathway = 'WP2512', systemCode = 'H')    #
    Wikipathways

into_luad<-read.table("~/intogen-LUAD-LUAD_TCGA-drivers-data.tsv"
    , sep = "\t", header = T)
into_luad<-as.vector(into_luad$SYMBOL)        #Intogen

genes = unique(c(hkGenes, wiki_cc, wiki_brca, into_brca ,wiki_
    luad, into_luad))
length(genes)
## [1] 1103
```

The next step is the intersection between the counting matrix columns and the gene vector previously obtained. A loss of 44 genes which were not represented in the counting matrix occurs. Besides, an object is created with the union between the gene counting matrix and the response variable.

```
i = intersect(colnames(data) , genes)
data_reduce = data[,i]
dim(data_reduce)
## [1] 1035        1059
```

To continue with the analysis, the next step is to estimate the normalization factors that will make the samples comparable. Since normalization is a common step between the two types of analyses, it is conducted before starting the actual analyses. In this case, we have opted for the TMM standardization, which is the default one for the edgeR package, but the function we call is not from that package, but from the tweeDEseq package. This is mainly for convenience purposes as—if the function of edgeR were used—the counting matrix would have to be previously transformed to a specific class. This way, only the class matrix data.frame is applied to the function and we create a new class object data.frame that will be useful for the subsequent analysis using Machine Learning.

```
data_tmm = normalizeCounts(data_reduce, method = 'TMM')
```

So far, the analysis corresponds to the pre-processing of the data. The two objects created 'data_tmm' and 'class' will be the ones with which we will perform the analysis.

Below are some graphs that help us observe how our data is distributed.

3.3.6 Graphics

A very important aspect in any data analysis is a correct visualization of data. Below there is a series of graphs that will help understand the distribution the data we are working with. First, Fig. 3.6 shows the average expression profile for each of the genes so that we can compare the two groups of patients as well as the effect that normalization has on each of the datasets.

Figure 3.7 is a way of representing the distance between the data. In this case, we do not focus on the variables, but on the observations (i.e. the patients). It can be observed therefore that the samples we are working with are actually well differentiated.

Finally, for this part of pre-processing and inspection of the data we present a PCA plot (Fig. 3.8). With this graph our focus is again on the variables. The high dimensionality of the problem was reduced to two dimensions to facilitate its interpretability. We can see that the sum of these two dimensions explains about 50% of the variance of the data.

3.3.7 Classical Statistical Analysis

3.3.7.1 Analysis Using the edgeR Package

In this section, a differential expression analysis will be performed using the edgeR package. This package assumes that the number of readings in each sample j assigned to a gene i is modeled through a BN distribution with two parameters, the mean μ_{ij} and the overdispersion parameter θ_{ij}.

$$Y_{ij} \sim BN(\mu_{ij}, \theta_{ij})$$

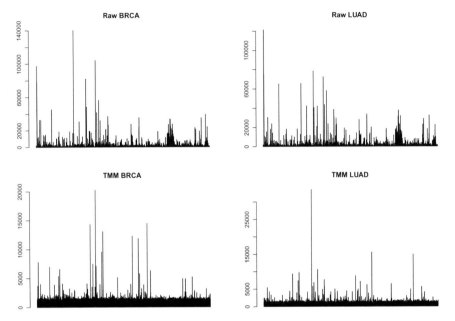

Fig. 3.6 Average expression profile for each of the genes. Expression profiles are differentiated for patients with breast cancer and patients with lung cancer. In addition, for each of the classes the profile can be observed before normalization (raw) and after normalization (TMM)

Y_{ij} corresponds to the non-negative whole number of readings in each sample j assigned to a gene i. The values of the mean and the overdispersion, in practice, are not known so we must estimate them from the data. Finally, using the exact test for the negative binomial distribution, differentially expressed genes are estimated.

Below, the code in R is shown and the steps to be followed for differential expression analysis using edgeR are explained.

3.3.7.2   Installation and Loading of the Package

For the installation of the edgeR package—hosted in the Bioconductor repository—we must connect to that repository, and using the biocLite () command, install it. Finally, we just have to load it in our session.

```
source("http://bioconductor.org/biocLite.R")
biocLite("edgeR")
library(edgeR)
```

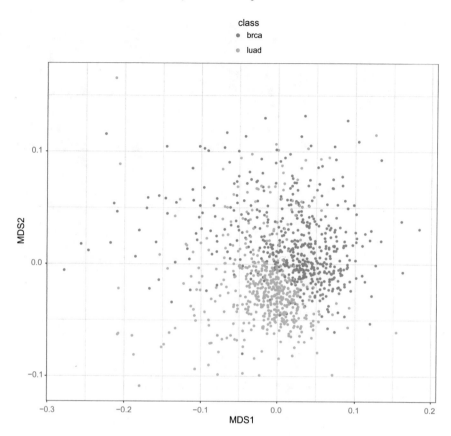

Fig. 3.7 Representation of multidimensional scaling where the distance in two dimensions between the observations can be seen. As shown in the legend, each color represents a class

3.3.7.3 Creating a DGEList Object

Once the entire working ecosystem is ready, a DGEList class object is created—specific to the package—to perform the analysis. For the creation of this class the DGEList function is used, providing the data of counts and the dependent variable. The transpose of the count matrix is performed since the class must have the genes by rows and the observations by columns.

```
d<- DGEList(counts=t(data_reduce), group=class)
```

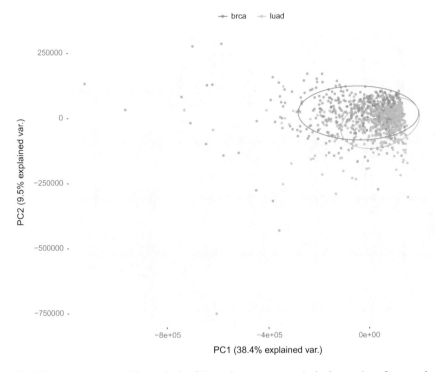

Fig. 3.8 Representation of the analysis of the main components. As in the previous figure, colors denote the type of class

3.3.7.4 The Common and Tagwise Dispersion for Each Gene is Calculated

As mentioned above, it is important to model the variability in the data appropriately. First, the function estimates the mean of the counts for each of the genes, called the size of the library. Then, dispersion is calculated, which is the parameter that determines how the variance is modeled in each gene. The variance function of each gene is

$$\sigma = \mu(1 + \theta * \mu),$$

the μ value being different for each gene. First, the common dispersion for all genes is calculated using the function estimateCommonDisp (), which assumes the same value θ for all genes. But for genes with a low level of expression, the estimated common dispersion entails a greater variability than what they actually have, so the tagwise dispersion is estimated using the estimateTagwiseDisp () function. In this case, from the common dispersion and the real variability of each gene, each of them will be assigned a specific dispersion of each gene. Therefore, to calculate these estimates, the common dispersion must have been estimated previously.

```
d<- estimateCommonDisp(d)
d<- estimateTagwiseDisp(d)
```

3.3.7.5 The DE of Genes in Each Group of Patients is Estimated by Statistical Test

Once we have the data modeled under the negative binomial distribution and the specific parameters of this distribution have been calculated, we can infer the differentially expressed genes through some statistical test. In this case, the edgeR package has the exactTest () function that performs paired tests for the differential expression between the two groups.

In the code shown below, the statistical test for the common dispersion and the gene-to-gene dispersion has been calculated, so that the differences of each model can be observed. In addition, the 10 genes that present a more significant value for each database are shown.

```
res.edgeR.common<- exactTest(d, pair=c("brca", "luad")
   , dispersion="common")
res.edgeR.tagwise<- exactTest(d, pair=c("brca", "luad"
   ), dispersion="tagwise")
topTags(res.edgeR.common)
## Comparison of groups:  luad-brca
##                logFC          logCPM        PValue      FDR
## GATA3      -6.522537     10.420014           0          0
## FGF10      -5.343317      2.630506           0          0
## ESR1       -5.323579     10.379560           0          0
## AR         -4.363861      6.375613           0          0
## DKK4        4.353374      2.033561           0          0
## PIGR        4.181836     11.353202           0          0
## FOLR1       4.072321      8.814847           0          0
## PNMT       -3.829508      4.889089           0          0
## MYB        -3.634474      7.941307           0          0
## TBX3       -3.403445      8.330207           0          0
topTags(res.edgeR.tagwise)

## Comparison of groups:  luad-brca
##                logFC          logCPM            PValue
                              FDR
## GATA3      -6.522141     10.420014       0.000000e+00
     0.000000e+00
## TBX3       -3.403195      8.330207       0.000000e+00
     0.000000e+00
## MYB        -3.634111      7.941307       9.514716e-320
     3.358695e-317
```

```
## ESR1       -5.323498     10.379560      9.695158e-302
   2.566793e-299
## TKT         1.525309     10.906745      3.079238e-258
   6.521826e-256
## MECOM        2.608538      7.719947      8.441187e-257
   1.489869e-254
## KIFC3        1.941571      7.623826      6.281716e-250
   9.503340e-248
## TFPI         3.131980      9.020959      1.716729e-248
   2.272519e-246
## AR          -4.359533      6.375613      7.940535e-235
   9.343363e-233
## CTSH         2.309607     10.693307      4.640301e-221
   4.914079e-219
```

3.3.7.6 Graphics

As mentioned above, once the analysis is done, it is necessary to visualize the results in different ways. First, Fig. 3.9 shows the number of genes that have presented a value below the proposed significance level ($\alpha = 0.0001$) both after calculating the common dispersion coefficient and tagwise.

On the other hand, an important factor that influences the number of differentially expressed genes is the variation among the samples. That is why the graph of the coefficient of biological variation (BCV) is shown in Fig. 3.9. The BCV is the square root of the negative binomial dispersion.

Finally, Fig. 3.10 shows the differentially expressed genes (infra or on), which gives us visual information on the two groups of patients.

3.3.8 Machine Learning Analysis

The different phases that can be followed in conducting an analysis of gene differential expression using automated learning techniques will be shown next.

3.3.8.1 Loading of Packages

The Machine Learning-based analysis will be carried out using the mlr package [4]. The clean, easy to use and flexible language of this package is a great advantage when writing code for these types of experiments. It consists of an interface of more than 160 basic algorithms (called learners) and includes meta-algorithms and model selection techniques to improve and extend the functionality of learner bases. The package is installed from the CRAN repository of R and then loaded into our working directory.

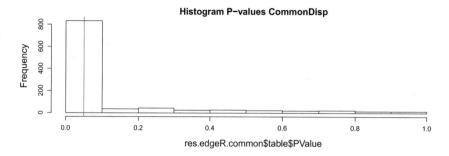

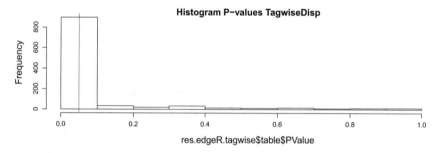

Fig. 3.9 Representation of the results obtained after the execution of the exact test after calculation of the common dispersion coefficient (above) and the tagwise dispersion coefficient (below). It is observed how most of the genes, in both histograms, present an extremely low p-value. The red line denotes the value $\alpha = 0.0001$

```
install.packages("mlr")
library(mlr)
```

3.3.8.2 Loading Data

The next step is to build the Task object of the mlr package using the makeClassif-Task() function. It is with this structure that the package correctly reads the data.

```
data_rc = cbind(data_reduce, class)
colnames(data_rc) = make.names(colnames(data_rc))
task = makeClassifTask(data = data_rc, target = "class")
```

3.3.8.3 Standardization and Elimination of Constant Variables

Machine Learning techniques, as we have mentioned above, are sensitive to the differences in the dimension of each of their variable. Consequently, it is necessary to standardize them to obtain variables with zero mean and standard deviation so that,

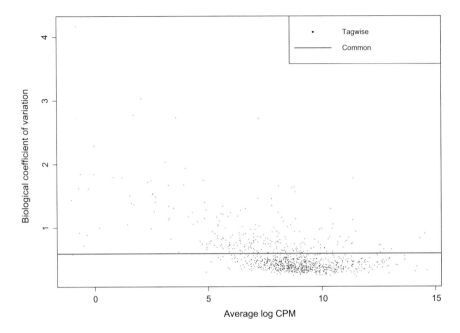

Fig. 3.10 Representation of the coefficient of biological variation

in this case, all the genes affect the classification in the same way. In addition, the mlr package has a function that eliminates the constant and/or correlated variables from the database.

```
task = normalizeFeatures(task)
task = removeConstantFeatures(task)
```

3.3.8.4   Feature Selection

In any ML-based experiment, the researcher must decide which data should be entered as input to the algorithm. To facilitate the work of the analysis algorithms, a previous filtering should be performed to maintain only those variables with a greater correlation with the dependent variable.

To this end, Feature Selection techniques can be used. These techniques do not alter the original representation of the variables. They just select a subset of them. Among the three modes of FS techniques available (filter, wrapped and embedded) a filter technique has been chosen. This, unlike the other two, is independent of the algorithm used, which facilitates the comparison of different algorithms under the same conditions and a better biological interpretation of the results. In this specific instance, the p-value between each gene and the dependent variable (type of cancer) is calculated and the genes are sorted by importance. Finally, all the tasks are grouped in a list for further analysis through a benchmark.

```
#Feature Selection
brca_fs_10<-filterFeatures(task, method = "kruskal.
    test", abs=10)
brca_fs_25<-filterFeatures(task, method = "kruskal.
    test", abs=25)
brca_fs_50<-filterFeatures(task, method = "kruskal.
    test", abs=50)
brca_fs_100<-filterFeatures(task, method = "kruskal.
    test", abs=100)

tasks = list(brca_fs_100, brca_fs_50, brca_fs_25, brca
    _fs_10)
```

3.3.8.5   Algorithm Construction/Hyperparameter Tuning

A crucial step in the training of ML algorithms is choosing the values of their internal parameters. One option is to conduct a search with different values for each of its parameters. By so doing, the general training of the model will be carried out with the values that have historically performed best, thus increasing the optimization and performance of the model. In this case, the decision was made to look for the optimal values of the 'lambda' and 'alpha' parameters for the Glmnet algorithm and 'mtry' and 'nodesize' for the RandomForest algorithm. Note that the 'ntree' parameter remained fixed. The search for the optimal parameters was carried out by means of a Grid control and a Holdout was used as an internal resampling technique ($\frac{2}{3}$ of the data will be used to train the models whereas a $\frac{1}{3}$ of the data will be used to validate the result and select the best one). This way, the adjustment of the parameters is tested without the risk of overtraining the models.

```
#Glmnet
psglmnet<-makeParamSet(
  makeDiscreteParam("lambda", values = c(0.0001,0.001,0.01,0.1,1)
    ),
  makeDiscreteParam("alpha", values= c
    (0,0.15,0.25,0.35,0.5,0.65,0.75,0.85,1))
)

ctrl<-makeTuneControlGrid()
inner<-makeResampleDesc("Holdout")

l<-makeLearner("classif.glmnet", predict.type = "prob")
lrn_glmnet<-makeTuneWrapper(l, resampling = inner, par.set =
    psglmnet, measures = acc, control=ctrl, show.info = T)

#RandomForest
psrf<-makeParamSet(
```

```
  makeDiscreteParam("mtry", values = c(5:13)),
  makeDiscreteParam("ntree", values= 1000L),
  makeDiscreteParam("nodesize", values = c(1:7))
)

l<-makeLearner("classif.randomForest", predict.type = "prob")
lrn_rf<-makeTuneWrapper(l, resampling = inner, par.set = psrf,
    measures = acc, control=ctrl, show.info = T)

lrns = list(lrn_glmnet, lrn_rf)
```

3.3.8.6   Resampling

For the general evaluation of the model, a 10 Repeated 10-fold-cross-validation was
used. It is necessary to study the deviation that occurs in the performance of the
models to verify that there has been no bias resulting from the data. If the error is
stable, we can be sure that the algorithms have been trained properly. A list of 5
elements is defined on the basis of the number of datasets used. The algorithms must
be compared in equal conditions in order for the conclusions to be valid.

```
outer_bmr <- list(
  makeResampleDesc('RepCV', reps=10, folds=10,
      stratify = T),
  makeResampleDesc('RepCV', reps=10, folds=10,
      stratify = T),
  makeResampleDesc('RepCV', reps=10, folds=10,
      stratify = T),
  makeResampleDesc('RepCV', reps=10, folds=10,
      stratify = T))
```

3.3.8.7   Benchmark Experiment

In a Benchmarking experiment, different learning methods are applied to one or
more databases to compare and rank the algorithms with respect to one or more
performance measures. It has the advantage that all algorithms are trained under the
same conditions. In addition, the possibility of being able to train the algorithms on
a large number of different datasets means that a later comparison is possible, not to
mention that the experiment is very easy to design.

```
bmr = benchmark(lrns, tasks, outer_bmr, measures =
      list(auc, acc), show.info = T, models = T)
```

3.4 Conclusions

As described throughout this chapter, a variety of approaches are available to conduct a differential analysis of the expression of genes that may be involved, for instance, in a certain type or stage of cancer. In this chapter, we have described the methodology required to perform two such approaches. First, a standard statistical analysis of a set of genetic expression data and secondly an approach based on automated learning algorithms and techniques.

To represent the analysis methodology, the transcriptome (mRNA) of two different groups of patients was sequenced. One group consisted of patients who had breast cancer and the other group consisted of patients with lung cancer. This type of analysis makes it possible to quantify the difficulty of the analyses that use all the possible genes and how—by resorting to a preprocessing of the genes according to different biological behaviors and also, by means of gene filtering techniques based on machine learning—the number of genes in the study can be drastically reduced. This results in more specific analyses that can be performed in time. 99% of the genes of the subset of 100 genes generated by a Kruskal test following an FS filter approximation are among those selected as differentially expressed by the conventional statistical approach (Fig. 3.11).

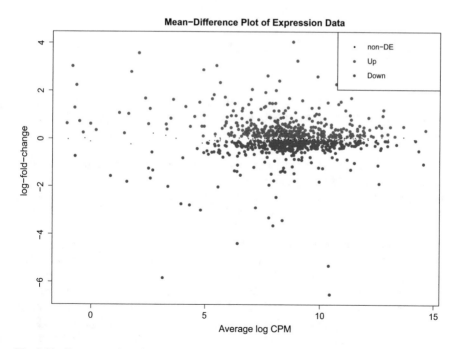

Fig. 3.11 Representation of genes that are differentially expressed following a color code: red if expressed at a higher level (up), blue if expressed at a lower level (down) and black if expressed without differences (non-DE)

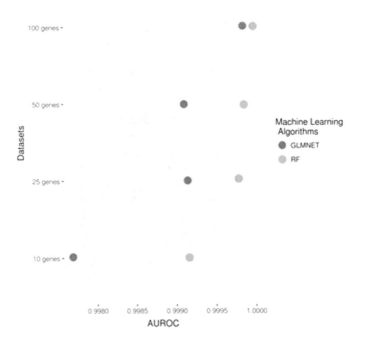

Fig. 3.12 Mean performance of GLMNET and RF using AUROC during 10 10-fold experiments and using different number of genes

Figure 3.12 shows the average performance of the 10 10-fold experiments in AUROC of the two techniques with the different datasets. As expected, both techniques have been shown to yield a very good result with a very small number of genes.

It is important to test now the performance obtained during all runs and see whether it has remained stable throughout the process. For this, a plot violin is presented in Fig. 3.13 in which shows that, although very good results were obtained with all the subsets, the smaller the number of genes, the greater the instability, especially with GLMNET.

The most differentially expressed genes were shown previously following a conventional approach. Now in Figs. 3.14 and 3.15 the most relevant genes identified by Random Forest and by GLMNET are shown. The value assigned by each technique is specific to it, but it is maintained in the corresponding units to assess the degree of importance of the variables in their original scale. The characteristic that led to its biological filtering in the initial preprocessing phase is shown in color code. In the event that one of the genes were identified in several databases of biological information, the identifiers will be concatenated.

As shown in Fig. 3.12, the RF results using 10 genes are very good from the beginning. This is important and, in our humble opinion, it is because the ten genes

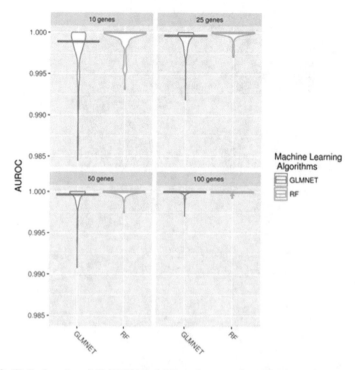

Fig. 3.13 Violin boxplot of GLMNET and RF performance in AUROC during the 10 10-fold experiments

in the subset and TKT, as shown in Fig. 3.14, are among of those identified in the Top 40 as the most important when using a subset of 100 genes. These genes are very informative for both a conventional statistical approach and for RF. In the TOP 40 represented in Fig. 3.12, there are 31 genes within the Top 40 of the conventional statistical approach as opposed to 9 genes that are not: MAP3K1, CDA, C2orf27A, CCND3, PRKCZ, FLT3, BUB3 and VEGFA.

As shown in Fig. 3.15, the FOLR1 and HYB genes are not in the TOP40 although they did appear in the set of the 10 most important in the Kruskal–Wallis test. In addition to this, it can be observed that, for this particular technique, there are only 16 matching genes in the Top 40 with the classical statistical approximation: GATA3, KIFC3, MECOM, PIK3R1, PIK3R3, PITPNM1, TBX3, TFPI, TKT, ANXA6, AR, SFN, MAP3K1, ODC1, MUC20 and ENO1.

These results show that the RF and classical statistical approaches are very similar in the identification of genes with discriminatory potential between both types of tumor. The results obtained by GLMNET are different in terms of the choice of genes, so it should be taken into account as a path for analyzing different genes. Finally, it is also interesting to note that the filtering or preprocessing of genes by

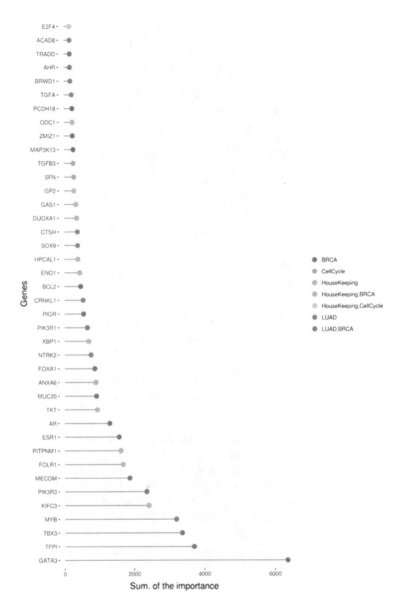

Fig. 3.14 Top 40 genes identified as the most important by RF in the 10 10-fold experiments with the subset of 100 gene data

their different biological functions makes it possible to characterize tumors on the basis of known and unknown genes, such as those related to the cell cycle or others.

The next step, which is beyond the scope of this chapter of the book, would be the analysis of the genes that differed between the three approaches and the

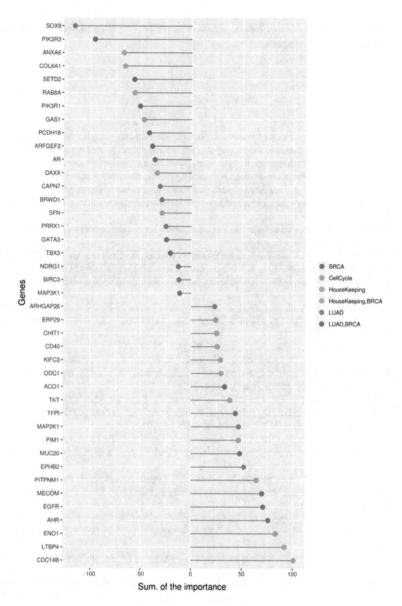

Fig. 3.15 Top 40 genes identified as most important by GLMNET in the 10 10-fold experiments with the subset of 100 gene data

characterization of their function to try to understand the why of their independent choice among the techniques.

References

1. F. Alonso-Atienza, J.L. Rojo-Álvarez, A. Rosado-Muñoz, J.J. Vinagre, A. García-Alberola, G. Camps-Valls, Feature selection using support vector machines and bootstrap methods for ventricular fibrillation detection. Expert. Syst. Appl. **39**(2), 1956–1967 (2012)
2. R. Alonso-Calvo, S. Paraiso-Medina, D. Perez-Rey, E. Alonso-Oset, R. van Stiphout, Y. Sheng, M. Taylor, F. Buffa, C. Fernandez-Lozano, A. Pazos et al., A semantic interoperability approach to support integration of gene expression and clinical data in breast cancer. Comput. Biol. Med. **87**, 179–186 (2017)
3. S. Anders, W. Huber, Differential expression analysis for sequence count data. Genome Biol. **11**(10), R106 (2010)
4. B. Bischl, M. Lang, L. Kotthoff, J. Schiffner, J. Richter, E. Studerus, G. Casalicchio, Z.M. Jones, mlr: machine learning in r. J. Mach. Learn. Res. **17**(1), 5938–5942 (2016)
5. A.L. Blum, P. Langley, Selection of relevant features and examples in machine learning. Artif. Intell. **97**(1–2), 245–271 (1997)
6. J.C. Brégains, J. Dorado, M. Gestal, J.A. Rodriguez, F. Ares, A. Pazos, Avoiding interference in planar arrays through the use of artificial neural networks. IEEE Antennas Propag. Mag. **44**(4), 61–65 (2002)
7. L. Breiman, Random forests. Mach. Learn. **45**(1), 5–32 (2001)
8. M. Esnaola, P. Puig, D. Gonzalez, R. Castelo, J.R. Gonzalez, A flexible count data model to fit the wide diversity of expression profiles arising from extensively replicated RNA-seq experiments. BMC Bioinform. **14**(1), 254 (2013)
9. C. Fernandez-Lozano, R.F. Cuiñas, J.A. Seoane, E. Fernandez-Blanco, J. Dorado, C.R. Munteanu, Classification of signaling proteins based on molecular star graph descriptors using machine learning models. J. Theor. Biol. **384**, 50–58 (2015)
10. C. Fernandez-Lozano, M. Gestal, C.R. Munteanu, J. Dorado, A. Pazos, A methodology for the design of experiments in computational intelligence with multiple regression models. PeerJ **4**, e2721 (2016)
11. C. Fernandez-Lozano, J.A. Seoane, M. Gestal, T.R. Gaunt, J. Dorado, A. Pazos, C. Campbell, Texture analysis in gel electrophoresis images using an integrative kernel-based approach. Sci. Rep. **6**, 19256 (2016)
12. F.J. Ferri, P. Pudil, M. Hatef, J. Kittler, Comparative study of techniques for large-scale feature selection, in *Machine Intelligence and Pattern Recognition*, vol. 16 (Elsevier, 1994), pp. 403–413
13. J. Friedman, T. Hastie, R. Tibshirani, Regularization paths for generalized linear models via coordinate descent. J. Stat. Softw. **33**(1), 1 (2010)
14. W. Huber, V.J. Carey, R. Gentleman, S. Anders, M. Carlson, B.S. Carvalho, H.C. Bravo, S. Davis, L. Gatto, T. Girke et al., Orchestrating high-throughput genomic analysis with Bioconductor. Nat. methods **12**(2), 115 (2015)
15. A.K. Jain, B. Chandrasekaran, 39 dimensionality and sample size considerations in pattern recognition practice. Handb. Stat. **2**, 835–855 (1982)
16. T.M. Khoshgoftaar, M. Golawala, J. Van Hulse. An empirical study of learning from imbalanced data using random forest, in *19th IEEE international conference on Tools with Artificial Intelligence, 2007. ICTAI 2007*, vol. 2 (IEEE, 2007), pp. 310–317
17. R. Kohavi, G.H. John, Wrappers for feature subset selection. Artif. Intell. **97**(1–2), 273–324 (1997)
18. W.H. Kruskal, W. Allen Wallis, Use of ranks in one-criterion variance analysis. J. Am. Stat. Assoc. **47**(260), 583–621 (1952)

19. Y. Liu, S. Tang, C. Fernandez-Lozano, C.R. Munteanu, A. Pazos, Y.-z. Yu, Z. Tan, H. González-Díaz, Experimental study and random forest prediction model of microbiome cell surface hydrophobicity. Expert. Syst. Appl. **72**, 306–316 (2017)
20. D.J. McCarthy, Y. Chen, G.K. Smyth, Differential expression analysis of multifactor RNA-Seq experiments with respect to biological variation. Nucl. Acids Res. **40**(10), 4288–4297 (2012)
21. P. McCullagh, J.A. Nelder, *Generalized Linear Models*, vol. 37 (CRC Press, 1989)
22. A. Mortazavi, B.A. Williams, K. McCue, L. Schaeffer, B. Wold, Mapping and quantifying mammalian transcriptomes by RNA-Seq. Nat. Methods **5**(7), 621 (2008)
23. A. Oshlack, M.D. Robinson, M.D. Young, From RNA-Seq reads to differential expression results. Genome Biol. **11**(12), 220 (2010)
24. M.D. Robinson, D.J. McCarthy, G.K. Smyth, edgeR: a Bioconductor package for differential expression analysis of digital gene expression data. Bioinformatics **26**(1), 139–140 (2010)
25. Y. Saeys, I. Inza, P. Larrañaga, A review of feature selection techniques in bioinformatics. Bioinformatics **23**(19), 2507–2517 (2007)
26. S. Salcedo-Sanz, G. Camps-Valls, F. Pérez-Cruz, J. Sepúlveda-Sanchis, C. Bousoño-Calzón, Enhancing genetic feature selection through restricted search and Walsh analysis. IEEE Trans. Syst. Man Cybern. Part C (Appl. Rev.) **34**(4), 398–406 (2004)
27. C. Soneson, M. Delorenzi, A comparison of methods for differential expression analysis of RNA-Seq data. BMC Bioinform. **14**(1), 91 (2013)
28. C. Strobl, A.-L. Boulesteix, A. Zeileis, T. Hothorn, Bias in random forest variable importance measures: Illustrations, sources and a solution. BMC Bioinform. **8**(1), 25 (2007)
29. C.W. Therrien, C.W. Therrien, *Decision, Estimation, and Classification: An Introduction to Pattern Recognition and Related Topics* (Wiley, New York, 1989)
30. W.G. Touw, J.R. Bayjanov, L. Overmars, L. Backus, J. Boekhorst, M. Wels, S.A.F.T. van Hijum, Data mining in the life sciences with random forest: a walk in the park or lost in the jungle? Brief. Bioinform. **14**(3), 315–326 (2012)
31. G. Tsiliki, C.R. Munteanu, J.A. Seoane, C. Fernandez-Lozano, H. Sarimveis, E.L. Willighagen, RRegrs: an R package for computer-aided model selection with multiple regression models. J. Cheminformatics **7**(1), 46 (2015)
32. J.D. Watson, F.H.C. Crick et al., Molecular structure of nucleic acids. Nature **171**(4356), 737–738 (1953)
33. J.N. Weinstein, E.A. Collisson, G.B. Mills, K.R. Mills Shaw, B.A. Ozenberger, K. Ellrott, I. Shmulevich, C. Sander, J.M. Stuart et al., Cancer genome atlas research network, the cancer genome atlas pan-cancer analysis project. Nat. Genet. **45**(10), 1113 (2013)
34. D. Zongker, A. Jain, Algorithms for feature selection: an evaluation, in *Proceedings of the 13th International Conference on Pattern Recognition, 1996*, vol. 2 (IEEE, 1996), pp. 18–22

Chapter 4
Machine Learning Approaches for Pap-Smear Diagnosis: An Overview

E. Karampotsis, G. Dounias and J. Jantzen

Abstract This chapter is a typical example of usage of Computational Intelligence Techniques-CI-Techniques (Machine Learning-Artificial Intelligence) in medical data analysis problems, such as optimizing the Pap-Smear or Pap-Test diagnosis. Pap-Smear or Pap-Test is a method for diagnosing Cervical Cancer (4th leading cause of female cancer and 2nd common female cancer in the women aged 14–44 years old), invented by Dr. George Papanicolaou in 1928 (Bruni et al. in Human papillomavirus and related diseases in the world [1]; Marinakis and Dounias in The Pap Smear Benchmark, Intelligent and Nature Inspired Approaches in Pap Smear Diagnosis, Special Session Proceedings of the NISIS—2006 Symposium [2]). According to Pap-Smear, specialized doctors collect a sample of cells from specific areas of cervical, observe (using microscope) specific cells of the above cell-sample and classify these cells into 2 general (Normal and Abnormal cells) and 7 individual categories/classes: Superficial squamous epithelial, Intermediate squamous epithelial, Columnar epithelial, Mild squamous non-keratinizing dysplasia, Moderate squamous non-keratinizing dysplasia, Severe squamous non-keratinizing dysplasia and Squamous cell carcinoma in situ intermediate. The ideal aim of this classification process is the early diagnosis of cervical cancer. Pap-Test was a time-consuming process and with considerable errors of observation, resulting in diagnosis with a high degree of uncertainty. Considering these problems, Data Analysis researchers in collaboration with specialized doctors have presented several successful approaches to the Pap-Smear diagnosis optimization problem using Computational Intelligence (CI) Techniques, whose results are acceptable to the medical community and have room for improvement. An equivalent effort was made by the researchers and students of Department of Automation of the Technical University of Denmark for the first time

E. Karampotsis · G. Dounias (✉) · J. Jantzen
Management & Decision Engineering Lab (MDE-Lab), Department of Financial and Management Engineering, University of the Aegean, 41 Kountouriotou Street, 82100 Chios, Greece
e-mail: g.dounias@aegean.gr

E. Karampotsis
e-mail: fmem1206@fme.agean.gr

J. Jantzen
e-mail: jj@aegean.gr

© Springer Nature Switzerland AG 2019
G. A. Tsihrintzis et al. (eds.), *Machine Learning Paradigms*,
Learning and Analytics in Intelligent Systems 1,
https://doi.org/10.1007/978-3-030-15628-2_4

in 1999 is the cornerstone of further Pap-Smear data analysis using CI-Techniques, which was then continued until nowadays by researchers and students of the Management and Decision Engineering Laboratory (MDE-Lab) of Technical Department of Financial and Management Engineering of the University of the Aegean (http://mde-lab.aegean.gr/downloads). This research focuses on the approach of the Pap-Smear Classification Problem with the use of CI-Techniques and has as an ideal goal the contribution of Artificial Intelligence to the optimization of medical diagnoses. In addition, the research conducted aims at diagnosing cervical cancer both at an early stage and at an advanced stage, improving the Pap-Smear classification process as well as the process of Feature Selection. The aforementioned research was based on two databases, called Old Data and New Data which consist of 500 and 917 single cell patterns respectively, described by 20 features. These data were collected by qualified doctors and cyto-technicians from the Department of Pathology of the Herlev University Hospital and are available at the web-page of MDE-Lab (http://mde-lab.aegean.gr/downloads). The CI-Techniques, that were used to build classifiers, both for the 2-class classification problem and for the 7-class classification problem, are machine learning algorithms, such as Adaptive Network-based Fuzzy Inference Systems, Artificial Neural Networks, k-Nearest Neighbor approaches, etc. Also, various algorithms, such as Fuzzy C-Means, Tabu Search, Ant Colony etc., were used to improve the operating processes, such as the training process or the feature selection approach. All these CI-Techniques are briefly presented in Sect. 3. Finally, the results of the application of the above CI-Techniques are presented in Sect. 4, and prove very satisfactory, fully accepted by the doctors of this particular field of medicine and with room for further improvement.

4.1 Introduction

A medical diagnosis is a vital issue, so must be as prompt and dependable as possible. In general, diagnosis of both a particular disease and its categories can be considered a nominal classification process according to various data (several factors or symptoms) derived from medical observations. This classification process requires a quick and accurate analysis of a large amount of data (nominal or/and numerical), that, in combination with the experience and knowledge of medical scientists, will bring more accurate diagnostics of a disease [3–8].

One solution to this classification problem is the application of computational intelligence techniques (CI-Techniques). Using these CI-Techniques, we can improve the above classification process and reduce the degree of uncertainty in medical diagnosis or/and decision.

CI-Techniques, such as Decision Trees, Artificial Neural Networks, Fuzzy Systems, Evolutionary Computing Techniques, Algorithms inspired by Nature, Intelligent Agents and so on, can improve one medical diagnosis or/and decision, integrating human logic into it [3].

Table 1 CI-Techniques and medicine: a conclusion table of some typical examples

Medical problems and research fields	Brief description of CI-Techniques applications
Breast cancer	Use of Bayesian networks to support the prompt detection of the breast cancer [31]
Electroencephalography (EEG)	A new classification of continuous EEG recordings based on a network of spiking neurons [32]
Atheromatosis (a frequent cause of stroke)	Use of the personalized reasoning mechanism for an intelligent medical e-learning system on atheromatosis [33]
Acute coronary syndromes	Use of online analytical process (OLAP) methodology for a more accurate risk assessment of developing acute coronary syndromes [34]
Cancer	Development of useful data-mining tools using CI-Techniques, such as Fuzzy Systems, that help the evolution of cancer research [35]
Texture characterization and classification of capsule-endoscopic images	Pattern recognition system based on Artificial Neural Networks [36]
Telemedicine service, patient monitoring and diagnosis, emergency management, online informing patients and so on	Development of mobile applications using web, mobile agents and Ci-Techniques (Artificial Neural Networks, Fuzzy Systems and etc.) [37–42]
Acid-base disturbances	A Hybrid Intelligent System that consists of a Fuzzy expert System that incorporates an Evolutionary Algorithm in an off-line mode and deals with diagnosis and treatment consultation of acid-base disturbances based on blood gas analysis data [43]
Bladder tumor classification	A combination of Fuzzy Cognitive Maps and Support Vector Machines to achieve better tumor malignancy classification [44]

There are many notable application cases of CI-Techniques on medical data and some of these are presented in Table 1. According to the following presentation of the successful applications of CI-Techniques in various fields of medicine, it is obvious that CI-Techniques have become an integral part of medical decision-making, opening up new horizons to the development of medical and artificial intelligence.

A corresponding example of the contribution of artificial intelligence to medicine is the improvement of Pap-Smear diagnosis, using CI-Techniques, which is presented in this chapter.

This chapter presents the implementation of some CI-Techniques on medical data which relate to the problem of cervical cancer and are derived from the observation and collection of specific features of various cell specimen images, that were collected and examined according to the Pap-Smear or Pap-Test process. The aim was to improve the diagnosis of cervical cancer improving the classification process of cells, that were collected during the Pap-Smear or Pap-Test.

The following sections of this chapter present a summary of some medical information about cancer, cervical cancer and Pap-Smear as well as the application of CI-Techniques to the Pap-Smear classification problem.

4.2   Cervical Cancer and Pap-Test

Cancer is a disease of cells, that are the building blocks of tissues of human organs. All types of cancer, besides leukemia (blood cancer), effect an uncontrolled growth of a group of cells within human body resulting in the appearance of the so-called cancerous tumor. A tumor (Benign or Malignant) can grow and spread to the healthy part of a tissue (cancer-infected) or to another part of the human body via bloodstream and lymphatic systems and can cause various problems to digestive, nervous and circulatory systems. Cancer can be classified into four different categories (Carcinoma, Sarcoma, Lymphoma and Leukemia) and its name comes from the organ, in which it first appeared [9].

Cervical cancer (Benign or Malignant) is predominantly sexually transmitted, it begins in cells on the surface of the cervix, and it sometimes can invade more deeply into the cervix and nearby tissues. The cervix, as a part of a woman's reproductive system, is on the pelvis and it is the lower, narrow part of the uterus.

The cervix connects the uterus to the vagina, makes mucus and during pregnancy, is tightly closed to help keep the baby inside the uterus. The symptoms of cervical cancer, which is in advanced stage, are the abnormal vaginal bleedings (bleeding that occurs between regular menstrual periods, after sexual intercourse, after douching or a pelvic exam, either going through menopause and menstrual periods, that last longer and are heavier than before), the increased vaginal discharge, the pelvic pain and the pain during intercourse. The diagnosis of cervical cancer may be accomplished by cervical exams, biopsies of tissue samples and lab tests. The most widely used laboratory method for diagnosing cervical cancer is the Pap-Smear [1, 9–15].

Pap-Smear was described for first time by Dr. George Papanicolaou in 1928 and is a microscopic examination of cells scraped from the cervix and is used to detect cancerous or pre-cancerous conditions of the cervix or other medical conditions. According to the Pap-smear process, a cell sample is taken, using a small brush, a cotton stick or wooden stick, from the uterine cervix and transferred onto a thin, rectangular glass plate (slide), where the specimen (smear) is stained using the Papanicolaou method to observe characteristics of cells more clearly in a micro-scope. The purpose of the smear screening, which is done by a cyto-technologist and/or cyto-pathologist, is to diagnose pre-malignant cell changes before they progress to cancer. Each slide may contain up to 300.000 columnar and squamous epithelial cells, that have been collected respectively from the upper and lower part of cervix (Fig. 1) [14–16].

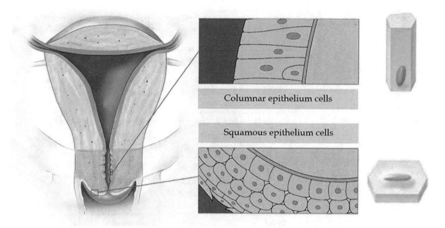

Fig. 1 Areas of columnar and squamous epithelium, columnar and squamous epithelium cells. *Source* Epithelium, Dr. Swati Patil, National Cancer Institute and JO's Cervical Cancer Trust

Dysplastic cells are cells that have undergone pre-cancerous changes, have larger and darker nuclei and have a tendency to cling together in large clusters. There are three different classes of Squamous dysplasia. The first class includes mild dysplastic cells, which have enlarged and light nuclei, the second class includes the moderate dysplastic cells, whose nuclei are darker and larger than the mild dysplastic cells and the third class, which is the last stage of precancerous changes and includes the severe dysplastic cells with large, dark and often de-formed nuclei (Fig. 1) [1, 9–15].

Pap-Smear is a method for the detection of cervical cancer in early stages with the ultimate purpose of treating it and deter its metastasis. As has been reported, cancer takes its name from the organ it has initially offended, when the cancer spreads to another part of body, then the tumor has the same cancer cells and the same name as the original tumor. Cervical cancer can spread to nearby tissues of pelvis and often makes metastasis in the lymph nodes, lungs, liver and bones. According to the point, where the cervical cancer has been detected, there are 5 stages of it, those are presented in Table 2 [1, 9–15].

One major problem in Pap-Smear diagnosis is the cells recognition and classification of Pap-Smear samples. This problem was called as Pap-Smear Classification Problem in its first approach, using CI-Techniques, that had been in 1999 from the researchers and students of Department of Automation of the Technical University of Denmark in cooperation with the doctors of Department of Pathology of the Herlev University Hospital [2, 16–25] (Fig. 2).

The above-mentioned research aims to help the optimization of Pap-Smear diagnosis using CI-Techniques on the Pap-Smear Classification Problem and could be considered as a characteristic example of AI contribution into Medical Data Analysis.

Table 2 Stages of cervical cancer [25]

Stages	Definition	Therapy	5-year-healing rates (%)
Stage 0	Carcinoma in situ	Conization	99
Stage 1	Restricted to cervix	Hysterectomy	75–90
Stage 2	Infiltration of vulva or parametrium	Radiation therapy (and operation)	50–70
Stage 3	Infiltration of >2/3 of vagina or parametrium (pelvis)	Radiation therapy only	30
Stage 4	Metastasis or infiltration of other organs	Radiation therapy only	0

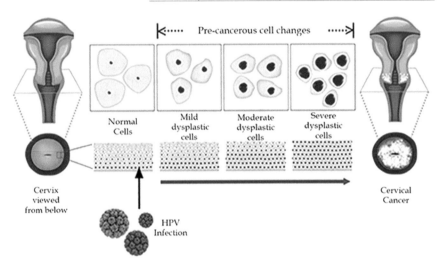

Fig. 2 Pre-cancerous cell changes. *Source* The JC School of Public Health and Primary Care, Faculty of Medicine, The Chinese University of Hong Kong (http://www.cuhk.edu.hk/sphpc/hpvselfsampling/en/cc-n-hpv.html)

4.3　The Pap-Smear Databases

The used databases were derived from the Department of Pathology of Herlev University hospital and are called "Old Data" and "New Data". These two databases, which, are respectively comprised of 500 and 917 images of different columnar and squamous epithelium cells, which were classified, manually, into 2 big cell classes (Normal and Abnormal) and into 7 smaller cell classes (Superficial Squamous, Intermediate Squamous, Columnar, Mild Dysplasia, Moderate Dysplasia, Severe Dysplasia and Carcinoma in Situ) and then 20 cell features resulted from the conversion of the images into numerical data using CHAMP and MATLAB software. All the above information is presented in the summary Tables 3, 4, 5 and 6 [2, 16–26].

4.3.1 A Basic Data Analysis of New Data

As the number of New Data is bigger than the number of Old Data, the database named New Data can be considered as a representative sample of Pap-Smear data. Under this hypothesis, a basic data analysis was performed on New data by Jan Jantzen and Georgios Dounias in 2006 [24].

This basic data analysis was based on the Mahalanobis distance norm [24] and provides numerical measures indicating how well the classes are separated. Observing the data, it seems that the classes can't be linearly separated. To achieve an indication of the degree of overlap, the distance between class centers was measured

Table 3 Cell features that were extracted using CHAMP and MATLAB

Nr.	Name	Feature	Description
1	Kerne_A	Nucleus area	Calculated by counting the corresponding pixels of the segmented picture
2	Cyto_A	Cytoplasm area	
3	K/C	Nucleus/cytoplasm ratio	Tells how small the nucleus area is compared to the area of the cytoplasm
4	Kerne_Ycol	Nucleus brightness	The average perceived brightness, that is a function of the colors wavelength
5	Cyto_Ycol	Cytoplasm brightness	
6	KerneShort	Nucleus shortest diameter	The shortest diameter a circle can have, when surrounding the whole object
7	KerneLong	Nucleus longest diameter	The biggest diameter a circle can have, when the circle is totally encircled of the object
8	KerneElong	Nucleus elongation	The elongation is calculated as the ratio between the shortest diameter and the longest diameter of the object
9	KerneRund	Nucleus roundness	The roundness is calculated as the ratio between the actual area and the area bound by the circle given by the longest diameter of the object
10	CytoShort	Cytoplasm shortest diameter	The shortest diameter a circle can have, when surrounding the whole object
11	CytoLong	Cytoplasm longest diameter	The biggest diameter a circle can have, when the circle is totally encircled of the object

(continued)

Table 3 (continued)

Nr.	Name	Feature	Description
12	CytoElong	Cytoplasm elongation	The elongation is calculated as the ratio between the shortest diameter and the longest diameter of the object
13	CytoRund	Cytoplasm roundness	The roundness is calculated as the ratio between the actual area and the area bound by the circle given by the longest diameter of the object
14	KernePeri	Nucleus perimeter	The length of the perimeter around the object
15	CytoPeri	Cytoplasm perimeter	
16	KernePos	Nucleus relative position	A measure of how well the nucleus is centered in the cytoplasm
17	KerneMax	Maxima in nucleus	This is a count of how many pixels is a maximum/minimum value inside of a 3 pixel radius
18	KerneMin	Minima in nucleus	
19	CytoMax	Maxima in cytoplasm	
20	CytoMin	Minima in cytoplasm	

Table 4 Classes and categories of New Data [20]

Class	Name	A short description	Category	Image
1	Superficial squamous epithelial	The shape of cells in Class 1 is flat and oval, their nucleus is very small and the analogy between nucleus and cytoplasm is small	Normal	
2	Intermediate squamous epithelial	The shape of cells in Class 2 cells is round, their nucleus is large and the analogy between nucleus and cytoplasm is small		
3	Columnar epithelial	The shape of cells in Class 3 is column-like, their nucleus is large and the analogy between nucleus and cytoplasm is medium		

(continued)

Table 4 (continued)

Class	Name	A short description	Category	Image
4	Mild squamous non-keratinizing dysplasia	The nucleus of cells in Class 4 is light and large and the analogy between nucleus and cytoplasm is medium	Abnormal	
5	Moderate squamous non-keratinizing dysplasia	The nucleus of cells in Class 5 is dark (as cytoplasm) and large and the analogy between nucleus and cytoplasm is large		
6	Severe squamous non-keratinizing dysplasia	The nucleus of cells in Class 6 is dark (as cytoplasm), large and deform and the analogy between nucleus and cytoplasm is large		
7	Squamous cell carcinoma in situ intermediate	The nucleus of cells in Class 7 is dark, large and deform and the analogy between nucleus and cytoplasm is very large		

Table 5 Presentation of Old Data [16]. Available at: http://mde-lab.aegean.gr/downloads

Class	Name	Category	Number of single cells
1	Superficial squamous epithelial	Normal 200 single cells	50
2	Intermediate squamous epithelial		50
3	Columnar epithelial		50
4	Parabasal squamous epithelial		50
5	Mild squamous non-keratinizing dysplasia	Abnormal 300 single cells	100
6	Moderate squamous non-keratinizing dysplasia		100
7	Severe squamous non-keratinizing dysplasia		100
Summary			500

Table 6 Presentation of New Data [20]. Available at: http://mde-lab.aegean.gr/downloads

Class	Name	Category	Number of single cells
1	Superficial squamous epithelial	Normal 242 single cells	74
2	Intermediate squamous epithelial		70
3	Columnar epithelial		98
4	Mild squamous non-keratinizing dysplasia	Abnormal 675 single cells	182
5	Moderate squamous non-keratinizing dysplasia		146
6	Severe squamous non-keratinizing dysplasia		197
7	Squamous cell carcinoma in situ intermediate		150
Summary			917

and compared with the variation. For instance, if two class centers are far from each other in the feature space, and the standard deviation within each class is small, then the separation is good. The Mahalanobis distance, which is relative to each class as it depends on the standard deviation existing within the class, takes the local variation into account by using the standard deviation as a yardstick [24].

Table 7 defines what we shall call the distance matrix, that is, a 7-by-7 table D of distances from each class to all the other class centers. The shortest distance in each column (Table 7) indicates the nearest neighbor of each class. This relationship between the classes is presented in Fig. 3, where it clearly seems that classes 1 and 2 are each other's neighbor, and the distance to class 4 is relatively large, indicating that classes 1 and 2 are lying separately from the other classes. Furthermore, there is a forward path from class 4 to class 5 to classes 6 and 7, the last two being mutual neighbors [24].

As it is observed in Table 7 and is seemed in Figs. 3 and 4, the normal class 3 represents a problem, since its nearest neighbor is the abnormal class 6, and it is quite far from classes 1 and 2. Also, for class 3 the distribution is about 4 (mean 4.2). The distance from class 4 to class 6 is 4.3, according this, it is expected an overlap between classes 3 and 6 and it will be difficult to separate them. Similarly, there is

Table 7 Distance matrix D. An element Dij is the Mahalanobis distance seen from class j to the center μi of class i. For a given class, the shortest distance to a center is marked with cyan color) [24]

Center	Class 1	Class 2	Class 3	Class 4	Class 5	Class 6	Class 7
μ_1	0.0	**6.0**	115.4	19.2	31.3	72.6	108.4
μ_2	**9.0**	0.0	76.5	12.1	19.6	47.6	73.1
μ_3	222.8	76.0	0.0	5.3	3.5	2.6	4.5
μ_4	197.9	45.9	18.8	0.0	2.6	9.7	17.8
μ_5	259.6	73.9	7.8	**1.7**	0.0	2.5	5.0
μ_6	328.7	107.7	**4.3**	4.5	**2.1**	0.0	**1.4**
μ_7	412.4	140.0	5.2	6.7	3.4	**1.1**	0.0

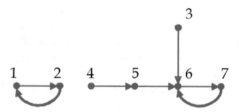

Fig. 3 Digraph of class neighbors. The numbered nodes are classes, and the arrows point from a class to its nearest class neighbor, identified by the shortest distance in the columns of matrix D [24]

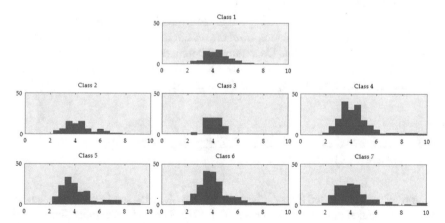

Fig. 4 Histograms of point distances (Mahalanobis) within each class. Few points are closer than distance 2 to their own center. All 20 features are used [24]

overlap between classes 3 and 7, but it is less. Furthermore, classes 4, 5, 6, and 7 are pairwise close, and all are rather dispersed, indicating overlap [24]. Doctors and cyto-technicians were aware of the overlap, and they were willing to assign a shared membership of one cell to several classes simultaneously.

4.4 The Used Methodologies

The main purpose was to optimize the Pap-Smear Diagnosis developing corresponding classifiers and optimizing their feature selection process. More specifically, the Pap-smear Classification Problem was dealt with in two phases.

First, when the 2-class problem was considered, we tried to develop intelligence systems, that could classify the Pap-Smear Data into 2 Categories (1-Normal and 2-Abnormal). Then, during the second phase, when the 7-class problem was considered, we tried to develop intelligence systems, that could classify the Pap-Smear Data into 7 Classes (1-Superficial Squamous Epithelial, 2-Intermediate Squamous Epithelial, 3-Columnar Epithelial, 4-Mild Squamous non-keratinizing Dysplasia, 5-Moderate Squamous non-keratinizing Dysplasia, 6-Severe Squamous non-keratinizing Dysplasia and 7-Squamous cell carcinoma in Situ intermediate).

The intelligence systems developed, were built on the basis of computational intelligence classification using both Old and New Data. In addition, different feature selection methods were used for improving the classification process.

The Training-Validation process of these systems, in some cases was supervised and in other unsupervised and was become using several Training-Validation methods such as k-fold cross validation, single split and swap random tests.

Finally, the performance measures, that were used during the validation process of the above intelligence systems, were Root Mean Squared Error (RMSE), False-negative rate ($F_N\%$), False-positive rate ($F_P\%$), Sensitivity, Specificity, and Positive and Negative Predictive Values (P_P and P_N).

The following paragraphs briefly present the CI-Techniques, that were used on the Pap-Smear Classification Problem, from the researchers and students of the Department of Automation, Technical University of Denmark as well as of the Management and Decision Engineering Laboratory (MDE-Lab) of the Department of Financial and Management Engineering, University of the Aegean.

4.4.1 Adaptive Network-Based Fuzzy Inference System (ANFIS)

Adaptive Network-based Fuzzy Inference System (ANFIS) is a fuzzy inference system implemented in framework of networks. ANFIS could be illustrated as an input-output network based both human knowledge (Fuzzy IF-THEN Rules) and simulated input-output data patterns. ANFIS could be used to (a) model non-linear functions, (b) identify non-linear components automatically in a control system, (c) predict a chaotic time-series, (d) automatic control or (e) signal processing [27].

4.4.2   Artificial Neural Networks

Artificial Neural Network (ANN) is a computational system inspired by the learning ability and structure of a biological brain. An ANN consists of a large number of very simple processing neuro-like processing elements and their weighted connections, with distributed representation of knowledge over the connections, that are acquired by network through a learning process. The main features of an ANN are its topology, learning and validation. ANNs are a major machine learning category, that have a huge application range in deferent domains of deferent science [28].

4.4.3   Heuristic Classification

Heuristic Classification is a classification method, which differs from the Simple, Certain and Statistical classification. The difference, between heuristic and simple/certain classifications, exploits uncertain knowledge and the difference with the statistical classification is that factors are estimated by experts and not calculated from a database. The heuristic classification model characterizes a form of knowledge and reasoning-patterns of familiar problem situations and solutions, heuristically related [29].

Given this and having in mind that human beings find it difficult to keep track of many individual assessments for a solution, the technique of step-by-step abstraction from observations to solutions is necessary, where the constellations of observations are combined to a single concept, which can be treated as one observation in assessing solutions. There are two types of abstraction (a) the compression of raw data to simple data abstractions, which is usually based on concept definitions, and (b) the derivation of partial solutions or solution classes, which is based on uncertain knowledge [29].

4.4.4   Minimum Distance Classification

According of the process of minimum distance classification, the training is made using the objects of known classes, each class is represented by its mean vector, the mean of the feature vectors for the object within the class is calculated and the new objects are classified by finding the closest mean vector [24].

4.4.5   Hard C-Means Clustering

Hard C-Means (HCM) or k-Means Clustering is a data-analysis method that treats observations of a dataset as objects based on locations and distance between various input data points and develops clusters characterized by their center points (Centroid) and the objects within each cluster remain as close as possible to each other but as far as possible from objects in other clusters. The coordinates of a cluster centroid are defined by the calculation of the average of each of the coordinates of the points of samples assigned to the cluster. HCM Clustering starts up by randomly choosing of clusters centroid, then all data is mapped to the nearest centroid and according of the objects' coordinates of each cluster, a new cluster centroid is defined with the ultimate goal of minimizing the distance between the coordinates of the objects and the centroid of each cluster [30].

4.4.6   Fuzzy C-Means Clustering

Fuzzy C-Means Clustering method is a clustering technique that allows one part of data to belonging into two or more clusters with associated membership degree values from 0 to 1. FCM algorithm is based on the minimization of an objective function called C-Means function. During the clustering process, the FCM clustering algorithm searches for optimal clustering centers and membership degrees by minimizing the objective function iteratively [30].

4.4.7   Gustafson and Kessel Clustering

The Gustafson and Kessel clustering technique, is a similar classification technique to FCM with the only difference that the GK uses the Mahalanobian distance measure while FCM uses the Euclidian distance measure. The objective function of GK is to minimize the distances from the cluster centers to the data points [18].

4.4.8   k-Nearest Neighborhood Classification

The k-Nearest-Neighborhood Classifier (KNN) is a simple classification method, that commonly based on the Euclidean distance between a test sample and the specified training samples. The KNN algorithm assigns class membership to a sample vector rather than assigning the vector of a particular class. The aim of this method is to assign membership as a function of the vector's distance from its AT-nearest Neighborhoods and those Neighborhoods' memberships in the possible classes [20].

4.4.9 Weighted k-Nearest Neighborhood Classification

The Weighted k-Nearest Neighborhood (WKNN) is a similar to KNN classification method with only deferent that the class of the k-nearest samples is weighted in the output. The weighting of each sample is inversely proportional with the distance between training and test samples. In this way the training sample that is the most distant from the test point, is weighted as a minimum [20].

4.4.10 Tabu Search

Tabu search (TS) was introduced by Glover [26] as a general iterative metaheuristic for solving combinatorial optimization problems and is form of local Neighborhood search where each solution S has an associated set of Neighborhoods N(S). A solution $S' \in N(S)$ can be reached from S by an operation called a move. Tabu Search moves from a solution to its best admissible Neighborhood, even if this causes the objective function to deteriorate. To avoid cycling, solutions that have been recently explored are declared forbidden or tabu for a number of iterations. The tabu status of a solution is overridden when certain criteria (aspiration criteria) are satisfied. It should be mentioned that, sometimes, intensification and diversification strategies are used to improve the search, where in the first case, the search is accentuated in the promising regions of the feasible domain and in the second case, an attempt is made to consider solutions in a broad area of the search space. Computational experience has shown that Tabu Search is a well-established approximation technique, which can compete with almost all known techniques and which, by its flexibility, can outperform many classic processes [22].

4.4.11 Genetic Programming

Genetic programming (GP), as an Evolutionary Algorithm (EA)—a subset of machine learning, is a domain-independent method that genetically breeds a population of computer programs to solve a problem. Generally, GP is applied in various domains where a direct search method cannot be applied due to the nature of the problem. Genetic algorithms demand the use of a coding scheme for a fixed-length string, which is a selection constraining the "size" of a solution. The application of genetic programming, follows much of the theory of genetic algorithms with a basic difference to the form of representation of the structures used, which resembles more to a variable-length tree, than a fixed string. In genetic programming, a population of random trees is initially generated, representing programs. Then, the genetic operations (crossover, mutation, reproduction and inversion) on these trees are performed. The most important operations of GP are the crossover and reproduction [17].

4.4.12   Ant Colony

An Ant Colony Optimization algorithm (ACO) is essentially a system based on agents which simulate the natural behavior of ants, including mechanisms of cooperation and adaptation The ants, in search of their nearest source of food, search randomly the surrounding area of their nest and as they move, a certain amount of pheromone is dropped on the ground, marking the path with a trail of this substance. It is remarkable to mention that the amount of pheromone depends on the distance, quantity and quality of the food source and as time passes, the pheromone starts to evaporate. Based on this process, a path is created, which becomes more attractive to other ants, when it is followed by more ants. It was observed that a short path to source of food is more possible to be followed by the ants with the result that the pheromone density remains high for a longer period of time and since ants prefer to follow trails with larger amounts of pheromone, eventually all the ants converge to the shorter path [23].

ACO approaches Combinatorial Optimization Problems (COPs) according to the search of a minimum cost path in a graph, like ants search a short path to food source. The best paths are found with the cooperation among ants in the colony, as only one ant has a rather simple behavior and will typically only find rather poor-quality paths on its own. An ACO algorithm consists of iterations, which are a number of cycles of solution construction. During each iteration a number of ants (which is a parameter) constructs complete solutions using heuristic information and the collected experiences of previous groups of ants. These collected experiences are represented by a digital analogue of trail pheromone which is deposited on the constituent elements of a solution [23].

4.5   The Pap-Smear Classification Problem

The presentation of methods below is made according to the sequence of their appearance in literature. More details of each approach are presented and can be found in [2, 16–25].

4.5.1   Classification with ANFIS

The ANFIS models of Pap-Smear Classification Problem, which was the first approach effort of Pap-Smear Classification Problem and was presented by Jens Byriel in 1999 [16], were developed to classify the Old Data (Table 5) into 2 classes (Normal and Abnormal) according to specific features of cells.

The system of the first ANFIS model, which, in this case, is called ANFIS_M, is a network with 34 adaptive parameters (27 premise and 7 consequent), 4 nominal inputs and 2 numerical outputs. The ANFIS_M is set to classify the cells, according to 7 IF_THEN rules, into normal (output = 1) and abnormal/dysplastic (output = 2) categories. Table 8 presents the basic features and Table 9 presents the classification rules of ANFIS_M [16].

The ANFIS_M developed and trained for 200 Epochs using the MATLAB Fuzzy toolbox. The validation of this model used Swap-random tests and the performance measures were the Root Mean Squared Error (RMSE) between actual and produced output value, False-negative (Normal Cells) rate (F_N%) and False-positive (Abnormal Cells) rate (F_P%) [16].

Tables 9 and 10 presents the classification rules of ANFIS_M and its performance during the classification process of Old Data.

A false negative ratio of 9% and a false positive ratio of 13% was achieved as a result. Also, acceding of the developing process of this model, it is observed that the model lagged behind to the classification of the Columnar Epithelial, Parabasal Squamous Epithelial, Mild Squamous non-keratinizing Dysplastic, Moderate Squamous non-keratinizing Dysplastic, Severe Squamous non-keratinizing Dysplastic cells. A

Table 8 Basic features of ANFIS_M [16]

Inputs	Nucleus area (Nominal)	Nucleus/cytoplasm ratio (Nominal)	Nucleus color (Nominal)	Cytoplasm color (Nominal)
Outputs	1 = Normal cell (Numerical)	2 = Abnor-mal/dysplastic cell (Numerical)	Outputs	1 = Normal cell (Numerical)
Rules	7 classification rules			

Table 9 Classification rules of ANFIS_M [16]

Rules	Nucleus area (Nominal)	Nucleus/cytoplasm ratio (Nominal)	Nucleus color (Nominal)	Cytoplasm color (Nominal)	Output
1	Small	Not small	Light	Dark	1
2	Small	Medium	Dark	Dark	1
3	Small	Small	Dark	Light	1
4	Large	Medium	Light	Light	2
5	Large	Not small	Light	Light	2
6	Large	Large	Light	Light	2
7	Large	large	Dark	Dark	2

Table 10 ANFIS performances after training [16]

RMSE	0.20
F_N %	9
F_P %	13

suggestion that could solve this problem, is to find ways to improve the structure of rules bases and a better distinguishing method of the above cells [16].

Another Fuzzy Inference System (FIS) was used to improve the performance of ANFIS application on Pap-Smear Classification Problem. This FIS is called GEN-FIS2 and generates a Sugeno-type FIS structure using subtractive clustering of used data by extracting a set of rules that models the data behavior [16].

The produced model of this approach, that, in this case, is called GENFIS_M, is presented in [27] and is set to classify the Old Data into 2 classes (Normal and Abnormal) according specifically features of cells.

GENFIS_M is a network with 111 adaptive parameters (108 premise and 3 consequent), 9 numerical inputs and 2 numerical outputs. The GENFIS_M is set to classify the cells, according 3 IF_THEN rules, into normal (output = 1) and abnormal/dysplastic (output = 2) categories using 9 cells features (Inputs), which emerged from the process of Simulated Annealing (SA) Feature Selection and are presented in Table 12 [16].

The GENFIS_M was developed and trained using the MATLAB Fuzzy toolbox. The validation of this model used Swap-random tests and the performance measures were the Root Mean Squared Error (RMSE) between actual and produced output value, False-negative (Normal Cells) rate ($F_N\%$), False-positive (Abnormal Cells) rate ($F_P\%$), Sensitivity-Specificity and Positive-Negative Predictive of the model [16] (Table 11).

Tables 12 and 13 present the inputs of GENFIS_M and its performance during the classification process of Old Data.

The performances of GENFIS_M show that the application of GENFIS2, using the feature selection process of Simulated Annealing (SA), produces a satisfactory model for the 2 classes classification problem of Pap-Smear.

To improve the above classification model (GENFIS_M), FCM clustering technique was applied on it, following the same training/validation process and doing feature selection according the Simulated Annealing (SA) Feature Selection.

The inputs and the performances of the new model, that, in this case, is called GENFIS_M1, are present in Tables 14 and 15.

In improving the above performances of GENFIS_M1, GK clustering technique was applied and a new model was produced, that, in this case, is called GENFIS_M2. The GENFIS_M2 was trained/validated with the same method of the previous mod-

Table 11 Basic features of GENFIS_M [16]	Inputs	9 numerical inputs that are presented in Table 12	
	Outputs	1 = Normal cell (Numerical)	2 = Abnormal/dysplastic cell (Numerical)
	Rules	3 classification rules	

els. The performances and the inputs of this model, that came from the application of the Simulated Annealing (SA) Feature Selection, are presented in Tables 16 and 17.

Finally, a Nearest Neighborhood Clustering technique (NNH) was applied on the same data and under the same training/validation conditions. The result of this effort

Table 12 The numerical inputs of GENFIS_M [16]

Nr.	Name	Feature
2	Cyto_A	Cytoplasm area
4	Kerne_Ycol	Nucleus brightness
6	KerneShort	Nucleus shortest diameter
9	KerneRund	Nucleus roundness
10	CytoShort	Cytoplasm shortest diameter
11	CytoLong	Cytoplasm longest diameter
14	KernePeri	Nucleus perimeter
16	KernePos	Nucleus relative position
17	KerneMax	Maxima in nucleus

Table 13 GENFIS_M performances after training [16]

RMSE	0.19
F_N %	0.7
F_P %	2.7
Sensitivity %	99.3
Specificity %	97.3
P_P %	98.3
P_N %	99.3

Table 14 The numerical inputs of GENFIS_M1 [16]

Nr.	Name	Feature
5	Cyto_Ycol	Cytoplasm brightness
6	KerneShort	Nucleus shortest diameter
9	KerneRund	Nucleus roundness
15	CytoPeri	Cytoplasm perimeter

Table 15 GENFIS_M1 performances after training [16]

RMSE	0.23
F_N %	7.0
F_P %	3.7
Sensitivity %	93.0
Specificity %	96.3
P_P %	97.7
P_N %	90.3

Table 16 The numerical inputs of GENFIS_M2 [16]

Nr.	Name	Feature
3	K/C	Nucleus/cytoplasm ratio
4	Kerne_Ycol	Nucleus brightness
5	Cyto_Ycol	Cytoplasm brightness
6	KerneShort	Nucleus shortest diameter
7	KerneLong	Nucleus longest diameter
8	KerneElong	Nucleus elongation
9	KerneRund	Nucleus roundness
10	CytoShort	Cytoplasm shortest diameter
12	CytoElong	Cytoplasm elongation
14	KernePeri	Nucleus perimeter
16	KernePos	Nucleus relative position
17	KerneMax	Maxima in nucleus
20	CytoMin	Minima in cytoplasm

Table 17 GENFIS_M2 performances after training [16]

RMSE	0.17
F_N %	1.0
F_P %	6.3
Sensitivity %	99.0
Specificity %	93.7
P_P %	96.0
P_N %	98.7

was to have a new classification model, which in this case is called GENFIS_M3. The performances and inputs, after the process of Simulated Annealing (SA) Feature Selection, of the GENFIS_M3 are presented in Tables 18 and 19.

In conclusion, this first approach of Pap-Smear Classification Problem with CI-Techniques, showed that for the 2-class classification problem, the best classification model was the GENFIS_M with $F_N = 0.7\%$ and $F_P = 2.7\%$ (Table 13).

However, observing and comparing the 3 clustering-based classifications (GENFIS_M1, GENFIS_M2 and GENFIS_M3), it seems that the NNH algorithm was better adapted to the shape of clusters in the database with $F_N = 1.7\%$ and $F_P = 3.0\%$.

4.5.2 Heuristic Classification Based on GP

In this approach given in [17], a hybrid computational intelligence model, that combines Genetic Programming and Heuristic Classification, is suggested. This proposed

Table 18 The numerical inputs of GENFIS_M3 [16]

Nr.	Name	Feature
1	Kerne_A	Nucleus area
3	K/C	Nucleus/cytoplasm ratio
4	Kerne_Ycol	Nucleus brightness
5	Cyto_Ycol	Cytoplasm brightness
6	KerneShort	Nucleus shortest diameter
8	KerneElong	Nucleus elongation
10	CytoShort	Cytoplasm shortest diameter
11	CytoLong	Cytoplasm longest diameter
12	CytoElong	Cytoplasm elongation
16	KernePos	Nucleus relative position
18	KerneMin	Minima in nucleus

Table 19 GENFIS_M3 performances after training [16]

RMSE	0.13
F_N %	1.7
F_P %	3.0
Sensitivity %	98.3
Specificity %	97.0
P_P %	98.0
P_N %	97.6

system uses a genetic programming process, in order to produce generalized classification rules between abnormal cells and each of the normal cell types. This proposal approaches the above classification problem from the following 3 directions: (a) the application of genetic programming itself, (b) the extraction of simple rules, comprehensible by humans and (c) the verification of existent medical indices based on existing data [17].

The used data, in this, cases, derive from the Old Data and consist of 450 patterns of 24 cell features (20 feature of Tables 3 and 4 new features) and 5 cell classes (Table 20). For the training and validation of this hybrid model the used data were divided into 2 equal sets (Training and Test Set) of 225 data patterns, as they are presented in Table 20. After the separation of the available data into training and test sets, the rules presented in Table 22 were extracted. Training and test sets are composed using the involved classes [17] (Tables 21 and 22).

The results of this approach to the Pap-Smear classification problem have been accepted by the expert doctors and could be useful for providing computer-based medical aid in the Pap-Smear diagnosis. This positive response by the medical experts leads to new research, which is related to the optimization of the genetic programming process with a view to its optimal application, both in the case of Pap-Smear Diagnosis and others similar cases.

Table 20 The New Classes and the Used Sets [17]

Old class	New Class	Name of New Class	Patterns for training set	Patterns of test set
1	1	CYL	25	25
2	2	PARA	25	25
3	3	INTRA	25	25
4	4	SYPORA	25	25
5, 6 and 7	5	DYSPLASIC	125	125

Table 21 Rules and separation intension. The classes shown for each rule, denote also the corresponding training and test set for that rule [17]

Rule	Separation class	Class to have separation from
1	CYL and PARA	INTRA and SUPRA
2	INTRA	SUPRA
3	INTRA	DYSPLASIC
4	CYL	PARA
5	SUPRA	DYSPLASIC
6	CYL	DYSPLASIC
7	PARA	DYSPLASIC

Table 22 The rules and their accuracy in test set of the hybrid model [17]

Nr.	Rule	Accuracy in unknown data (%)
1	If KerneY > CytoMin and KernePeri < 28 and KerneMax > CytoShort then cell class is Intra (class 3)	90.42
2	If KerneY > CytoMin and KernePeri < 28 and KerneMax $\leq$ CytoShort then cell class is Dysplasia (classes 5, 6, 7)	90.42
3	If KerneY > CytoMin and KernePeri > 28 and CytoLong $\leq$ Kerne_A then cell class is Supra (class 4)	97.96
4	If KerneY > CytoMin and KernePeri > 28 and CytoLong > Kerne_A then cell class is Dysplasia (classes 5, 6, 7)	97.96
5	If KerneY < CytoMin and (KernePos $\geq$ CytoRund * CytoRund or CytoRund * CytoRund $\geq$ KerneElong) and Kerne_A > KerneMax/Kerne_Ycol then cell class is Para (class 2)	88.46

(continued)

Table 22   (continued)

Nr.	Rule	Accuracy in unknown data (%)
6	If KerneY < CytoMin and (KernePos ≥ CytoRund * CytoRund or CytoRund * CytoRund ≥ KerneElong) and Kerne_A ≤ KerneMax/Kerne_Ycol then cell class is Dysplasia (classes 5, 6, 7)	99.17
7	If KerneY < CytoMin and KernePos < CytoRund * CytoRund and CytoRund * CytoRund < KerneElong and Kerne_A > 112 then cell class is Cyl (class 1)	66.35
8	If KerneY < CytoMin and KernePos < CytoRund * CytoRund and CytoRund * CytoRund < KerneElong and Kerne_A < 112 then cell class is Dysplasia (classes 5, 6, 7)	91.28

4.5.3   Classification Using Defuzzification Methods

This Pap-Smear classification effort, which was presented by Erik Martin in 2003 [18], approaches the 2-class classification problem using defuzzification methods and based on HCM, FCM and GK clustering.

The final models, which are presented in reference [18] and in this case are called DF_HCM, DF_FCM and DF_GK, are a result of several experimental application-tests of HCM/FCM and GK on Old and New Data following supervised clustering process and were trained/validated according k-fold cross validation.

As has been reported, the used data were both the Old and New (Tables 5 and 6) and the performances (Overall Error, F_P, F_N) of the above models, which are presented in Tables 23, 24, 25, 26, 27 and 28 [18].

The DF_HCM and DF_FCM emerged following the next steps: (a) Determining the number of clusters used. This was made possible by measuring the error with 2-fold cross-validation with 20 reruns, for a series of clusters and using all the cell features (Table 4). (b) Calculation of the error of application of HCM and FCM clustering. This was obtained using 10-fold cross validation with 20 reruns and without feature selection and (c) according to the results of previous steps and applied Simulated Annealing (SA) Feature Selection method, the classification models DF_HCM and DF_FCM were resulted. The DF_GK came out following a process similar to the above [18].

Tables 23, 24, 25, 26, 27 and 28 present the cell features, that were used in every classification case after the feature selection process, and the models' performances according the process of 10-fold cross validation [18].

Table 23 The cells features that were used in DF_HCM [18]

Old Data	Selected features for DF_HCM		New Data
Nr.	Name	Feature	Nr.
1	Kerne_A	Nucleus area	1
–	Cyto_A	Cytoplasm area	2
3	K/C	Nucleus/cytoplasm ratio	3
4	Kerne_Ycol	Nucleus brightness	4
5	Cyto_Ycol	Cytoplasm brightness	5
6	KerneShort	Nucleus shortest diameter	6
7	KerneLong	Nucleus longest diameter	7
–	KerneElong	Nucleus elongation	8
11	CytoLong	Cytoplasm longest diameter	11
12	CytoElong	Cytoplasm elongation	12
13	CytoRund	Cytoplasm roundness	–
–	KernePeri	Nucleus perimeter	14
–	CytoPeri	Cytoplasm perimeter	15
16	KernePos	Nucleus relative position	16
–	KerneMax	Maxima in nucleus	17
18	KerneMin	Minima in nucleus	18
–	CytoMax	Maxima in cytoplasm	19
–	CytoMin	Minima in cytoplasm	20

Table 24 Performances of DF_HCM [18]

Supervised clustering	Overall Error %	F_P %	F_N %
Old Data	8.77	18.81	5.24
New Data	7.86	17.12	4.55

Table 25 The cells features that were used in DF_FCM [18]

Old Data	Selected features for DF_FCM		New Data
Nr.	Name	Feature	Nr.
–	Kerne_A	Nucleus area	1
2	Cyto_A	Cytoplasm area	–
3	K/C	Nucleus/cytoplasm ratio	3
4	Kerne_Ycol	Nucleus brightness	4
5	Cyto_Ycol	Cytoplasm brightness	5

(continued)

Table 25   (continued)

Old Data	Selected features for DF_FCM		New Data
Nr.	Name	Feature	Nr.
6	KerneShort	Nucleus shortest diameter	6
7	KerneLong	Nucleus longest diameter	7
–	KerneElong	Nucleus elongation	–
–	KerneRund	Nucleus roundness	–
10	CytoShort	Cytoplasm shortest diameter	10
11	CytoLong	Cytoplasm longest diameter	–
12	CytoElong	Cytoplasm elongation	–
–	CytoRund	Cytoplasm roundness	–
–	KernePeri	Nucleus perimeter	14
15	CytoPeri	Cytoplasm perimeter	15
16	KernePos	Nucleus relative position	–
–	KerneMax	Maxima in nucleus	17
18	KerneMin	Minima in nucleus	18
–	CytoMax	Maxima in cytoplasm	19
20	CytoMin	Minima in cytoplasm	20

Table 26   Performances of DF_FCM [18]

Supervised clustering	Overall Error %	F_P %	F_N %
Old Data	1.64	2.02	1.38
New Data	6.10	13.89	3.29

Table 27   The cells features that were used in DF_GK [18]

Old Data	Selected features for DF_GK		New Data
Nr.	Name	Feature	Nr.
1	Kerne_A	Nucleus area	1
2	Cyto_A	Cytoplasm area	2
3	K/C	Nucleus/cytoplasm ratio	–
4	Kerne_Ycol	Nucleus brightness	4
5	Cyto_Ycol	Cytoplasm brightness	5

(continued)

Table 27 (continued)

Old Data	Selected features for DF_GK		New Data
Nr.	Name	Feature	Nr.
6	KerneShort	Nucleus shortest diameter	6
7	KerneLong	Nucleus longest diameter	7
8	KerneElong	Nucleus elongation	8
9	KerneRund	Nucleus roundness	9
10	CytoShort	Cytoplasm shortest diameter	10
11	CytoLong	Cytoplasm longest diameter	–
12	CytoElong	Cytoplasm elongation	12
–	CytoRund	Cytoplasm roundness	13
14	KernePeri	Nucleus perimeter	14
15	CytoPeri	Cytoplasm perimeter	15
16	KernePos	Nucleus relative position	16
17	KerneMax	Maxima in nucleus	17
18	KerneMin	Minima in nucleus	18
–	CytoMax	Maxima in cytoplasm	19
–	CytoMin	Minima in cytoplasm	20

Table 28 Performances of DF_GK [18]

Supervised clustering	Overall Error %	F_P %	F_N %
Old Data	2.89	4.46	1.77
New Data	6.10	13.89	3.29

4.5.4 Direct and Hierarchical Classification

In the previous paragraph, the classification problem has been to decide between the normal and abnormal cells (2-class problem). In this paragraph, direct and Hierarchical Classification are presented for all the diagnoses of the cells. According to the results of the previous approach in the 2-class problem, the best clustering method proved to be the FCM. In this approach in the 7-class classification problem using the directed and hierarchical classification methods, the only used clustering method was the FCM [18].

The used databases were both the Old and New Data. The selected features, the clusters number and the fuzzy exponent were modified in the same manner as earlier when facing the 2-class problem. The validation process followed in these classifications, was the k-fold cross validation [18] (Fig. 5 and Tables 29, 30, 31, 32 and 33).

Classification results obtained for the old data, shows that the accuracy for all the normal cell diagnoses is above 90% for the hierarchical classifiers. The direct classifiers have an almost 5% worse accuracy for each of the normal cell diagnoses. The poor separation between the abnormal cells was expected, since this is also one of the hardest classifications for cyto-technicians. Classification results with the new data, shows a good separation of the superficial and intermediate normal cells from the abnormal. But the Columnar epithelial cells had a very poor separation from the abnormal cells. Compared to the separation between the columnar epithelial cells and the abnormal cells in the old data, this is surprising. The pictures in the new

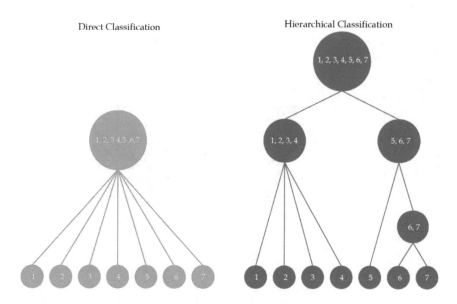

Fig. 5 Direct and hierarchical classification of 7 classes classification problem [18]

Table 29 (Un)supervised direct classification of Old Data-parameters and confusion matrix [18]

Unsupervised direct classification of Old Data-parameters and confusion matrix

Features	1, 2, 3, 4, 5, 6, 7, 8, 11, 12, 14, 16						
Fuzzy exponent q = 1.5							
Number of clusters = 50							
Classes	COL (%)	PAR (%)	INT (%)	SUP (%)	MIL (%)	MOD (%)	SEV (%)
COL	86.4	0.0	0.0	0.0	0.2	3.9	9.5
PAR	4.0	92.0	0.0	0.0	0.0	3.3	0.7
INT	0.0	0.0	84.6	13.6	1.8	0.0	0.0
SUP	0.0	0.0	13.3	86.7	0.0	0.0	0.0
MIL	1.0	2.45	0.0	0.0	55.05	24.25	17.25
MOD	1.0	0.05	0.0	0.0	26.6	39.7	32.65
SEV	0.55	1.25	0.0	0.0	8.3	24.85	65.05

Supervised direct classification of Old Data-parameters and confusion matrix

Features	1, 2, 3, 4, 5, 6, 7, 10, 11, 16						
Fuzzy exponent q = 1.4							
Number of clusters = 100							
Classes	COL (%)	PAR (%)	INT (%)	SUP (%)	MIL (%)	MOD (%)	SEV (%)
COL	85.8	0.0	0.0	0.0	0.0	0.7	13.5
PAR	4.0	92.0	0.0	0.0	0.0	2.3	1.7
INT	0.0	0.4	88.6	10.0	1.0	0.0	0.0
SUP	0.0	0.0	6.0	94.0	0.0	0.0	0.0
MIL	0.95	1.1	0.0	0.0	61.45	25.6	10.9
MOD	1.0	0.0	0.0	0.0	22.6	56.95	19.45
SEV	0.65	0.9	0.0	0.0	6.5	31.75	60.2

Table 30 (Un)supervised direct classification of New Data-parameters and confusion matrix [18]

Unsupervised direct classification of New Data-parameters and confusion matrix

Features	1, 2, 3, 4, 5, 6, 7, 8, 12, 16						
Fuzzy exponent q = 1.25							
Number of clusters = 100							
Classes	COL (%)	PAR (%)	INT (%)	SUP (%)	MIL (%)	MOD (%)	SEV (%)
COL	85.64	14.36	0	0	0	0	0
PAR	9.43	87.86	1.28	1.43	0	0	0
INT	0	0.87	52.33	6.79	13.55	20.96	5.50
SUP	0	0.50	2.06	62.42	20.91	12.98	1.13
MIL	0	0	4.27	32.50	32.11	18.94	12.18
MOD	0	0	11.16	12.53	15.04	34.81	26.46
SEV	0	0	4.63	2.27	9.13	35.43	48.54

Supervised direct classification of New Data-parameters and confusion matrix

Features	1, 2, 3, 4, 5, 6, 7, 10, 11, 14, 16						
Fuzzy exponent q = 1.26							
Number of clusters = 100							
Classes	COL (%)	PAR (%)	INT (%)	SUP (%)	MIL (%)	MOD (%)	SEV (%)
COL	89.57	10.43	0	0	0	0	0
PAR	9.21	89.79	0	1.0	0	0	0
INT	0	0	55.48	7.73	11.28	23.16	2.35
SUP	0	0.93	1.40	79.09	9.40	9.15	0.0
MIL	0	0	5.02	41.24	26.29	21.66	5.79
MOD	0	0	8.14	11.15	10.65	41.47	28.59
SEV	0	0	3.3	2.97	7.36	21.87	64.5

Table 31 (Un)supervised hierarchical classification of Old Data-parameters and confusion matrix [18]

Unsupervised hierarchical classification of Old Data-parameters and conf. matrix

Nodes	Chosen features				q	Clusters
Node 1234567	1, 2, 3, 4, 5, 6, 7, 10, 11, 12, 15, 16, 17, 18, 20				q = 1.25	100
Node 1234	All				q = 1.17	100
Node 12	3, 4, 5, 9, 10, 12, 15, 16, 17, 18, 19				q = 1.6	100
Node 34	1, 4, 7, 12, 14				q = 1.76	6
Node 567	1, 2, 3, 4, 5, 7, 9, 10, 11, 12, 13, 14, 15, 16, 17, 18				q = 1.27	40
Node 67	1, 2, 3, 4, 5, 7, 11, 14, 15, 16, 17, 19				q = 1.5	28

Classes	SUP (%)	INT (%)	COL (%)	MIL (%)	MOD (%)	SEV (%)	CIS (%)
SUP	92.76	7.24	0	0	0	0	0
INT	6.07	92.36	0.79	0.78	0	0	0
COL	0	0	51.48	0.99	10.11	33.1	4.32
MIL	0	0	0.72	67.36	24.38	6.82	0.72
MOD	0	0	2.21	31.28	44.86	17.76	3.89
SEV	0	0.03	6.15	6.61	11.71	47.56	27.94
CIS	0	0	2.27	1.13	5.87	45.07	45.66

Supervised hierarchical classification of Old Data-parameters and conf. matrix

Nodes	Chosen features				q	Clusters
Node 1234567	2, 3, 4, 5, 6, 7, 10, 11, 12, 15, 16, 18, 20				q = 1.2	50
Node 1234	All				q = 1.17	50
Node 12	1, 2, 3, 4, 5, 6, 7, 8, 10, 12, 13, 14, 15, 16, 17, 18, 19, 20				q = 1.38	50
Node 34	3, 4, 5, 6, 7, 8, 10, 12, 14				q = 1.27	1
Node 567	1, 2, 3, 4, 5, 7, 9, 10, 11, 12, 13, 14, 15, 16, 17, 18				q = 1.31	12
Node 67	1, 2, 3, 4, 5, 7, 11, 14, 15, 16, 17, 19				q = 1.57	50

Classes	SUP (%)	INT (%)	COL (%)	MIL (%)	MOD (%)	SEV (%)	CIS (%)
SUP	93.53	6.47	0	0	0	0	0
INT	10.21	87.79	1.21	0.79	0	0	0
COL	0	0	66.81	1.86	5.79	18.71	6.83
MIL	0	0	0.71	68.09	24.86	5.13	1.21
MOD	0	0	1.98	28.99	46.02	17.37	5.64
SEV	0	0.48	6.39	6.65	8.38	48.19	29.91
CIS	0	0	2.53	2.67	4.6	34.1	56.1

Table 32 (Un)supervised hierarchical classification of New Data-parameters and confusion matrix [18]

Unsupervised hierarchical classification of Old Data-parameters and conf. matrix

Nodes	Chosen features	q	Clusters
Node 1234567	1, 2, 3, 4, 6, 7, 10, 12, 13, 14	$q = 1.13$	100
Node 123	All	$q = 1.74$	30
Node 12	5, 7, 11, 14	$q = 1.9$	60
Node 4567	2, 3, 5, 6, 10, 15	$q = 1.25$	96
Node 45	1, 4, 5, 7, 10, 14, 16, 17, 19	$q = 1.3$	96
Node 67	1, 4, 5, 7, 10, 14, 16, 17, 19	$q = 1.3$	80

Classes	SUP (%)	INT (%)	COL (%)	MIL (%)	MOD (%)	SEV (%)	CIS (%)
SUP	92.76	7.24	0	0	0	0	0
INT	6.07	92.36	0.79	0.78	0	0	0
COL	0	0	51.48	0.99	10.11	33.1	4.32
MIL	0	0	0.72	67.36	24.38	6.82	0.72
MOD	0	0	2.21	31.28	44.86	17.76	3.89
SEV	0	0.03	6.15	6.61	11.71	47.56	27.94
CIS	0	0	2.27	1.13	5.87	45.07	45.66

Supervised hierarchical classification of Old Data-parameters and conf. matrix

Nodes	Chosen features	q	Clusters
Node 1234567	1, 3, 4, 5, 6, 7, 10, 14, 15, 17, 18, 19, 20	$q = 1.2$	50
Node 123	All	$q = 1.8$	8
Node 12	1, 5, 7, 8, 11, 14, 20	$q = 1.9$	3
Node 4567	1, 2, 3, 4, 5, 10, 13, 17, 19	$q = 1.4$	50
Node 45	1, 2, 4, 5, 6, 10, 14, 16, 18, 19, 20	$q = 1.27$	49
Node 67	1, 2, 4, 5, 6, 10, 14, 16, 18, 19, 20	$q = 1.25$	48

Classes	SUP (%)	INT (%)	COL (%)	MIL (%)	MOD (%)	SEV (%)	CIS (%)
SUP	93.53	6.47	0	0	0	0	0
INT	10.21	87.79	1.21	0.79	0	0	0
COL	0	0	66.81	1.86	5.79	18.71	6.83
MIL	0	0	0.71	68.09	24.86	5.13	1.21
MOD	0	0	1.98	28.99	46.02	17.37	5.64
SEV	0	0.48	6.39	6.65	8.38	48.19	29.91
CIS	0	0	2.53	2.67	4.6	34.1	56.1

Table 33 Classification accuracy and average number of selected features [18]

Classification accuracy and average number of selected features		
Accuracy of direct classification		
	Unsupervised (%)	Supervised (%)
Old Data	72.5	77.0
New Data	57.67	63.74
Accuracy of hierarchical classification		
	Unsupervised (%)	Supervised (%)
Old Data	79.74	80.52
New Data	63.15	66.65
Average number of selected features		
	Unsupervised	Supervised
Old Data	11.8	13.6
New Data	7.6	10.2

database has been verified by a cyto-technician and no problems with diagnoses of the cells was found [18].

4.5.5 Classification Using Feed-Forward Neural Network

This approach of Pap-Smear Classification Problem, that was presented in [19], suggests a feed forward neural network based on both LMAM (Levenberg-Marquardt with Adaptive Momentum) and OLMAM (Optimized Levenberg-Marquardt with Adaptive Momentum) training algorithms.

In this approach, two deferent databases, coming from Old Data, were used. The first database consists of 500 samples each of which is represented by 20 numerical attributes and the second database consists of 500 samples each of which is represented by 9 numerical attributes. The nine (9) cell features that came from process of Feature Selection during the classification with ANFIS (Sect. 5.1/Table 12). Each of these databases was divided into training and test sets using 90% and 10% of the samples respectively using 10-folds cross-validation [19].

The best ANN-LMAM for the 2 classes classification problem is that has 10 hidden nodes (1 Hidden Layer) and 20 inputs (all cell features) and the best ANN-OLMAM for the 2 classes classification problem is that has 9 hidden nodes (1 Hidden Layer) and 20 inputs (all cell features), with corresponding accuracy in the test process 98.42 and 98.86% [19] (Table 34).

Table 34 ANN models for 2 class classification problem [19]

ANN-LMAM	20 inputs—1 Hidden Layer (10 nodes)—1 output (normal and abnormal)	Testing accuracy 98.42%
ANN_OLMAM	20 inputs—1 Hidden Layer (10 nodes)—1 output (normal and abnormal)	Testing accuracy 98.86%

4.5.6 Nearest Neighborhood Classification Based on GP Feature Selection

This approach, that was presented in 2006 [2] and based on both the previous presented and similar works, like the effort made by Norup in 2005 [21], concerns the feature selection problem of Pap-Smear Classification Problem and suggests the application of GA feature selection under the scepter of nearest Neighborhood classification.

In this case, Old Data and New Data were used to approach both the 2 classes and 7 classes classification, using the above hybrid CI method. The Old Data, as the New Data, were separated into two data sets (Training Set and Test Set) according to the process of k-fold cross validation and for k = 2, 3, 4, 5, 10 and 20 [2].

As it has already been mentioned two approaches that use different classifiers, the 1-NN, the K-NN and the WK-NN, are used. The first approach is called the GEN-1NN, the second is called GEN-KNN and the third is called GEN-WKNN. In GEN-WKNN and in GEN-KNN the value of k is changed dynamically depending on the number of iterations. The parameter settings for the genetic based metaheuristic are: (a) Population size equal to 1000, (b) Number of generations equal to 50, (c) Probability of crossover equal to 0.8 and (d) Probability of mutation equal to 0.25 [2].

Tables 35, 36, 37, 38 and 39 present the performance of each model and the future selection made. Specially, Table 35 presents the errors (the Root Mean Square Error, $F_N\%$ and $F_P\%$, and the Overall Error) for the proposed algorithms in the 2-class classification problem for both data bases. The proposed algorithms give, for example, in the 2-fold Cross Validation, very good results as the RMSE is between 0.219 and 0.243 the New Data and is between 0.089 and 0.108 for the Old Data. In this case, in the New Data set although GEN-1NN has the best performance based on the RMSE, the $F_P\%$ and the OE%, the value of $F_N\%$ is not as good as in the other method, meaning is equal to 2.667 in GEN-1NN while for GEN-KNN is equal to 1.333. As the value of s in the k-fold Cross Validation is increasing the results are improved taking into account the values of the four performance measures. For example, the best solution in the 10-fold Cross Validation has in the New Data the RMSE equal to 0.127 and in the old data set has the RMSE equal to 0 while in the

20-fold Cross Validation in the new data set the RMSE has been improved and is equal to 0.022 and in the Old Data in all methods all the performance measures are equal to 0. All methods give very good results and these results are almost identical for all methods. However, a higher performance is observed for the GEN-1NN. It should, also, be noted that the value of the most significant performance measure, the $F_N\%$, is always smaller than 2.669 and especially in 10-fold and 20-fold Cross Validation is smaller than 1 and for a number of methods is equal to 0 in the new data set while in the Old Data is smaller than 1 in all methods independent of the selected cross validation [2].

Table 35 present the errors (the Root Mean Square Error and the Overall Error) for the proposed algorithms in the 2-class classification problem for both data bases (as it has already been mentioned the $F_N\%$ and the $F_P\%$ do not make sense in the 2-class classification problem. the proposed algorithms give very good results as the RMSE is between 0.219 and 0.243 for the New Data and between 0.089 and 0.108 for the Old Data. As the value of s in the k-fold Cross Validation is increasing the results are improved taking into account the values of the four performance measures. It can be observed that all methods give very good results and these results are almost identical for all methods. However, a higher performance is observed for the GEN-1NN [2].

Table 37 presents the errors (the Root Mean Square Error and the Overall Error) for the proposed algorithms in the 7-class classification problem for both databases. It is observed that the proposed algorithms give very good results as the RMSE is between 0.998 and 1.020 for the New Data and between 0.860 and 0.911 for the Old Data. From this table it can be observed that all methods give very good results and these results are almost identical for all methods. However, a higher performance is observed for the GEN-1NN [2].

Tables 36 and 38 present the average number of used features in every model for both Old Data and New Data. According to these tables, it is observed that in both classification problems (2-class and 7-class classification problem), about half of 20 cell features were used in every model of this approach.

Finally, Table 39 presents the times that each feature was selected in the optimal solutions of all presented models. The three most important features are the 5th feature (Cytoplasm brightness) as it was selected totally 408 times, the 3rd feature (K/C ratio (Size of nucleus relative to cell size) as it was selected totally 366 times, and the 4th feature (Nucleus brightness) as it was selected totally 328 times. Instead, the five less important features were the 9th feature (Nucleus roundness) as it was selected totally only 154 times, the 8th feature (Nucleus elongation) as it was selected totally 170 times and the 12th feature (Cytoplasm elongation) as it was selected 216 times [2].

According to the above tables, that present the results of this classification effort improving the feature selection process using GA, it is observed that the high performance of the proposed algorithm in searching for a reduced set of features (in almost all cases less than 50% of all features are used) with high accuracy and in achieving excellent classification of Pap-smear cells both in 2 classes and in 7 classes.

Table 35 Models performances on 2 classes classification problem of Pap-Smear classification problem [2]

2 classes Cl. Pr.	New Data				Old Data			
Model	RMSE	$F_N\%$	$F_P\%$	OE%	RMSE	$F_N\%$	$F_P\%$	OE%
2-fold cross validation (50% training set—50% test set)								
GEN–1NN	0.218828	2.669306	10.74380	4.798737	0.107967	1	1.5	1.20
GEN–KNN	0.235818	1.333556	17.35537	5.561502	0.089443	0	2	0.8
GEN–WKNN	0.242629	1.927028	16.94215	5.889013	0.094868	0.666667	1	1
Model								
3-fold cross validation (50% training set—50% test set)								
GEN–1NN	0.189321	1.62963	9.084362	3.598343	0.077460	0.333333	1.002563	0.600005
GEN–KNN	0.192821	1.62963	9.907407	3.816208	0.081155	1	0.995025	0.998004
GEN–WKNN	0.197085	1.333333	11.1677	3.925855	0.072957	0.666667	0.995025	0.798403
Model								
4-fold cross validation (50% training set—50% test set)								
GEN–1NN	0.194807	2.073648	8.674863	3.816214	0.031623	0	1	0.4
GEN–KNN	0.19909	1.038145	12.39071	4.033606	0.031623	0	1	0.4
GEN–WKNN	0.194807	2.073648	8.674863	3.816214	0.022361	0	0.5	0.2
Model								
5-fold cross validation (50% training set—50% test set)								
GEN–1NN	0.176904	1.481482	7.874150	3.164053	0.04	0	1	0.4
GEN–KNN	0.183068	0.740741	10.76531	3.380851	0	0	0	0

(continued)

Table 35 (continued)

2 classes Cl. Pr.	New Data				Old Data			
	RMSE	$F_N\%$	$F_P\%$	OE%	RMSE	$F_N\%$	$F_P\%$	OE%
GEN-WKNN	0.177229	0.888889	9.532313	3.163459	0.02	0.333333	0	0.2
Model	10-fold cross validation (50% training set—50% test set)							
GEN–1NN	0.126588	0.741879	4.95	1.853798	0	0	0	0
GEN–KNN	0.152297	0	9.1	2.398471	0	0	0	0
GEN-WKNN	0.159971	0	9.95	2.618251	0	0	0	0
Model	20-fold cross validation (50% training set—50% test set)							
GEN–1NN	0.022116	0	1.25	0.326087	0	0	0	0
GEN–KNN	0.073721	0	4.102564	1.086957	0	0	0	0
GEN-WKNN	0.076775	0	4.519231	1.195652	0	0	0	0

Table 36 Results of the algorithms (average number of features used) for the 2-classes classification problem [2]

	2-fold	3-fold	4-fold	5-fold	10-fold	20-fold
Model	Old Data					
GEN–1NN	11.5	8.33	10	11.2	10.3	9.85
GEN–KNN	13.5	11.33	9.25	10	9.8	10.7
GEN-WKNN	11.5	12	10.75	9.6	10.1	9.75
Model	New Data					
GEN–1NN	10	10	11	11	10.2	9.4
GEN–KNN	8.5	12	10.75	9.2	10.5	9.4
GEN-WKNN	9.5	11.33	11.25	10	9.9	10.55

4.5.7 Nearest Neighborhood Classification Using Tabu Search for Feature Selection

This approach, that was presented in 2006 [22], concerns the feature selection problem of Pap-Smear Classification Problem and suggests the application of Tabu Search (TS) for feature selection under the scepter of nearest Neighborhood classification.

In this case, Old and New Data were used to approach both the 2-class and 7-class classification problems, using the above hybrid CI method. The Old Data, as the New Data, were separated into two data sets (Training Set and Test Set) according the process of k-fold cross validation and for $k = 2, 3, 4, 5, 10$ and 20 [22].

Table 40 presents the errors (the Root Mean Square Error, the Error for Old and New Data, $F_N\%$ and $F_P\%$, and the Overall Error) for the proposed algorithms in the 2 classes classification problem for both databases. It is observed that the proposed algorithms give very good results as the RMSE is between 0.260 and 0.284 for the New Data and between 0.128 and 0.154 for the Old Data. In this case, in the New Data although Tabu-1NN has the best performance based on the RMSE, the $F_P\%$ and the OE%, the value of $F_N\%$ is not as good as in the other methods, meaning is equal to 4.741 in Tabu-1NN while for Tabu-W3NN is equal to 0.741. From this table it can be observed that all methods give very good results and these results are almost identical for all methods. However, a higher performance is observed for the Tabu-1NN. It should, also, be noted that the value of the most significant performance measure, the $F_N\%$, is always smaller than 4,741 and especially in 10-fold and 20-fold Cross Validation is smaller than 1 and for a number of methods is equal to 0 [22] (Table 41).

Table 42 presents the errors (the Root Mean Square Error and the Overall Error) for the proposed algorithms in the 7-class problem for both databases. It is observed that the proposed algorithms give very good results as the RMSE is between 1.042 and 1.117 for the New Data and between 0.970 and 1.297 for the Old Data. However, a higher performance is observed for the Tabu-1NN [22].

Table 37 Models performances on 7 classes classification problem of Pap-Smear classification problem [22]

7 classes Cl. Pr.	New Data				Old Data			
	RMSE	$F_N\%$	$F_P\%$	OE%	RMSE	$F_N\%$	$F_P\%$	OE%
Model	2-fold cross validation (50% training set—50% test set)							
GEN–1NN	1.019975	5.125534	0.883509	1.4	1.019975	5.125534	0.883509	1.4
GEN–KNN	0.9989	6.543083	0.910777	2	0.9989	6.543083	0.910777	2.
GEN–WKNN	1.011858	6.107591	0.860093	1.4	1.011858	6.107591	0.860093	1.4
Model	3-fold cross validation (50% training set—50% test set)							
GEN–1NN	0.910197	4.362299	0.803111	1.798812	0.910197	4.362299	0.803111	1.798812
GEN–KNN	1.000221	5.779492	0.758028	1.598009	1.000221	5.779492	0.758028	1.598009
GEN–WKNN	1.024441	5.671631	0.770776	1.20001	1.024441	5.671631	0.770776	1.20001
Model	4-fold cross validation (50% training set—50% test set)							
GEN–1NN	0.893888	4.687678	0.697431	0.4	0.893888	4.687678	0.697431	0.4
GEN–KNN	0.932513	5.124359	0.658152	0.8	0.932513	5.124359	0.658152	0.8
GEN–WKNN	0.951439	4.905544	0.684904	0.8	0.951439	4.905544	0.684904	0.8
Model	5-fold cross validation (50% training set—50% test set)							
GEN–1NN	0.894883	4.253386	0.682463	0.8	0.894883	4.253386	0.682463	0.8
GEN–KNN	0.871883	4.034212	0.635022	0.4	0.871883	4.034212	0.635022	0.4

(continued)

Table 37 (continued)

7 classes Cl. Pr.	New Data				Old Data			
	RMSE	$F_N\%$	$F_P\%$	OE%	RMSE	$F_N\%$	$F_P\%$	OE%
GEN–WKNN	0.875094	3.926111	0.615672	0.4	0.875094	3.926111	0.615672	0.4
Model	10-fold cross validation (50% training set—50% test set)							
GEN–1NN	0.782835	3.054228	0.499766	0	0.782835	3.054228	0.499766	0
GEN–KNN	0.761857	3.27162	0.529694	0.2	0.761857	3.271620	0.529694	0.2
GEN–WKNN	0.790183	3.059006	0.500224	0	0.790183	3.059006	0.500224	0
Model	20-fold cross validation (50% training set—50% test set)							
GEN–1NN	0.624011	1.852657	0.34175	0	0.624011	1.852657	0.34175	0
GEN–KNN	0.66072	1.637681	0.432793	0	0.66072	1.637681	0.432793	0
GEN–WKNN	0.67809	1.958937	0.372304	0	0.67809	1.958937	0.372304	0

Table 38 Results of the algorithms (average number of features used) for the 7 classes classification problem [22]

	2-fold	3-fold	4-fold	5-fold	10-fold	20-fold
Model	Old Data					
GEN–1NN	11.5	10.7	8.75	9.6	10.4	9.4
GEN–KNN	12.5	11.7	13	13.6	11.1	10.3
GEN-WKNN	13.5	12	13.3	12.6	11.1	9.5
Model	New Data					
GEN–1NN	10	10.33	10	9.2	11.5	9.5
GEN–KNN	8	9.66	9	12	10.2	9.45
GEN-WKNN	9	11	11	12.6	10.4	10.5

Table 39 Results of the algorithms (times each feature was selected) [22]

Feature	2 classes classification problem				7 classes classification problem			
	New Data		Old Data		New Data		Old Data	
	Times Sel.	Average %	Times Sel.	Average %	Times Sel.	Average %	Times Sel.	Average %
1	**80**	**60.60606**	68	51.51515	79	59.84848	63	47.72727
2	66	50	65	49.24242	68	51.51515	70	53.0303
3	**92**	**69.69697**	71	53.78788	**117**	**88.63636**	86	**65.15152**
4	**93**	**70.45455**	57	43.18182	**84**	**63.63636**	94	**71.21212**
5	**92**	**69.69697**	109	82.57576	**94**	**71.21212**	113	**85.60606**
6	69	52.27273	69	52.27273	51	38.63636	74	56.06061
7	90	68.18182	**77**	**58.33333**	**83**	**62.87879**	74	56.06061
8	37	28.0303	69	52.27273	33	25	31	23.48485
9	28	21.21212	50	37.87879	40	30.30303	36	27.27273
10	62	46.9697	69	52.27273	73	55.30303	71	53.78788
11	71	53.78788	54	40.90909	67	50.75758	**86**	**65.15152**
12	56	42.42424	60	45.45455	43	32.57576	57	43.18182
13	48	36.36364	54	40.90909	63	47.72727	69	52.27273
14	76	57.57576	49	37.12121	**88**	**66.66667**	61	46.21212
15	61	46.21212	**73**	**55.30303**	68	51.51515	57	43.18182
16	64	48.48485	**105**	**79.54545**	38	28.78788	**109**	**82.57576**
17	61	46.21212	53	40.15152	60	45.45455	56	42.42424
18	64	48.48485	67	50.75758	67	50.75758	63	47.72727
19	57	43.18182	**72**	**54.54545**	57	43.18182	74	56.06061
20	65	49.24242	59	44.69697	73	55.30303	62	46.9697

Table 40 Results of the algorithms in the 2 classes classification problem [22]

2 classes Cl. Pr.	New Data				Old Data			
Model	RMSE	$F_N\%$	$F_P\%$	OE%	RMSE	$F_N\%$	$F_P\%$	OE%
2-fold cross validation (50% training set—50% test set)								
TS–1NN	0.260023	4.74119	12.39669	6.761186	0.140705	1.333333	3	2
TS–3NN	0.26204	2.37257	19.42149	6.870594	0.154377	2.666667	2	2.4
TS–5NN	0.264153	2.075396	20.66116	6.979527	0.128387	1.666667	2	1.8
TS–8NN	0.27024	0.889769	25.20661	7.3068	0.133956	1.333333	2.5	1.8
TS–10NN	0.284046	0.889769	28.09917	8.069564	0.140705	1	3.5	2
TS–W3NN	0.278197	0.74184	27.27273	7.743005	0.154377	1.666667	3.5	2.4
TS–W5NN	0.266172	2.076712	21.07438	7.088697	0.140705	2	2	2
TS–W8NN	0.268251	0.740084	25.20661	7.197629	0.138438	2	2	2
TS–W10NN	0.266172	0.88933	24.38017	7.088697	0.138438	2.333333	1.5	2
Model								
3-fold cross validation (50% training set—50% test set)								
TS–1NN	0.213638	2.222222	11.14712	4.579449	0.132853	1.666667	1.997588	1.798812
TS–3NN	0.240132	2.666667	14.44959	5.779135	0.138942	2.333333	1.500075	1.998413
TS–5NN	0.228482	1.777778	14.86111	5.233758	0.113938	1	1.997588	1.398408
TS–8NN	0.230258	0.888889	17.74691	5.342334	0.114771	0	4.975124	1.996008
TS–10NN	0.229904	0.740741	18.15844	5.341976	0.145854	0.666667	4.492688	2.198014
TS–W3NN	0.237495	3.111111	12.79835	5.670917	0.132878	0.666667	3.505201	1.800014
TS–W5NN	0.230784	2.962963	11.98045	5.343048	0.139765	2	1.997588	1.998413

(continued)

Table 40 (continued)

2 classes Cl. Pr.	New Data				Old Data			
	RMSE	$F_N\%$	$F_P\%$	OE%	RMSE	$F_N\%$	$F_P\%$	OE%
TS–W8NN	0.230784	2.666667	12.80864	5.343048	0.120027	1.333333	1.99005	1.598009
TS–W10NN	0.223177	1.925926	13.62654	5.015536	0.089353	0.666667	1.99005	1.197605
Model	4-fold cross validation (50% training set—50% test set)							
TS–1NN	0.208316	2.371267	9.911202	4.361116	0.107967	0.666667	2	1.2
TS–3NN	0.230544	2.369505	13.62705	5.342225	0.121066	1	2.5	1.6
TS–5NN	0.230444	2.225099	14.05738	5.342225	0.092713	1	1.5	1.2
TS–8NN	0.229618	1.038145	17.34973	5.3427	0.101975	0.333333	3	1.4
TS–10NN	0.263987	1.333122	22.71858	6.978356	0.092713	0.666667	2	1.2
TS–W3NN	0.240204	2.074528	16.09973	5.778906	0.131443	0.666667	3.5	1.8
TS–W5NN	0.223262	2.520957	11.98087	5.015189	0.092713	1.333333	1	1.2
TS–W8NN	0.23288	2.373028	14.05055	5.451396	0.115074	1	2	1.4
TS–W10NN	0.230372	1.781312	15.28005	5.3427	0.092713	1	1.5	1.2
Model	5-fold cross validation (50% training set—50% test set)							
TS–1NN	0.203104	2.518518	8.690476	4.14469	0.074641	0.666667	1.5	1
TS–3NN	0.211221	1.333333	13.23129	4.470777	0.074641	1.333333	0.5	1
TS–5NN	0.213751	1.62963	12.81463	4.579473	0.074641	1	1	1
TS–8NN	0.226317	1.037037	16.55612	5.125327	0.094641	1.333333	1	1.2
TS–10NN	0.221381	1.037037	15.71429	4.906747	0.094641	0.666667	2	1.2

(continued)

Table 40 (continued)

2 classes Cl. Pr.	New Data				Old Data			
	RMSE	$F_N\%$	$F_P\%$	OE%	RMSE	$F_N\%$	$F_P\%$	OE%
TS–W3NN	0.228462	1.481481	15.7398	5.234616	0.108284	1.333333	2	1.6
TS–W5NN	0.216151	1.925926	12.38946	4.688762	0.06	1	1	1
TS–W8NN	0.213264	1.777778	12.43197	4.58066	0.074641	1	1	1
TS–W10NN	0.213758	1.037037	14.48129	4.580066	0.094641	1	1.5	1.2
Model	10-fold cross validation (50% training set—50% test set)							
TS–1NN	0.164553	1.18525	7.45	2.835643	0.042426	0.666667	0.5	0.6
TS–3NN	0.197351	0.443371	13.66667	3.927377	0.02	0.333333	0.5	0.4
TS–5NN	0.197349	0.298507	14.08333	3.927377	0.028284	0.666667	0	0.4
TS–8NN	0.223651	0.149254	18.61667	5.016722	0.028284	0	1	0.4
TS–10NN	0.218049	0.149254	17.78333	4.798137	0.028284	0	1	0.4
TS–W3NN	0.225876	0.296313	18.61667	5.125418	0.028284	0	1	0.4
TS–W5NN	0.197683	0.59482	13.26667	3.927377	0.02	0.666667	0	0.4
TS–W8NN	0.208206	0.298507	15.73333	4.363354	0.028284	0.666667	0	0.4
TS–W10NN	0.208206	0.298507	15.73333	4.363354	0.034142	0.333333	1	0.6
Model	20-fold cross validation (50% training set—50% test set)							
TS–1NN	0.090369	0.748663	3.717949	1.52657	0	0	0	0
TS–1NN	0.151264	0.151515	9.51923	2.615942	0	0	0	0

(continued)

Table 40 (continued)

2 classes Cl. Pr.	New Data				Old Data			
	RMSE	$F_N\%$	$F_P\%$	OE%	RMSE	$F_N\%$	$F_P\%$	OE%
TS–3NN	0.151818	0	9.51923	2.507246	0.01	0.333333	0	0.2
TS–5NN	0.19207	0	14.45513	3.811594	0	0	0	0
TS–8NN	0.18905	0	14.07051	3.705314	0.01	0	0.5	0.2
TS–10NN	0.182943	0	13.17308	3.487923	0	0	0	0
TS–W3NN	0.156694	0.147059	10.32051	2.835749	0	0	0	0
TS–W5NN	0.156694	0	10.73718	2.835749	0	0	0	0
TS–W8NN	0.160979	0	10.73718	2.833333	0	0	0	0

Tables 41 and 43 present the average number of used features in every model for both Old Data and New Data. according to these tables, it is observed that in both classification problems (2-class and 7-class classification problem), about half of 20 cell features were used in every model of this approach [22].

Finally, Table 44 presents the times that each feature was selected in the optimal solutions of all presented models. As it can be seen, the three most important features are the 3rd feature (N/C ratio (Size of nucleus relative to cell size)) as it was selected totally 1138 times, the 5th feature (Cytoplasm brightness) as it was selected totally 1090 times, and the 4th feature (Nucleus brightness) as it was selected totally 962 times. The three less important features are the 9th feature (Nucleus roundness) as it was selected totally only 503 times, the 8th feature (Nucleus elongation) as it was selected totally 576 times, and the 15th feature (Cytoplasm perimeter) [22].

Table 41   Results of the algorithms (average number of features used) for the 7-classes classification problem [22]

Model	2-fold	3-fold	4-fold	5-fold	10-fold	20-fold
	New Data					
TS-1NN	6.5	12	9.5	11.2	10.9	10.1
TS–3NN	10.5	11	8.25	9	11.2	10.25
TS-5NN	10	11	10.25	10	10.9	9.55
TS–8NN	9	11.33	9.5	10.2	10	9.3
TS–10NN	9	10.67	9.25	10.2	9.6	9.5
TS–W3NN	10.5	12.33	11	9.2	10.2	9.75
TS–W5NN	11.5	12	10	9.4	9.8	9.9
TS–W8NN	11	12	9.5	9	10.2	9.35
TS–W10NN	9	12.33	8.5	12.2	11.1	9.55
TS-1NN	Old Data					
TS–3NN	10.5	11	10.75	10	9.5	7.4
TS-5NN	11	10.33	10.5	12.2	10.5	9.4
TS–8NN	11	11	11.5	12.2	9.6	9.25
TS–10NN	13.5	13	11.75	12.8	10.5	9
TS–W3NN	14	14.67	11.5	9.6	9.6	8.3
TS–W5NN	9.5	10.67	12.25	10.8	9.8	9.4
TS–W8NN	11	10.33	11.25	13.4	9.7	9.85
TS–W10NN	11	11.33	11	12	10.5	9.15
TS-1NN	13	13.67	12	11.6	10.8	8.85

Table 42 Results of the algorithms in the 7 classes classification problem [22]

7 classes Cl. Pr.	New Data				Old Data			
	RMSE	$F_N\%$	$F_P\%$	OE%	RMSE	$F_N\%$	$F_P\%$	OE%
Model	2-fold cross validation (50% training set—50% test set)							
TS–1NN	1.097725	6.979527	0.970155	2.4	1.097725	6.979527	0.970155	2.4
TS–3NN	1.107526	7.743481	1.297558	4.8	1.107526	7.743481	1.297558	4.8
TS–5NN	1.075632	7.307037	1.049504	2.8	1.075632	7.307037	1.049504	2.8
TS–8NN	1.07402	7.088459	1.076184	3.2	1.07402	7.088459	1.076184	3.2
TS–10NN	1.064957	8.615416	1.032272	2.4	1.064957	8.615416	1.032272	2.4
TS–W3NN	1.116924	8.615654	1.191824	3.2	1.116924	8.615654	1.191824	3.2
TS–W5NN	1.093116	7.52514	1.067658	2.4	1.093116	7.52514	1.067658	2.4
TS–W8NN	1.069031	8.180162	1.015378	2.6	1.069031	8.180162	1.015378	2.6
TS–W10NN	1.0416	6.761186	1.027794	2.8	1.0416	6.761186	1.027794	2.8
Model	3-fold cross validation (50% training set—50% test set)							
TS–1NN	0.998415	5.234115	0.901895	2.397614	0.998415	5.234115	0.901895	2.397614
TS–3NN	1.11935	7.851352	1.05101	3.399226	1.11935	7.851352	1.05101	3.399226
TS–5NN	1.104866	6.978463	0.965862	2.796816	1.104866	6.978463	0.965862	2.796816
TS–8NN	1.106326	6.871674	0.926921	2.199216	1.106326	6.871674	0.926921	2.199216
TS–10NN	1.116397	7.087753	0.870997	2.199216	1.116397	7.087753	0.870997	2.199216
TS–W3NN	1.126345	7.634559	0.948671	2.799221	1.126345	7.634559	0.948671	2.799221
TS–W5NN	1.114004	6.869174	0.902844	2.397614	1.114004	6.869174	0.902844	2.397614

(continued)

Table 42 (continued)

7 classes Cl. Pr.	New Data				Old Data			
	RMSE	$F_N\%$	$F_P\%$	OE%	RMSE	$F_N\%$	$F_P\%$	OE%
TS–W8NN	1.103898	6.434873	0.851216	1.99721	1.103898	6.434873	0.851216	1.99721
TS–W10NN	1.129295	7.633844	0.923595	2.198014	1.129295	7.633844	0.923595	2.198014
Model	4-fold cross validation (50% training set—50% test set)							
TS–1NN	0.962652	5.233529	0.818468	1.4	0.962652	5.233529	0.818468	1.4
TS–3NN	1.070987	6.65132	0.922433	2.6	1.070987	6.65132	0.922433	2.6
TS–5NN	1.042917	6.870609	0.818359	1.6	1.042917	6.870609	0.818359	1.6
TS–8NN	1.022539	6.652269	0.802134	1.4	1.022539	6.652269	0.802134	1.4
TS–10NN	1.070661	7.087526	0.794862	1.2	1.070661	7.087526	0.794862	1.2
TS–W3NN	1.043391	6.541675	0.98445	3.4	1.043391	6.541675	0.98445	3.4
TS–W5NN	1.030383	6.215588	0.851283	1.8	1.030383	6.215588	0.851283	1.8
TS–W8NN	1.034986	7.088475	0.844032	2	1.034986	7.088475	0.844032	2
TS–W10NN	1.022443	6.215113	0.822172	1.6	1.022443	6.215113	0.822172	1.6
Model	5-fold cross validation (50% training set—50% test set)							
TS–1NN	0.977585	4.909717	0.828779	1.2	0.977585	4.909717	0.828779	1.2
TS–3NN	0.996275	5.779876	0.838493	1.4	0.996275	5.779876	0.838493	1.4
TS–5NN	1.00079	6.432644	0.743451	1.6	1.00079	6.432644	0.743451	1.6
TS–8NN	0.975546	5.887978	0.727123	0.8	0.975546	5.887978	0.727123	0.8
TS–10NN	0.966245	5.669399	0.731373	1.2	0.966245	5.669399	0.731373	1.2

(continued)

Table 42 (continued)

7 classes Cl. Pr.	New Data				Old Data			
	RMSE	$F_N\%$	$F_P\%$	OE%	RMSE	$F_N\%$	$F_P\%$	OE%
TS–W3NN	1.017272	5.88976	0.856501	2	1.017272	5.88976	0.856501	2
TS–W5NN	1.000737	5.99608	0.814177	1.4	1.000737	5.99608	0.814177	1.4
TS–W8NN	0.979248	6.105963	0.776908	1.6	0.979248	6.105963	0.776908	1.6
TS–W10NN	0.980239	5.125921	0.741778	1	0.980239	5.125921	0.741778	1
Model	10-fold cross validation (50% training set—50% test set)							
TS–1NN	0.905763	4.252269	0.669753	0.6	0.905763	4.252269	0.669753	0.6
TS–3NN	0.929426	5.017917	0.760973	1.4	0.929426	5.017917	0.760973	1.4
TS–5NN	0.908669	4.363354	0.727018	1.4	0.908669	4.363354	0.727018	1.4
TS–8NN	0.916278	4.36216	0.722991	1.4	0.916278	4.36216	0.722991	1.4
TS–10NN	0.911791	4.359771	0.680138	0.8	0.911791	4.359771	0.680138	0.8
TS–W3NN	0.912102	5.019111	0.753263	0.6	0.912102	5.019111	0.753263	0.6
TS–W5NN	0.933394	4.910416	0.672604	1	0.933394	4.910416	0.672604	1
TS–W8NN	0.922104	4.363354	0.663337	1	0.922104	4.363354	0.663337	1
TS–W10NN	0.910576	4.142379	0.659465	1	0.910576	4.142379	0.659465	1
Model	20-fold cross validation (50% training set—50% test set)							
TS–1NN	0.776461	3.048309	0.487488	0	0.776461	3.048309	0.487488	0
TS–3NN	0.855476	4.140097	0.658098	0.4	0.855476	4.140097	0.658098	0.4

(continued)

Table 42 (continued)

7 classes Cl. Pr.	New Data				Old Data			
	RMSE	$F_N\%$	$F_P\%$	OE%	RMSE	$F_N\%$	$F_P\%$	OE%
TS-5NN	0.821645	3.485507	0.595209	1	0.821645	3.485507	0.595209	1
TS–8NN	0.821604	4.031401	0.625383	0.6	0.821604	4.031401	0.625383	0.6
TS–10NN	0.82703	3.705314	0.627029	0.4	0.82703	3.705314	0.627029	0.4
TS–W3NN	0.88653	4.036232	0.644507	0.2	0.88653	4.036232	0.644507	0.2
TS–W5NN	0.817966	2.949275	0.599635	0.2	0.817966	2.949275	0.599635	0.2
TS–W8NN	0.832524	3.821256	0.575656	0.2	0.832524	3.821256	0.575656	0.2
TS–W10NN	0.806831	3.594203	0.592631	0.4	0.806831	3.594203	0.592631	0.4

Table 43 Results of the algorithms (average number of features used) for the 7-classes classification problem [22]

	2-fold	3-fold	4-fold	5-fold	10-fold	20-fold
Model	New Data					
TS-1NN	12.5	12.67	10.5	11.6	10.5	10.15
TS–3NN	9.5	10	9.5	9.6	10.6	11.3
TS-5NN	9	9.33	9.25	9.2	11.2	9.9
TS–8NN	8.5	8	9.5	11.4	10.1	9.85
TS–10NN	8.5	8.33	8.25	11.8	8.7	9.8
TS–W3NN	10.5	12	13	10	11.7	10.8
TS–W5NN	9	9.5	10.25	9	11.1	11.05
TS–W8NN	8.5	10	10	11.4	10.8	10.2
TS–W10NN	9	7.67	10.75	11.6	9.8	10.3
Model	Old Data					
TS-1NN	10	13.7	11.3	10.2	10	9.15
TS–3NN	12	8.67	12.3	13.6	10.9	10.1
TS-5NN	14	13	13.3	13.4	11.2	9.9
TS–8NN	12.5	13	12.3	12	11.3	9.4
TS–10NN	11	9.33	13.3	13.8	11	10.4
TS–W3NN	12.5	11.7	10.3	10	10.8	10.3
TS–W5NN	11.5	11.7	11.3	10	11.3	9.25
TS–W8NN	14	11	14	12.4	12	10
TS–W10NN	13	13.3	12.5	12.8	11	10.2

Table 44 Results of the algorithms (times each feature was selected) [22]

Feature	2 classes classification problem				7 classes classification problem			
	New Data		Old Data		New Data		Old Data	
	Times Sel.	Average %	Times Sel.	Average %	Times Sel.	Average %	Times Sel.	Average %
1	**245**	**61.87**	**248**	**62.63**	229	57.83	193	48.74
2	189	47.73	212	53.54	197	49.75	212	53.54
3	**282**	**71.21**	**241**	**60.86**	**346**	**87.37**	**269**	**67.93**
4	**248**	**62.63**	211	53.28	**238**	**60.10**	**265**	**66.92**
5	**242**	**61.11**	**287**	**72.47**	242	61.11	319	80.56
6	203	51.26	226	57.07	182	45.96	210	53.03
7	**260**	**65.66**	206	52.02	230	58.08	224	56.57
8	123	31.06	168	42.42	135	34.09	150	37.88
9	78	19.70	165	41.67	120	30.30	140	35.35

(continued)

Table 44 (continued)

Feature	2 classes classification problem				7 classes classification problem			
	New Data		Old Data		New Data		Old Data	
	Times Sel.	Average %	Times Sel.	Average %	Times Sel.	Average %	Times Sel.	Average %
10	183	46.21	150	37.88	210	53.03	212	53.54
11	150	37.88	201	50.76	187	47.22	224	56.57
12	197	49.75	188	47.47	153	38.64	217	54.80
13	177	44.70	160	40.4	**238**	**60.10**	**256**	**64.65**
14	228	57.58	106	26.77	218	55.05	170	42.93
15	175	44.20	158	39.9	187	47.22	151	38.13
16	158	39.90	**271**	**68.43**	152	38.38	**326**	**82.32**
17	219	55.30	168	42.42	188	47.47	166	41.92
18	214	54.04	214	54.04	223	56.31	202	51.01
19	192	48.48	170	42.93	191	48.23	187	47.22
20	202	51.01	227	57.32	219	55.30	191	48.23

4.5.8 Nearest Neighborhood Classification Using ACO for Feature Selection

This approach presented in 2006 [23], concerns the feature selection problem of Pap-Smear Classification Problem and suggests the application of an algorithm based on the Ant Colony Optimization to improve the feature selection process under the scepter of nearest Neighborhood classification.

In this case, two approaches that use different classifiers, the 1-NN and the WK-NN, are used. Both in two approaches, Old Data and New Data were used for both the 2 classes and 7 classes classification problems. The Old Data, as the New Data, were separated into two data sets (Training Set and Test Set) according to the process of k-fold cross validation and for k = 2, 3, 4, 5, 10 and 20 [22]. The first approach is called the ACO-1NN, and the other is called ACO-WKNN. In ACO-WKNN the value of k is changed dynamically depending on the number of iteration and the parameter settings for the ACO based metaheuristic are: (a) the number of ants used is equal to the number of features (20) because in the initial iteration each ant begins from a different feature and (b) the number of iterations that each ant constructs a different solution, based on the pheromone trails, is equal to 50 [23].

Table 45 presents the errors (the Root Mean Square Error, the Error for Old and New Data, $F_N\%$ and $F_P\%$, and the Overall Error) for the proposed algorithms in the 2-class classification problem for both databases. It is observed that the proposed algorithms give very good results as the RMSE is between 0.2189 and 0.249 for the New Data set and 0.108 for the Old Data. ACO-1NN has the best performance based on the RMSE, the $F_P\%$ and the OE%, the value of $F_N\%$ is not as good as in the other

Table 45 Results of the algorithms in the 2 classes classification problem [23]

2 classes Cl. Pr.	New Data				Old Data			
	RMSE	$F_N\%$	$F_P\%$	OE%	RMSE	$F_N\%$	$F_P\%$	OE%
Model	2-fold cross validation (50% training set—50% test set)							
ACO-1NN	0.218828	2.669306	10.7438	4.798737	0.107967	1	1.5	1.2
ACO-WKNN	0.249235	1.776465	18.59504	6.216285	0.107967	0.666667	2	1.2
Model	3-fold cross validation (50% training set—50% test set)							
ACO-1NN	0.19486	1.925926	9.089506	3.816565	0.088144	0.666667	1.002563	0.799606
ACO-WKNN	0.207896	1.777778	11.57922	4.362299	0.098828	0.333333	1.997588	0.999206
Model	4-fold cross validation (50% training set—50% test set)							
aACO-1NN	0.188547	1.776028	8.674863	3.597874	0.031623	0	1	0.4
ACO-WKNN	0.20535	2.222457	9.918033	4.251946	0.031623	0.333333	0.5	0.4
Model	5-fold cross validation (50% training set—50% test set)							
ACO-1NN	0.1711	1.481481	7.040816	2.945474	0.048284	0.333333	1	0.6
ACO-WKNN	0.183286	1.185185	9.532313	3.382038	0	0	0	0
Model	10-fold cross validation (50% training set—50% test set)							
aACO-1NN	0.126588	0.739684	4.933333	1.853798	0	0	0	0
ACO-WKNN	0.159971	0.445566	8.7	2.618251	0	0	0	0
Model	20-fold cross validation (50% training set—50% test set)							
ACO-1NN	0.014744	0.147059	0.416667	0.217391	0	0	0	0
ACO-WKNN	0.069403	0	4.102564	1.086957	0	0	0	0

Table 46 Results of the algorithms (average number of features used) for the 2-classes classification problem [23]

	2-fold	3-fold	4-fold	5-fold	10-fold	20-fold
Model	New Data					
ACO-1NN	10	11	8.5	9.8	9.7	9.7
ACO-WKNN	**10**	**10.33**	**11**	**10.8**	**10.3**	**9.4**
Model	Old Data					
ACO-1NN	10	9	7.75	11.4	10.5	8.9
ACO-WKNN	14.5	11.67	11.25	11.4	10.4	11.25

method, meaning is equal to 2.669 in ACO-1NN while for ACO-WKNN is equal to 1.776. From this table it can be observed that all methods give very good results and these results are almost identical for all methods. However, a higher performance is observed for the ACO-1NN. It should, also, be noted that the value of the most significant performance measure, the $F_N\%$, is always smaller than 2.669306 and especially in 10-fold and 20-fold Cross Validation is smaller than 1 and for a number of methods is equal to 0 in the new data set while in the old data is smaller than 1 in all methods independent of the selected cross validation [23].

Table 46 presents the average number of features that each algorithm selects for both data bases. It can be observed that in all cases the number of features selected are fewer than the total number of features used in order to describe a cell. For the New Data, the minimum average number of features used is equal to 8.5. For the Old Data, the minimum average number of features used is equal to 7.75 [23].

Table 47 presents the errors (the Root Mean Square Error, the Error for Old and New Data, $F_N\%$ and $F_P\%$, and the Overall Error) for the proposed algorithms in the 2 classes classification problem for both databases. It is observed that the proposed algorithms give very good results as the RMSE is between 1.021 and 1.030 for the New Data set and between 0.899 and 0.915 for the Old Data. From this Table it can be observed that all methods give very good results and these results are almost identical for all methods. However, a higher performance is observed for the ACO-1NN.

Table 48 presents the average number of features that each algorithm selects for both data bases. It can be observed that in all cases the number of features selected are fewer than the total number of features used in order to describe a cell. For the New Data, the minimum average number of features used is equal to 9. For the Old Data, the minimum average number of features used is equal to 8.25 [23].

Finally, Table 49 presents the times that each feature was selected in the optimal solutions of all presented models. As it can be seen, the three most important features are the 5th feature (Cytoplasm brightness) as it was selected totally 247 times, the 4th feature (Nucleus brightness) as it was selected totally 233 times and the 3rd feature (N/C ratio (Size of nucleus relative to cell size)) as it was selected totally 226 times. The three less important features are the 9th feature (Nucleus roundness) as it was selected totally only 100 times, the 8th feature (Nucleus elongation) as it was selected

Table 47 Results of the algorithms in the 7 classes classification problem [23]

7 classes Cl. Pr.	New Data				Old Data			
	RMSE	$F_N\%$	$F_P\%$	OE%	RMSE	$F_N\%$	$F_P\%$	OE%
Model	2-fold cross validation (50% training set—50% test set)							
ACO-1NN	1.030028	5.780556	0.898933	1.6	1.030028	5.780556	0.898933	1.6
ACO-WKNN	1.020786	6.652729	0.914948	1.8	1.020786	6.652729	0.914948	1.8
Model	3-fold cross validation (50% training set—50% test set)							
ACO-1NN	0.910197	4.362299	0.819103	1.39961	0.910197	4.362299	0.819103	1.39961
ACO-WKNN	1.060677	5.234473	0.807685	1.398408	1.060677	5.234473	0.807685	1.398408
Model	4-fold cross validation (50% training set—50% test set)							
aACO-1NN	0.893888	4.687678	0.704141	0.4	0.893888	4.687678	0.704141	0.4
ACO-WKNN	0.968416	5.779381	0.710849	0.8	0.968416	5.779381	0.710849	0.8
Model	5-fold cross validation (50% training set—50% test set)							
ACO-1NN	0.887065	4.362675	0.710722	1.2	0.887065	4.362675	0.710722	1.2
ACO-WKNN	0.911689	4.909123	0.664113	0.4	0.911689	4.909123	0.664113	0.4
Model	10-fold cross validation (50% training set—50% test set)							
aACO-1NN	0.77879	2.943144	0.524292	0	0.77879	2.943144	0.524292	0.77879
ACO-WKNN	0.812723	3.823459	0.540925	0	0.812723	3.823459	0.540925	0.812723
Model	20-fold cross validation (50% training set—50% test set)							
ACO-1NN	0.638976	2.070048	0.366199	0	0.638976	2.070048	0.366199	0.638976
ACO-WKNN	0.70262	2.181159	0.393758	0	0.70262	2.181159	0.393758	0.70262

Table 48 Results of the algorithms (average number of features used) for the 7-classes classification problem [23]

	2-fold	3-fold	4-fold	5-fold	10-fold	20-fold
Model	New Data					
ACO-1NN	9	10.33	10	9.2	9.2	9.7
ACO-WKNN	10.5	9.67	10	9	10.9	9.4
Model	Old Data					
ACO-1NN	11.5	10.33	9.75	10.6	10.4	8.25
ACO-WKNN	13	10.33	12	11.2	9.33	9.2

Table 49 Results of the algorithms (times each feature was selected) [23]

Feature	2 classes classification problem				7 classes classification problem			
	New Data		Old Data		New Data		Old Data	
	Times Sel.	Average %	Times Sel.	Average %	Times Sel.	Average %	Times Sel.	Average %
1	**52**	**59.09091**	51	**57.95455**	**60**	**68.18182**	50	56.818
2	48	54.54545	41	46.59091	48	54.54545	44	50
3	**58**	**65.90909**	47	53.40909	**65**	**73.86364**	**56**	**63.636**
4	**63**	**71.59091**	48	54.54545	**58**	**65.90909**	**64**	**72.727**
5	**64**	**72.72727**	**53**	**60.22727**	**63**	**71.59091**	**67**	**76.136**
6	43	48.86364	36	40.90909	35	39.77273	43	48.864
7	**57**	**64.77273**	46	52.27273	53	60.22727	48	54.545
8	26	29.54545	45	51.13636	24	27.27273	21	23.864
9	21	23.86364	31	35.22727	23	26.13636	25	28.409
10	38	43.18182	41	46.59091	41	46.59091	42	47.727
11	40	45.45455	40	45.45455	39	44.31818	**55**	**62.5**
12	36	40.90909	38	43.18182	24	27.27273	34	38.636
13	33	37.5	43	48.86364	34	38.63636	29	32.955
14	45	51.13636	35	39.77273	49	55.68182	30	34.091
15	46	52.27273	55	62.5	44	50	40	45.455
16	41	46.59091	**70**	**79.54545**	32	36.36364	**68**	**77.273**
17	36	40.90909	41	46.59091	43	48.86364	32	36.364
18	41	46.59091	52	59.09091	39	44.31818	30	34.091
19	39	44.31818	47	53.40909	30	34.09091	40	45.455
20	40	45.45455	**53**	**60.22727**	49	55.68182	35	39.773

totally 116 times and the 12th feature (Cytoplasm elongation) as it was selected 132 times [23].

4.5.9 Minimum Distance Classifier

This approach presented in 2006 [24], is about both the 2-class and the 7-class classification problem and suggests the using of the minimum diastase classification (Table 50).

The results of this approach prove that the application of this CI-Technique is more efficient than comparable techniques, such as Least Squares, which had been presented in 2005 by Jonas Norup [20], as it is presented in Table 51, which shows the results of these two methods (Least Squares and Minimum Distance) after the proceeds of 10-fold cross validation.

Also, according to the confusion matrix (Table 50), which was obtained from the application of the above CI-Technique to New Data, there is a little overlapping between classes 1 and 2, which is to be expected, and also between classes 6 and 7. Furthermore, class 3 is confused with classes 4, 5, 6, and 7, and mostly with class 6. This is as predicted by the class Neighborhood relationship according to the basic data analysis presented in Sect. 3 [23].

Table 50 Average confusion matrix C. An element Cij is the number of objects of true class j estimated as class i. The diagonal is the number of correctly classified samples of the test data (91 objects, 10-fold, 500 reruns) [24]

Estimate	Class 1	Class 2	Class 3	Class 4	Class 5	Class 6	Class 7
$\widehat{C_1}$	6.6	0.6	0.0	0.0	0.0	0.0	0.0
$\widehat{C_2}$	0.7	6.1	0.0	0.0	0.0	0.0	0.0
$\widehat{C_3}$	0.0	0.0	6.2	0.1	0.4	1.1	0.3
$\widehat{C_4}$	0.0	0.1	0.3	14.7	5.1	2.3	0.4
$\widehat{C_5}$	0.0	0.1	0.9	2.4	6.7	2.8	1.9
$\widehat{C_6}$	0.0	0.0	2.0	0.9	1.8	10.8	6.6
$\widehat{C_7}$	0.0	0.0	0.2	0.0	0.5	2.6	5.7

Table 51 Performance. Comparison of percent misclassifications in the test set. The format of the numbers is: means' standard deviation [24]

Classifier	2 classes	7 classes	Comment
Least squares	6.4 ± 1.9	42.9 ± 4.7	Least squares
Minimum distance	6.2 ± 2.4	37.7 ± 4.8	Mahalanobis D.
Minimum distance	15.5 ± 3.6	15.5 ± 3.6	Euclidian D.

4.6 Conclusion and Future Work

This chapter is an overview of the approaches of the Pap-Smear Classification Problem using CI-Techniques. This problem arises from the need to optimize the diagnosis of cervical cancer through the Pap-Smear method.

The Pap-Smear Classification problem was approached, from the following three sides: (a) the sub-problem of 2-class classification (Normal and Abnormal) (b) the sub-problem of 7-class classification (Superficial squamous epithelial, Intermediate squamous epithelial, Columnar epithelial, Mild squamous non-keratinizing dysplasia, Moderate squamous non-keratinizing dysplasia, Severe squamous non-keratinizing dysplasia and Squamous cell carcinoma in situ intermediate) and (c) the sub-problem of optimization of the previous classifications, improving their functional processes using different feature selection methods and clustering techniques.

The aforementioned research was based on two databases, which are called Old Data and New Data and consist respectively of 500 and 917 single cell patterns/20 features.

The CI-Techniques, that were used to build classifiers, both for the 2-class classification problem and for the 7-class classification problem, are machine learning algorithms, such as Adaptive Network-based Fuzzy Inference System, Artificial Neural Networks, k-Nearest Neighborhood etc. Also, various algorithms, such as Fuzzy C-Means, Tabu Search, Ant Colony etc., were used to improve the functional processes of every classifier. All these CI-Techniques are presented, briefly, in Sect. 4.

According to the following table, which presents, briefly, the results of these approaches and comes from [21], and also according to the results of the latest classification efforts made in 2006, it could be concluded that the performance of the proposed classification models are satisfactory but with room for further improvement.

Specifically, it is observed that, for the 2-class classification problem, almost all of the CI Techniques present a fairly good adaptation to the existing data of the problem but for the 7-class classification, their performances are not so satisfied. This may be a result of the data problems, such as the confusion and overlap of some classes, as it was observed according to the basic statistical analysis carried out in [24], and also into the number and the type of the cells' features used as inputs at the classifiers (Table 52).

Finally, it would be interesting perhaps to investigate whether the machine approach performs competitively to human beings in the specific diagnostic task. In any case, if a machine is able to find a considerable percentage of the normal cells with zero false negatives, then these could be discarded, and the human work would be considerably less. So, it might be better after all to look for high certainty rather than high performance in future research.

Table 52 Classification accuracy results, for separating: (a) between normal and abnormal cells (2 classes classification problem), and (b) between all 7 diagnostic classes (7 classes classification problem); accuracy refers to the best performance obtained on test (new) data [21]

CI-Techniques used for Pap-Smear classification	Classification accuracy for 2 classes classification problem	Classification accuracy for 7 classes classification problem
Hard C-Means	94–96	72–80
Fuzzy C-Means	96–97	72–77
Gustafson-Kessel clustering method	89–96	<75
Feature selection and clustering	90–97	<75
ANFIS neuro-fuzzy classification	96	<75
Nearest neighborhood classification	96	<75
Inductive machine learning	<75	<75
Genetic programming	89	81
Second order neural network	99	<75

References

1. L. Bruni, L. Barrionuevo-Rosas, G. Albero, B. Serrano, M. Mena, D. Gómez, J. Muñoz, F.X. Bosch, S. de Sanjosé, Human papillomavirus and related diseases in the world, Summary Report 27 July 2017, ICO/IARC Information Centre on HPV and Cancer (HPV Information Centre) (2017). http://www.hpvcentre.net/statistics/reports/XWX.pdf
2. Y. Marinakis, G. Dounias, Pap smear diagnosis using a hybrid intelligent scheme focusing on genetic algorithm based feature selection and nearest neighborhood classification, in *The Pap Smear Benchmark, Intelligent and Nature Inspired Approaches in Pap Smear Diagnosis, Special Session Proceedings of the NISIS—2006 Symposium*, 15–24, November 29–December 1 2006, Puerto de la Cruz, Tenerife, Spain (Spain, 2006), pp. 15–24
3. E. Rakus-Andersson, L.C. Jain, Computational intelligence in medical decisions making, in *Recent Advances in Decision Making*. Studies in Computational Intelligence, vol. 222 (Springer, Berlin, Heidelberg, 2009), pp. 145–159. https://doi.org/10.1007/978-3-642-02187-9_9
4. F. Lemke, J.-A. Müller, Medical data analysis using self-organizing data mining technologies, in *Systems Analysis Modelling Simulation*, vol. 43 (Taylor & Francis Group, London, 2010), pp. 1399–1408. https://doi.org/10.1080/02329290290027337
5. A. Tsanas, M.A. Little, P.E. McSharry, A methodology for the analysis of medical data, in *A Methodology for the Analysis of Medical Data*. Handbook of Systems and Complexity in Health, vol. 1 (Springer, New York, 2013), pp. 113–125. https://doi.org/10.1007/978-1-4614-4998-0_7
6. E. López-Rubio, D.A. Elizondo, M. Grootveld, J.M. Jerez, R.M. Luque-Baena, Computational intelligence techniques in medicine, in *Computational and Mathematical Methods in Medicine*, vol. 2015 (Hindawi, 2015), pp. 37–47. http://dx.doi.org/10.1155/2015/196976
7. A.N. Ramesh, C. Kambhampati, J.R.T. Monson, P.J. Drew, Artificial intelligence in medicine, in *Annals of the Royal College of Surgeons of England*, vol. 86 (PMC, 2004), pp. 334–338. http://doi.org/10.1308/147870804290

8. C.K. Reddy, C.C. Aggarwal, *Healthcare Data Analytics*, 1st edn. (CRC Press, Taylor & Francis Group, 2015). ISBN 978-1-4822-3211-0
9. R.A. Weinberg, *The Biology of Cancer*, 2nd edn. (Taylor & Francis Group, Garland Science, 2014). ISBN 978-0-8153-4219-9
10. A. González Martín, Molecular biology of cervical cancer, in *Clinical and Translational Oncology*, vol. 9 (Springer, Milan, 2007), pp. 347–354. https://doi.org/10.1007/s12094-007-0066-8
11. J.G. De la Garza-salazar, F. Morales-Vasquez, A. Meneses-García, *Cervical Cancer* (Springer, Switzerland, 2017). https://doi.org/10.1007/978-3-319-45231-9
12. J. Mothoneos, Understanding cervical cancer, a guide for women with cancer, their families and friends, in *Cancer Council Australia Cancer Council SA*, vol. 13 (Cancer Council Australia, 2017). ISBN 978-1-925651-03-4
13. R. Sankaranarayanan, J.W. Sellors, Colposcopy and treatment of cervical intraepithelial neoplasia, World Health Organization—International Agency for Research on Cancer (IARC), Lyon (2003). ISBN 9283204123
14. V. Mehta, V. Vasanth, C. Balachandran, Pap smear, in *Indian Journal of Dermatology, Venereology and Leprology*, vol. 75 (Wolters Kluwer Medknow Publications, 2009), pp. 214–216. https://doi.org/10.4103/0378-6323.48686
15. P. Pisani, R.J. Black, P. Pisani, M.T. Valdivieso, A.B. Miller, N.E. Day, M. Kallio, A.B. Miller, N.E. Day, H. Moller, P. Lauriola, E. Magliola, L. Bonelli, E. Rossi, C. Gustavino, M. Ferreri, M.R. Giovagnoli, C. Midulla, M.E. Boon, S. Beck, J.A. Knottnerus, N. Day, G. Douglas, E. Farney, E. Lynge, J. Philip, G.P. Vooijs, M. Confortini, A. Biggeri, A. Russo, The pap test process. Leonardo Da Vinci Project—Cytotrain (2000). http://www.apof.eu/ZAMBIA2/RobertoL/THE%20PAP%20TEST%20PROCESS.pdf
16. J. Byriel, Neuro-fuzzy classification of cells in cervical smears. MSc. thesis, Department of Automation, Technical University of Denmark. Lyngby, Denmark, 1999
17. A. Tsakonas, G. Dounias, J. Jantzen, B. Bjerregaard, A hybrid CI approach combining genetic programming and heuristic classification for Pap-Smear diagnosis, in *Presented in "Hybrid CI Methods in Medicine" session*, EUNITE-01, Tenerife, Spain, December 13–14, 2001, also published in G. Dounias, D.A. Linkens (eds.), Adaptive Systems and Hybrid Computational Intelligence in Medicine, pp. 123–132. Joint Publication of the University of the Aegean and EUNITE, The European Network on Intelligent Technologies for Smart Adaptive Systems (2001). ISBN 960-7475-19-4
18. E. Martin, Pap-Smear classification. MSc. thesis, Department of Automation, Technical University of Denmark. Lyngby, Denmark, 2003
19. N. Ampazis, G. Dounias, J. Jantzen, Pap-Smear classification using efficient second order neural network training algorithms, in *Methods and Applications of Artificial Intelligence*. SETN 2004. Lecture Notes in Computer Science, vol. 3025 (Springer, Berlin, Heidelberg, 2004), pp. 230–245. https://doi.org/10.1007/978-3-540-24674-9_25
20. J. Norap, Classification of Pap-Smear data by transudative neuro-fuzzy methods. MSc thesis, Department of Automation, Technical University of Denmark. Lyngby, Denmark, 2005
21. G. Dounias, B. Bjerregaard, J. Jantzen, A. Tsakonas, N. Ampazis, G. Panagi, E. Panourgias, Automated identification of cancerous smears using various competitive intelligent techniques, in *Oncology Reports*, vol. 15 (2006), pp. 1001–1006. https://doi.org/10.3892/or.15.4.1001
22. Y. Marinakis, G. Dounias, Nearest neighborhood based pap smear cell classification using tabu search for feature selection, in *The Pap Smear Benchmark, Intelligent and Nature Inspired Approaches in Pap Smear Diagnosis, Special Session Proceedings of the NISIS—2006 Symposium*, 25–34, November 29–December 1 2006, Puerto de la Cruz, Tenerife, Spain (Spain, 2006), pp. 25–34
23. Y. Marinakis, G. Dounias, Nature inspired intelligent techniques for pap smear diagnosis: ant colony optimization for cell classification, in *The Pap Smear Benchmark, Intelligent and Nature Inspired Approaches in Pap Smear Diagnosis, Special Session Proceedings of the NISIS—2006 Symposium*, 35–45, November 29–December 1, 2006, Puerto de la Cruz, Tenerife, Spain (Spain, 2006), pp. 35–45

24. J. Jantzen, G. Dounias, Analysis of Pap-Smear data, in *NISIS 2006, The Pap Smear Benchmark, Intelligent and Nature Inspired Approaches in Pap Smear Diagnosis, Special Session Proceedings of the NISIS—2006 Symposium*, 25–34, November 29–December 1 2006, Puerto de la Cruz, Tenerife, Spain (Spain, 2006), pp. 5–14

25. Y. Marinakis, M. Marinaki, G. Dounias, C. Zopounidis, Metaheuristic algorithms in medicine, the Pap-Smear cell classification problem, *Book of Abstracts of the ECO-Q Management and Quality in Health Care*, 30–31 March 2007, Chania, Greece (presentation) (Greece, 2007)

26. F. Glower, Tabu search-part I. ORSA J Comput. **3**, 190–206 (1989). https://pubsonline.informs. org/doi/abs/10.1287/ijoc.1.3.190

27. J. Jang, ANFIS: adaptive-network-based fuzzy inference system, in *IEEE Systems, Man, and Cybernetics Society*, vol. 23 (IEE, 1993), pp. 665–685. https://doi.org/10.1109/21.256541

28. S. Haykin, *Neural Networks and Learning Machines*, 3rd edn. (Pearson Education, Inc. Upper Saddle River, New Jersey, 2009). ISBN 978-0-13-147139-9

29. F. Puppe, Heuristic classification, in *Systematic Introduction to Expert Systems* (Springer, Berlin, Heidelberg, 1993), pp. 131–148. https://doi.org/10.1007/978-3-642-77971-8_15

30. B.N. Prasad, M. Rathore, G. Gupta, T. Singh, Performance measure of hard c-means, fuzzy c-means and alternative c-means algorithms, in *International Journal of Computer Science and Information Technologies (IJCSIT)*, vol. 7 (2016), pp. 878–883. http://ijcsit.com/docs/ Volume%207/vol7issue2/ijcsit2016070297.pdf

31. M. Velikova, P.J.F. Lucas, N. Ferreira, M. Samulski, N. Karssemeijer, A decision support system for breast cancer detection in screening programs, in *Proceedings of the 18th European Conference on Artificial Intelligence*, ECAI 2008, vol. 178 (IOS Press, Amsterdam, 2008), pp. 658–662. http://doi.org/10.3233/978-1-58603-891-5-658

32. P. Goel, H. Liu, D. Brown, A. Datta, Spiking neural network based classification of task-evoked EEG signals, in *Knowledge-Based Intelligent Information and Engineering Systems*. KES 2006. Lecture Notes in Computer Science, vol. 4251 (Springer, Berlin, Heidelberg, 2006), pp. 825–832. https://doi.org/10.1007/11892960_99

33. K. Kabassi, M. Virvoua, G.A. Tsihrintzis, Y. Vlachos, D. Perrea, Specifying the personalization reasoning mechanism for an intelligent medical e-learning system on atheromatosis: an empirical study, in *Intelligent Decision Technologies*, vol. 2 (IOS Press, 2008), pp. 179–190. https://doi.org/10.3233/IDT-2008-2304

34. H. Kostakis, B. Boutsinas, D.B. Panagiotakos, L.D. Kounis, A computational algorithm for the risk assessment of developing acute coronary syndromes, using online analytical process methodology, in *International Journal of Knowledge Engineering and Soft Data Paradigms*, vol. 1 (Inderscience Enterprises Ltd, 2008), pp. 85–99. https://doi.org/10.1504/IJKESDP.2009. 021986

35. F. Menolascina, R.T. Alves, S. Tommasi, P. Chiarappa, M. Delgado, V. Bevilacqua, G. Mastronardi, A.A. Freitas, A. Paradiso, Fuzzy rule induction and artificial immune systems in female breast cancer familiarity profiling, in *Knowledge-Based Intelligent Information and Engineering Systems*. KES 2007. Lecture Notes in Computer Science, vol. 4694 (Springer, Berlin, Heidelberg, 2007), pp. 830–837. https://doi.org/10.1007/978-3-540-74829-8_101

36. V.S. Kodogiannis, J.N. Lygouras, T. Pachidis, An intelligent decision support system in wireless-capsule endoscopy, in *Intelligent Techniques and Tools for Novel System Architectures*. Studies in Computational Intelligence, vol. 109 (Springer, Berlin, Heidelberg, 2008), pp. 259–275. https://doi.org/10.1007/978-3-540-77623-9_15

37. E. Kang, Y. Im, U. Kim, Remote control multi-agent system for u-healthcare service, in *Agent and Multi-Agent Systems: Technologies and Applications*. KES-AMSTA 2007. Lecture Notes in Computer Science, vol. 4496 (Springer, Berlin, Heidelberg, 2007), pp. 36–644. https://doi. org/10.1007/978-3-540-72830-6_66

38. C.W. Jeong, D.H. Kim, S.C. Joo, Mobile collaboration framework for u-healthcare agent services and its application using PDAs, in *Agent and Multi-Agent Systems: Technologies and Applications*. KES-AMSTA 2007. Lecture Notes in Computer Science, vol. 4496 (Springer, Berlin, Heidelberg, 2007), pp. 747–756. https://doi.org/10.1007/978-3-540-72830-6_78

39. R.M.A. Mateo, L.F. Cervantes, H.K. Yang, J. Lee, Mobile agents using data mining for diagnosis support in ubiquitous healthcare, in *Agent and Multi-Agent Systems: Technologies and Applications*. KES-AMSTA 2007. Lecture Notes in Computer Science, vol. 4496 (Springer, Berlin, Heidelberg, 2007), pp. 795–804. https://doi.org/10.1007/978-3-540-72830-6_83

40. M. Tentori, J. Favela, M. Rodriguez, Privacy-aware autonomous agents for pervasive healthcare, *IEEE Intelligent Systems*, vol. 21 (IEEE, 2006), pp. 55–62. https://doi.org/10.1109/MIS.2006.118

41. S.G. Nejad, R. Martens, R. Paranjape, An agent-based diabetic patient simulation, in *Agent and Multi-Agent Systems: Technologies and Applications*. KES-AMSTA 2008. Lecture Notes in Computer Science, vol. 4953 (Springer, Berlin, Heidelberg, 2008), pp. 832–841. https://doi.org/10.1007/978-3-540-78582-8_84

42. J. Koleszynska, GIGISim—the intelligent telehealth system: computer aided diabetes management—a new review, in *Knowledge-Based Intelligent Information and Engineering Systems*. KES 2007. Lecture Notes in Computer Science, vol. 4692 (Springer, Berlin, Heidelberg, 2007), pp. 789–796. https://doi.org/10.1007/978-3-540-74819-9_97

43. C. Koutsojannis, I. Hatzilygeroudis, Fuzzy-evolutionary synergism in an intelligent medical diagnosis system, in *Knowledge-Based Intelligent Information and Engineering Systems*. KES 2006. Lecture Notes in Computer Science, vol. 4252 (Springer, Berlin, Heidelberg, 2006), pp. 1313–1322. https://doi.org/10.1007/11893004_166

44. E. Papageorgiou, G. Georgoulas, C. Stylios, G. Nikiforidis, P. Groumpos, Combining fuzzy cognitive maps with support vector machines for bladder tumor grading. in *Knowledge-Based Intelligent Information and Engineering Systems*. KES 2006. Lecture Notes in Computer Science, vol. 4251 (Springer, Berlin, Heidelberg, 2006), pp. 515–523. https://doi.org/10.1007/11892960_63

Part II
Learning and Analytics in Intelligent Power Systems

Chapter 5
Multi-kernel Analysis Paradigm Implementing the Learning from Loads Approach for Smart Power Systems

Miltiadis Alamaniotis

Abstract The future of electric power grid infrastructure is strongly associated with the heavy use of information and learning technologies. In this chapter, a new machine learning paradigm is presented focusing on the analysis of recorded electricity load data. The presented paradigm utilizes a set of multiple kernel functions to analyze a load signal into a set of components. Each component models a set of different data properties, while the coefficients of the analysis are obtained using an optimization algorithm and more specifically a simple genetic algorithm. The overall goal of the analysis is to identify the data properties underlying the observed loads. The identified properties will be used for building more efficient forecasting tools by retaining those kernels that are higher correlated with the observed signals. Thus, the multi-kernel analysis implements a "learning from loads" approach, which is a pure data driven method avoiding the explicit modeling of the factors that affect the load demand in smart power systems. The paradigm is applied on real world nodal load data taken from the Chicago metropolitan Area. Results indicate that the proposed paradigm can be used in applications where the analysis of load signals is needed.

5.1 Introduction

The future of electric power is associated with the extensive utilization of advanced information technologies [1]. Generation, collection and processing of information will be the cornerstone in developing a modern infrastructure whose operation will attain maximum efficiency for both suppliers and consumers [2]. The envisioned coupling of information and data technologies with power delivery systems is known as the "smart grid" [3]. A more general term that contains the entire power system from end to end—generation, transmission and distribution—is the term "smart power"

M. Alamaniotis (✉)
Department of Electrical and Computer Engineering, University of Texas at San Antonio, UTSA Circle 1, San Antonio, TX 78249, USA
e-mail: malamaniotis@gmail.com; miltos.alamaniotis@utsa.edu

© Springer Nature Switzerland AG 2019
G. A. Tsihrintzis et al. (eds.), *Machine Learning Paradigms*,
Learning and Analytics in Intelligent Systems 1,
https://doi.org/10.1007/978-3-030-15628-2_5

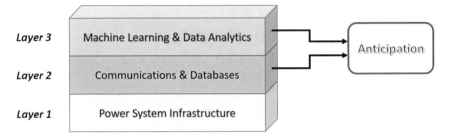

Fig. 1 The notion of smart power systems and the role of anticipation

systems" [4]. Smart power systems fully integrate information technologies with physical infrastructure, thus, providing a complex man made cyber-physical system.

The cyber-physical nature of smart power systems is clear from their architecture—as depicted in Fig. 1—that is comprised of three main layers [5]. The first layer contains the physical electrical infrastructure, i.e., the electrical power system. The other two layers are associated with the cyber-space. In particular, layer 2 contains those technologies pertained to communication and data storage (databases), while layer 3 contains machine learning techniques for data processing and decision making. The two cyber layers play crucial role in implementing smart power systems. The synergism of these two layers will utilize communications and collected data to regulate power flow, shape demand and ensure grid reliability at all times [6]. Furthermore, the cyber layers will enable the extensive use of anticipation in the management of the power system. Notably, anticipation has been identified as an essential component for the efficient management of the power grid from a demand-size point of view [7]. An example, among many visions, is the *Energy Internet* [1, 2], proposed by CIMEG [8], which models the electric grid as demand-driven system, where load forecasting plays a crucial role [9]. Electric load anticipation is a challenging task since energy consumption is based on a set of time-varying and difficult to accurately predict factors, such as the weather conditions and human habits [10].

The diverse and complex infrastructure of power systems hosts demanding and data intensive activities that need to be independently monitored for forecasting nodal load demand [11]. Data analytics attracts increasing attention as a tool that may be used for drawing conclusions from observed historical load demand data and subsequently making predictions over the future demand. Therefore, several approaches have been developed for analyzing load time series data and subsequently making predictions over the future load demand. The vast majority of the methods adopt tools from the statistics and machine learning library. In [12], a method for electric load analysis using a feedforward neural network is introduced, while in [13] a neural network is used for the load flow analysis in power systems. Neural networks have also found use in synergism with other tools: a combination of neural networks with wavelets is introduced in [14], with immunity lion algorithm in [15], with particle swarm optimization in [16], with fuzzy inference in [17], with principal component analysis in [18], with Jaya algorithm in [19], and genetic algorithms in

[20]. Other methods proposed for load data analysis were based on multifractals [21], Kalman filtering [22], Cuckoo search algorithm [23], singular spectral analysis [24], support vector regression [25], regression trees [26], Gaussian processes [27], independent component analysis [28], and relevance vector machines [29]. Notably, there is a plethora of methods that have been presented for load analysis and then utilized for load anticipation.

Despite the vast amount of method that have been proposed, the complexity of the load data still remains high and exhibit high uncertainty. Inference making driven by load data imposes significant difficulties to scientists and engineers. Thus, there is still room for more sophisticated methods that will push the envelope in the area of load data utilization. In this chapter, a new machine learning paradigm is proposed that is applied to load signal analysis. In particular, the proposed paradigm utilizes a set of kernel machines [30]—kernel modeled Gaussian Processes—to analyze a single load signal into several components. Each component is taken with a single kernel machine, which is equipped with a different kernel function [31]. The signal components that are close to the analyzed signal are retained and the rest are rejected. It should be noted that this new paradigm adopts multiple kernels to implement the "learning from loads" approach, which had been previously used for detection of cyber-attacks [32]. The contribution of the current work contains:

– A new multi-kernel paradigm for load signal analysis in power grids,
– A new approach for load data representation,
– An approach for selecting kernels in making prediction making,
– Application of the multi-kernel paradigm in real data taken from the Chicago metropolitan area.

The roadmap of the chapter is as follows: in Sect. 2 the kernel modeled Gaussian processes are briefly introduced, while in Sect. 3 the multi-kernel learning paradigm is described. Section 4 presents the results taken by applying the paradigm to real world data, and Sect. 5 concludes the papers by summarizing its main points.

5.2 Background

In this section, the Gaussian processes from a machine learning perspective are introduced. Initially the section gives a brief description of kernel machines and subsequently presents the derivation of Gaussian processes with respect to kernel functions.

5.2.1 Kernel Machines

Machine learning refers to models and methods that learn from data to solve problems in an automatic way. In general, machine learning is used for regression and classification problems: regression refers to prediction making of variables that take values in a continuous range, while classification refers to problems where unknown patterns are assigned labels taken from a specific discrete set. A preeminent set of methods in machine learning library that are utilized for both regression and classification problems is the *kernel machines* [30]. Kernel machines are analytical models that are expressed with the aid of a kernel function.

A kernel function, or simply known as kernel, denoted by $k(x, x)$ is any valid analytical function that is written as an expression of the dual formula which is given below [25]:

$$k(x_1, x_2) = f(x_1)^T f(x_2)^T \qquad (1)$$

where $f(x)$ is a mathematical function and x_1, x_2 represent the inputs (either scalars or vectors). The recasting of a function using the formula in (1) is known as the kernel trick [30].

5.2.2 Gaussian Processes

The Gaussian (or Normal) distribution is a well-defined and widely used probability distribution which describes a random variable. It is expressed as a function of two parameters, namely, the mean and the variance. Likewise, a Gaussian process (GP) is a collection of random variables that jointly follow a Gaussian distribution. A Gaussian process is also comprised of two parameters, namely, the mean and the covariance function. Furthermore, GP can be used for classification and regression problems; in the latter case it is known as Gaussian process regression and is denoted as GPR.

In order to derive the GPR framework, it is helpful to start from the simple linear regression that takes the form:

$$y(x) = w^T \varphi(x) \qquad (2)$$

with w denoting linear weights, and $\varphi(x)$ the basis functions. In the next step, a prior distribution over the weights is assigned that follows a Gaussian with zero mean and variance $\sigma^2 I$:

$$p(w) = N(w|0, \sigma^2 I) \qquad (3)$$

with I being the identity matrix. Considering a matrix form then the formula in Eq. (2) takes the form:

$$y = \Phi w \qquad (4)$$

where the matrix Φ is populated with elements of the form $\Phi_{nm} = \varphi_n(x_m)$. It should be mentioned that since y is a linear combination of Gaussian variables it follows also a Gaussian distribution. By taking into consideration the prior distribution, i.e., Eq. 3, then the parameters of the Gaussian distribution are evaluated as:

$$E[y] = E[\Phi w] = \Phi E[w] = 0 \tag{5}$$

$$Cov[y] = E[yy^T] = E[\Phi ww^T \Phi^T] = \Phi E[ww^T]\Phi^T = \sigma^2 \Phi \Phi^T. \tag{6}$$

It should be noted that the last term of Eq. 6 can be denoted as K which is known as the Gram matrix [31]:

$$K = \sigma^2 \Phi \Phi^T \tag{7}$$

with the elements of the Gram matrix being expressed as a function of a kernel $k(x_n, x_m)$ as given below:

$$K_{nm} = \sigma^2 \phi(x_n)^T \phi(x_m) = k(x_n, x_m). \tag{8}$$

Thus, the output of the linear regression y follows a Gaussian distribution with zero mean and variance being equal to K:

$$p(y) = N(y|0, K) \tag{9}$$

where, a careful observation of Eq. 9 reveals that the distribution of y depends on the kernel that is part of the Gram matrix.

Lastly, to derive the GPR framework, a population of N training datapoints with known targets $\mathbf{t}$ for known inputs $\mathbf{x}$, is required. Then, a new input, which does not belong to the training set, is denoted as x_{N+1} and has a respective target t_{N+1}. Then, the joint distribution between the N training datapoints and the new input x_{N+1} follows a Gaussian distribution as well. By utilizing the joint Gaussian distribution, it has been shown in [33, 34] that GPR gives a predictive distribution whose parameters are computed by:

$$m(x_{N+1}) = \mathbf{k}^T K_N^{-1} \mathbf{t}_N \tag{10}$$

$$\sigma^2(x_{N+1}) = k - \mathbf{k}^T K_N^{-1} \mathbf{t}_N \tag{11}$$

where $\mathbf{K}_N$ represents the covariance matrix of the N training datapoints, $\mathbf{k}$ is a vector whose entries are computed as the covariances between the new $N + 1$ point and the training N points, and k is the scalar value [30]. The dependence of the predictive distribution on the kernel form, accounts for a high flexibility since it allows the modeler to select different types of kernels and vary the result [35].

5.3 Multi-kernel Paradigm for Load Analysis

5.3.1 Problem Statement

In this section the proposed multi-kernel paradigm is presented and its individual steps are described. The overall goal of the paradigm is to utilize the various kernel types to analyze the load signal into a set of various components and then select the components that are closer to the original load signal.

The objective of the multi-kernel paradigm is to learn from observed load data and use what it learnt to analyze the load signal. The underlying idea is the expansion of a load signal into several components, by using multiple kernel machines. Each kernel machine is equipped with a different kernel function, where each kernel models a different set of data properties. Therefore, the use of multiple kernels at the same time allows the creation of a pool of various data properties.

In power systems, recorded load patterns exhibit high fluctuation and depend upon various dynamically varying factors to name a few: weather conditions, day of the week (e.g., weekend), season of the year, and special days (e.g., Christmas day). Capturing the dynamic behavior of all those factors may be done in an indirect way: by identifying the data properties underlying the observed datasets [36]. Therefore, in the present paradigm the various kernels are employed to capture and model the underlying properties of the dynamic factors. In addition, it allows a data driven modeling of load factors without explicitly knowing the real factors that influence the load signal.

5.3.2 Multi-kernel Paradigm

The underlying idea of the proposed framework is the analysis of the observed load signal into a set of components. Each component is computed by a single kernel machine, hence, providing, for N kernels a set of N components. The block diagram of the proposed paradigm is depicted in Fig. 2, where the individual steps are explicitly given.

Initially, the load signal under study is acquired and forwarded for processing; the signal spans a specific time period—anything from minute to week long time intervals. Next, the load signal is sampled at uniform intervals; for instance, for hourly data, 24 samples will be collected, with every sample representing the load demand at each hour. The sampling frequency depends on the modeler and the specifics of the application at hand. However, it should be noted that the sampling process should be uniform no matter the interval length at hand.

In the next step, the sampled data are put together to form the training dataset. To make it clear, the training dataset coincides with the sampled data. Then, the formed training set is forwarded to the next stage, which contains a set of kernel machines, and more specifically five Gaussian processes. As seen in Fig. 2, each of the Gaussian

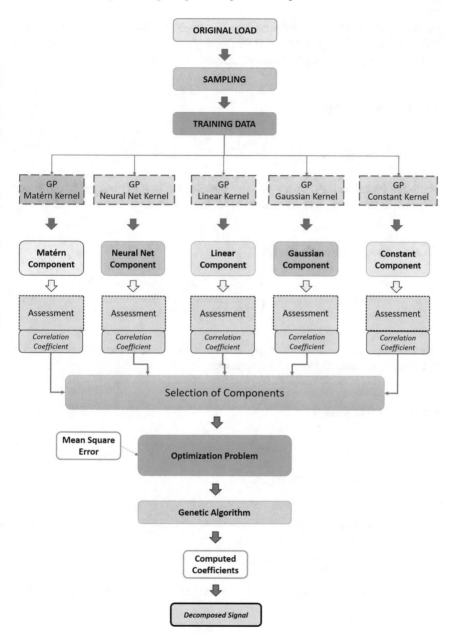

Fig. 2 Block diagram of the proposed multi-kernel paradigm

process is equipped with a different kernel, namely, the Matérn kernel, the Neural Network kernel, the Linear kernel, the Gaussian kernel, and the Constant kernel. The analytical forms of the aforementioned kernels are given below.

5.3.2.1 Matérn Kernel

$$k(\mathbf{x}_1, \mathbf{x}_2) = \left(\frac{2^{1-\theta_1}}{\Gamma(\theta_1)}\right)\left[\frac{\sqrt{2\theta_1}|\mathbf{x}_1 - \mathbf{x}_2|}{\theta_2}\right]^{\theta_1} K_{\theta_1}\left(\frac{\sqrt{2\theta_1}|\mathbf{x}_1 - \mathbf{x}_2|}{\theta_2}\right) \tag{12}$$

where the parameters θ_1, θ_2 are positive defined, while $K_{\theta_1}()$ is a modified Bessel function. In this work, it is convenient to set the first parameter equal to 3/2.

5.3.2.2 Neural Network Kernel

$$k(\mathbf{x}_1, \mathbf{x}_2) = \theta_0 \sin^{-1}\left(\frac{2\tilde{\mathbf{x}}_1^T \Sigma \tilde{\mathbf{x}}_2}{\sqrt{\left(1 + 2\tilde{\mathbf{x}}_1^T \Sigma \tilde{\mathbf{x}}_1\right)\left(1 + 2\tilde{\mathbf{x}}_2^T \Sigma \tilde{\mathbf{x}}_2\right)}}\right) \tag{13}$$

where $\tilde{\mathbf{x}} = (1, x_1, \ldots, x_D)^T$ is an augmented input vector, Σ the covariance matrix of the input vectors, and θ_0 a scale parameter.

5.3.2.3 Linear Kernel

$$k(x_1, x_2) = \theta_1 x_1^T x_2 \tag{14}$$

that has a single scale parameter, i.e., θ_1.

5.3.2.4 Gaussian Kernel

$$k(x_1, x_2) = \exp\left(-\|x_1 - x_2\|^2/2\sigma^2\right) \tag{15}$$

which contains one parameter σ^2 that expresses the variance of the training data.

5.3.2.5 Constant Kernel

$$k(x_1, x_2) = \frac{1}{\theta_0} \tag{16}$$

where parameter θ_0 is a scale factor.

The parameters of the aforementioned kernels are evaluated in the training phase. Therefore, the parameter values are adjusted to the load signals, i.e., 'learn from signals'. In other words, the Gaussian processes learn the load at hand via parameter evaluation.

Once the training phase is completed, then the five Gaussian process models are utilized for load analysis. The analysis is conducted as following:

- The GP utilizes the training data, where each sample point is stamped with a time point.
- The GP provides an output for each of the time point in the sampled data.
- Lastly, the sampled data are replaced by the GP outputs.

The above analysis implicitly exhibits that the GP output uses the training data to output a set of new values at the timepoints where the training data were acquired. To make it clearer, the GP replaces the training data with a new dataset that has been generated with the use of the respective kernel. The above analysis process is performed by each of the GP models, thus, providing a set of five signals.

In the next step, the computed components are being assessed. The assessment contains the computation of the correlation coefficient among the component and the sampled data. The analytical form of the correlation coefficient is given below:

$$cc = \frac{\sum_{i=1}^{N} (x_i - \bar{x})(y_i - \bar{y})}{\sqrt{\sum_{i=1}^{N} (x_i - \bar{x})} \sqrt{\sum_{i=1}^{N} (y_i - \bar{y})}} \tag{17}$$

for the two signals x and y, with $\bar{x}$, $\bar{y}$ being the mean values of the signals respectively. The correlation coefficient takes values in the interval $[-1\ 1]$ where 1 denotes absolute similarity and -1 denotes absolute dissimilarity between the two signals.

In the next step, the computed correlation coefficient values are forwarded to the selection module together with the analysis components. The objective of the selection module is to identify the data properties underlying the original signal by retaining those components that are close to the original signal and reject the rest. Selection is performed by:

- Retain those components whose correlation coefficient is equal to or above 0.7 (Threshold in Fig. 3),

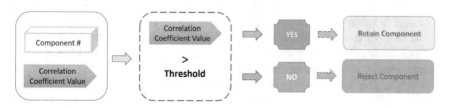

Fig. 3 Selection process of the components in the selection module

– Reject those components whose correlation coefficient is less than 0.7 (Threshold in Fig. 3).

with the selection process to be depicted in Fig. 3 as well. At this point it should be noted that selection of the threshold is done empirically and is based on previous experience [36].

Component selection is followed by the formulation of a linear assembly of the selected components as given by:

$$E = \alpha_1 C_1 + \cdots + \alpha_n C_n \tag{18}$$

where E is the estimated load, n expresses the population of components, $C_i\ i = 1, \ldots, n$ stand for the components, and $\alpha_i, i = 1, \ldots, n$ are the linear coefficients. In Eq. (18), the linear coefficients express the contribution of coefficients in the estimated load. Evaluation of the coefficients is performed via an optimization problem. Getting into details, the optimization problem is formulated with the aid of the mean square error (MSE), which is defined as the difference between the estimated load E, i.e., Eq. (18), and the original signal. Denoting the original signal as L, then the MSE is given by:

$$MSE = \frac{1}{N} \sum_{i=1}^{N} (L(i) - E(i))^2 \tag{19}$$

while the single objective optimization problem takes the following form:

$$\operatorname*{minimize}_{\alpha_1 \ldots \alpha_n} \left\{ \frac{1}{N} \sum_{i=1}^{N} (L(i) - E(i))^2 \right\}$$
$$\text{subject to } \{\alpha_1, \ldots, \alpha_n \geq 0\}$$
$$\text{where to } E = \alpha_1 C_1 + \cdots + \alpha_n C_n \tag{20}$$

with the unknown coefficients being imposed to the constraint to be zero or positive. This constraint stems from the fact that the coefficients express the contribution of the component in the spectrum; contributions cannot be negative.

Solution to the optimization problem in Eq. (20) is sought using a simple genetic algorithm [37]. In this work, the genetic algorithm has an initial population of 30 solutions, while the mutation probability is set equal to 0.01. The identified solution, i.e., a set of values for coefficients $c_1, \ldots, c_n$, is the final solution of the problem.

At this point, it should be emphasized the applicability of the paradigm: the optimal linear assembly that consists of the computed coefficients and the respective components, may be used for making predictions over the next interval demand. In this application, the general idea is that by analyzing the most recent past, the near future may be predicted given that the observed data properties will be dominate the near future as well. Overall, the presented paradigm is not restricted to load prediction but can be used in various applications pertained to load signals.

5.4   Results

In this section, the presented multi-kernel paradigm is tested on a set of real world data taken from the metropolitan Chicago area [35]. In particular, the datasets include load data taken from a grid node and are expressed in the scale of Megawatts (MW). In the next section, a demonstration case will be explicitly given, while further results are briefly provided in Sect. 4.2.

5.4.1   Problem Statement

In this problem, a step by step case is provided. In the first step, a day long load demand signal is obtained. The signal is sampled at 1-h intervals, and hence, 24 samples are acquired; the 24-sample signal is depicted in Fig. 4.

In the next step, the signal depicted in Fig. 4 is set as the training dataset and is forwarded to the Gaussian process models for training. Then, each GPR model is trained and the kernel parameters are evaluated. Once the training is completed, the GPR models are used to provide an output that is the model's component for that respective day (i.e., Fig. 4). The output signal of each GPR model are provided in Fig. 5.

Determination of components is followed by computing the correlation coefficients between the components in the Fig. 5 and the original sampled signal in the Fig. 4. The coefficient values are presented in Table 1. Next, the correlation coefficients are forwarded to the selection module and are compared to the predefined

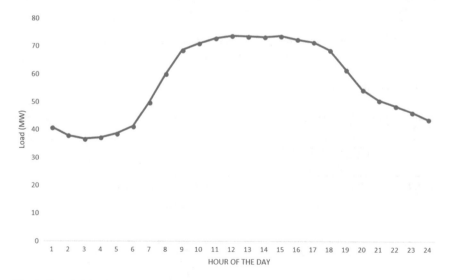

Fig. 4 Sampled day long load signal

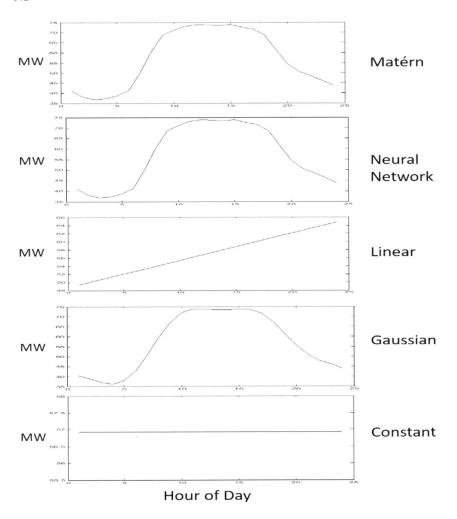

Fig. 5 GPR computer components

Table 1 Correlation coefficient values between computed components (Fig. 5) and sampled signal (Fig. 4)

	Kernel—Component				
	Matérn	Neural net	Linear	Gaussian	Constant
Original signal	1.0	1.0	0.3294	0.9984	0.1

threshold, whose value is equal to 0.7. The results of comparison indicate that the three kernels that are selected are, namely, the *Matérn*, *Neural Net* and *Gaussian* kernels, whose correlation values are 1.0, 1.0 and 0.9984 and lay above the 0.7 threshold. Thus, the respective components are retained, while the components of the linear and the constant kernel are rejected.

In the next step, the three selected components are put together to form the linear assembly given by:

$$E = \alpha_1 C_M + \alpha_{NN} C_{NN} + \alpha_G C_G \tag{21}$$

where C_M, C_{NN}, and C_G denote the components of the Matérn, Neural Net and Gaussian components. The contributions of the coefficients are unknown, and thus, the optimization problem is formulated as shown below:

$$\underset{\alpha_1, \alpha_2, \alpha_3}{\text{minimize}} \left\{ MSE = \frac{1}{24} \sum_{i=1}^{24} (L(i) - E(i)) \right\}$$

$$\text{subject to } \{\alpha_1, \alpha_2, \alpha_3 \geq 0\}$$

$$\text{where to } E = \alpha_1 C_M + \alpha_1 C_M + \alpha_n C_n \tag{22}$$

whose solution is identified with a genetic algorithm. By applying the genetic algorithm to the problem of Eq. (22), the following solution is taken:

$$\alpha_1 = 0.440945$$
$$\alpha_2 = 0.094488$$
$$\alpha_3 = 0.472441 \tag{23}$$

and therefore, the final linear assembly becomes:

$$L = 0.440945 \cdot C_M + 0.094488 \cdot C_{NN} + 0.472441 \cdot C_G \tag{24}$$

which expresses the expansion of the original signal (denoted as L) into three components. Thus, the above assembly consists of the final output of the multi-kernel paradigm.

At this point it should be noted that the linear assembly of Eq. (24) may be used in applications where inference making pertained to load management. For instance, it may be used to find factors related to patterns of interest of the power system operation.

5.4.2 *Further Results*

In this section, a set of load signals is analyzed and the computed linear assemblies are obtained. The signals under study are sampled at different rates, going beyond the hourly sampling that was adopted in the previous subsection. The set of four signals are depicted in Fig. 6, while the obtained linear assemblies are given in Table 2.

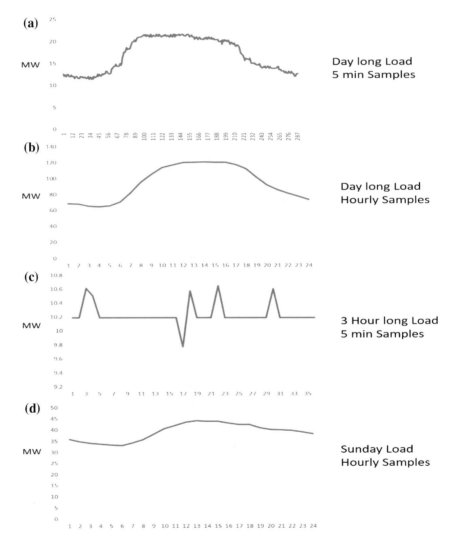

Fig. 6 Load data curves for testing: **a** weekday long hourly sampled, **b** 5-min day long sampled, **c** 5 min 3 h long sampled, and **d** Sunday hourly sampled load data

Load	Computed linear assembly
Table 2 Obtained linear assemblies for load data of Fig. 6	
Day long load 5 min samples	$L = 0.283465 \cdot C_M + 0.370079 \cdot C_{NN} + 0.346557 \cdot C_G$
Day long load hourly samples	$L = 0.401575 \cdot C_M + 0.425197 \cdot C_{NN} + 0.188976 \cdot C_G$
3 h long load 5 min samples	$L = 1 \cdot C_{Const}$
Sunday load hourly samples	$L = 0.291339 \cdot C_M + 0.685039 \cdot C_{NN} + 0.0001 \cdot C_{Lin} + 0.023622 \cdot C_G$

The first case of Fig. 6 contains a day long signal that has been sampled every five minutes (this is a weekday). Therefore, the signal exhibits variance and higher fidelity. The analysis performed regarding this first load provides a linear assembly comprised of three components. The three components as shown in Table 2 are the Matérn, Neural Net and Gaussian components, while the respective computed coefficients are also given in the right column of Table 2.

The second case contains a day long load signal that is sampled every one hour (this is also a weekday). Therefore, it contains a 24 value signal as shown in Fig. 6b. The multi-kernel paradigm analyzes the signal and provides a linear assembly that is presented in Table 2; the obtained assembly is comprised of three components— Matérn, Neural Net and Gaussian—with the respective coefficients to be given in the second column of Table 2. Notably, cases 1 and 2 provide the same set of components, though the fact that the two signals have different sampling rate. However, a more careful observation of Fig. 6a, b show that the two signals are very similar. Therefore, the fact that are represented by the same components confirms the validity of our approach: similar patterns, which share similar properties, are analyzed and represented by the same components (but contributions, i.e., coefficients are may be different).

In the third case, which contains the signal depicted in Fig. 6c, the load is recorded every five minutes for a period of three hours. Essentially, the load signal exhibits a constant behavior except for some cases that take other values. This behavior is also identified by the multi-kernel analysis: the computed assembly contains only one components, namely, the constant component whose coefficient is equal to 1.

In the last case, the day long load of a Sunday is taken into consideration (in other words this is the load of a day taken from a weekend). The analysis conducted by the presented paradigm provides a four-factor linear assembly that is given in the last row of Table 2. The obtained assembly consists of the following components: Matérn, Neural Net, Linear and Gaussian. It should be stated that the load signal in this case is taken from the same power grid node as the signal in cases 1 and 2. Comparing those cases with the current one, it is noticed that the weekend signal contains an extra component—the Linear component. Hence, it is concluded that the Sunday signal is different than the weekday—an observation that validates the general truth of load patterns that weekdays and weekend exhibit different load trends-, and that conclusion is strongly supported by the analysis provided by the multi-kernel paradigm.

Overall, it is concluded from the above studied cases, that the multi-kernel paradigm is able to analyze and express the load signal into a set of components. Its efficiency is supported by our findings: similar load patterns are expressed with the same set of components, while load signals with different trends are expressed with different set of components.

5.5   Conclusion and Future Work

In this work, a new machine learning paradigm that implements the "learning from loads" approach for analyzing load signals in smart power systems is presented. The essential components of the presented paradigm are a set of intelligent tools, and more specifically, the kernel modeled Gaussian processes and genetic algorithms. The basic idea is to utilize a set of Gaussian processes equipped with different kernels and then represent the load as a linear assembly of components provided by the Gaussian processes. Each kernel represents a set of data properties, and thus, the presented paradigm implements a data-driven approach for load signal analysis.

Representation of load signals with the kernel obtained components promotes utilization of load data in smart grid applications, such as load anticipation. The paradigm successfully analyzed the studied load signals independently of their length and the type of sampling. Furthermore, load signals that were representing the similar signal were expressed with linear assemblies that shared the same number of factors and same type of components; this was strongly supported by results obtained on the set of real world load data taken from the Chicago area.

In the future, research will be follow in two main directions. The first one is the use of a wider number of kernels for analyzing the signals, i.e., going beyond the 5 kernels employed in the current chapter. The second direction will focus on integrating the presented paradigm with critical applications in smart power systems, such as load management and cybersecurity.

References

1. L.H. Tsoukalas, R. Gao, Inventing energy internet. The role of anticipation in human-centered energy distribution and utilization, in *SICE Annual Conference, 2008* (IEEE, 2008, August), pp. 399–403
2. M. Alamaniotis, R. Gao, L.H. Tsoukalas, Towards an energy internet: a game-theoretic approach to price-directed energy utilization, in *International Conference on Energy-Efficient Computing and Networking* (Springer, Berlin, Heidelberg, 2010, October), pp. 3–11
3. J. Momoh, *Smart Grid: Fundamentals of Design and Analysis*, vol. 63 (Wiley, 2012)

4. M. Alamaniotis, L.H. Tsoukalas, Layered-based approach to virtual storage for smart power systems, in *2013 Fourth International Conference on Information, Intelligence, Systems and Applications (IISA)* (IEEE, 2013, July), pp. 1–6
5. M. Alamaniotis, L.H. Tsoukalas, Utilization of virtual buffer in local area grids for electricity storage in smart power systems, in *Power Symposium (NAPS), 2017 North American* (IEEE, 2017, September), pp. 1–6
6. K. Moslehi, R. Kumar, A reliability perspective of the smart grid. IEEE Trans. Smart Grid **1**(1), 57–64 (2010)
7. M. Alamaniotis, N. Gatsis, L.H. Tsoukalas, Virtual budget: integration of electricity load and price anticipation for load morphing in price-directed energy utilization. Electr. Power Syst. Res. **158**, 284–296 (2018)
8. L.H. Tsoukalas, R. Gao, From smart grids to an energy internet: assumptions, architectures and requirements, in *Third International Conference on Electric Utility Deregulation and Restructuring and Power Technologies, 2008. DRPT 2008* (IEEE, 2008, April), pp. 94–98
9. L.H. Tsoukalas, Modeling the grid as a customer-driven system: emerging challenges and opportunities for self-healing infrastructures, in *Power Engineering Society Winter Meeting, 2001*, vol. 1 (IEEE, 2001, January), pp. 158-vol
10. M. Alamaniotis, A. Ikonomopoulos, L.H. Tsoukalas, Evolutionary multiobjective optimization of kernel-based very-short-term load forecasting. IEEE Trans. Power Syst. **27**(3), 1477–1484 (2012)
11. M. Alamaniotis, A. Nasiakou, R. Fainti, L.H. Tsoukalas, Leaky bucket approach implementing anticipatory control for nodal power flow management in smart energy systems, in *PES Innovative Smart Grid Technologies Conference Europe (ISGT-Europe), 2016 IEEE* (IEEE, 2016, October), pp. 1–6
12. F. Cavallaro, Electric load analysis using an artificial neural network. Int. J. Energy Res. **29**(5), 377–392 (2005)
13. W.L. Chan, A.T.P. So, L.L. Lai, Initial applications of complex artificial neural networks to load-flow analysis. IEE Proc. Gener. Transm. Distrib. **147**(6), 361–366 (2000)
14. T. Senjyu, Y. Tamaki, H. Takara, K. Uezato, Next day load curve forecasting using wavelet analysis with neural network. Electr. Power Compon. Syst. **30**(11), 1167–1178 (2002)
15. Y. Li, Y. Huang, M. Zhang, Short-term load forecasting for electric vehicle charging station based on niche immunity lion algorithm and convolutional neural network. Energies **11**(5), 1253 (2018)
16. S. Yufei, J. Chuanwen, PSO clustering analysis and Elman neural network for short-term electrical load forecasting. Wseas Trans. Circuits Syst. **4**(8), 1002 (2005)
17. M. Tamimi, R. Egbert, Short term electric load forecasting via fuzzy neural collaboration. Electr. Power Syst. Res. **56**(3), 243–248 (2000)
18. X.C. Guo, Z.Y. Chen, H.W. Ge, Y.C. Liang, Short-term load forecasting using neural network with principal component analysis, in *Proceedings of 2004 International Conference on Machine Learning and Cybernetics, 2004*, vol. 6 (IEEE, 2004, August), pp. 3365–3369
19. P. Singh, K.K. Mishra, P. Dwivedi, Enhanced hybrid model for electricity load forecast through artificial neural network and Jaya algorithm, in *2017 International Conference on Intelligent Computing and Control Systems (ICICCS)*, (IEEE, 2017, June), pp. 115–120
20. R.R. de Aquino, O.N. Neto, M.M. Lira, A.A. Ferreira, K.F. Santos, Using genetic algorithm to develop a neural-network-based load forecasting, in *International Conference on Artificial Neural Networks* (Springer, Berlin, Heidelberg, 2007, September), pp. 738–747
21. X. Yuan, B. Ji, Y. Yuan, Y. Huang, X. Li, W. Li, Multifractal detrended fluctuation analysis of electric load series. Fractals **23**(02), 1550010 (2015)
22. S.H. Jo, S. Son, S. Lee, J.W. Park, Kalman-filter-based multilevel analysis to estimate electric load composition. IEEE Trans. Industr. Electron. **59**(11), 4263–4271 (2012)
23. X. Zhang, J. Wang, K. Zhang, Short-term electric load forecasting based on singular spectrum analysis and support vector machine optimized by Cuckoo search algorithm. Electr. Power Syst. Res. **146**, 270–285 (2017)

24. K. Afshar, N. Bigdeli, Data analysis and short term load forecasting in Iran electricity market using singular spectral analysis (SSA). Energy **36**(5), 2620–2627 (2011)
25. G.F. Fan, L.L. Peng, W.C. Hong, F. Sun, Electric load forecasting by the SVR model with differential empirical mode decomposition and auto regression. Neurocomputing **173**, 958–970 (2016)
26. B. Gładysz, D. Kuchta, Application of regression trees in the analysis of electricity load. Bad.A Oper. I Decyz. **4**, 19–28 (2008)
27. M.Alamaniotis, S. Chatzidakis, L.H. Tsoukalas, *Monthly Load Forecasting Using Kernel Based Gaussian Process Regression* (2014)
28. H. Liao, D. Niebur, Exploring independent component analysis for electric load profiling, in *Proceedings of the 2002 International Joint Conference on Neural Networks, 2002. IJCNN'02*, vol. 3 (IEEE, 2002), pp. 2144–2149
29. M. Alamaniotis, D. Bargiotas, L.H. Tsoukalas, Towards smart energy systems: application of kernel machine regression for medium term electricity load forecasting. SpringerPlus **5**(1), 58 (2016)
30. C.M. Bishop, Pattern recognition and machine learning (information science and statistics) (Springer, New York, 2006)
31. B. Scholkopf, A.J. Smola, *Learning with Kernels: Support Vector Machines, Regularization, Optimization, and Beyond* (MIT Press, 2001)
32. M. Alamaniotis, L.H. Tsoukalas, Learning from loads: an intelligent system for decision support in identifying nodal load disturbances of cyber-attacks in smart power systems using Gaussian processes and fuzzy inference, in *Data Analytics and Decision Support for Cybersecurity* (Springer, 2017), pp. 223–241
33. C.E. Rasmussen, C. K. Williams, *Gaussian Process for Machine Learning* (MIT Press, 2006)
34. D.J.C. Mackay, Introduction to Gaussian processes, in *Neural Networks and Machine Learning*, ed. by C.M. Bishop (Springer, Berlin, 1998)
35. M. Alamaniotis, A. Ikonomopoulos, L.H. Tsoukalas, Probabilistic kernel approach to online monitoring of nuclear power plants. Nucl. Technol. **177**(1), 132–145 (2012)
36. M. Alamaniotis, A. Ikonomopoulos, L.H. Tsoukalas, A Pareto optimization approach of a Gaussian process ensemble for short-term load forecasting, in *2011 16th International Conference on Intelligent System Application to Power Systems (ISAP)* (IEEE, 2011, September), pp. 1–6
37. M.D. Vose, *The Simple Genetic Algorithm: Foundations and Theory*, vol. 12 (MIT Press, 1999)

Chapter 6
Conceptualizing and Measuring Energy Security: Geopolitical Dimensions, Data Availability, Quantitative and Qualitative Methods

John A. Paravantis, Athanasios Ballis, Nikoletta Kontoulis and Vasilios Dourmas

> *It is even probable that the supremacy of nations may be determined by the possession of available petroleum and its products.*
> —Calvin Coolidge, US President (1872–1933)
> ...*"patterns of oil peace" would be drawn up by governments because oil was too serious an affair to be left to oilmen...*
> —Frankel [24]
> *Without data, you're just another person with an opinion.*
> —W. Edwards Deming (1900–1993)

Abstract Energy is an important geopolitical driver, and energy security is an emerging field with growing interest in its measurement. This Chapter is a guide to energy security research that aims to estimate a quantitative energy security index with a geopolitical focus, by providing an in-depth dynamic geopolitical look into the history, evolution, dimensions, data, estimation, taxonomy, and forecasts of energy security. Discussion is complemented with examples from the area of the Southeastern Mediterranean and the Middle East. The Chapter commences with documenting the state of the art of the research literature on energy security. In particular, the definitions, the dimensions, the associated geopolitical issues, the policies, and the quantitative simple indicators and complex indexes of energy security are examined. An alternative to a full panel data approach for research that aims to calculate a quantitative energy security index would be to focus on specific milestone time periods such as the first oil crisis of the 1970s, the first Gulf War of 1990–91, the

J. A. Paravantis (✉) · N. Kontoulis
Department of International and European Studies, University of Piraeus, Piraeus, Greece
e-mail: jparav@unipi.gr

A. Ballis
Department of Transportation Planning and Engineering, National Technical University of Athens, Athens, Greece

V. Dourmas
Department of Environmental and Natural Resources Management, University of Patras, Patras, Greece

© Springer Nature Switzerland AG 2019 149
G. A. Tsihrintzis et al. (eds.), *Machine Learning Paradigms*,
Learning and Analytics in Intelligent Systems 1,
https://doi.org/10.1007/978-3-030-15628-2_6

Russian-Ukrainian disputes (2005–09), and the present time. The Chapter discusses how empirical geographical, energy, socioeconomic, environmental and geopolitical data could be collected, and a novel geopolitical energy security index could be developed with reference to appropriate statistical techniques. A Cluster Analysis of the data and security index values of each milestone time period could provide a dynamic taxonomy of countries, based on their energy security profile, which could be depicted on geopolitical maps. Of particular interest would be to see how countries shifted from one cluster to another over time. Following Cluster Analysis, it is suggested that a case study analysis of the energy security profile of key countries (such as Germany, Russia, Iran, Iraq, Saudi Arabia, the United States, Canada, Venezuela, China, India, and Japan) could complement the quantitative approach, and provide an in-depth appreciation of countries, the energy security issues they face, the policies they adopt to address them, and how well the obtained clustering and energy security values capture this knowledge. It is also proposed that any energy security index research include a small number of interviews with energy experts from the government, the academia, and the professional arena. It is advised that these interviews include open-ended questions on the concept of energy security, its evolution through the milestone time periods, the energy security index, the clusters, the geopolitical maps, and energy security policies. Finally, it is proposed that any energy security index research be complemented with forecasts that take into account socioeconomic and geopolitical data, the energy security index values, the dynamic taxonomy, as well as information gleaned from the interviews.

6.1   Preamble

Energy has been essential throughout human history, crucial for economic development and human security. According to the latest BP Statistical Review of World Energy [8], global primary energy consumption increased by 1% in 2016, following a growth of 0.9% in 2015 and 1% in 2014. Yet, energy is an economic, ill-distributed and expensive good, subject to price fluctuations, with repercussions in many domains of life. With energy being the *precondition of all commodities, a basic factor equal with air, water, and earth*" (E. F. Schumacher, Nobel laureate economist, 1977), energy security is paramount to human security [77], and has become an increasingly popular concept for policy makers, entrepreneurs and academics. Energy security can be viewed as a public good for societies, and its insufficient provision may be associated with disruptions of the supply of oil, gas and electricity, and severe consequences for societies, economies and individuals. Consequently, energy security is a political issue combining the concepts of high and low politics.

Concerns about energy security first arose in the early 1970s in Europe, Japan and the US, as the first oil crises uncovered the vulnerability of developed economies to oil price shocks. The International Energy Agency (IEA) was created in 1974 by the countries of the Organization for Economic Cooperation and Development

(OECD), so as to promote energy security among its member countries, through collective response to physical disruptions of energy supplies (e.g. by holding stocks equivalent to at least 90 days of net oil imports). Energy security became a matter of national security for many developed countries in the aftermath of the oil shocks of 1973. The late and great International Relations theorist Kenneth Waltz argued that there is a continuity of strategies of the major western states in energy geopolitics, and he predicted that the oil shocks, provoked by the Arab export embargo, would not cause a change of power in the West [83]. The current global attention to energy security is mostly explained by the new emerging giants of the world economy and their rising energy demand.

Energy security is a field dominated by a traditional approach to security, and means different things to different countries. The differentiation is based on their geographical location, their natural resource endowment, the status of their international relations, their political system, their economic disposition [55], and their ideological views and perceptions [58]. Approaches to energy security may differ between countries, depending on the structure of the energy system and historical experiences. This is observed in the various strategies chosen by the different member states within the European Union (EU), e.g. the degree of reliance on Russian gas and the diverse historical experiences from the Cold War that have led to different approaches to energy security ([51], as cited by Johansson [40]).

Energy security is argued to be part and parcel of security, and thus it should be high on the policy agenda of countries across the world. The Copenhagen School of security studies has described a four-level approach to international politics: the international (system), the regional (sub-system), the national (unit), and the internal (sub-unit) [16]. Energy security needs to be investigated at these four levels: globally, to ensure adequacy of resources; regionally, to ensure that networking and trade can take place; at a country level, to ensure national security of supply; and finally at a consumer level, to ensure that consumer demand is satisfied.

Yet, as it will become evident in the rest of this Chapter, part of the challenge is (1) to conceptualize energy security in a concrete way, and (2) measure it and formulate a quantitative energy security index that would be helpful in solving geopolitical problems. In this respect, this Chapter aims to be of use as an in-depth guide to: conceptualizing energy security; collecting data on various energy indicators; synthesizing a wide array of energy related variables and indicators into a geopolitical energy security index; using such an index to develop a dynamic geopolitical taxonomy of countries over time; examining key countries in depth; utilizing experts to assess gained knowledge; and compiling energy security forecasts.

Some of the background research leading to this Chapter has been published [67].

6.1.1 Structure of Chapter

The rest of this Chapter is structured as follows:

- Section 2 reviews the energy security research literature, with subsections covering the definitions, the history, the geopolitics, the dimensions, and the indexes of energy security.
- Section 3 presents in an organized manner a number of important research questions that were extracted from the energy security literature; these research questions could guide research aiming to develop an energy security index as well as suggest specific tasks that could be achieved with its use.
- Section 4 achieves a number of objectives, while at the same time outlining appropriate statistical techniques for addressing the research questions presented previously, and presenting examples from the Southeastern Mediterranean and the Middle East. Some of these tasks are: list sources for collecting data on a wide variety of energy security indicators; discuss historical time periods that could be focused upon; dissect energy security into components that should be accounted for in any attempt to measure it; discuss the role of Cluster Analysis (CA) in establishing groups of countries with a similar energy security level, and observe how countries change from one group to another, and how regional hotspots are formed and moved over time; examine case studies of key countries, to complement a quantitative with a qualitative approach; conduct (a small number of) key interviews with energy (security) experts to firm up the concept of any new geopolitical energy security index; and use all these tools to develop energy security forecasts.
- Finally, Sect. 5 wraps up the Chapter by presenting a few closing comments.

6.2 Review of Energy Security Literature

The research literature on energy security is organized around the concept of security, which, whether approached from a realist or a liberalist standpoint, is based on similar ontological and epistemological assumptions. Energy security is understood to be closely linked with national security, and greatly influenced by fossil fuels, particularly oil. It is also closely related to microeconomic and macroeconomic developments [15].

Energy security may be divided into security of supply, and security of demand [55] or into physical security (uninterrupted supply), price security, and geopolitical security. Much of the literature relates the term energy security to the security of supply. According to Kruyt et al. [47], energy security is often used as a synonym for the security of energy supply, particularly by researchers adopting an economic perspective [38, 40, 42, 49]. According to IEA [33], energy security is related to the security of supply and its physical availability. Even if energy security is a term which includes the economic competitiveness of energy, its environmental sustainability and a wide variety of geopolitical issues, the literature usually focuses on the uninterrupted physical availability of energy at an affordable price [64].

6.2.1 Brief History of Energy Security

Energy security has played an important role in the 20th century. According to Yergin [90], Winston Churchill believed that oil supply security was essential to fuel his army during the First World War, and was an important concern for Germany and Japan as they invaded the Soviet Union (USSR) and Indonesia during the Second World War. During these wars, energy security was often implicitly used as a synonym for national security.

According to 2013 United Nations (UN) energy statistics, the world energy demand increased from 1676 million tons of oil equivalent (Mtoe) in 1950 to 4197 Mtoe in 1969 [18]. The dominance of oil was assured by its liquid form, which made it the only viable fuel for transportation and the emerging automobile market: solid fuels had to be burned to raise steam, and only then could they drive an engine [24]. Globally, economic growth, improved living standards, motorization, and electrification pushed energy demand in all sectors, and the international energy trade increased from 331 Mtoe in 1950 to 1513 Mtoe in 1969. Oil exporting countries formed the Organization of Petroleum Exporting Countries (OPEC) in part to address the distribution of wealth derived from oil exports. In that period, security of energy supply was not a priority in many developing countries because companies supplied cheap oil and, thus, stability. At the same time, in many developing countries the majority of the population did not have access to modern energy. In 1969, 72% of the electricity was consumed in OECD countries that had only 20% of the world total population.

It was during the 1970s that energy security arose as a problem in the research literature. In the first oil crisis of 1973, oil embargoes by the Organization of Arab Petroleum Countries (OAPEC) shook the oil importing countries to the core, while the second oil crisis shot up international oil prices above $30 per barrel, which amounted to $100 per barrel in 2015 values [18]. The year 1974 was a milestone for the energy security concept, and as a response to the 1973 oil embargo, the OECD established the IEA. During that period, international energy security still largely meant oil security. According to IEA [36], in 1979 oil shared as much as 86% of the world energy trade, and the Middle East supplied 58% of the internationally traded oil. Consequently, by the end of 1970, energy security was a high priority issue on the policy agenda, in view of its significance for the entire economy.

The concern for energy security was somewhat reduced in the 1980s, as a result of the supply expansion and lower demand for energy. Global oil imports decreased by 25% during the first half of the decade [19]. Oil was in part replaced by natural gas and nuclear energy, especially for power generation. According to IEA [36], while the world energy demand increased by 20% per year during the 1980s, the share of oil shrank from 42 in 1980 to 37% in 1989 for primary energy supply, and from 20 in 1980 to 12% in 1989 for power generation.

The early 1990s were marked by the First Gulf War (https://history.state.gov/departmenthistory/short-history/firstgulf) and the collapse of the Soviet Union (https://history.state.gov/milestones/1989-1992/collapse-soviet-union). Energy security gained momentum after that, when global resources became scarce, in the face of a growing global demand for energy. At the time of the First Gulf War (1990–1991) and the dissolution of the Soviet Union (1991), new concepts emerged, and the concern for energy security began to gain prominence in global discourse [90]. According to 2013 UN energy statistics (https://unstats.un.org/unsd/energy/yearbook/2013.htm), power demand in non-OECD countries grew much faster than that of the OECD countries, although 63% of electricity was still consumed within OECD. Global warming issues became gradually institutionalized throughout the decade. The Kyoto Protocol was signed in 1997, and was the first international treaty to outline emission targets for combating global climate change. Implementation required participating members to create policies and measures to reduce and offset domestic emissions, and increase absorption of greenhouse gases.

As mentioned by Yergin [91] and Hancock and Vivoda [31], the energy security issue re-emerged in the 2000s, driven by the rising demand in Asia, disruptions of gas supplies in Europe, and the pressure to decarbonize energy systems. In 2005 the Russian federation cut down the supply of gas to Ukraine on the premise that they were not ready to accept the new prices. As a result, the supply of gas towards western European was also shortened. The Russian-Ukrainian crises of 2006 and 2009 showed that the main supplier of the EU was not only unreliable, but capable of using energy resources as a geopolitical weapon as well. A more recent revival of interest in energy security was stirred by high oil prices (in the period up to 2008), as well as geopolitical supply tensions [33]. At the same time, terror attacks led to the wars in Afghanistan and Iraq. The Arab Spring and the Islamic State created further tensions and instability. Finally, in 2011 the Fukushima nuclear accident revived important questions about the safety of nuclear energy.

So, the meaning of energy security expanded over time. Deviating gradually from the origin of the wording that suggested stable energy flow, the fair price element was added in the 1970s and 1980s. Recently, energy security became closely entangled with other energy policy problems, such as providing equitable access to modern energy, and mitigating climate change [27].

6.2.2 Defining Security and Energy Security

When one tries to define energy security, the question of what the term security means in general emerges. The Walzian notion of security asserts that the issue of security stems from the anarchical structure of society [83]. Neoclassical realism, as expressed by Morgenthau [63] and Klare [44], considers energy as a geopolitical tool to pursue state goals, and adds an element of hard power with overlapping military, political and economic dimensions.

According to Winzer [84], the attempts to define energy security have varied over time because of challenges in the realm of energy policy. Different countries may define energy security differently, in relevance to their own energy situation, and their vulnerability to energy supply disruptions. Countries not only differ in their definition of energy security, but also in the way they address energy security challenges [55].

Some authors [33, 91] point out that the term of energy security is not clearly defined; consequently there is no universal concept of energy security. Taking into consideration the absence of a clear definition [4, 47], energy security has become an umbrella term for different policy goals [84]. A multiplicity of concepts and dimensions enter the realm of energy security. Chester [12] aptly described energy security as "*slippery*" and "*polysemic*". The diversity of definitions is shaped by the perspective and nature of different countries, and their place in the energy chain and the complex global energy system.

Countries may be divided into three groups: producers/exporters, who wish to ensure reliable demand for their commodities; consumers, who commonly aim towards diversity of energy supply to maximize their security; and transit states, who are the essential bridges connecting producers/exporters with their markets [55]. Given the above differentiation among countries, there are two important concepts for energy security: security of supply (for the consumers), and security of demand (for the exporters). For energy exporting countries, the security of demand is equally important to the security of supply [40].

According to Sovacool [74], there are at least 45 different definitions of energy security that share a great deal of similarity among them, and lead to difficulties in terms of the operationality of the concept. Ang et al. [3] identified 83 energy security definitions in the literature. As pointed out by Cherp and Jewell [11], a classic definition of energy security is provided by Yergin [89], who visualized energy security as the assurance of "*adequate, reliable supplies of energy at reasonable prices,*" adding a geopolitical component by qualifying that this assurance must be provided "*in ways that do not jeopardize national values or objectives*". Yergin's definition identifies "*national values and objectives*" as the assets that safeguard energy security.

The IEA, a pioneer institution in energy security and the most important multinational platform, defines energy security as the "*uninterrupted availability of energy sources at an affordable price,*" and considers it to have a long-term and a short-term aspect. The IEA has restated the definition through the years to characterize energy security as the adequate, affordable and reliable supply of energy. Long-term energy security relates to "*timely investments to supply energy in line with economic developments and environmental needs*". Short-term energy security relates to "*the ability of the energy system to react promptly to sudden changes in the supply-demand balance*" [35, 39, 43].

In 2000, the European Commission [21] referred to energy supply security as "*the uninterrupted physical availability of energy products on the market, at a price that is affordable for all (private and industrial) consumers, while respecting environmental concerns and looking towards sustainable development*". This was an extension of the IEA definition, involving the inclusion of environmental and sustainability issues.

Definitions of energy security [4, 21, 33, 91] mainly use the term availability to imply stable and uninterrupted supply of energy. Other authors [41, 85] use the term reliability for energy infrastructure. Accessibility has been at the center of the energy security debates and policy approaches into the 21st century [45]. So, as an extension to the original IEA definition of energy security, the Asia Pacific Energy Research

Centre (APERC [4] highlighted the four As of availability, affordability, accessibility (to all), and acceptability (from a sustainability standpoint), and defined energy security as "*the ability of an economy to guarantee the availability of the supply of energy resources in a sustainable and timely manner with the energy price being at a level that will not adversely affect the economic performance of the economy*". Cherp and Jewell [11] compared the 4 As to the five As of access to health care (availability, accessibility, accommodation, affordability and acceptability). The first two As (availability and affordability) constitute the classic approach to energy security (20th century), while the latter two (accessibility and acceptability) reflect certain contemporary concerns (21st century) e.g. fuel poverty and global climate change.

6.2.3 Energy Security Since the 20th Century

Chester [12] and Vivoda [82] highlight the polysemic and multi-dimensional nature of energy security. Chester suggested that the term has risen highly on the policy agenda of governments because there is a complex system of global markets, a vast cross-border infrastructure network, and a small group of primary energy suppliers. Manson et al. [57] have described the energy security as a dynamic concept, with a perspective that depends on the time frame analyzed.

Energy security studies have changed in scope and focus over time, evolving from classic political economy studies of oil supplies for industrialized democracies, to a research field addressing a much wider range of energy sectors and energy security challenges [11]. The sources that Cherp and Jewell cite show that, in the 1970s and 1980s, energy security signified the stable supply of cheap oil, and threats to energy security included threats of embargoes, and price manipulation by exporters. Contemporary energy security studies: go beyond OECD oil importers; include nations at all levels of developments (that extract, import and export); address perspectives on energy security of non-state actors, including, individual regions, utilities, consumers, and global production networks; examine a variety of energy sources and carriers; address environmental issues such as climate change; address equity issues, mainly providing equitable access to energy; and address a wider range of threats.

Chester [12] described the energy regime after the Second World War, highlighting some important characteristics. During that time, there was a growing dominance of fossil fuels, particularly of Middle Eastern oil, which in 2008 accounted for one third of world energy use. Oil was cheap until the oil shocks of the 1970s, effectively triggered by the OPEC. As a consequence, energy security became synonymous with the need to reduce dependence on oil consumption.

The 1970s economic crisis took place at a time when production was internationalized against a backdrop of rampant inflation, persistently high unemployment, growing government expenditures, declining productivity, shrinking revenues, and smaller profits margins. Accepted management strategies failed to restore the growth of the 1950s and 1960s. In the 1980s, the need for the introduction of greater competition and less government involvement was strongly advocated for the electricity, telecommunications, and gas grids, so the energy markets took the path towards restructuring and liberalization. Monopolies were broken up, and new pricing schemes were introduced. This restructuring was promoted by the OECD, the World Bank, the International Monetary Fund (IMF), and international trading agreements.

In the meantime, nuclear energy was developed rapidly. By the end of the 1970s, 25 countries were generating electricity from nuclear power (including the US, the UK, France, Germany, and the former Soviet Union). The public acceptance of nuclear energy was hit by the Three Mile Island (USA 1979) and the Chernobyl (USSR 1986) accidents. The growth of nuclear energy was slowed down by high capital costs, lengthy construction times, problems with decommissioning older nuclear plants and disposing their radioactive waste, reactor safety, and concerns about the potential use to produce nuclear weapons [54]. By 2006, nuclear energy accounted for around 15% of the world's electricity generation, compared to 41% for coal, and 20% for gas. Interestingly, the current climate change debate has rekindled interest in nuclear energy as an energy source that is low in carbon dioxide emissions (more concerns will be discussed in Sect. 4.3.4 of this Chapter).

Consequently, the emergence of nuclear power reduced the role and geopolitical significance of coal. While oil had become the dominant fuel by 1930 mainly due to its importance in transport, it replaced coal as the industry's primary energy source in the 1950s. It is worth pointing out that these two energy sources are very different economically [24]: on the one hand, while locating coal seams is relatively easy (i.e. cheap), bringing coal to the surface is very costly; on the other hand, while locating oil is very difficult (with one fourth to 95% of exploratory holes failing to yield oil), once located oil rises to the surface under its own steam. In the meantime, the advent of supertankers and the development of pipeline networks lowered the price of oil and natural gas. Gas accounts for about a quarter of global energy consumption [86], a dominance which has been catalyzed by the development of a global market for Liquefied Natural Gas (LNG). The geopolitical problem is that natural gas resources are concentrated in a handful of countries, with the former Soviet Union and the Middle East holding about three quarters of known world reserves [8].

Presently, developing nations are characterized by escalating energy demands. Of these, China and India ("*Chindia*") have emerged as major geopolitical players, key energy consumers, and major energy producers. Interestingly, China's growth has taken place hand in hand with a rapid growth in China's coal consumption, and an increase in its share of coal in world energy use. As of 2017, the cost of coal was comparatively low, as oil and natural gas prices had escalated and remained high. The political instability in supplier countries, rapidly increasing oil prices, the increasing frequency of disruptions to gas supplied from Russia to Europe, terrorist attacks, and extreme weather events (e.g. the Hurricane Katrina that hit the Gulf coast of the US in 2005) often followed by electricity blackouts, constitute additional energy security risks.

6.2.4 Energy Security and Geopolitics

The consequences of the two oil crises of the 1970s uncovered the degree of vulnerability and dependence on fossil fuels of the industrialized Western world. Conant and Gold [13] made the first systematic study of energy issues, from a geopolitical perspective, asserting that "*the country having control over resources will control those who rely on the resources, which will lead to a profound transformation of International Relations*".

Energy geopolitics gained momentum after the 1990s, when global resources (mainly fossil fuels) became scarce in the face of a growing world demand for energy. These changes in the global energy geopolitical situation after the Cold War were documented by Mitchell et al. [61]. The term "*new geopolitics*" was used to reflect factors such as the end of the Cold War, the transformation of the international energy trade by the Russian oil and gas resources, and the increase of the importance of natural gas and related technology developments.

Despite the fact that the concept of energy security has varied across different time periods, it remains central to the study of geopolitics and the global agenda. Esakova [20] presented the relationships in the energy sector, as interdependencies between energy producers and energy importers. Due to the interdependence of energy markets and energy prices, energy producers are also highly interconnected among themselves [40]. The same applies to energy importers, who are often forced to cooperate in order to ensure collectively their energy security. Exporting and energy importing states normally have contradictory interests: the first group is interested in the security of energy demand, while the latter seek the security of energy supply.

The concept of energy security is central to the study of geopolitics. Vivoda [82] pointed out the following issues that have made energy security an emerging area of focus in international relations: high energy prices; the increasing demand for geographically concentrated resources; the threat of resource scarcity and depletion in the foreseeable future; and concern for the likely social and political effects of climate change.

Citing several literature sources, Johansson [40] suggested that ongoing or potential conflicts over small islands or water areas (such as the South China Sea and the Arctic Region) may be attributed to the existence of valuable energy resources. Furthermore, it is pointed out that US military activities in the Middle East (such as in Iraq) may be motivated by the Carter Doctrine, enunciated by President Jimmy Carter on January 23rd, 1980: "*An attempt by any outside force to gain control of the Persian Gulf region will be regarded as an assault on the vital interests of the United States of America, and such an assault will be repelled by any means necessary, including military force.*" (http://www.presidency.ucsb.edu/ws/?pid=33079).

Johansson [40] has also argued that the development of large-scale solar power in North Africa for import to the EU, may create dependencies with security policy implications for Europe. Furthermore, Johansson has suggested that the presence of abundant energy and natural resources in poor countries may create regional insecurity, a phenomenon known as the resource curse.

Citing 2009 BP data, Vivoda [82] argued that the following 10 largest regional economies consume 61% of the world's energy, producing 54% of the world's Gross Domestic Product (GDP), and emitting 66% of the world's carbon dioxide (CO_2): United States, Canada, Mexico, Russia, China, India, South Korea, Japan, Australia and Indonesia. These data show a major geopolitical disconnect: the world's top four oil consumers (China, Japan, India and the US) account for 42% of the world's oil demand, but control only 4% of global oil reserves. Vivoda asserted that energy security concerns among these four major powers have the potential to cause power competition so intense that may translate into open confrontation, especially in the Asia-Pacific region, the world's fastest growing energy consumer and a strategically vulnerable region of paramount importance for global stability and development. Vivoda also observed that the world's top ten regional economies are characterized by very different energy efficiency and carbon intensity values, citing as examples: Japan being four times as efficient as Russia, and four times as efficient as China; the economies of Russia, China and India being five to six times as carbon intensive as Japan; and the inefficient use of energy (with a high carbon intensity), creating energy-related pollution in Russia, China and India. Consequently, the growth of energy use in the Asia-Pacific region, particularly in China and India, is likely to have profound impacts on the energy landscape, and may have major geopolitical consequences.

6.2.5 Dimensions of Energy Security

The historical discourse on energy security has shown that it encompasses a number of dimensions. According to Kruyt et al. [47], the bulk of the recent literature on energy security seeks to classify energy security concerns into dimensions, even if this classification has been criticized for lacking transparency, being systemically unjustified, and arbitrary [10].

The IEA considers energy security to have three components: reliable and uninterrupted supply; affordable and competitive supply; and accessible or available supply (http://www.iea.org/topics/energysecurity/subtopics/whatisenergysecurity).   Furthermore, the IEA has proposed the following characteristics for energy security measurement: physical availability that accounts for geopolitical energy security; pipe-based import dependence; power system reliability; and market power [12]. A study by the U.S. Chamber of Commerce [81] presented four dimensions of energy security: geopolitical (energy imports, particularly from politically unstable regions); economic (high energy intensity and trade imbalances); reliability (adequacy and reliability of infrastructure); and environmental (related to the carbon intensity of energy systems).

In the words of Cherp and Jewell [10]: "*there are three perspectives on energy security, namely those of sovereignty, robustness and resilience*". Alhajji [1] differentiated between six dimensions of energy security: economic, environmental, social, foreign policy, technical and security. Vivoda [82] listed seven salient energy security dimensions (environment, technology, demand side management, socio-cultural or political factors, human security, international elements like geopolitics, and the formulation of energy security policy) and 44 attributes of energy security.

Sovacool and Mukherjee [77] considered energy security to comprise five dimensions that may be broken down into 20 components, as follows:

- *availability*: security of supply and production, dependency, and diversification;
- *affordability*: price stability, access and equity, decentralization, and low prices;
- *technology development*: innovation and research, safety and reliability, resilience, energy efficiency, and investment;
- *sustainability* (i.e. environmental component): land use, water, climate change, and air pollution;
- *regulation*: governance, trade, competition, and knowledge of sound regulation.

To this end, Sovacool and Mukherjee have assembled 320 simple indicators and 52 complex indexes of energy security.

In a study for the evaluation of energy security in the Asia-Pacific region, Sovacool [74] listed 20 dimensions of energy security identified by experts. For each dimension (availability, dependency, diversification, decentralization, innovation, investment, trade, production, price stability, affordability, governance, access, reliability, literacy, resilience, land use, water, pollution, efficiency, and greenhouse gas emissions) a number of metrics and indicators was presented.

Putting energy security dimensions into perspective, Radovanović et al. [69] pointed out that it is not possible to develop a unique methodology of assessing energy security because each country has a different wealth of energy resources. The use of these resources differs in: the type and intensity at different points of development; climate; geopolitical position; demographic indicators; economic growth; and strategic priorities, which depend on the historical, social and political social conditions [12]. Furthermore, Radovanović et al. argued that all countries try to improve their energy security by increasing energy efficiency, improving the stability of energy systems, reducing energy vulnerability, and increasing self-sufficiency.

Key energy security models, indicators and indexes are reviewed next.

6.2.5.1 Energy Security Models

The *IEA Model of Short-Term Energy Security* (MOSES) [39] is designed to analyze short-term energy security in IEA member countries, and it functions as a tool to understand the energy security profiles, and identify energy policy priorities. The MOSES model looks at resource adequacy, diversity, flexibility, asset performance, and sustained emergency events [43]. MOSES uses 35 indicators each of which relates to one of the dimensions and is meant to indicate a "*level of risk*" or the "*adequacy of resilience*" of the energy sources under investigation [35, 39].

MOSES has produced some interesting results. Energy security, in terms of diversity of supply of crude oil and oil products, is the highest in Denmark, Estonia, France and Italy. In terms of natural gas, Hungary, Ireland, Finland and Sweden are at the greatest risk. Relatively lower risk is reported for Austria, Luxembourg, Greece, Slovakia and Czech Republic, and the lowest for Denmark, Italy, Estonia, France and the United Kingdom. When it comes to natural gas, the same research took storage capacities (expressed as a percentage of annual demand) into account for the assessment of energy security. Results indicated that Estonia, Switzerland, Luxembourg, Norway, Finland and Sweden are at the greatest risk in this respect. Medium level of risk was reported by the United Kingdom, the Netherlands, Germany, Italy, Spain, France and Czech Republic. The lowest risk is reported by Austria, Hungary and Slovakia. When it comes to coal, not a single European Union country has more than 60% of coal from its own reserves, so the risk was determined indirectly, by considering the adequacy of importing infrastructure. The highest level of risk in this regard is reported by Finland and Switzerland [69].

The *Risk of Energy Availability*: *Common Corridors for Europe Supply Security* model (REACCESS) [71] is a unique world model that links three sub-models: the Pan European TIMES multi-regional model (PET36); the global multi-regional TIMES Integrated Assessment Model (TIAM-World); and the REACCESS Corridor (RECOR) model, representing the technical-economic details of all energy corridors, which bring energy from all resource-rich locations to consuming regions. REACCESS is a large partial equilibrium model of the global energy system representing 51 regions: the 36 countries of PET36 plus the 15 regions of TIAM-World remaining after the EU region is excluded from consideration.

6.2.5.2 Energy Security Indicators and Indexes

The conceptualization and formulation of energy security dimensions is the first step for an analysis of energy security. These dimensions must be complemented by indicators which should relate to the dimensions, and seek to quantify the identified energy security risks and concerns [10]. The aggregation of indicators into composite indexes allows a comparison between energy security risks and policy trade-offs.

The interest in measuring energy security results not only from its rising prominence, but also from its increasing complexity. Many indicators are available in the literature, based on the perspective of the user [64]. The literature on the indicators of energy security is quite extensive [28, 33, 49, 53, 78] and may be a useful tool for monitoring, measuring, and evaluating the current and future effects of energy security on the economy, the society, and the environment. Indicators for energy security are necessary to link the concept with model-based scenario analyses in the context of addressing policy issues related to affordable energy and climate change [47].

Chester [12] suggested that there are quantitative and qualitative approaches to the measurement of energy security. Threats to energy security are short-term (operational), and long-term (related to adequacy of sources, transit, storage and delivery). The literature reviewed by Chester suggests that the quantifiable energy security indicators have the potential of being analytically helpful, and are necessary to assess the consequences of alternative development scenarios.

Various studies [43, 69, 77, 78] have proposed a wide variety of energy security indexes, either to compare performance among countries or to track changes in a country's performance over time. In these studies, some indicators are first identified based on specific considerations or theoretical framework. This is followed by data collection, normalization, weighting, and aggregation of the chosen indicators to give one or more composite energy security indexes. A quick review of these studies has shown that there are large variations in the choice of indicators [3]. Radovanović et al. [69] applied Principal Component Analysis (PCA) to assess the impact of individual indicators on an energy security index and found energy intensity, GDP per capita, and carbon intensity to have the greatest impact on energy security (more on this in Sect. 4.4).

Sovacool [74] defined an index with 20 dimensions and 200 attributes. In a subsequent work, Sovacool and Mukherjee [77] reduced the number of dimensions to five and the number of attributes to 20. Sovacool et al. [78] applied the index to a set of countries, and found that Japan had the highest energy security index among the 18 countries considered. The impact of the Fukushima nuclear accident on Japan's energy system and economy, hints at the (often unexpected) difficulties that may be encountered when attempting to construct robust energy security indexes.

Ang et al. [3] reviewed 53 studies that dealt with energy security indicators, and found the number of indicators examined to vary from a few to more than 60. About two-thirds of the studies employed no more than 20 indicators. Their research identified that there are two major types of studies that use energy security indicators: those that deal with performance over time, and those that compare performance among countries, with no significant difference in the number of indicators used.

Turning to specific indexes, the *Herfindahl-Hirschmann Index* determines the degree of a certain country's dependence on a certain supplier, and may be used as an indicator that indirectly points to the energy security of a country [69]. The *Supply/Demand Index for the long-term security of supply* (SD Index) [73] has been designed on the basis of expert assessments on all possible relevant aspects of the security of supply, and covers demand, supply, conversion, and transport of energy in the medium to long-term [47]. It is a composite indicator (i.e. an index) that comprises 30 individual indicators, and considers the characteristics of demand, supply and transport [69]. According to Kruyt et al. [47], the basic difference with other indicators, is that the SD Index attempts to grasp the entire energy spectrum, including conversion, transport, and demand (taking into account that a decrease in energy use lowers the overall impact of supply disruptions).

The *Oil Vulnerability Index* (OVI) [30] is an aggregated index of oil vulnerability, based on seven indicators: ratio of value of oil imports to GDP; oil consumption per unit of GDP; GDP per capita; oil share in total energy supply; ratio of domestic reserves to oil consumption; and exposure to geopolitical oil supply concentration risks, measured by net oil import dependence, diversification of supply sources, political risk in oil-supplying countries, and market liquidity. According to Radovanović et al. [69], the OVI is a comprehensive composite indicator that manages to consider economic indicators, import dependence, and political stability.

The *Vulnerability Index* [26] is a composite indicator which considers five indicators: energy intensity; energy import dependency; ratio of energy-related carbon emissions to the total primary energy supply (TPES); electricity supply vulnerability; and lack of diversity in transport fuels [69]. The six-factor *Risky External Energy Supply* [49] is entirely supply-oriented, and considers solely the level of diversification, with particular emphasis given to the assessment of transport safety of energy generating products [69].

The *Aggregated Energy Security Performance Indicator* (AESPI) [59] has been developed by considering 25 individual indicators representing social, economic, and environmental dimensions. The indicator (essentially an index) ranges from zero to 10, and requires time series data for its estimation. The advantages of AESPI is that, it not only assists in knowing the past energy security status of a country, but also helps in assessing the future status considering the energy policies and plans, thus enabling monitoring the impacts of policies.

The *Socio-economic Energy Risk* is a composite index that considers the following indicators: energy source diversification, energy resource availability and feasibility, energy intensity, energy transport, energy dependence, political stability, market liquidity, and the GDP [69].

The *US Energy Security Risk Index* [81] is an index based on 83 individual indicators assessing geopolitical indicators, economic development, environmental concerns and reliability [69].

The *Energy Development Index* (EDI) [39] is composed of four indicators, each of which captures a specific aspect of potential energy poverty: per capita commercial energy consumption (which serves as an indicator of the overall economic development of a country); per capita electricity consumption in the residential sector (which serves as an indicator of the reliability of, and consumer's ability to pay for, electricity services); share of modern fuels in total residential sector energy use (which serves as an indicator of the level of access to clean cooking facilities); and the share of population with access to electricity. This index was intended as a simple composite measure of the progress of a country or region in its transition to modern fuels, and of the degree of maturity of its energy end use [32].

The *Energy Security Index* is composed of two indicators (ESI_{price}, ESI_{volume}) that measure the energy security implications of resource concentration, from the viewpoint of both price and physical availability [33]. ESI_{price} is a composite measure of the diversification of energy sources and suppliers, and the political stability of exporting countries, while ESI_{volume} is a measure of the level of dependence of natural gas imports.

In what constitutes an interesting concept, the *"energy trilemma"* is defined as balancing the trade-offs between three major energy goals, namely energy security, economic competitiveness, and environmental sustainability [3]. The dimensions of energy trilemma are defined by WEC [85] as follows:

- *Energy security*: Effective management of primary energy supply from domestic and external sources, reliability of energy infrastructure, and ability of energy providers to meet current and future demand.

- *Energy equity*: Accessibility and affordability of energy supply across the population.
- *Environmental sustainability*: Encompasses the achievement of supply and demand-side energy efficiency, and the development of energy supply from renewable and other low-carbon sources.

The Energy Trilemma Index, formerly known as the *Energy Sustainability Index*, was first introduced in 2009, ranking close to 90 countries. This ranking has been expanded to include 130 countries and greater detail about the performance of countries on the specific trilemma dimensions by adding a balance score, and an index watch list to indicate countries that are expected to display trend changes in the next few years [3]. The Index 2.0 methodology uses a set of 34 indicators and approximately 100 data sets to rank countries on their trilemma performance (compared to 23 indicators and 60 data sets in the previous index methodology [85]).

Finally, the *Energy Architecture Performance Index* (EAPI) was proposed in 2010 by the World Economic Forum (WEF), and was modified the next year into the *Energy Sustainability Index* [87]. EAPI is a composite index based on a set of indicators divided into three basic categories (energy security, energy equity, and environmental sustainability), the so-called *Energy Trilemma Index* [69].

More on some of these indexes this will be graphed and discussed in Sect. 4.4.

6.3 Methodology

In a seminal work, O'Neill [65] discusses social scientist authors setting up game theoretic models, but failing to follow through with the mathematics, thus failing to use game theoretic tools to advance their research. O'Neill calls this approach *"proto-game theory"*. In a similar vein, much energy security research addresses the development of energy security indexes, but fails to achieve much with them (beyond e.g. sorting countries according to the value of some energy security index) in way of solving real geopolitical problems.

According to Ang et al. [3], research on energy security may follow either a qualitative or a quantitative approach, with the former relying on theory and geopolitical considerations, and the latter focusing on the analysis of mostly numerical indicators [92]. This Chapter explains how these two approaches may be combined by outlining methods that would allow the formulation and estimation of an energy security index (with a geopolitical slant) by: analyzing various types of data; resorting to CA and tracking clusters of countries over time; analyzing case studies; and incorporating the opinions of energy experts through interviews. Such approaches would help solve geopolitical problems and develop forecasts, taking into account the diversity of country interests over space and time, and the polysemic nature of energy security that has risen near the top of the agenda of the international community.

6.3.1 Research Questions

The tasks that a researcher would have to pursue in an endeavor to develop and use a geopolitical energy security index are formulated and presented below in the form of grouped research questions.

1. *Literature*: What is the state of the art of the research literature on the topic of energy security? How has the definition of energy security changed over time? What are the dimensions of energy security? What has the role of energy security been in the foreign policy of energy-exporting, energy-importing, and energy-transition countries? What geopolitical issues has it been mostly associated with? Are there important aspects related to energy security that current research methods have failed to capture?

2. *Milestones*: Which specific time periods could be regarded as important milestones in the history of energy security? What were the socioeconomic and geopolitical characteristics at these times? How was energy security conceptualized at each milestone time period, and what specific data should be taken into account? While continuous annual data could be collected for the estimation of an energy security index (which would be quite demanding), the authors of this Chapter believe that the establishment of a few distinct milestones that were associated with important geopolitical events would be more fruitful, for better calibration of the results and a more focused (and thus efficient) utilization of the energy security index.

3. *Data*: What (online) sources may be used for collecting data for the development of an energy security index? What kind of data are contained in these sources? What data types may be collected? Which areas of the world or specific countries appear to not be well covered, and at which time periods? What kind of analyses may be run with these different types of data? What problems should one be aware of? What statistical software is suggested for these analyses?

4. *Index*: How may an energy security index be formulated and estimated, based on the data collected in the context of the previous research question? Based on the published literature, what issues might hinder the development of an energy security index geared towards geopolitical considerations, and how can they be overcome? What socioeconomic and geopolitical data would be required to estimate an energy security index? How may different types of data be of use in addressing specific types of research questions? What would the values of an energy security index (annual or at the milestone time periods) be for most countries of the world? How have they changed over time? Beyond observing how energy security is associated (perhaps causally) to major geopolitical events, how could the milestone time periods be used otherwise to bring added value to the research? How do other energy indicators and indexes compare, e.g. which aspects are better covered, and which aspects are not well addressed by them?

5. *Clusters*: Having formulated and estimated an energy security index, it is proposed that one of the most useful things that could be done with it, would be to cluster analyze the countries of the world across different time periods according

to the values of the energy security index and possibly some complementary socioeconomic and geopolitical indicators that were left out of the index; this, to the knowledge of the authors of this Chapter, has not been attempted in the published literature. Related questions might include: Could the countries of the world be organized fruitfully into similar groups, based on various socioeconomic, geopolitical characteristics, and the values of the energy security index? How many such groups exist at each milestone time period (or annual time series)? Have countries moved from one group to another over time? Depicting these findings on a dynamic map (that would involve the use of a Geographical Information System or GIS software tool), which were the geopolitical hotspots from an energy security perspective (at each milestone or time period), and how have they changed and moved over time? Which energy policies have been pursued by energy import countries, energy export countries, and energy-transition countries, to address the issues of the energy security challenges they face? How are such energy policies related to the foreign policy of these countries? Can any insights be drawn as to why countries under comparable international conditions prioritize energy security differently (as observed in Sect. 2)?

6. *Cases*: The results of any quantitative analysis could be complemented with the study of individual cases. To this end, selected key countries could be analyzed as case studies that could showcase the application of the entire approach in geopolitical issues. Different countries could be selected at different milestones, in part aided by the clustering of countries, and other specifics of the statistical analyses. A full longitudinal statistical analysis could also be employed, although this would be more demanding in data (as will be seen in Sect. 4.2). Questions that could be addressed during the examination of case studies include: How has the participation of these key countries in energy security clusters changed from one milestone time period to another, i.e. have they shifted from one group to another over time? Have the energy security related policies of these countries changed over time? Does the involvement of an energy security index shed light on the analysis of these policies, and does it help determine the efficacy of energy security policies over the milestone time periods (or a complete time series)?

7. *Experts*: To add more perspective from a qualitative approach, an energy security researcher could turn to a group of energy experts in order to seek answers to questions such as: How may a small group of energy security experts be defined (i.e. what is an "energy security" expert)? What criteria should be used for selecting and including individuals to the group? What would the resulting identity of these experts (including government officials, academics and professionals) be? What would their opinion be of the concept of energy security, perhaps with emphasis on the milestone time periods (which would be easier to poll compared to registering their view for a complete time series)? How would they judge the energy security policies that the case-study countries have followed over time? Finally, what would their opinion of the energy security index be, and would they agree with the findings that would have resulted from its usage?

8. *Forecasts*: The final issue to address in an energy security research could be the issue of forecasts. A research work that would have developed an energy security

index and responded to the previous questions, would probably want to address the following issues: How may energy security forecasts be developed? How are the energy security index values expected to vary in the foreseeable future, and what are the corresponding geopolitical implications? How may such forecasts take into account global climate change, and any surrounding uncertainties?

Attention now shifts to a detailed discussion of the research tasks that would have to be undertaken to answer some of the aforementioned questions.

6.4 Analyses and Results

This section details the analyses that would have to be run to address the research questions outlined in the previous section.

Research question 1 (*Literature*) has been addressed in Sect. 2 of this Chapter, so this section commences with research question 2 (*Milestones*), i.e. the task of defining appropriate key time periods for energy security research.

6.4.1 Milestone Time Periods

Vivoda [82] has argued that the definition and dimensions of energy security are dynamic and evolve as circumstances change over time. Furthermore, energy security regimes tend to be transformed or even weaken over a period of time. Therefore, taking time into account is important in energy security research.

To address the second research question (*Milestones*), it is recommended that an energy security researcher focus on a small number of distinct and well defined milestones that were characterized by important energy events in the post-1970s world (the energy world before 1970 was vastly different; [24]). An examination and analysis of such milestones should look at the socioeconomic and geopolitical dynamics that culminated at those times.

Four such milestones are proposed and described in the following sections: the first oil crisis of 1973–74, the Gulf War of 1990–91, the period of the Russia–Ukraine gas disputes, and the present (2018, at the time of writing).

6.4.1.1 Milestone 1: First Oil Crisis (1973–74)

The first oil crisis was caused by a number of measures that were coordinated by the members of the Organization of Arab Petroleum Exporting Countries (OAPEC), and were directed at the countries that had supported Israel in the Yom Kippur War (1973).

The year 1974 has been a milestone for the energy security concept. As a response to the 1973 oil embargo, the Organization for Economic Cooperation and Development (OECD) established the IEA. During that period, the concept of international energy security still largely meant oil security. Back then, it was all Middle Eastern oil: IEA [36] data show that oil shared as much as 86% of the world energy trade in 1979, while the Middle East supplied 58% of the internationally traded oil.

This is, without doubt, a very important first milestone that should be taken into consideration in energy security research that aims to develop a geopolitical energy security index.

6.4.1.2   Milestone 2: Gulf War (1990–91)

Energy geopolitics and the concept of energy security gained momentum after the 1990s, when global resources became scarce in the face of a growing world demand for energy. Concurrently, the first Gulf War (1990–1991) and the dissolution of the Soviet Union (1991) took place (as mentioned previously), new concepts emerged, and concern for energy security began to gain prominence in the world's discourse [90].

As mentioned, global warming issues were gradually institutionalized throughout the decade, while the signing of the Kyoto Protocol in 1997 marked the starting point of countries considering the concept of energy security coupled with environmental protection and sustainable development.

6.4.1.3   Milestone 3: Russia–Ukraine Gas Dispute (2005–9)

The Russia-Ukraine gas disputes refer to a number of periods of tension that occurred around the beginning of the 21st century, against a constant backdrop of rather complicated intergovernmental relations between Russia and Ukraine.

Energy security has played an important role in Russian-Ukrainian relations since Ukraine became independent in 1991. The two major crises that stick out in the bilateral gas relationship occurred in 2006 and 2009. During both of these disputes, Russia halted the supply of gas through Ukraine, which is the basic energy-transition country between Russian gas and the EU.

Ukraine is a key factor in the global energy arena because half of Russia's total gas exported to Europe is transmitted through the Ukrainian pipeline system [37]. As a result of the above disputes, several European countries lost gas supplies, and the matter of diversifying the energy sources and transportation routes emerged as an important concern in the EU. The most severely affected countries were in Southeastern Europe, most of which relied completely on Russia for gas, and at the time did not have sufficient alternative energy sources [68].

The 2009 dispute is considered the most serious of all. The crisis of 2009 became a battle of self-image for both Russia and Ukraine in Europe; Russia's reputation as a supplier to Europe and Ukraine's reputation as a transit country were seriously damaged. The 2009 dispute ended with an agreement, where the heads of Gazprom and Naftogaz (the national oil and gas company of Ukraine) signed supply and transit contracts covering the ten-year period 2009–19.

The milestone of Russia-Ukraine gas disputes is important from an energy security perspective. The security of gas supply has a totally different meaning than the security of supply of oil. Gas is mainly sold on the basis of long-term bilateral contracts, and shipped through dedicated pipelines, which often cross several countries; with oil, the main issue is that there is no global market for an interchangeable product [50].

6.4.1.4   Milestone 4: Present (2018)

Presently, energy technologies keep advancing, the awareness of climate change and sustainability issues are on the increase, and the facets of energy security are being reshaped (as evidenced by the review of the literature). The global attention to energy security may be explained by the new emerging giants of the world economy (China and India included) and their rapidly rising energy demand.

As the geopolitical changes that commenced in the 20th century continue in the 21st century, and constitute a basic dynamic component of the concept of energy security, the present time (2018) is proposed as an appropriate final milestone that should be considered.

6.4.2   Data

To adopt a marketing approach to data science (as it bears to research in the field of energy security), there are primary data, secondary data, and marketing intelligence [56]. Primary data are originated by a researcher for the specific purposes of a particular research task, may require a longer time frame to collect, and are often more expensive to generate. Secondary data are usually of a quantitative nature, are collected for purposes other than those of the researcher that accesses them, and are oftentimes used by many researchers. Marketing intelligence may be defined as *"qualified observation of events and developments in the marketing environment"* [56] and may, for instance, refer to expert opinions in the case of energy security. Such marketing intelligence usually is of a qualitative nature, suffers from poor structure, and since it is opinion based may be characterized by varying credibility. It is much easier to evaluate the accuracy of secondary data than marketing intelligence [29], i.e. the opinion of experts. While marketing intelligence is used systematically by companies such as Microsoft, IBM and Coca Cola, it is also employed by companies in the field of energy, such as General Electric [56]. The rest of this section is devoted to secondary data, although mention of primary data will be made later on (when discussing expert interviews in Sect. 4.7).

Secondary data are easily accessible (usually online), and may be obtained quickly and inexpensively. Yet, with secondary data, a researcher may not be aware or information such as the margin of error or the response rate (in cases where secondary data were collected with a survey). In fact, a variety of accuracy issues plague all types of secondary data [29]: observational errors; approximations and roundoffs; selectivity bias created by nonresponse (for data that were collected by survey); biases created by different sampling methods; a high level of aggregation (often due to confidentiality). Therefore, the reliability, dependability and generalizability of data may be questionable [56]. Furthermore, secondary data may not be current enough (for the purposes of the research at hand). Nevertheless Malhotra and Birks [56] advise that secondary data be examined as an option before deciding where primary data must be generated and collected. Accuracy may be a particular concern with dynamic global phenomena that affect geopolitics and energy security. Multiple sources of secondary data may be sought and compared, to overcome issues of accuracy and possibly fill in missing cases.

Sources of secondary data include published materials, online and offline (electronic) databases, and syndicated services [56]. Researchers should try to obtain secondary data from original rather than acquired (surrogate) sources (i.e. sources that reproduce data obtained from other secondary sources). Many variables may be included in such databases, and it is very important to be clear on their definition and units of measurement. Occasionally, a researcher may need to convert a variable from one system of units to another, or change its time frame (which may not always be feasible).

There is an accelerating tendency for secondary data to be available in electronic format including the Internet and offline electronic sources such as CD or DVD-ROMS [56], with physical media perhaps being phased out over time. The Internet in particular has emerged as the most extensive source of secondary data. While not all online secondary data are free, the Internet is replete with such collections that include diverse energy related data, including supranational entities (like the European Union), national and government sources (like the Hellenic Statistical Authority), international organizations (like the International Energy Agency or the United Nations), national organizations (like the CIA) and businesses (like British Petroleum). There are even thematic lists, directories and indexes of secondary sources, including reputable periodical publications and newspapers, e.g. *The Financial Times Index* or the *Foreign Affairs* list of energy related articles (https://www.foreignaffairs.com/topics/energy?cid=int-gna&pgtype=hpg).

International trade, national accounts, demography, population, socioeconomic data, industry, energy, environment, and other information are all available to an energy security researcher, in a variety of online secondary sources [56]. All an energy security researcher has to do is locate and select the data required, taking into consideration the expertise, credibility, reputation and trustworthiness of the data provider. Such a list of reputable and inclusive list of free online sources of global secondary data related to the many dimensions of energy security is displayed in Table 1, sorted alphabetically (with the exception of Eurostat, as a source of predominantly regional data, i.e. on European Union member and affiliated countries).

Table 1 Online sources of energy data

Source	Contents	Link
British Petroleum (BP)	The BP *Statistical Review of World Energy* contains data on: primary energy; oil; natural gas; coal; nuclear energy; hydroelectricity; renewable energy; electricity and CO_2 emissions	https://www.bp.com/en/global/corporate/energy-economics/statistical-review-of-world-energy.html
Central Intelligence Agency (CIA)	*The CIA World Factbook* contains data on: geography; people and society; government; economy; communications; transportation; military and security; terrorism and transnational issues. Especially on energy issues, it contains data on: electricity; crude oil; refined petroleum; natural gas; and CO_2 emissions from energy consumption. Country information much qualitative information in each category, so this source would be a good complement to case studies	https://www.cia.gov/library/publications/the-world-factbook/
International Energy Agency (IEA)	The IEA *Atlas of Energy* contains data on: CO_2 emissions from fuel combustion; electricity; oil; coal; natural gas and renewables. In addition, it contains energy indicators and energy balance data	http://energyatlas.iea.org/#!/topic/DEFAULT
Organisation for Economic Co-operation and Development (OECD)	The *OECD database* contains data on: agriculture; development; economy; education; energy; environment; finance; government; health; innovation and technology; jobs and society. Energy data include: CO_2 emissions; natural gas; oil; coal; renewables; electricity; energy prices and taxes; and world energy statistics	http://www.oecd.org/statistics/listofoecddatabases.htm; https://data.oecd.org/searchresults/?r=%2Bf%2Ftype%2Findicators&r=%2Bf%2Ftopics_en%2Fenergy
United Nations Development Program (UNDP)	The *Human Development Reports* contains data on: health; education; income/composition of resources; inequality; gender; poverty; work; employment and vulnerability; human security; trade and financial flows; mobility and communication; environmental sustainability and demography. The Human Development Report Office releases five indices each year: the Human Development Index (HDI), the Inequality-Adjusted Human Development Index (IHDI), the Gender Development Index (GDI), the Gender Inequality Index (GII), and the Multidimensional Poverty Index (MPI)	http://hdr.undp.org/en/countries

(continued)

Table 1 (continued)

Source	Contents	Link
United Nations (UN) Data	Contains data on most countries of the world, including: general information; economic indicators; major trading partners; social indicators; environment and infrastructure indicators; and some energy data	http://data.un.org/en/index.html; https://unstats.un.org/unsd/energy/default.htm
World Bank Open Data	Contains many development data sets on: agriculture and rural development; aid effectiveness; climate change; economy and growth; education; environment; energy and mining; external debt; financial sector; gender; health; infrastructure; poverty; private sector; public sector; science and technology; social development; social protection and labor; and trade and urban development	https://data.worldbank.org; https://data.worldbank.org/indicator?tab=all; https://data.worldbank.org/country
World Data Atlas	These data contain data on: agriculture; climate statistics; demographics; economy; education; environment; food security; foreign trade; health; land use; national defense; poverty; research and development; telecommunication; tourism; transportation; water and world rankings. Data include indicators on total energy, electricity, renewables, oil; gas and coal	https://knoema.com/atlas; https://knoema.com/atlas/topics/Energy
World Nuclear Association	Contains profiles on countries that use nuclear power, including: energy policy; nuclear facilities; research and development; regulatory framework and non-proliferation. Profiles include much textual and qualitative information, so this source would be a great complement to case studies	http://www.world-nuclear.org/information-library/country-profiles.aspx

The majority of the data to be used for the estimation of an energy security index is expected to be of a secondary nature, although some primary (mostly qualitative) data may be of use as well (as in the case of case study and energy expert approaches that will be discussed in Sects. 4.6 and 4.7).

Turning to the types of data, three major categories are generally used in various statistical and econometric analyses: (1) cross-sectional, (2) time series, and (3) pooled data, including balanced and unbalanced panel or longitudinal data [29].

Cross-sectional data are data that are collected at the same time, e.g. a survey of households in an area, a census of the population in a country, or a Gallup poll among the countries of the European Union. As Gujarati and Porter [29] mention, cross-sectional data may suffer from heterogeneity, and may require appropriate mathematical transformations to correct.

Time-series data may be collected daily (e.g. stock prices), weekly (e.g. various economic figures such as the money supply), monthly (e.g. the inflation rate, the unemployment rate or the Consumer Price Index), quarterly (e.g. the Gross Domestic Product or GDP of a country), annually (e.g. the government budget), every five years (such as the census of manufacturers) or even every 10 years (such as the population census in a country) [29]. Some data may be available both quarterly and annually, such as the GDP or consumer expenditure. The analysis of time-series data presents special problems in econometric analyses, mostly because time series must be tested for stationarity (i.e. no upward or downward trending and steady variance over time) prior to any analysis such as regression modeling; if found to be nonstationary, they must be corrected with appropriate mathematical techniques, depending on the type of nonstationarity established (for a practical example, see [66]).

Pooled data combine the characteristics of cross-sectional and time-series data, as for every time unit there is an entire cross-sectional data set (instead of a single observation), e.g. the annual GDP of European Union countries over the last 10 years. If the same cross-sectional units are surveyed over time, the pooled data are called panel data; if all the units have the same number of observations, the panel data are balanced, otherwise the panel data are unbalanced [29]. The terms panel and longitudinal data are used interchangeably, although it may be argued that the concept of panel data analysis is broader [25].

Variables used in econometric research (like most related to energy security), belong to the following four broad measurement scales (in decreasing amount of useful information contained in them: ratio, interval, ordinal and nominal [29]):

1. With *ratio* scale data, both the ratio of two values, and the distance between them are meaningful. Such data may be ordered (in an ascending or descending) manner, so comparisons between their values may be made, while the value of zero denotes the absence of any quantity of such data meaningfully. All in all, ratio-scaled data are numeric data that contain the most information. Personal or family income or GDP are examples of ratio scaled data.

2. *Interval* scaled data are also numeric data with values that may be ordered (in an ascending or descending manner), but while the distance between two of their values is meaningful, their ratio is not, i.e. they lack a meaningful definition of the location of the value of zero. Time is such a data type, and taking years as an example, it would be meaningless to claim that 2000 AD is twice as big as 1000 AD. Temperature scales also belong to this data type, and it is easy to realize that different scales (such as Celsius, Fahrenheit and Kelvin) place the location of zero degrees at a rather arbitrary point.

3. Although *ordinal* data may not necessarily be numeric, they can certainly be ordered (in an ascending or descending manner) but this is pretty much all one can do with them. Age classes (e.g. 0–17, 17–35, 35–50, over 50 years of age), income classes (e.g. low, medium and high) or grading systems (such as A, B, C, D, and F for fail in the US grading system).

4. Finally, *nominal* scale data contain the least amount of information, and just represent different values of a variable that cannot even be ordered in a meaningful manner. Gender (e.g. male, female) and marital status (e.g. single, married, divorced, separated) are examples of nominal scaled data.

It is worth making a special mention to dummy variables that are used in statistical and econometric analyses to indicate the presence or absence of a certain quality or characteristic. In the case of energy security research, one may use a dummy variable to indicate, for instance, the membership of a country in OPEC or the IEA, with the value of one indicating membership, and the value of zero indicating nonmembership. Dummy variables may also be employed to represent a nominal variable with more than one categories. For instance, if a researcher has arrived at a clustering scheme that groups countries into three energy security clusters, then one may employ two (i.e. one less than the number of clusters or categories) dummy variables to indicate the membership of each observation into a cluster. Assuming that these two dummy variables were named D2 and D3, here is how it would work: if both D2 and D3 were zero, the corresponding observation would belong to the first cluster; if D2 was equal to one, and D3 was equal to zero, then the corresponding observation would belong to the second cluster; finally, if D2 equalled zero and D3 equalled one, the corresponding observation would belong to the third cluster. It is important to remember that if the number of categories of a nominal variable equals N, then one need only define $N - 1$ dummy variables to represent them; in fact, if all N dummy variables were defined and included in a multiple regression model, perfect multicollinearity would prevent the model from being estimated.

Although the issue of which software tools to use in energy security research (that includes analysis of secondary data with the aim of estimating an energy security index) is rather subjective, some advice may be useful. The authors of this Chapter use *Microsoft Excel* exclusively for data entry and manipulation, and find Excel great because almost all statistical packages may open Excel files. *Minitab*, in the opinion of the authors, is hard to beat for graphing, basic statistical analyses and multiple regression modeling (http://www.minitab.com). The freely available *SAS University Edition* (https://www.sas.com/en_us/software/university-edition.html) is outstanding for carrying out specific analyses, such as Poisson and Negative Binomial regression of count data [2]. Finally, *IBM SPSS* (https://www.ibm.com/analytics/spss-statistics-software) is a great tool for many types of multivariate statistical analysis, including Two-Step Cluster Analysis that can accommodate a mixture of quantitative and qualitative data [62].

6.4.3 Measuring Energy Security

Having discussed the data that are usually involved in energy security research, attention now turns to the specific socioeconomic and geopolitical data that should be collected, in order to compute an energy security index.

As it has probably become evident from the literature review, the measurement of energy security is not a straightforward issue, as there is uncertainty over the definition of the concept [92]. Different approaches to measuring energy security have been developed in the literature [69], although even the simplest definition (*"the uninterrupted availability of energy sources at an affordable price"* by the IEA) illustrates how complex can any attempt of measurement be. Despite the multitude of dimensions, indicators, and models, the concept of energy security (especially its geopolitical role) is relatively novel, and there are no well-replicated scales for measuring it [14].

Measuring energy security helps remove ambiguity, but encounters numerous complications. As an example, energy security oftentimes focuses on specific energy sources, and talk is made of oil security or coal security. In fact, if one centers on certain aspects of energy supply (such as electricity supply, nuclear power, or gasoline powered automobiles), developing and least developed countries are automatically excluded from consideration since they have poor electricity networks, limited nuclear power units, and non-motorized forms of transport. Another way to go about it, is to consider isolated quantitative variables (such as electricity supply, percent of population with access to electricity, or energy intensity) that are often sectoral in nature, e.g. household energy consumption or share of commercial energy in total energy use. Alternatively, one may estimate composite indicators, such as industrial efficiency, transportation productivity, and environmental quality [77], although these are nontrivial to estimate in a meaningful and accurate way.

Cherp and Jewell [11] have suggested that energy security be conceptualized as an instance of a more general construct of security. A good basis for approaching security has been provided by the seminal work of Baldwin [5], who argued that military, economic, social and environmental instances of security constitute different forms of security, not fundamentally different concepts. Cherp and Jewell suggested that Baldwin makes a distinction between *"acquiring new values,"* which is not related to security, and *"protecting existing values,"* which lies at the heart of the concept of security. The latter approach in the literature focuses on the dynamics of separate indicators, and offers a rather limited possibility for comparative assessment of multiple countries in term of energy security through quantitative analysis [92].

Secondary literature sources are used typically for the estimation of an energy security index. Collected socioeconomic and geopolitical data at a country (or region) level and each milestone time period may be entered in a single Excel file that will essentially be a flat database. Further to socioeconomic and geopolitical data, this flat database may include simple indicators, functions and combinations of simple indicators, and composite indexes collected from the Internet or the literature (even by direct communication with the authors of the original source, if necessary). A composite index is formed when individual indicators are compiled into a single index, on the basis of an underlying model of energy security. The analysis of these empirical data must commence with graphical depiction and numeric measures of location and dispersion. Such a descriptive analysis of the collected data will help concretize the theoretical framework and provide a wealth of information for case studies.

Table 2 Countries (and regions with a state-like status) of the Southeast Mediterranean and the Middle East

Country	Label
Bahrain	Bah
Cyprus	Cyp
Djibouti	Dji
Egypt	Egy
Eritrea	Eri
Greece	Gre
Iran	Irn
Iraq	Irq
Israel	Isr
Jordan	Jor
Kuwait	Kuw
Lebanon	Leb
Libya	Lib
Oman	Oma
Qatar	Qat
Saudi Arabia	Sau
Somalia	Som
Sudan	Sud
Syria	Syr
Turkey	Tur
United Arab Emirates	UAE
West Bank	WB
Yemen	Yem

In the next sections, an attempt is made to organize and discuss briefly thoughts and notions, mostly drawn from literature findings, as to the nature of the data that need to be collected and organized as a researcher tries to measure the conceptual components of an energy security index. Some examples from the area of the Southeast Mediterranean and Middle East are included, with country (and region) abbreviations used for labeling points on graphs as in Table 2.

6.4.3.1 Geographical Components

According to Correlje and van der Linde [15], the geographic distribution of oil, gas and coal reserves should be considered in energy security research. Furthermore, the location of a country in an oil producing area such as the Middle East, Russia, the Caspian, the Persian Gulf, and Africa, should be represented (possibly by dummy variables). Considering that even though the Middle East has vast gas resources, its political climate is not much better than Russia and the Caspian Sea as far as investments are concerned, it could be that interaction terms of location and political climate may be useful in energy security research.

Variables that are of a geographic nature, may be found in the sources of Table 1, and may be of use in energy security research include: area of a country, broken down into land and sea parts (in km^2); land boundary and coastline of a country (in km); percent of the area of a country that is agricultural or forestry; and dummy variables indicating the location of a country, e.g. in the Middle East or Africa.

To look at a few examples, Fig. 1 shows that the area of Southeast Mediterranean and the Middle East contains four large countries that appear to the left of the chart: Saudi Arabia, Sudan, Libya, and Iran. The area also contains quite a few very small countries such as Djibouti, Israel, Kuwait, Qatar, Lebanon, Cyprus, the region of the West Bank, and Bahrain.

Looking at the coastline of countries in the Southeast Mediterranean and the Middle East affords an entirely different picture, as shown in Fig. 2. Greece and Turkey have the longest coastlines by far, while Jordan and the West Bank are entirely landlocked. The length of a coastline may act as a surrogate variable for access to marine resources, so it may prove to be important in energy security research.

Figure 3 charts the percent of the area of a country in Southeast Mediterranean and the Middle East that is agricultural land. It may be seen that Sudan is an exclusively agricultural country; 70–80% of the area of Saudi Arabia, Syria, Eritrea, Djibouti and Somalia is agricultural; and Greece and Lebanon are very similar in that just over 63% of their area is agricultural land. The rest of the countries have less than half of

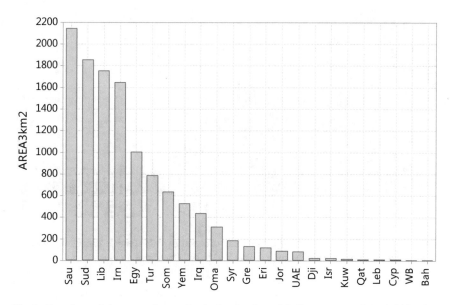

Fig. 1 Bar chart of the area of countries in the Southeast Mediterranean and the Middle East (AREA3km2: area in thousand km^2)

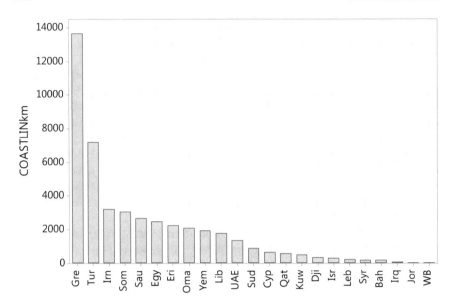

Fig. 2 Bar chart of the coastline of countries in the Southeast Mediterranean and the Middle East (COASTLINkm: length of coastline in km)

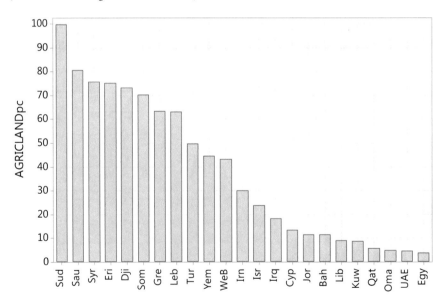

Fig. 3 Bar chart of the percent of agricultural land of countries in the Southeast Mediterranean and the Middle East (AGRICLANDpc: agricultural land in percent of total area)

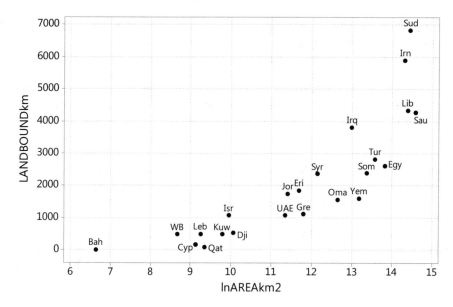

Fig. 4 Scatterplot of the land boundary of countries in the Southeast Mediterranean and the Middle East as a function of their area (LANDBOUNDkm: land boundary of a country in km; lnAREAkm2: natural logarithm of the area of a country in km^2)

their area devoted to agriculture, with some energy resource rich countries (such as Kuwait, Qatar, Oman, and the United Arab Emirates) having very little agricultural land.

Finally, a look at data transformations that may have to be carried out is presented in Fig. 4, where the size of the land boundary in km (LANDBOUNDkm) is shown to be nonlinearly related to the natural logarithm of the area of a country in km^2 (lnAREAkm2). The natural logarithm is preferred to decimal logarithms because it allows a more natural interpretation of the coefficients in regression models in terms of percentage changes and elasticities [79].

While some of the variables that were examined graphically in this section may not be strongly related to the concept of an energy security index, they could be useful in a case study approach to energy security (which is discussed in Sect. 4.6). The graphs also hint at possible clusters of countries with similar geographical characteristics [22], and in this respect may guide (in part) a Cluster Analysis.

6.4.3.2 Socioeconomic Components

Thinking hard about economic variables that should be taken into account in an energy security index should be an important part of such research.

Turning to literature findings on socioeconomic characteristics, Corner et al. [14] argue that high concern about energy security (as a foreign policy issue) was found in a summary of international public opinion (by the World Council on Foreign Relations), with significant support for energy conservation and investment in renewable energy sources. The same authors cited a UK national survey that indicated a high degree of support for government subsidies of renewable energy technologies as well as attaining energy independence. Of the renewable energy sources, solar and wind power appear to be perceived as being able to deliver reliable and secure energy [72].

Probably the most important element that is included in all definitions of energy security is the availability of energy to the economy [47]. As a consequence, many countries consider energy security to be a major priority of their energy policies. Energy security is often used implicitly as a synonym for the security of supply, particularly by researchers adopting an economic perspective, such as Kruyt et al. [47] and Johansson [40]. So, energy security is closely related to micro and macroeconomic developments [15].

A minimum supply of energy is essential for the functioning of the economy, thus energy security is and will continue to be an issue [48]. The cost of energy is an important factor in the inflation rate as well as in the competitive position of the economy of a country [15]. On the relationship between energy prices and energy consumption, Radovanović et al. [69] have argued that when energy prices increase, the consumption does not decline, while when energy prices decrease, energy consumption increases significantly. On the affordability of energy prices, Deese [17], as cited by Cherp and Jewell [11], writes that affordable prices do not cause "*severe disruptions of normal social and economic activity*".

Consuming countries wish to keep energy prices low, as long as this does not remove incentives for the development of supply); producing countries gain when prices are high, as long as they do not cause big reductions in demand [40]. Imports from OECD and non-OECD countries could be tallied and involved in the economic components of the energy security index [15]. Investments as well as Foreign Direct Investments (FDS) in the oil, gas and coal industry could also be taken into account [15].

The existence and number of independent bilateral agreements with producer countries could be indicated by a dummy or other quantitative variables [15], possibly including quantitative measures of economic instability. As a related thought, in countries like Nigeria, there are conflicts over economic rents and oil exploitation, which in turn cause political and social rivalry [15] that should be accounted for in energy security research.

Socioeconomic variables that could be found at a country level in the sources of Table 1 include: population and percent of urban population; life expectancy; health expenses, physicians, and hospital beds; education expenses and literacy rate; unemployment and inflation rate; GDP and GDP per capita, in current or Purchasing Power Parity (PPP) terms; house and government consumption as percent of GDP; imports and exports as percent of GDP; the contribution of agriculture, industry and services to the GDP (as percentages); size of labor force; percent of the population that is below the poverty line; as well as taxes and public debt as percent of the GDP.

To consider some examples, do bigger countries tend to have a larger population? It turns out that, to get a near linear relationship for countries in the Southeast

Mediterranean and the Middle East, one has to take the natural logarithm of both variables, as shown in Fig. 5. The physical meaning of such a log-log relationship is this: the slope of a regression line passing through the points of the scatterplot of Fig. 5 (not shown) would be equal to the percentage change in population (expressed in thousands) every time the area (expressed in km²) was increased by one percentage point [79]. Incidentally, Fig. 5 shows the three largest and most populated countries in the area at the top right of the scatterplot: Egypt, Turkey, and Iran. Similarly, the scatterplot reveals that Israel, Lebanon, Kuwait, West Bank, Qatar, Cyprus and Djibouti are countries that have relatively similar area and population characteristics (although it should be kept in mind that the axes are in logarithm units, so points that appear near may in fact have very different numerical values).

Figure 6 shows the percentage of the population living in urban areas for countries in the Southeast Mediterranean and the Middle East. It is shown that Kuwait and Oman are almost exclusive urbanized, many countries have an urban population ranging from about two thirds to 90%, while countries like Somalia, Egypt, Eritrea, Yemen and Sudan have an urban population that is less than 50% of the total.

As another example, the level of taxation in countries in the Southeast Mediterranean and the Middle East is graphed in Fig. 7. It may be concluded that Libya, Greece and Kuwait draw 45–50% of their GDP from taxes, which one may argue constitutes a hindrance to investment and innovation. Cyprus, Djibouti, Eritrea, Iraq and Oman draw around 30–35% of their GDP from taxation. Israel, Saudi Arabia

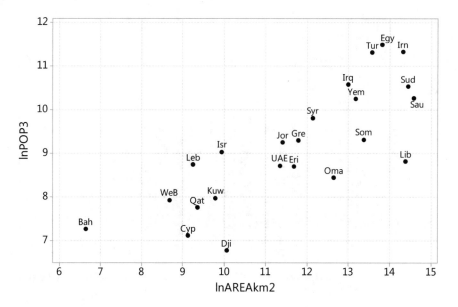

Fig. 5 Scatterplot of the population of countries in the Southeast Mediterranean and the Middle East as a function of their area (lnPOP3: natural logarithm of population in thousands; lnAREAkm2: natural logarithm of area in km²)

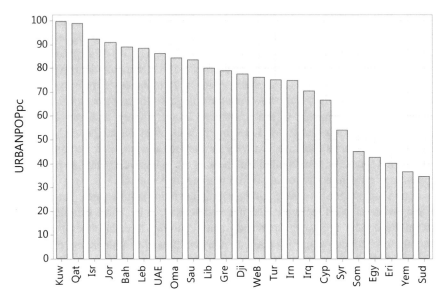

Fig. 6 Bar chart of percent of population residing in urban areas of countries in the Southeast Mediterranean and the Middle East (URBANPOPpc)

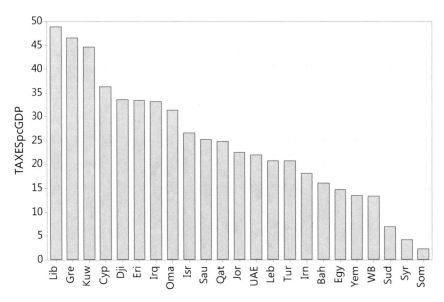

Fig. 7 Bar chart of taxation as percentage of GDP of countries in the Southeast Mediterranean and the Middle East (TAXESpcGDP)

and Qatar draw around one quarter of their GDP from taxation. One could go making analogous observations about groups of countries with similar levels of taxation.

Exports minus imports expressed as percent of GDP is graphed in the bar chart of Fig. 8. Interestingly, oil and gas rich countries (such as the United Arab Emirates, Bahrain, Iraq, Qatar, Saudi Arabia, Kuwait and Oman) are characterized by a surplus of exports. On the other hand, one sees countries struggling with their economies and political situation (such as the West Bank, Yemen and Djibouti) to have a large deficit of exports (compared to their imports). So, this particular indicator may be an interesting metric in the sense as it appears to be associated with energy exports of countries in the Southeast Mediterranean and the Middle East.

How is average life expectancy (of countries of Southeast Mediterranean and the Middle East) related to the number of physicians per population? An answer is given in Fig. 9, which shows that: Somalia has the lowest life expectancy (below 55 years of age); Yemen, Djibouti and Sudan have a life expectancy of about 65 years of age (although Sudan has a much larger number of physicians per population); and most other countries are clustered together, with a life expectancy ranging from below 75 to less than 85 years of age. Israel has the highest life expectancy, and it achieves this without the near double number of doctors per population that Greece has.

Turning to technology diffusion, as a last example in the socioeconomic section, Fig. 10 shows that there is a rather weak linear relationship between the Internet users per 1000 inhabitants and the cellular phone subscribers per 1000 inhabitants of the countries in Southeast Mediterranean and the Middle East. It is interesting to note that energy rich countries, such as Kuwait, the United Arab Emirates, Bahrain,

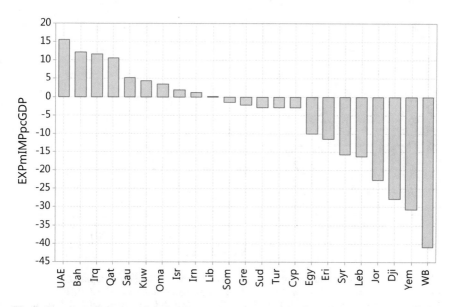

Fig. 8 Bar chart of exports minus imports as a percentage of the Gross Domestic Product (GDP) of countries in the Southeast Mediterranean and the Middle East

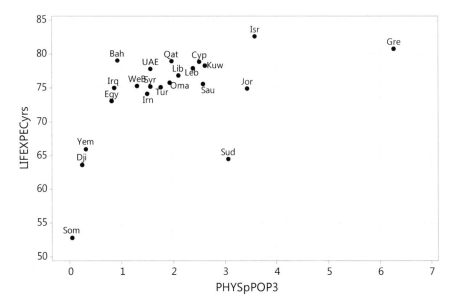

Fig. 9 Scatterplot of the life expectancy of countries in Southeast Mediterranean and the Middle East as a function of the number of physicians per population (LIFEXPECyrs: average life expectancy in years for the entire population, including both genders; PHYSpPOP3: physicians per 1000 inhabitants)

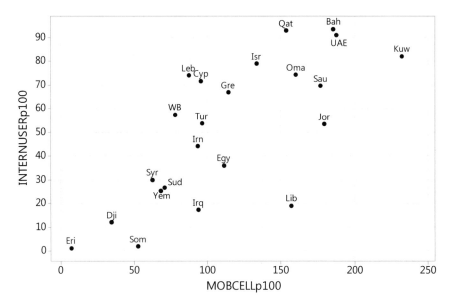

Fig. 10 Scatterplot of the number of internet users of countries in Southeast Mediterranean and the Middle East as a function of the number of mobile/cellular phone subscribers (INTERNUSERp100: Internet users per 100 inhabitants; MOBCELLp100: mobile/cellular phone subscribers 1000 inhabitants)

Oman, Qatar and Saudi Arabia have over 1.5 cellular subscriptions per individual—in fact Kuwait has over two. Bahrain, Qatar and the United Arab Emirates have over 90 Internet users per 100 people. At the other end, countries like Djibouti, Somalia and Eritrea have very low usage of Internet. As a general observation though, cellular phone diffusion is (much) bigger than Internet usage, so perhaps a (large) part of Internet access is now carried via cellular connections (especially in certain countries).

6.4.3.3 Institutional and Political Components

Turning to institutional and political components, organizations that should be considered in the context of energy security include the World Trade Organization (WTO), IEA, OPEC, and IMF. Relevant regional free trade organizations include the European Union (EU), the North American Free Trade Agreement (NAFTA), the Southern Common Market (MERCOSUR), and the Southern African Development Community (SADC) [15].

Although the US maintains a transatlantic relationship with the EU, cooperating on issues of energy security, at the same time it discourages independent EU energy initiatives, e.g. the Euro-Arab dialog that took place in the 1970s. Therefore, political difficulties should be taken into account in the development of an energy security index, with possible interactions with oil producing country [15].

The ability of a country to develop its own means of deterrence, and the actual deterrence capacity both relate to energy security. The case of Iraq, a central player in the global energy landscape, makes one think that an indicator variable showing whether a country is a fragile or failed state, could be incorporated in the political component of energy security research. Competition for scarce resources between consumer countries could cause conflict, so there is a risk that key producing regions slip into chaos—this should also be accounted for.

Delving deeper to the political component of an energy security index, the presence of a centralist political structure, lack of effective political institutions, and a weak civil society are also indicator variables that could be considered. The presence of elites, and the concentration of bigger amounts of wealth in smaller portions of the society (measured by the corresponding Gini coefficient), should also be considered. Taking corruption into account is important because large oil and gas revenues generate political pressures and corruption [15].

Institutional and political variables that could be found at a country level in the sources of Table 1 include: type of government; years of being an independent state; and the Gini coefficient for family income, which along with the percent of households with income at the lowest and highest 10% could act of surrogate variables for the effectiveness of democratic institutions.

For countries in the Southeast Mediterranean and the Middle East the type of government is tabulated in Table 3.

One could pool similar categories and come up with a more summary description, such as 10 republics and democracies, seven monarchies, one theocratic republic, and

Table 3 Type of government in countries of the Southeast Mediterranean and the Middle East (CIA Word Factbook characterization)

Type of government	Number of countries	Average years independent
Absolute monarchy	3	168
Parliamentary republic	3	120.3
Presidential republic	3	62
Federal parliamentary republic	2	73
In transition	2	48.5
Constitutional monarchy	1	48
Constitutional monarchy (emirate)	1	58
Federation of monarchies	1	48
Parliamentary constitutional monarchy	1	73
Parliamentary democracy	1	71
Presidential republic (highly authoritarian)	1	73
Semi-presidential republic	1	42
Theocratic republic	1	40

one authoritarian state presenting as a presidential republic. Looking at the years of independence per government type (rightmost column of Table 3), one may conclude that absolute monarchies and republics tend to be the countries enjoying the longest status as independent states; on the other hand, there are countries that have been in transition for decades.

6.4.3.4 Energy Components

Coming to potential energy components of an energy security index, clean coal and safe nuclear power will continue to play an important transitional role in the energy mix [9] and help offset dependence on imported gas [15]. So the contribution of each energy source to the energy mix of each country should be taken into account. Strategic coal, oil and gas reserves [6] as well as the ratio of reserves to production of the present and near future, should also be taken into consideration. In particular, the oil and gas consumption growth, and the investments to construct new or replace existing production capacity should be considered in the analysis.

Given that the US purchases crude oil from over 60 different countries [15], a work researching energy security should keep track of the number of countries that a country purchases oil, gas, and coal from. This adds a significant degree of complexity, but (together with the fact that much of the global oil production originates in countries that are characterized by internal instability) hints at the need of possibly defining and using energy fragility variables.

The incorporation of a nuclear energy component in an energy security index seems obvious since nuclear power can be an economical and reliable way of generating large amounts of base load electricity without producing CO_2 emissions. With nuclear power, the marginal cost of electricity is low, but construction, decommissioning, and waste disposal costs are high [85]. A total of 31 countries already have nuclear power and a number of them, including India and China, are looking to build new nuclear power units, while 20 countries are reported to be looking to develop nuclear power [88].

As pointed out by Corner et al. [14], the long-time association of nuclear energy with destroying power and world-making, sets nuclear apart from other technological systems in the mind of the public, what Masco [60] called the "*nuclear uncanny*". In the words of Winzer [84]: "*For some people the goal of energy security is the reliable provision of fuels and the role of nuclear energy is one of enhancing security. For others, energy security is concerned with a reduction of hazards from accidents and proliferation and the expansion of the nuclear industry is a potential threat to energy security*".

Corner et al. [14] argue that public attitudes towards nuclear energy constitute a reluctant acceptance [7] that should be interpreted against the following background: (a) high levels of awareness and concern about climate change, coupled with (b) an emerging concern about energy security (which remains poorly defined). Nuclear energy has been viewed as a potential method of addressing these two issues since the late 1990s [80], yet there is a clear preference for renewable energy over fossil fuels, as sources for electricity generation.

A comprehensive summary of up-to-date objections to nuclear power was made by Levite [52]. In his informed opinion, one could cite the following reasons for supporting nuclear power: ubiquity of fuel; exceptional form of energy for use as a base loader of the electrical grid; environmental benefits (mainly reduction of CO_2 emissions); manageable proliferation risks; improving operational performance record; safety standards improving since Chernobyl; and alternative energy sources also plagued by limitations and risks. Yet, Levite asserts that nuclear power is now in free fall, mainly due to the following shortcomings: lack of standardization; absence of harmony in (the update of) core regulatory requirements; cost management and financing challenges; enduring challenges of handling spent fuel; proliferation risks, to be added to safety and security issues; decommissioning overhang; and industry mentality and corporate myopia. Levite furthers his arguments against nuclear power by arguing that the problems that plague nuclear power now occur against a backdrop of: abundance of alternative energy supplies (such as shale, wind and solar); rapidly declining costs of the renewables (especially wind turbines and solar panels) that also possess better political appeal; significant improvements in storage capacity and affordability; daunting nuclear decommissioning overhang; political paralysis in confronting the storage challenge; a meltdown in the US nuclear leadership; and failure to come up with working industry solutions.

As expected, a wide variety of energy variables may be found in the sources of Table 1. These include: crude oil exports and imports (in barrels per day or bpd); refined oil production and consumption (in bpd); refined oil exports and imports (in bpd); gas production and consumption (in m^3); gas exports and imports (in m^3); electricity production and consumption (in kWh); electricity exports and imports (in kWh); installed capacity for electricity production (in kW); the contribution of each energy source (including coal, oil, gas, nuclear, hydro, solar, wind, geothermal, biofuel, and waste) expressed in GWh as well as percentage of the energy mix. Energy infrastructure variables may be found as well, such as gas and oil pipeline network (in km), although in general these data are incomplete and not up to date. Finally, pertinent socioeconomic variables such as population without electricity, and percent of electrification at urban centers, rural areas, and country average, may also be found in those sources.

The electricity energy mix of most countries in the Southeast Mediterranean and the Middle East is presented in Table 4. Quite a few observations may be made on the data presented there. For one, it may be argued that countries with a more varied energy mix (such as Greece and Turkey) would tend to be more energy secure. Also, there are unlikely countries whose energy resource endowment is such that energy security would appear to be easy to achieve, such as Sudan (with two thirds of its energy coming from hydro sources). Countries that use a single energy source predominantly (such as Cyprus, Eritrea, and Lebanon, all of which use oil almost exclusively) would have an energy security that would depend strongly on the availability of oil imports (and their own production of oil). In particular, countries that use (imported) natural gas almost exclusively would clearly have an energy security that would also depend strongly on its continued physical availability and affordability. Naturally, one should be careful with the data, e.g. countries like Qatar or Oman use natural gas almost exclusively because they have an abundance of it, and are exporting countries.

In the case of many EU countries (not shown on the table) that import the majority of the natural gas from Russia, this would mean that they would remain energy secure as long as Russia continues to supply its European market smoothly. Situations such as the Ukrainian crises (that were discussed in Sect. 4.1.3) or other unforeseen geopolitical considerations that would affect Russia, would also affect its European (and other) gas customers, more so in the case of small countries (like Greece) and less so in the case of bigger players (such as Germany) that are able to secure better terms. It is this very concern that the EU is currently trying to address with its much touted project of Energy Union, in a manner that would safeguard the energy security of its members and combat climate change, which the EU regards as a priority.

A final look at an energy variable is given by Fig. 11, which charts the energy self-sufficiency of the countries of the Southeast Mediterranean and the Middle East. On the one hand, countries that appear very secure from a self sufficiency point of view include: Qatar and Kuwait (with a self sufficiency between 450 and 500%); Iraq (with a self sufficiency between 350 and 400%); and the UAE, Oman and Saudi Arabia (with a self sufficiency around 300%). On the other hand, countries that appear quite insecure from a self sufficiency point of view include: Lebanon, Jordan, and Cyprus (with a near zero self-sufficiency); Greece, Turkey and Israel (with a self sufficiency less than 50%); and Syria (with a self-sufficiency of around 50%).

Table 4 Electricity energy mix of countries in the Southeast Mediterranean and the Middle East (percent of GWh; no data on Djibouti, Somalia, and the West Bank)

Country	Coal	Oil	Gas	Biofuel	Waste	Nuclear	Hydro	Geo-thermal	Solar PV	Solar thermal	Wind
Bahrain	0.0	0.0	100.0	0.0	0.0	0.0	0.0	0.0	0.0	0.0	0.0
Cyprus	0.0	91.2	0.0	1.1	0.0	0.0	0.0	0.0	2.8	0.0	4.9
Egypt	0.0	21.0	70.7	0.0	0.0	0.0	7.4	0.0	0.1	0.0	0.7
Eritrea	0.0	99.5	0.0	0.0	0.0	0.0	0.0	0.0	0.5	0.0	0.0
Greece	42.6	10.9	17.5	0.4	0.2	0.0	11.9	0.0	7.5	0.0	8.9
Iran	0.2	14.4	79.3	0.0	0.0	1.0	5.0	0.0	0.0	0.0	0.1
Iraq	0.0	71.9	24.4	0.0	0.0	0.0	3.7	0.0	0.0	0.0	0.0
Israel	45.8	0.7	51.6	0.1	0.0	0.0	0.0	0.0	1.7	0.0	0.0
Jordan	0.0	50.6	48.4	0.0	0.0	0.0	0.3	0.0	0.0	0.0	0.6
Kuwait	0.0	63.6	36.4	0.0	0.0	0.0	0.0	0.0	0.0	0.0	0.0
Lebanon	0.0	97.4	0.0	0.0	0.0	0.0	2.6	0.0	0.0	0.0	0.0
Libya	0.0	46.3	53.7	0.0	0.0	0.0	0.0	0.0	0.0	0.0	0.0
Oman	0.0	2.6	97.4	0.0	0.0	0.0	0.0	0.0	0.0	0.0	0.0
Qatar	0.0	0.0	100.0	0.0	0.0	0.0	0.0	0.0	0.0	0.0	0.0
Saudi Arabia	0.0	44.2	55.8	0.0	0.0	0.0	0.0	0.0	0.0	0.0	0.0
Sudan	0.0	35.5	0.0	0.0	0.0	0.0	64.5	0.0	0.0	0.0	0.0
Syria	0.0	29.1	68.6	0.0	0.0	0.0	2.3	0.0	0.0	0.0	0.0
Turkey	29.1	0.8	37.9	0.5	0.0	0.0	25.6	1.3	0.1	0.0	4.5
United Arab Emirates	0.0	1.2	98.5	0.0	0.0	0.0	0.0	0.0	0.0	0.2	0.0
Yemen	0.0	60.2	39.8	0.0	0.0	0.0	0.0	0.0	0.0	0.0	0.0

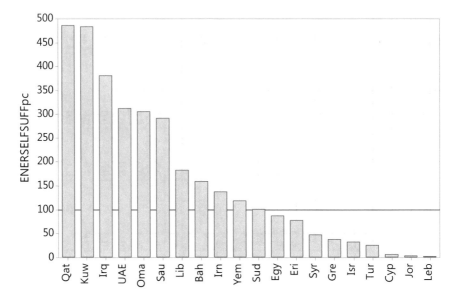

Fig. 11 Bar chart of energy self sufficiency expressed as a percentage for countries of the Southeast Mediterranean and the Middle East (reference line drawn at 100% energy self sufficiency)

Part of the challenge in defining (and calculating the values of) an energy security index is exactly this: how does one put together these opposing influences, e.g. variety in the energy mix versus self-sufficiency, in order to construct an index that would be a useful tool in analyzing geopolitical situations.

6.4.3.5 Environmental Components

Turning to environmental variables, although air pollution became a big concern in industrialized countries during the 1950s, 1960s and 1970s, global warming and awareness of climate change remained low. In the 1980s, the World Health Organization (WHO) established the Intergovernmental Panel on Climate Change (IPCC) to provide a scientific view of climate change. In 1997, the Kyoto Protocol was signed to set binding obligations on industrialized countries to reduce emissions of greenhouse gases. In 2011 a new value was included in the basic definition of energy security by IEA, the "*respect for environmental concerns*".

Although there is little doubt that environmental concerns should be accounted for, in a definition of energy security, the way to formalize their inclusion is not as obvious and may, in fact, be fruitful to address this task via qualitative approaches (e.g. with case studies and expert interviews, mentioned in Sects. 4.6 and 4.7 correspondingly).

Environmental (including geomorphological and ecological) variables that are found in the sources of Table 1 include: land and water area of a country (in km^2); land boundary and coastline of a country (in km); agricultural and forest land in a

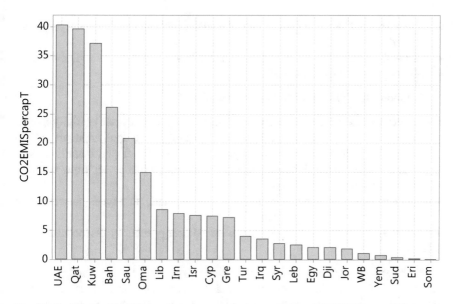

Fig. 12 Carbon dioxide (CO_2) emissions per capita in metric tons (CO2EMISpercapT) for countries of the Southeast Mediterranean and the Middle East

country (in percentages); contribution of fossil fuels and renewable energy sources to electrification in a country (in percentage); and CO_2 emissions of a country (in metric tons).

This last variable, CO_2 emissions per capita (in metric tons), is depicted in Fig. 12. It may be concluded that fossil-fuel producing countries such as the United Arab Emirates, Qatar and Kuwait, are major producers of CO_2 emissions on a per capita basis (between 35 and 50 tons per capita). Libya, Iran, Israel, Cyprus, and Greece are remarkably similar in their CO_2 emissions per capita (between five and 10 tons per capita). Finally, poor countries such as Somalia, Eritrea and Sudan have near zero CO_2 emissions per capita.

6.4.4 Creating a Geopolitical Energy Security Index

According to Sovacool [74], creating an energy security index involves: conceptualizing energy security; putting the existing definitions in the proper frame; collecting and consolidating data on previous energy security metrics; and scoring the results, so that a better index is obtained. The evaluation of such a new index may be based on the following criteria:

- *Suitability*: How well does the index measure the relevant aspects of the energy security framework?

- *Transparency*: How transparent and objective is the index? Is expert judgment required?
- *Availability of data*: Are sufficient and robust data available to compile the index?
- *Ability of forecast* (self-evident).

  Common shortcomings of indices include [76]:

- *Topical focus*, i.e. studies centering on pressing concerns (such as electricity supply, nuclear power, and automobiles) instead of striving to be applicable not only in developed countries, but also in developing or least developed countries (with patchy and incomplete electricity networks, limited nuclear power units, and non-motorized forms of transport).
- *Scope and coverage* as many indices are sector-specific, and focus on security of supply rather than security of demand.
- *Continuity*, as very few studies have assessed energy security performance over time.

Motivated by the above observations, an energy security index should attempt to include as many countries as possible, certainly all member states of the UN, taking into account the full complexity of the global environment, where energy decisions are taken at the political level [24] and global geopolitical consequences unfold. Such an energy security index should take into consideration both security of supply and demand, and it should attempt to measure energy security across a range of political systems and geopolitical priorities. Finally, an energy security index should provide the element of continuity by assessing energy security during the milestone time periods (if not along a complete time series).

Turning to the scope and nature of an energy security index, such research should strive to address Sovacool's [75] remark that *"metrics for energy security will always have this tension between comprehensiveness and understandability, but the most important aspect of any energy security index is for it to be simple, meaningful, and accessible – usability is better than perfectibility."*

The creation of a composite energy security index would normally include the following activities [3]: the energy security definition is framed; any specific focus areas are set; the appropriate simple indicators and selected; the required data are collected; the simple indicators are normalized; the normalized indicators are weighted; and the weighted and normalized simple indicators are aggregated into a (composite) index. The normalization step may resort to a minimum-maximum scale, the distance to a reference, or other standardization techniques; the weighting may use equal weights, recognize the relative importance of different fuel types, use PCA to sort out overlapping information, obtain expert opinions, or resort to Data Envelopment Analysis.

Further to what has been presented in previous sections, common simple indicators for developing an energy security index have included [46]: the reserve/production ratio, which shows the remaining amount of unexploited resources a country possesses; the Total Primary Energy Supply (TPES), which accounts for the indigenous production from primary sources, the import/export balance, the availability in international marine bunkers, and any stock changes; the reserve/consumption ratio, which provides an indication of how much energy a country may use, if in a pinch; and the reserve margin of generation capacity.

The development of a new quantitative energy security index with an emphasis on geopolitical considerations could be done by collecting quantitative data, categorical data, and qualitative data and indicators that highlight the causative connection between energy security and geopolitical events. The steps of normalization, weighting, and aggregation should take into consideration the (previously outlined) state of the art in the research literature. A significant contribution would be achieved if such an energy security indicator reflected—in a novel way—the energy security realities across time, geopolitical priorities, and political systems.

An example of such a research work is provided by Radovanović et al. [69], who used the following indicators to construct an energy security index (ESI): energy intensity (EI); final energy consumption (FEC); Gross Domestic Product per capita (GDPpc); share of nuclear and renewable energy (SRN); energy dependence (ED); and carbon intensity (CI). PCA was used to determine the coefficients in the following final equation of their energy security index:

$$ESI = 20 \times (EI) + 20 \times (FEC) + 10 \times (GDPpc) + 20 \times (SRN) - 20 \times (ED) - 10 \times (CI)$$

In a newer effort, Radovanović et al. [70] used an expanded list of variables (adding energy dependence, electricity prices, and sovereign credit rating) to estimate values of a Geo-Economic Energy Security Index for many European countries from 2004 to 2013. Such panel data may be used for further statistical analysis, especially in reference to case studies and expert opinions.

Turning to specific energy and environment-related indexes, the 2018 Environmental Performance Index (EPI, a collaborative effort of the Yale Center for Environmental Law and Policy, the Center for International Earth Science Information Network of Columbia University, and the World Economic Forum) was used to rank 180 countries on 24 performance indicators across ten issue categories covering "*environmental health and ecosystem vitality*" (https://epi.envirocenter.yale.edu/downloads/epi2018policymakerssummaryv01.pdf). The EPI has been touted as a scorecard highlighting "*leaders and laggards in environmental performance*". Low scores of the EPI may indicative a need for, e.g. cleaning up air quality, protecting biodiversity, or reducing greenhouse gas emissions.

Values of the 2018 EPI for the countries of the Southeast Mediterranean and the Middle East are plotted against the United Nations Human Development Index in Fig. 13. It may be seen that countries like Eritrea, Djibouti and Sudan score low on both indexes. On the other hand, countries like Israel, Greece, Cyprus and Qatar appear to be clustered together at the highest value range of both indexes. Quite a few other countries are located in the middle range of both indexes, with countries like Iraq doing relatively well on the HDI, but poorly on the EPI.

Another index of interest is the INFORM global risk index, a tool for understanding the risk of humanitarian crises and disasters (http://www.inform-index.org/Portals/0/InfoRM/2018/INFORM2018-WithCovers.pdf). It is composed of three

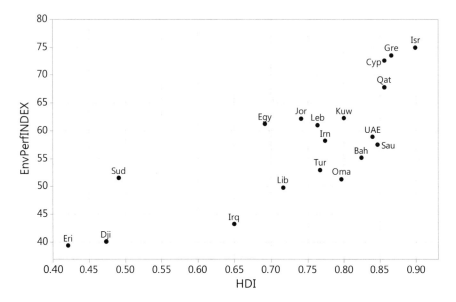

Fig. 13 Scatterplot showing the Environmental Performance Index (EnvPerfINDEX) as a function of the Human Development Index (HDI) for countries of the Southeast Mediterranean and the Middle East

dimensions: hazards and people's exposure to them; vulnerability; and lack of coping capacity (i.e. lack of resources to help people cope with the hazards). INFORM uses 50 different indicators to create a risk profile for every country, ranging from zero to 10, where: values less than two indicate very low risk; values from 2 to 3.5 indicate low risk; values from 3.5 to 5 indicate medium risk; values from five to 6.5 indicate high risk; and values over 6.5 (and less than 10, its maximum) indicate very high risk.

The Environmental Performance Index is plotted against the INFORM index for countries of the Southeast Mediterranean and the Middle East in Fig. 14. As expected, they appear to be negatively correlated: high values of the Environmental Performance Index appear to be associated with low values of the INFORM index and vice versa. Quite a few interesting observations may be made. First, Israel, Greece and Cyprus are clustered together at the top and middle-left of the figure, indicating that these three countries are similar in having a good level of environmental performance and relatively low global risk. On the other hand, countries like Djibouti, Eritrea, Iraq, Libya and Sudan, combine high levels of global risk with low levels of environmental performance. Energy producing countries such as Qatar, Kuwait, the United Arab Emirates, Saudi Arabia, Bahrain and Oman have reasonably low global risk and relatively average environmental performance.

A last look is shown by Fig. 15, where energy self-sufficiency (expressed as a percentage) is plotted against carbon dioxide emissions per capita. The data exhibit a moderately linear relationship, with more countries clustering at the lower left of

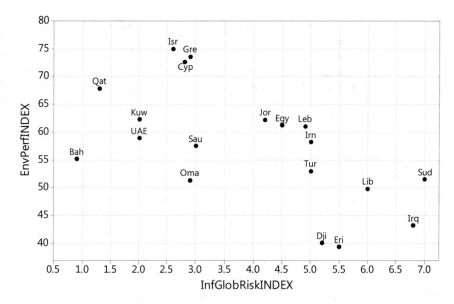

Fig. 14 Scatterplot showing the Environmental Performance Index (EnvPerfINDEX) as a function of the Inform Global Risk Index (InfGlobRiskINDEX) for countries of the Southeast Mediterranean and the Middle East

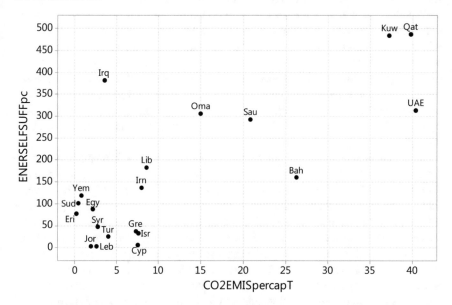

Fig. 15 Scatterplot showing the energy self-sufficiency as a percentage (ENERSELFSUFFpc) as a function of carbon dioxide (CO_2) emissions per capita in metric tons (CO2EMISpercapMt) for countries of the Southeast Mediterranean and the Middle East

the scatterplot. It is evident that countries with a very high self sufficiency, such as Qatar and Kuwait, are also characterized by high CO_2 emissions per capita. On the other hand, there is a very wide variation of CO_2 emissions for countries that are the mid range of self sufficiency, e.g. the United Arab Emirates, Saudi Arabia, Oman, and Iraq. Poor countries such as Eritrea and Sudan have near zero emissions of CO_2 per capita.

6.4.5   Using Cluster Analysis

The aim of this important step of the analysis in an energy security research work will be to produce country classifications by: forming groups of similar countries at each milestone time period; examining how countries change classification by moving from one group to another over time; and observing how energy security profiles of entire regions (and hotspots) have changed from the 1970s until today.

To formulate the groups, CA [22] may be run on a variety of socioeconomic and geopolitical variables as well as values of the formulated energy security index. Although it is expected that empirical degrees of freedom considerations, such as those proposed by Formann [23], as cited by Mooi and Sarstedt [62], will limit the number of variables that may be employed by CA, the aim would be to derive well-defined cluster solutions, which will constitute an innovative energy security taxonomy of countries over time. PCA may be employed to extract a small number of Principal Components out of (significantly more) well correlated variables, and thus manage to have more variables represented (even indirectly) during CA. Fuzzy clustering techniques or complexity dynamics may also be taken into consideration. Data quality and missing data consideration along with a priori expectations (as to which variables should characterize the profile of a country) could be taken into consideration, and may help with the final selection of variables to be used for clustering.

The classification that will result from the CA will provide information from which insight and knowledge may be gained as to: (a) the significance of the energy security index; (b) the factors that put a country in a specific cluster; and (c) the policies that are correlated with any changes in the classification of a country. Qualitative information that were left out of the CA (e.g. by not being represented by dummy variables) may also be associated with the classification, in an effort to gain additional understanding of the components of energy security over time. The clusters of each milestone time period may also be depicted on global or regional maps of energy security (also proposed by Vivoda [82]), the configuration and dynamics which, may generate additional geopolitical insight and highlight the location and movement of energy security hotspots over time.

Vivoda [82] has stressed the importance of constructing energy security typologies, i.e. clusters, which (it is asserted) are better than lists of variables and complex interrelations. The construction of typologies allows for the classification of countries, e.g. developing countries that are net energy importers, such as China and

India, have a significant degree of similarity across many dimensions and attributes of energy security, and are structurally different from developed energy importers (such as Japan and South Korea) and energy exporters (such as Russia and Australia).

6.4.6 Looking at Case Studies of Key Countries

The estimation of an energy security index and the results of CA could be complemented usefully by a case study analysis of a relatively small number of key countries, possibly including: Germany, Russia, Iran, Iraq, Saudi Arabia, the United States, Canada, Venezuela, Nigeria, China, India and Japan. If a researcher decided to concentrate one's efforts on the Southeast Mediterranean and the Middle East, plenty of ideas are present in the material present in previous sections; in particular, examining Turkey, Israel, Greece and Cyprus would be very important.

Different countries may be examined at different milestones, if this is recommended by the produced classification or grouping of countries, especially the calculated values of the energy security index at the milestone time periods. Since such key countries would be analyzed in depth, it would be interesting to see if they moved from one energy security cluster to another over time, and what policies were associated with such moves.

Since case studies may examine a broad spectrum of (quantitative and qualitative) energy security concerns, they may help explain why countries under comparable international conditions prioritize energy security differently. In this fashion, a case study approach would complement nicely the quantitative methods discussed in previous sections.

6.4.7 Carrying Out Interviews of Energy Experts

Conducting interviews with energy experts is a research method that has been combined with other methods in the field of energy security [77].

The interviewees could range from academic and policy experts on energy security to public officials from different government departments and bodies. Snowball sampling could be employed to augment the set of experts selected initially, i.e. at the end of each interview, experts could be asked to nominate other potential interviewees (and help provide contacts and access).

The interviews could be based on a questionnaire of open-ended questions, and would be used to complement quantitative data analyses, and provide more texture and depth to the knowledge that will have been gained thus far. The qualitative data produced by the interviews could be of use both from a theoretical and an empirical perspective. By construction, the theoretical approach will have a normative ambition, and will strengthen the comprehension of the main issues at stake. In particular, the interviewees could be polled on their opinion of the concept of energy security, with emphasis on the milestone time periods. The interviewees could also be asked to critique the energy security index and judge the classification, energy security policies, and geopolitical energy hotspots depicted on the dynamic maps throughout the milestone time periods.

The degree of involvement of expert opinions in the developed index would help provide a reality check and ensure its transparency [74].

6.4.8 Forecasting Energy Security

An energy security index research could be complemented by energy security forecasts.

The use of formal statistical methods would be contingent on data availability. If the socioeconomic and geopolitical energy security data were available on a yearly basis (which, at this point and with the experience gained in collecting such data by the authors of this Chapter, appears nontrivial) then formal modes (such as time series regression combined with atheoretical methods such as ARIMA) could be used to produce forecasts. If only data for the examined milestone time periods were available, forecasts would have to be developed based on a qualitative approach (and with a greater involvement of the experts).

Taking into account global climate change and energy mix transitions would be of particular importance in developing robust energy security forecasts. It would be particularly interesting to see how the energy security index values affect the forecasts, see how the state of energy security affairs is forecast to vary in the foreseeable future, and gain an understanding as to what geopolitical implications are likely to occur.

Importantly, energy security could be examined in the context of storylines, such as the *Markets and Institutions* (idealistic scenario) and the *Regions and Empires* (realistic scenario) examined by Correlje and van der Linde [15]: in a *Markets and Institutions* world, Russia becomes integrated in the EU market, while in a *Regions and Empires* world, Russia develops its own empire. Depending on which storyline prevails, the EU may need to transform itself [15] from an economically driven project into a project that is driven by geopolitics and the strategic use of state and economic power (possibly at the cost of significant political hardship and internal conflict).

6.5 Closing Comments

An extensive examination of the current energy security indicators and indexes in the literature has shown the differentiation among studies, and the need to integrate additional indexes in the framework for the analysis of geopolitical problems.

Many previous studies on energy security have focused on the security of supply. This Chapter has shown how future energy security research could add to the state of the art by combining quantitative and qualitative approaches in a way that is innovative and characterized by a more holistic geopolitical perspective.

The outcomes of such research could, in the long run, help address the issues of energy security in the years to come that may be characterized by energy transition, economic uncertainty and environmental hardship.

Acknowledgements The authors of this Chapter thank Drs. T. Nadasdi and S. Sinclair for their online Spell Check Plus (http://spellcheckplus.com), which was used for proofing the entire text.

References

1. A.F. Alhajji (2007). What is energy security? Definitions and concepts. Middle East Economic Survey, L, 45
2. P.D. Allison, *Logistic Regression Using SAS: Theory and Application*, 2nd edn. (USA, SAS Institute, 2012)
3. B.W. Ang, W.L. Choong, T.S. Ng, Energy security: definitions, dimensions and indexes. Renew. Sustain. Energy Rev. **42**, 1077–1093 (2015)
4. Asia Pacific Energy Research Center (APERC), A quest for energy security in the 21st century: Resources and constraints. Asia Pacific Energy Research Center, Japan (2007). Retrieved from http://aperc.ieej.or.jp/file/2010/9/26/APERC_2007_A_Quest_for_Energy_Security.pdf
5. D.A. Baldwin, The concept of security. Rev Int Stud **23**, 5–26 (1997). Retrieved from https://www.princeton.edu/~dbaldwin/selected%20articles/Baldwin%20(1997)%20The%20Concept%20of%20Security.pdf
6. A. Ballis, T. Moschovou, L. Dimitriou, Analysis of gas pipeline system from a European energy security perspective. Int. J. Decis. Support Syst. (forthcoming)
7. K. Bickerstaff, I. Lorenzoni, N.F. Pidgeon, W. Poortinga, P. Simmons, Reframing nuclear power in the UK energy debate: nuclear power, climate change mitigation, and radioactive waste. Public Underst. Sci. **17**, 145–169 (2008)
8. British Petroleum (BP), *BP Statistical Review of World Energy*. 66th edn. (PB, London, UK, 2017). Retrieved from https://www.bp.com/content/dam/bp/en/corporate/pdf/energy-economics/statistical-review-2017/bp-statistical-review-of-world-energy-2017-full-report.pdf
9. R. Caputo, Hitting the wall: a vision of a secure energy future, in *Synthesis Lectures on Energy and the Environment: Technology, Science and Society,* Lecture #3, (Morgan and Claypool Publishers, 2009)
10. A. Cherp, J. Jewell, The three perspectives on energy security: intellectual history, disciplinary roots and the potential for integration. Curr. Opin. Environ. Sustain. **3**(4), 202–212 (2011)
11. A. Cherp, J. Jewell, The concept of energy security: beyond the four As. Energy Policy **75**, 415–421 (2014)
12. L. Chester, Conceptualising energy security and making explicit its polysemic nature. Energy Policy **38**, 887–895 (2010)
13. M. Conant, F. Gold, *The geopolitics of energy* (Westview Press, Boulder, Colorado, USA, 1978)
14. A. Corner, D. Venables, A. Spence, W. Poortinga, C. Demski, N. Pidgeon, Nuclear power, climate change and energy security: exploring British public attitudes. Energy Policy **39**, 4823–4833 (2011)
15. A. Correlje, C. van der Linde, Energy supply security and geopolitics: a European perspective. Energy Policy **34**, 532–543 (2006)

16. J. De Wilde, Security levelled out: the dominance of the local and the regional, in *New Forms of Security: Views from Central, Eastern and Western Europe*, ed. by P. Dunay, G. Kardos, A. Williams (Aldershot, Dartmouth, 1995), pp. 88–102
17. D.A. Deese, Energy: economics, politics, and security. Int. Secur. **4**(3), 140–153 (1979)
18. Energy Charter Secretariat, International energy security: common concept for energy producing, consuming and transit (2015). Retrieved from http://www.energycharter.org/fileadmin/DocumentsMedia/Thematic/International_Energy_Security_2015_en.pdf
19. Energy Information Administration (EIA), Annual energy review 2011. U.S. Department of Energy, Washington (2012). Retrieved from https://www.eia.gov/totalenergy/data/annual/pdf/aer.pdf
20. N. Esakova, *European energy security: analysing the EU-Russia energy security regime in terms of interdependence theory* (Springer, Frankfurt, Germany, 2012)
21. European Commission (EC), *Towards a European Strategy for the Security of Energy Supply* (Office for Official Publications of the European Communities, Green Paper, Luxembourg, 2000)
22. B.S. Everitt, S. Landau, M. Leese, D. Stahl, *Cluster Analysis,* 5th edn. (King's College Wiley Series in Probability and Statistics, London, UK, 2011)
23. A.K. Formann, *Die Latent-Class-Analyse: Einführung in die Theorie und Anwendung* (Beltz, Weinheim, Germany, 1984)
24. P.H. Frankel, *Essentials to Petroleum: A Key to Oil Economics.* New edition, 2nd impression (Frank Cass, London, 1976)
25. E.W. Frees, *Longitudinal and panel data: analysis and application in the social sciences* (Cambridge University Press, UK, 2004)
26. E. Gnansounou, Assessing the energy vulnerability: case of industrialized countries. Energy Policy **36**, 3734–3744 (2008)
27. A. Goldthau, Governing global energy: existing approaches and discourses. Curr. Opin. Environ. Sustain. **3**(4), 213–217 (2011)
28. D. Greene, Measuring energy security: can the United States achieve oil independence?". Energy Policy **38**, 1614–1621 (2010)
29. D.N. Gujarati, D.C. Porter, *Basic Econometrics*, 5th edn. (Boston, McGraw-Hill Irwin, 2009)
30. E. Gupta, Oil vulnerability index of oil-importing countries. Energy Policy **36**, 1195–1211 (2008)
31. K.J. Hancock, V. Vivoda, International political economy: a field born of the OPEC crisis returns to its energy roots. Energy Res. Soc. Sci. **1**, 206–216 (2014)
32. International Atomic Energy Agency (IAEA), *Energy Indicators for Sustainable Development: Guidelines and Methodologies* (Vienna, 2005). Retrieved from http://www-pub.iaea.org/MTCD/publications/PDF/Pub1222_web.pdf
33. International Energy Agency (IEA), *Energy Security and Climate Policy: Assessing Interactions* (IEA/OECD, Paris, 2007)
34. International Energy Agency (IEA), *World Energy Outlook 2010* (International Energy Agency, Paris, 2010)
35. International Energy Agency (IEA), *Measuring Short-Term Energy Security* (OECD/IEA, Paris, France, 2011). Retrieved from https://www.iea.org/publications/freepublications/publication/Moses.pdf
36. International Energy Agency (IEA), *World Energy Outlook 2013* (OECD/IEA, Paris, France, 2013). Retrieved from https://www.iea.org/publications/freepublications/publication/WEO2013.pdf
37. International Monetary Fund (IMF) (2014). *Russian Federation: 2014 Article IV Consultation—Staff Report; Informational Annex; Press Release.* IMF Country Report No. 14/175. Retrieved from https://www.imf.org/external/pubs/ft/scr/2014/cr14175.pdf
38. J.C. Jansen, A.J. Seebregts, Long-term energy services security: what is it and how can it be measured and valued? Energy Policy **38**(4), 1654–1664 (2010)
39. J. Jewell, *The IEA Model of Short-Term Energy Security (MOSES): Primary Energy Sources and Secondary Fuels.* Working Paper (OECD/IEA, Paris, France, 2011)

40. B. Johansson, A broadened typology on energy and security. Energy **53**, 199–205 (2013)
41. E. Jun, W. Kim, S.H. Chang, The analysis of security cost for different energy sources. Appl. Energy **86**(10), 1894–1901 (2009)
42. J.H. Keppler, *International Relations and Security of Energy Supply: Risks to Continuity and Geopolitical Risks* (University of Paris-Dauphine, Paris, France, 2007). Retrieved from http://www.europarl.europa.eu/RegData/etudes/etudes/join/2007/348615/EXPO-AFET_ET(2007)348615_EN.pdf
43. E. Kisel, A. Hamburg, M. Harm, A. Leppiman, M. Ots, Concept for energy security matrix. Energy Policy **95**, 1–9 (2016)
44. M. Klare, *Rising Powers, Shrinking Planet, the New Geopolitics of Energy* (Metropolitan Books, New York, 2008)
45. S.D. Kopp, Politics, *Markets and EU Gas Supply Security. Case Studies of the UK and Germany* (Springer, Berlin, Germany, 2014). Retrieved from http://www.springer.com/gp/book/9783658083236
46. K. Koyama, I. Kutani, *Study on the Development of an Energy Security Index and an Assessment of Energy Security for East Asian Countries* (Economic Research Institute for ASEAN and East Asia (ERIA), 2012). Retrieved from http://www.eria.org/RPR-2011-13.pdf
47. B. Kruyt, D.P. van Vuuren, H.J.M. de Vries, H. Groenenberg, Indicators for energy security. Energy Policy **37**, 2166–2181 (2009)
48. X. Labandeira, B. Manzano, Some economic aspects of energy security. Econ. Ener. WP09/2012 (2012). Retrieved from https://eforenergy.org/docpublicaciones/documentos-de-trabajo/WP092012.pdf
49. C. Le Coq, E. Paltseva, Measuring the security of external energy supply in the European Union. Energy Policy **37**, 4474–4481 (2009)
50. N. Lefèvre, Measuring the energy security implications of fossil fuel resource concentration. Energy Policy **38**(4), 1635–1644 (2010)
51. M. Leonard, N. Popescu, *A Power Audit of the EU-Russia Relations* (European Council on Foreign Relations, London, UK, 2007)
52. A. Levite, Nuclear policy challenges revisited: the governmental and industrial dimensions. Presentation made in the *Energy: Strategy, Law and Economics* graduate program, (Department of International and European Studies, University of Piraeus, Greece, May 5 2018)
53. A. Löschel, U. Moslener, D.T.G. Rübbelke, Indicators of energy security in industrialised countries. Energy Policy **38**, 1665–1671 (2010)
54. R.J. Lovering, A. Yip, T. Nordhaus, Historical construction costs of global nuclear power reactors. Energy Policy **91**, 371–382 (2016)
55. G. Luft, A. Korin, Energy security: in the eyes of the beholder, in *Energy Security Challenges for the 21st Century: A Reference Handbook*, ed. by G. Luft, A. Korin (Praeger Security International, Santa Barbara, California, USA, 2009)
56. N.K. Malhotra, D.F. Birks, *Marketing Research: An Applied Approach*. Updated 2nd European Edition (Prentice Hall Financial Times, Harlow, England, 2006)
57. A. Manson, B. Johansson, L.J. Nilsson, Assessing energy security: an overview of commonly used methodologies. Energy **73**, 1–14 (2014)
58. A. Marquina, The southeast–southwest European energy corridor, in *Energy Security: Visions from Asia and Europe*, ed. by A. Marguina (Palgrave Macmillan, UK, 2008), pp. 54–68
59. J. Martchamadol, S. Kumar, An aggregated energy security performance indicator. Appl. Energy **103**(C), 653–670 (2013)
60. J. Masco, *The Nuclear Borderlands: The Manhattan Project in Post-Cold War New Mexico* (Princeton University Press, Princeton, New Jersey, USA, 2006)
61. J. Mitchell, P. Beck, M. Grubb, *The new geopolitics of energy* (The Royal Institute of International Affairs, London, UK, 1996)
62. E. Mooi, M. Sarstedt, *A Concise Guide to Market Research: The Process, Data, and Methods Using IBM SPSS Statistics* (Springer, Berlin/Heidelberg, Germany, 2011)
63. H. Morgenthau, *Politics Among Nations: The Struggle for Power and Peace*, 6th edn. (Knopf, New York, USA, 1985)

64. K. Narula, S. Reddy, Three blind men and an elephant: the case of energy indices to measure energy security and energy sustainability. Energy **80**, 148–158 (2015)
65. B. O'Neill, Game theory and the study of the deterrence of war, in *Perspectives on Deterrence*, ed. by P. Stern, R. Axelrod, R. Jervis, R. Radner (Oxford University Press, 1989), pp. 134–156
66. J.A. Paravantis, D. Georgakellos, Trends in energy consumption and carbon dioxide emissions of passenger cars and buses. Technol. Forecast. Soc. Chang. **74**(5), 682–707 (2007)
67. J.A. Paravantis, N. Kontoulis, A geopolitical approach to conceptualizing and measuring energy security. Arch. Econ. Hist./Αρχείον Οικονομικής Ιστορίας, **XXIX/1/2017**, 41–67 (2017)
68. S. Pirani, H. Stern, K. Yafimava, *The Russo-Ukrainian Gas Dispute of January 2009: A Comprehensive Assessment.* Oxford Institute for Energy Studies, NG27, February 2009. Retrieved from https://www.oxfordenergy.org/wpcms/wp-content/uploads/2010/11/NG27-TheRussoUkrainianGasDisputeofJanuary2009AComprehensiveAssessment-JonathanSternSimonPiraniKatjaYafimava-2009.pdf
69. M. Radovanović, S. Filipović, D. Pavlović, Energy security measurement—a sustainable approach. Renew. Sustain. Energy Rev. **68**, 1020–1032 (2017)
70. M. Radovanović, S. Filipović, V. Golušin, Geo-economic approach to energy security measurement—principal component analysis. Renew. Sustain. Energy Rev. **82**(P2), 1691–1700 (2018)
71. REACCESS (2011), Risk of Energy Availability Common Corridors for Europe Supply Security (REACCESS). Summary Report (draft). Retrieved from http://reaccess.epu.ntua.gr/LinkClick.aspx?fileticket=ez26yrScOcg%3D&tabid=721
72. D.M. Reiner (2006), *EPRG Public Opinion Survey on Energy Security: Policy Preferences and Personal Behaviour.* Retrieved from http://www.eprg.group.cam.ac.uk/wp-content/uploads/2008/11/eprg0706.pdf
73. M. Scheepers, A.J. Seebregts, J.J. De Jong, J.M. Maters, *EU Standards for Security of Supply* (Energy Research Center (ECN)/Clingendael International Energy Programme, Hague, Netherlands, 2007)
74. B.K. Sovacool, Evaluating energy security in the Asia Pacific: towards a more comprehensive approach. Energy Policy **39**, 7472–7479 (2011)
75. B.K. Sovacool, The methodological challenges of creating a comprehensive energy security index. Energy Policy **48**, 835–840 (2012)
76. B.K. Sovacool, An international assessment of energy security performance. Ecol. Econ. **88**, 148–158 (2013)
77. B.K. Sovacool, I. Mukherjee, Conceptualizing and measuring energy security: a synthesized approach. Energy **36**, 5343–5355 (2011)
78. B.K. Sovacool, I. Mukherjee, I.M. Drupady, A.L. D'Agostino, Evaluating energy security performance from 1990 to 2010 for eighteen countries. Energy **36**, 5846–5853 (2011)
79. A.H. Studenmund, B.K. Johnson, *Using Econometrics: A Practical Guide*, 7th edn. (Boston, Pearson, 2017)
80. T. Teravainen, M. Lehtonen, M. Martiskainen, Climate change, energy security and risk—debating nuclear new build in Finland, France and the UK. Energy Policy **39**, 3434–3442 (2011)
81. U.S. Chamber of Commerce. *Index of U.S. Energy Security Risk: Assessing America's Vulnerabilities in a Global Energy Market* (U.S. Chamber of Commerce, Washington D.C., USA, 2010)
82. V. Vivoda, Evaluating energy security in the Asia-Pacific region: a novel methodological approach. Energy Policy **38**, 5258–5263 (2010)
83. K. Waltz, *Theory of International Politics.* Addison-Wesley Series in Political Science, (Addison-Wesley, Reading, Massachusetts, USA, 1979)
84. C. Winzer, Conceptualizing energy security. Energy Policy **46**, 36–48 (2012)
85. World Energy Council (WEC), *World Energy Trilemma I 2016. Defining Measures to Accelerate the Energy Transition* (WEC, London, UK, 2016). Retrieved from https://www.worldenergy.org/wp-content/uploads/2016/05/World-Energy-Trilemma_full-report_2016_web.pdf

86. World Energy Council (WEC), *World Energy Resources. Natural Gas 2016*. (WEC, London, UK, 2016). Retrieved from https://www.worldenergy.org/wp-content/uploads/2017/03/WEResources_Natural_Gas_2016.pdf
87. World Energy Forum (WEF), Global energy architecture performance index report 2015. Switzerland (2014). Retrieved from http://www3.weforum.org/docs/WEF_GlobalEnergyArchitecture_2015.pdf
88. World Nuclear Association, *Emerging Nuclear Energy Countries* (September 2017). Retrieved from http://www.world-nuclear.org/information-library/country-profiles/others/emerging-nuclear-energy-countries.aspx
89. D. Yergin, Energy security in the 1990s. Foreign Aff. **67**(1), 110–132 (1988)
90. D. Yergin, *The Prize: The Epic Quest for Oil, Money, and Power* (Simon and Schuster, New York, USA, 1991)
91. D. Yergin, Ensuring energy security. Foreign Aff. **85**(2), 69–82 (2006)
92. S. Zeng, D. Streimikiene, T. Baležentis, Review of and comparative assessment of energy security in Baltic States. Renew. Sustain. Energy Rev. **76**, 185–192 (2017)

Part III
Learning and Analytics in Performance Assessment

Chapter 7
Automated Stock Price Motion Prediction Using Technical Analysis Datasets and Machine Learning

Apostolos Meliones and George Makrides

Abstract The purpose of this paper is to study, develop and evaluate a prototype stock motion prediction system using Technical Analysis theory and indicators, in which the process of forecasting is based on Machine Learning. The system has been developed following the cloud computing paradigm consisting of a backend application in Google's API Hosting Cloud and using an Android front end using inspiring, contemporary styles of tools and libraries and harmonized with modern technology trends. The focus of the paper is on system implementation and process automation. Due to the increased difficulty level of the task, the traditionally heavyweight stocks of the bank sector of the Athens Stocks Exchange market have been used in system modeling, as the attempt to generalize the model for all stocks would make our effort impossible. A use case with the famous US S&P 500 index has been also tested. We conclude with a discussion on the optimization of the accuracy of such systems.

Keywords Stock price motion prediction · Machine learning · Support vector machines (SVM) · Adaptive boost · Technical analysis · Cloud computing · Google App Engine · Android

7.1 Introduction

The rapid growth and deepening of the modern stockbroking environment which is the effect of several factors, from the entrance of new dynamic enterprises to the stock market, through the derivatives market, to the internationalization of economies, enforces the application of modern investment tools in stocks portfolio management. Many such tools rely on mathematical analysis and modeling techniques and require high performance computations and expert knowledge, thus being prohibitive for smaller investors. To this end, recently several studies have been performed for the application of machine learning algorithms in this area for the analysis and pre-

A. Meliones (✉) · G. Makrides
Department of Digital Systems, University of Piraeus, Piraeus, Greece
e-mail: meliones@unipi.gr

© Springer Nature Switzerland AG 2019   207
G. A. Tsihrintzis et al. (eds.), *Machine Learning Paradigms*,
Learning and Analytics in Intelligent Systems 1,
https://doi.org/10.1007/978-3-030-15628-2_7

diction of changes in stock prices (as well as indices etc.). The need to use intelligent trading systems which assist in predicting the stock prices values and decision making regarding investment strategies, is ever-increasing. Such portfolio optimization is nowadays the focus of joint efforts of investment theory and machine learning.

The fast variation of stock prices due to the inherent nature of the financial sector and the effect of unknown political and social factors (e.g. elections, rumors etc.) necessitates the close monitoring of market evolution and decision making using automated intelligent trading systems. A smart investor will predict a stock price and act appropriately. Although the investor experience cannot be replaced, an accurate prediction algorithm is always a useful tool that may yield high profits. The last phrase contains the keyword "accuracy", emphasizing a direct relationship between prediction performance and the generated profit.

The accuracy of prediction and trading systems is often questioned by a segment of the scientific community, as only a few research studies are published. Nevertheless, this work aims at developing and validating an automated stock price motion prediction system using machine learning technology. The major task of the proposed work is the design and implementation of a stock price change prediction algorithm. The proposed system will analyze a dataset of technical analysis indicators. The dataset will be properly modified to be valid input to the machine learning algorithm. The machine learning component relies on the support vector machines (SVM) algorithm, which is further optimized using an adaptive boost algorithm. The system backend has been implemented on a PaaS cloud computing environment. User access to the system is through an Android front-end client application. It must be noted that the emphasis of the client application is not on the user interface (UI), which is out of the scope of this work.

7.2 Technical Analysis Synopsis

Technical analysts claim it is possible to predict the future prices in a stock market through appropriate analysis of a stock price chart. In other words, prediction of a stock price is feasible based on past prices without considering other financial variables and data concerning enterprises and sectors of the economy, unlike what fundamental analysis claims. Unlike many academics who are skeptic against technical analysis, a lot of keen professionals keep using it as an indivisible part of their analysis [1, 2]. The authors in [3] ascertain that most modern survey reports demonstrate that technical analysis can lead to successful predictions of stock price motion. Both technical and fundamental analysts do not embrace the random walk theory which claims that the variance in a stock price is caused in a random way, thus rejecting the underlying effective market theory. Although stockbroking information shows up in a random way, market reaction is not instant: smart investors will react first, followed by the great mass of smaller investors [4].

Technical analysis tools/indicators are mathematical calculations involving a stock price and/or a stock trade volume. They can provide indications of the start of a new bullish or bearish stock price trend, its speed and momentum, maturity and finish, and possible reversal. In the ensuing, the major technical analysis indicators which were used in the present study are briefly introduced. A detailed description of the basic and prevailing technical analysis indicators, explained in most books on technical analysis, is out of the scope of this paper.

Simple Moving Average—SMA: An arithmetic moving average calculated by adding recent closing prices and then dividing that by the number of time periods in the calculation average. Short-term averages can act as levels of support when the price experiences a pullback.

Weighted Moving Average—WMA: It is calculated by multiplying the given price by its associated weighting and totaling the values. The denominator of the WMA is the sum of the number of price periods as a triangular number.

Exponential Moving Average—EMA: Exponentially weighted moving averages react more significantly to recent price changes than a simple moving average, which applies an equal weight to all observations in the period.

Moving Average Crossover: A popular use for moving averages is to develop simple trading systems based on moving average crossovers. A trading system using two moving averages would give a buy signal when the shorter (faster) moving average advances above the longer (slower) moving average.

Relative Strength: It is a ratio of a stock price performance to a market average (index) performance.

Bollinger Bands: A type of statistical chart characterizing the prices and volatility over time. It is plotted two standard deviations away from a simple moving average. Because standard deviation is a measure of volatility, when the markets become more volatile, the bands widen; during less volatile periods, the bands contract.

Momentum: A speed of movement indicator designed to identify the speed (or strength) of price movement. It compares the most recent closing price to a previous closing price (can be the closing price of any time frame).

Alexander's Filter—ALF: It shows the percentage growth or decline of a stock price. Like other momentum indicators, the market needs to be trending up or down to get the most reliable signals.

Moving Average Convergence/Divergence—MACD: A trend-following momentum indicator that shows the relationship between two moving averages of prices. The MACD is calculated by subtracting the 26-day EMA from the 12-day EMA.

Money Flow Index—MFI: An oscillator that uses both price and volume to measure buying and selling pressure. MFI is also known as volume-weighted RSI (see below).

Price Oscillator: A momentum oscillator that measures the difference between two moving averages as a percentage of the larger moving average.

Commodity Channel Index—CCI: It can be used to identify a new trend or warn of extreme conditions. CCI measures the current price level relative to an average price level over a given period of time.

Relative Strength Index—RSI: A momentum oscillator that measures the speed and change of price movements. RSI is considered overbought when above 70 and oversold when below 30.

Stochastics—K%D: Momentum indicator that shows the location of the close relative to the high-low range over a set number of periods. As a rule, the momentum changes direction before price.

Williams %R: Momentum indicator that measures overbought and oversold levels. It is commonly used to find entry and exit points in the market.

Accumulation–Distribution Index (ADL): A volume based indicator designed to measure underlying supply and demand. It accomplishes this by trying to determine whether traders are actually accumulating (buying) or distributing (selling).

Chaikin Oscillator: It measures the accumulation/distribution line of MACD. It is calculated by subtracting a 10-day EMA of the accumulation/distribution line from a three-day EMA of the accumulation/distribution line. This highlights the momentum implied by the accumulation/distribution line.

Price Rate of Change—ROC: It measures the percent change in price from one period to the next.

7.3 Machine Learning Component

A comparison between several Machine Learning algorithms (such as Support Vector Machines—SVM, neural networks etc.) applied in the field of financial prediction [5] highlights the C-SVM algorithm enhanced with Adaptive Boost as the best performing algorithm in this study. The comparison refers to a certain performance metric, usually termed as accuracy, defined as the percentage of correct predictions regarding the problem hypothesis across a specific test data set. Test results in this study demonstrate an accuracy of 64.32% ± 3.99%.

Another research work compares the RNN, SVM and APMA ML algorithms and concludes that SVM demonstrates a superior performance requiring at the same time less training data [6]. Based on these results, we decided to implement the C-SVM algorithm enhanced with adaptive boost as the machine learning component of our system.

7.3.1 SVM Algorithm

SVM training algorithms are often used in classification problems. The concept behind SVM is to create a hyperplane which maximizes the separation distance between samples. They perform well in classification problems with no need to embed problem knowledge, while at the same time their learning capability is independent of feature space dimensions. SVM based classification systems are quite popular nowadays because of their robustness, efficiency and response speed, as well as their ability to create nonlinear decision planes, rendering the solving of many real-world training problems which cannot be handled by linear models computationally feasible.

SVMs transform their input space into a higher dimension feature space which classifies the data set in two groups by a hyperplane maximizing the margin, i.e. the distance between the hyperplane and a few close points called support vectors, based on a decision function. The optimization problem in SVMs is convex, which means there is a single solution. This is an important difference between support vector machines and neural networks, which often converge in local minima.

In our study, we used the more robust Soft Margin SVM algorithm [7], also known as C-SVM, which handles better classification problems against the hard margin SVM and can tolerate the existence of isolated misclassified data using a slack variable. The tuning parameter C effectively controls how much misclassification the algorithm will allow. When C is small, the classifier allows only a small amount of misclassification. The support vector classifier will have low bias but may not generalize well and have high variance. We may overfit the training data if our tuning parameter is too small. If C is large, the number of misclassifications allowed has been increased. This classifier would generalize better but may have a high amount of bias. When the tuning parameter is zero, there can be no misclassification and we have the hard margin classifier. Table 1 summarizes the Short Margin SVM algorithm.

In the first part of the algorithm, a unique vector a that maximizes function W(a) is calculated taking into account the existing constraints. The parameter C, which is included in these calculations, limits the values of a_i elements. In the second part of the algorithm, based on the calculated values for a_i, variables λ, b and γ are determined. More specifically, to calculate b and γ, we need to select a certain value for i and j, as well as, by extension, for x_i and x_j such that the corresponding a_i and a_j multiplied by the y_i and y_j tags respectively, are greater than $-C$ and less than 0 regarding the former, and greater than 0 and less than C regarding the latter. The kernel function k of objects x and x′ is the inner product of the projection of the two objects in the higher dimension space. Efficient separability of non-linear regions is possible using non-linear kernel functions (e.g. polynomial, radial, sigmoid).

7.3.2   Adaptive Boost Algorithm

The efficiency of a model which is produced by a ML algorithm is determined by the size and quality of the training set as well as the proper fit of the algorithm, which, in general, are both hard to determine. An alternative approach aims at increasing the reliability of a ML system through the combined use of two or more models towards the successful classification of an unknown problem instance. This methodology of

Table 1 Summary of Short Margin SVM algorithm

Input	Training set $S = \{(x_1, y_1), \ldots, (x_l, y_l)\}$, $\delta > 0$, $C \in [1/l, \infty]$. y_i tag value is 1 or -1 if x_i belongs to the 1st or 2nd class respectively
Optimization process under constraints	Find the solution α^* to the optimization problem: $W(\alpha) = -\sum_{i,j=1}^{l} a_i a_j y_i y_j k(x_i, x_j)$ $\sum_{i=1}^{l} y_i a_i = 0$, $\sum_{i=1}^{l} a_i = 1$ and $0 \leq \alpha_i \leq C$, $i = 1, \ldots, l$
1	$\lambda^* = 1/2(\sum_{i,j=1}^{l} a_i^* a_j^* y_i y_j k(x_i, x_j))^{1/2}$
2	Select i, j such that $-C < a_i^* y_i < 0 < a_j^* y_j < C$, $i = 1, \ldots, l$
3	$b^* = -\lambda^*(\sum_{k=1}^{l} a_k^* y_k k(x_k, x_j) + \sum_{k=1}^{l} a_k^* y_k k(x_k, x_j))$
4	$\gamma^* = 2\lambda^* \sum_{k=1}^{l} a_k^* y_k k(x_k, x_j) + b^*$
5	$f(.) = \text{sgn}(\sum_{j=1}^{l} a_j^* y_j k(x_j, .) + b^*)$
6	$w = \sum_{j=1}^{l} y_j a_j^* \varphi(x_j)$
Output	Weight vector w, optimal solution a^*, margin γ^*, classification function f based on hyperplane

machine learning, which is termed Meta-Learning, was successfully applied in our system.

More specifically, the adaptive boosting algorithm proposed in [8] was used in our system. The training algorithm is initiated assigning equal weights on all training samples and creates a classifier for the training data set. In the ensuing, it recalculates the sample weights based on the classifier performance. The weights of successfully classified samples are decreased, and vice versa for the misclassified samples. This creates a data set including easy and hard to classify samples with smaller and larger weights respectively. In the next iteration, the classifier is created and sample weights are recalculated, focusing on the successful classification of hard cases. Following the weights update task, weights are normalized such that their total sum remains the same. After all iterations, the final classification is made. Table 2 illustrates the Ada-Boost algorithm in pseudo-code.

D_t is a distribution, an assignment of a probability to each example. L should be sensitive to the distribution. If not, we can take a bootstrap sample using the distribution. If $e_t \geq 0.5$ then T is set to $t - 1$ and the algorithm breaks from loop. After line 7, sum of D_{t+1} for misclassified examples equals the sum of correctly classified examples, which is 1/2. This forces h_{t+1} to be different from h_t. Classification is by plurality voting with base classifier h_t getting $\ln\beta_t$ votes.

Table 2 Summary of Ada-Boost algorithm

Adaboost(examples, L, T)

1. $D_1(i) \leftarrow 1/m$ for each of m examples
2. for $t \leftarrow 1$ to T
3. $h_t \leftarrow$ apply L to examples using Distribution D_t
4. $e_t \leftarrow$ sum of $D_t(i)$ for examples misclassified by h_t
5. $\beta_t \leftarrow (1-e_t)/e_t$
6. $D_{t+1}(i) \leftarrow D_t(i)\beta_t$ if h_t misclassifies example i
        $D_{t+1}(i) \leftarrow D_t(i)$ otherwise
7.   Normalize D_{t+1} so it sums to 1
8. return $h_1, …, h_t$ and $\beta_1, …, \beta_T$

7.4   System Implementation

7.4.1   System Structure

The application backend has been developed on Google's PaaS (Platform as a Service) cloud computing environment, known as Google App Engine (GEA) [9]. The obvious choice of using a compute cloud to execute efficiently the training process was not seamlessly available at the time of system development, therefore the GAE alternative, even if not being the best choice, was implemented and evaluated. GAE is a platform for hosting and developing internet applications on datacenters administered by Google. Using GAE is free of charge up to a certain amount of used computational resources. Exceeding this limit imposes additional cost for computation and storage use, depending on user needs. It is worth mentioning that the leading position of Google in the internet search engines market is not directly replicated in the cloud computing sector, a proof of the fierce competition between the leading companies in this market. A problem of the adopted PaaS model is that the developed application relies on the specific framework, complicating the porting of the application to another cloud service provider if necessary. The parameterization of the backend code differs between free of charge and commercial (payment required) applications. Basic configurations refer to the resources supporting the applications and the functional features of the activated instances (see Table 3). Furthermore, the way the application is scaled is important, as it prescribes how the system reacts to required resource fluctuations. It must be noted that GAE is not suggested for high performance computations, as it enforces a 10-min time limitation for the completion of a computational task. Our application required the use of B8 instance class during testing. GAE supports an object-oriented database (Datastore API) and file storage system (Blobstore API).

Table 3 Various GAE instance types with their characteristics and costs	Instance class	Memory limit (MB)	CPU limit	Cost per hour per instance
	B1	128	600 MHz	$0.05
	B2	256	1.2 GHz	$0.10
	E4	512	2.4 GHz	$0.20
	B4_1G	1024	2.4 GHz	$0.30
	B8	1024	4.8 GHz	$0.40
	F1	128	600 MHz	$0.05
	F2	256	1.2 GHz	$0.10
	F4	512	2.4 GHz	$0.20
	F4_1G	1024	2.4 GHz	$0.30

As already stated, the massive stock data that we used in our implementation refer to a heavyweight stock of ASE's FTSE20, namely the share of the National Bank of Greece (ETE:ATH). Figures 1 and 2 illustrate the system details involving the training and prediction processes.

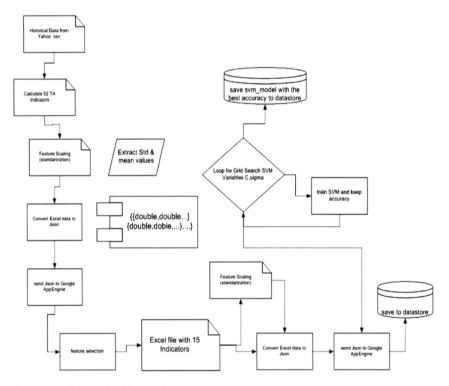

Fig. 1 Flowchart of classifier training process

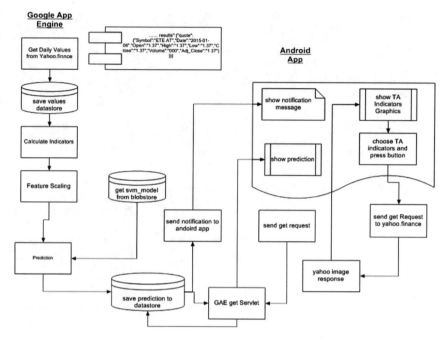

Fig. 2 Flowchart of prediction process

Figure 1 illustrates the classifier training workflow ranging from getting the csv file including the basic values required for the calculation of technical analysis values, to the storing of the SVM model in Google AppEngine in txt format. The SVM model represents the knowledge acquired by the system through the training process, which allows making predictions on unknown data. The intermediate steps comprise specific data pre-processing tasks, such as feature selection, feature scaling, training as well as storing specific data in GAE's datastore so that they can be used by subsequent tasks.

Figure 2 illustrates the steps of the prediction process which implements the daily predictions in a monthly timeframe. These involve all the tasks from the automatic acquisition of the daily share values which are used to calculate the new technical analysis indicator values referring to the last date, being the input of the prediction algorithm, through the prediction of the stock motion in a monthly timeframe, to the notification of the prediction in the front-end application.

Figure 3 summarizes the aforementioned tasks, which are analyzed in detail in the ensuing.

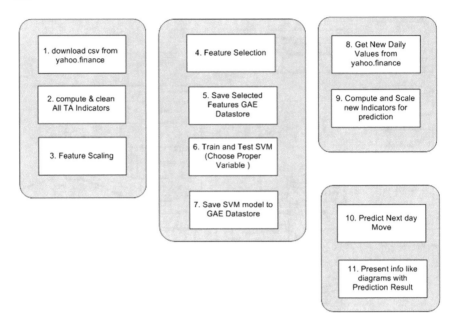

Fig. 3 System tasks

7.4.2 *Training Data Set*

As already stated in Sect. 3, the definitive prerequisite for efficiency of any machine learning algorithm is the credibility and appropriate processing of the training data set.

7.4.2.1 Source of Training Data

The data set used for training and validation of the machine learning algorithm is available through the Yahoo Finance webpage [10], as illustrated in Fig. 4.

Data was acquired in csv format. Initially, it included 3500 values spanning historic daily values since the year 2000. Using an excel sheet the values were classified in columns titled Date, Open, High, Low, Close, Volume and Adjusted Close, as illustrated in Fig. 5.

Since the formed data set of 3500 values may not be adequate for efficient training of a machine learning algorithm, the training data set was enriched with similar data of the remaining bank stocks of the FTSE20 index (Alpha Bank, Eurobank and Piraeus Bank), assuming all the bank sector blue chips demonstrate a similar stockbroking behavior. Along this data set, several technical analysis indicators were selected and used to train the proposed system. These technical analysis indicators are illustrated

Fig. 4 Yahoo Finance data source

A	B	C	D	E	F	G	H
	Date	Open	High	Low	Close	Volume	
	03-01-00	73.86	76.34	73.77	74.71	49300	
	04-01-00	73.98	73.98	70.99	71.81	60800	
	05-01-00	66.78	70.4	66.78	69.01	44000	
	06-01-00	68.98	68.98	68.98	68.98	0	
	07-01-00	69.65	70.25	68.22	68.71	37800	
	10-01-00	69.11	69.68	68.41	68.65	20800	
	11-01-00	68.61	68.61	67.1	67.43	14000	
	12-01-00	66.88	72.6	62.92	66.79	20400	
	13-01-00	67.02	67.92	66.47	67.02	19700	
	14-01-00	67.98	69.49	67.71	68.77	35200	
	17-01-00	70.16	70.85	69.04	69.37	42100	
	18-01-00	69.49	69.49	68.13	68.61	18200	
	19-01-00	65.93	67.65	65.72	65.99	46800	
	20-01-00	66.48	67.05	65.42	65.66	34500	
	21-01-00	65.81	66.41	65.54	65.84	15300	
	24-01-00	66.05	67.32	65.15	66.2	40000	
	25-01-00	65.03	66.3	64.37	65.39	26100	
	26-01-00	66.22	66.37	64.35	64.35	29100	

Fig. 5 Excel sheet with initial data for ETE share

Table 4 Calculated technical analysis indicators

Open	EMA 50	MOMENTUM 10	%K 7	EMA 26	Slope 20 High
High	ALF 5	MFI 14	%K 8	MACD	
Low	ALF 20	Upper Bollinger	%K 9	MACD EMA 9	
Close	ALF 50	Lower Bollinger	%K 10	CCI	
Volume	RSI 14	%R 6	PROC 12	High price average 2	
SMA 5	MOMENTUM 3	%R 7	PROC 13	Low price average 2	
SMA 20	MOMENTUM 4	%R 8	PROC 14	Slope 3 High	
SMA 50	MOMENTUM 5	%R 9	PROC 15	Slope 4 High	
EMA 5	MOMENTUM 8	%R 10	WCL	Slope 5 High	
EMA 20	MOMENTUM 9	%K 6	EMA 12	Slope 10 High	

in Table 4. Both the calculation formulas and value calculations of the indicators were implemented in an excel sheet.

7.4.2.2 Data Set Pre-processing

The training data set was properly processed to allow a better performance of the machine learning algorithm. The data set pre-processing comprises the following tasks.

Standardization: The calculation of the real values of the technical analysis indicators is followed by a feature scaling process. In our case the standardization scaling method has been applied using the Matlab environment. The mean value is subtracted from each value in the data set and the new value is subsequently divided by the standard deviation value.

Tag pre-processing: In this task, we calculated and tagged the fall or rise of the share price within a calendar month (22 working days of the stock market), with a 0 or 1 value respectively.

Imputation of missing values: This task refers to the handling of missing values from the data set. For various reasons, a data set may miss certain values that could be null/blank, no available or coded with certain placeholders. Such missing values may hinder a machine learning algorithm. One approach in resolving this issue is to drop from the data set the entire rows or columns including the missing values, thus dropping/losing possibly valuable data, though incomplete. A better handling is to estimate the missing values through the rest values, using the average or median value, or the value listed more often in the row or column of the missing value.

7.4.3　Selection of Machine Learning Algorithm and Implementation

We have compared three machine learning algorithms using the Matlab environment. Our modeling categorizes our problem as a classification one and therefore the algorithms tested are Logistic Regression, ANN (Artificial Neural Networks) and SVM (Support Vector Machines). Table 5 depicts the comparative performance of these algorithms.

The tested algorithms originate from Stanford University Coursera Machine Learning course [11] and might not be optimized for performance. Test results are in accordance with the literature review in Sect. 3. Therefore, we adopted the C-SVM algorithm enhanced with Adaptive Boost. Our implementation was supported by Java ML [12], Java Encog [13] and Java TA [14] frameworks/libraries for machine learning and technical analysis indicators respectively.

7.4.3.1　Feature Selection

The outcome of the data set pre-processing is a new data set which, unless their values have been properly modified, demonstrates the exact same characteristics with the initial data set. Next task is feature selection, alternatively known as feature subset selection (FSS). This is a method for selecting a subset of the feature set towards a more efficient model. The term "efficient" should be comprehended as follows: During field application of a machine learning algorithm, feature quantity is often large enough, including features that either lack useful information or are interrelated between each other. This task can reduce the number of features enhancing both the

Table 5 Performance of machine learning algorithms

Method	Accuracy on the training set	Accuracy on the test set	Remarks
Logistic regression	61.464596	59.833333	
Neural networks	65.739813	63.366667	Input layer: 12, hidden layer: 30, iterations: 1000
SVM	66.34	64.233333	C:1, σ:1

accuracy and timely performance of the model. A widespread selection method is the association of each feature with the model output.

In our case, the aforementioned selection was implemented using the Java ML library. Another Java application was developed using Java XL in order to read data from the excel file and convert it to a JSONObject, which is sent to the GAE (cloud environment) computations backend through a HTTP Post Request. Figure 6 depicts the JSONObject format including the indicator values sent to the backend.

A GAE Servlet receiving the call (made at URL/featureselection) calculates the association score of each feature against the tag output using the JavaML library, depicted in Fig. 7.

Out of the 51 variables and technical analysis indicators, the 15 entities with the greater score were selected, which are depicted in Table 6.

{"data": {[CLOSE[0],CLOSE[1],…CLOSE[N]},{OPEN[0],OPEN[1],…,OPEN[N]]…..[
SLOPE[0],SLOPE[1],…,SLOPE[N]],[OUT[0],OUT[1]….,OUT[N]]}

Fig. 6 Format of JSONObject sent to the backend, where N is the number of data items

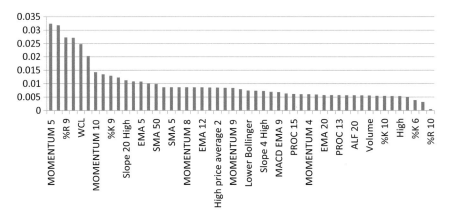

Fig. 7 Association score of technical analysis indicators

Table 6 Outcome of the feature selection process

Low	MOMENTUM 10	PROC 12
SMA 50	%R 7	PROC 14
EMA 5	%R 9	WCL
EMA 50	%K 8	Low price average 2
MOMENTUM 5	%K 9	Slope 20 High

7.4.3.2   Selection of C and γ Variables

In the ensuing, proper values for the C and γ parameters of the SVM algorithm must be selected. When implementing a SVM classifier a large scale greedy search is recommended for finding C and γ values that optimize the accuracy of the algorithm. In other words, selecting proper values for C and γ requires the iteration of the training process of the SVM algorithm under several parameter combinations, each time evaluating the performance of the algorithm and identifying the pair which has a better impact on the performance of the algorithm. Table 7 depicts this process for several C and γ pairs, considering the entire set of technical analysis indicators instead of those extracted by the feature selection process.

7.4.3.3   SVM Model Training

Following the selection of proper values for C and γ variables, the 15 features extracted through the feature selection process are conveyed to the GAE API. Our data set, comprising 13,500 samples, is split into two parts, a Training and Testing

Table 7 Accuracy performance of SVM algorithm as per C and γ values

C	γ	Accuracy on test dataset
0.1	0.001	0.6165
0.1	0.01	0.645
0.1	0.1	0.637
0.1	1	0.665
0.1	10	0.665
1	0.001	0.6395
1	0.01	0.6345
1	0.1	0.6345
1	*1*	*0.672*
1	10	0.6665
10	0.001	0.655
10	0.01	0.623
10	0.1	0.6545
10	1	0.6715
10	10	0.667
100	0.001	0.645
100	0.01	0.6275
100	0.1	0.6575
100	1	0.6685
100	10	0.667

data set, comprising 11,500 and 2000 values respectively, aiming at testing the accuracy of the algorithm on an unknown data set. The training process is executed on the backend and then the calculated SVM model is stored in GAE's Blobstore in txt format. Additionally, the mean and standard deviation values used in the standardization process are stored in the backend, as they are required for scaling the unknown data extracted daily through the Yahoo Finance API and used to produce daily predictions. Besides, the latest dataset values must be stored in the backend datastore, as they are required in the calculations of certain technical analysis indicators used for estimating the daily variables which determine the monthly prediction.

7.4.3.4   Daily Indicators Calculations

Having stored all the required values in the application backend, we can now proceed in the next step, the initial goal, of daily stock motion predictions. The required daily values are retrieved via a Get HTTPRequest call to the Yahoo Finance API URL, depicted in Fig. 8.

In the above URL we have highlighted the share symbol, and the start and end dates, which must be the same for retrieval of daily values. The reply to this request is a JSONObject including the asked values, depicted in Fig. 9.

The asked values are parsed through the JSONObject and stored in the datastore. Using these values and those of the previous days, which are already stored, the technical analysis indicators are calculated for the current day. These are consequently scaled via the standardization process (using the stored mean and std values) to be

https://query.yahooapis.com/v1/public/yql?q=select%20*%20from%20yahoo.finance.historical
data%20where%20symbol%20%3D%20%22ETE.AT%22%20and%20startDate%20%3D%20%2220
15-01-06%22%20and%20endDate%20%3D%20%222015-01-
06%22&format=json&diagnostics=true&env=store%3A%2F%2Fdatatables.org%2Falltableswithk
eys&callback=

Fig. 8   URL Yahoo API for daily values

...... results":{"quote":{"Symbol":"ETE.AT","Date":"2015-01-
06","Open":"1.37","High":"1.37","Low":"1.37","Close":"1.37","Volume":"000","Adj_Cl
ose":"1.37"}}}}

Fig. 9   Reply from Yahoo Finance API

acceptable inputs to the machine learning algorithm. Finally, using the SVM model stored in the blobstore and other required data retrieved from the database, the stock motion prediction for the next month can be reported.

All tasks described in this section must take place every day. Therefore, instead of calling the corresponding Servlets via using a timer, we employed the cron jobs provided by GAE, illustrated in Fig. 10. The executed tasks are summarized below:

- Cron Job—every day at 19.00
- Java—TA lib
- Feature scaling
- Prediction every day—Cron job at 19.30
- Load SVM model from Blobstore for prediction.

7.4.4 Android Client Application

As obvious, it is not feasible to develop the whole application on a mobile device, due to the limited computational power of such devices. Therefore, the mobile client application is only a visualization front-end which reports the last prediction along with the corresponding date. This is implemented through a HTTP Get request to the backend, as illustrated in Fig. 10. Furthermore, the mobile client, besides reporting the latest stock motion prediction, enables the user to display technical analysis

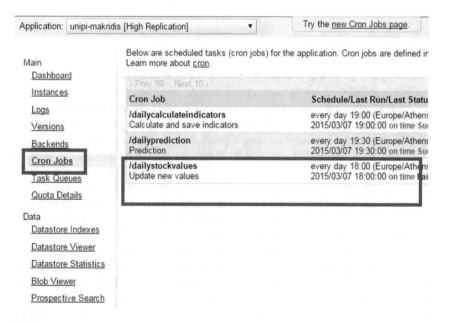

Fig. 10 GAE cron jobs

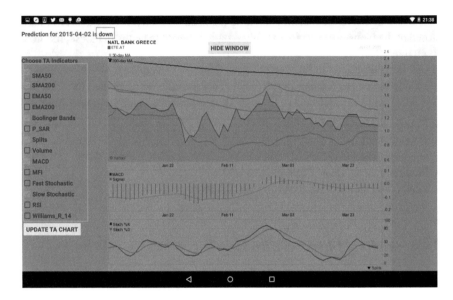

Fig. 11 Android App layout

diagrams selecting from the most popular technical analysis indicators, as illustrated in Fig. 11. Again, the corresponding implementation relies on using the Yahoo API.

7.5   System Evaluation

This work demonstrates an adequate stock price motion prediction capability of the applied machine learning methodology. Machine learning algorithms, when trained with big datasets, seem promising for predicting with a higher accuracy the motion of stock prices. However, any profits that result from a prediction system achieving 65–70% accuracy could well be zeroed because of stock trading costs.

In this work, the training dataset is populated with ASE FTSE20 bank stock prices from the year 2000 up to the year 2015. The prediction accuracy achieved by the proposed system on the training and test set is 70.8% and 64.5% respectively. The execution times on GAE of the dominating system tasks are approximately reported—Training SVM model (including saving): 7500 ms; Feature selection: 5000 ms; Daily TA index calculations: 2000 ms.

Because of the large long-term fall and disdain of the Greek ASE stock market index, and especially its bank sector, due to the severe long-lasting Greek financial crisis, we further tested our system with the famous US S&P 500 index. The training dataset included daily prices (Open, High, Low, Close) starting from the year 2008 up to the year 2016. These values were used to calculate several technical analysis indicators and their buy and sell signals. In this test, besides the technical analysis indicators, we also considered and included a few fundamental analysis indicators in the feature set. The created dataset is depicted in Table 8.

In the ensuing, we evaluated the score of the indicators of the feature set, which is illustrated in Fig. 12.

The profit generated by the applied buy-and-sell strategy based on the system predictions was evaluated using Metastock for backtesting, illustrated in Fig. 13. The predictions on the test set regarding rise or fall of the S&P 500 signaled corresponding buy and sell trading actions (without stop loss or profit take) closing the corresponding positions after 30 days. Trading expenses have not been taken into account in the calculations.

Table 8 S&P 500 use case dataset

High_Open	Low_Open	Close_Open	1_day_diff_close	2_day_diff_close	3_day_diff_close	Technical analysis
EMA 5	EMA 20	EMA 50	RSI 14	SMA 14RSI	MACD	
MACD EMA 9	SMA 5	SMA 20	SMA 50	%K	%D	
20–80%K	20–80%D	K-D	Cross EMA5-EMA20	Cross EMA20-EMA50	RSI	Buy-sell signals
MACD cross	MACD	20 High-low	3 High-low	Cross SMA5-SMA20	Cross SMA20-SMA50	
Sum						
Dividend S&P	Inflation rate	cpi	Long term unemployment	Unemployment	p/e ratio S&P	Fundamental analysis indicators
Earnings S&P	Treasury rate					

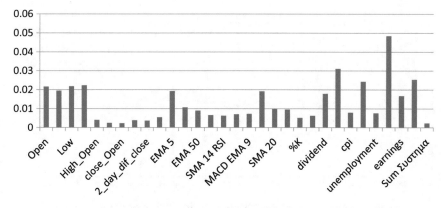

Fig. 12 S&P 500 use case feature selection

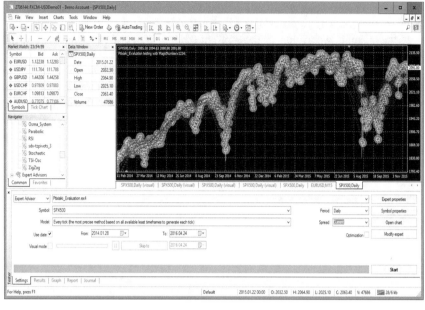

Period:	2014.01.28 – 2016.04.24				
Initial deposit:	10000.00				
Total net profit:	15850.44	Gross profit:	23490.07	Gross loss:	-7639.63
Total trades:	500	Short positions won:	159 (66.04%)	Long positions won:	341 (72.43%)
		Profit trades:	352 (70.40%)	Loss trades:	148 (29.60%)
		Largest profit trade:	249.15	Largest loss trade:	-227.55
		Average profit trade:	66.73	Average loss trade:	-51.62
	Max consecutive	wins (profit in money):	32 (3301.91)	losses (loss in money):	16 (-664.70)
	Average	consecutive wins:	8	consecutive losses:	3

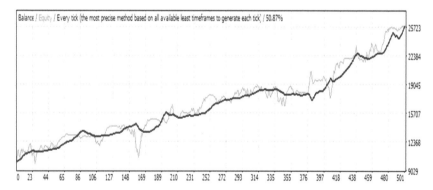

Fig. 13 Profit calculation in Metastock for buy-sell strategy following system predictions

7.6 Conclusions and Future Work

Our attempt to come up with a functional and reliable stock price motion prediction system relying on machine learning technology, which is a difficult problem to cope with, even for big financial organizations involving abundant development teams of keen professionals, clearly demonstrates that such systems can be helpful in the decision making and trading regarding stock market investments. Apart from the fact that the innovation behind the theoretical approach is not new, the implementation and deployment of such systems is pretty much challenging and interesting. Furthermore, it should be noted that only a few machine learning projects have been implemented in Java.

Both in theoretical and technical aspects, future work can evolve the presented system using fundamental economy indicators besides technical analysis (as demonstrated in our S&P 500 trial), as well as taking advantage of relevant big data from social networks, such as Twitter, and news, using text and emotional analysis tools, leading to more accurate predictions. Another improvement can be expected through the automatic analysis of the time series of technical analysis indicators. Expert analysts do not apply technical analysis based only on individual indicator values, but on certain formations in their time series diagrams. Time series analysis comprises three basic functions: (1) Application of distance metrics to determine the similarity between different time series diagrams; (2) Analysis of the structure of the time series to determine and even classify its behavior; (3) Use of time series diagrams for prediction of future values. An automated image recognition component for time series analysis is expected to greatly enhance the performance of machine learning based stock price motion prediction systems.

References

1. J.D. Schwager. *Getting Started in Technical Analysis* (Wiley, 1999). ISBN 9780471295426
2. M.P. Taylor, H. Allen, The use of technical analysis in the foreign exchange market. J. Int. Money Financ. **11**(3), 304–314 (1992)
3. C.H. Park, S.H. Irwin, What do we know about the profitability of technical analysis? J. Econ. Surv. **21**(4), 786–826 (2007)
4. E.F. Fama, Random walks in stock market prices. Financ. Anal. J. **21**(5), 55–59 (1965)
5. D. Brown, R. Jennings, On technical analysis. The Rev. Financ. Stud. **2**(4), 527–551 (1989)
6. U. Thissen, R. van Brakel, A.P. de Weijer, W.J. Melssen, L.M.C. Buydens, Using support vector machines for time series prediction. Chemometr. Intell. Lab. Syst. **69**(1–2), 35–49 (2003)
7. C. Cortes, V.N. Vapnik, Support-vector networks. Mach. Learn. **20**(3), 273–297 (1995)
8. Y. Freund, R.E. Schapire, Experiments with a new boosting algorithm, in *Machine Learning: Proceedings of the 13th International Conference* (1996), pp. 148–156
9. Google App Engine, https://cloud.google.com/appengine/
10. Yahoo Finance, http://finance.yahoo.com/
11. Machine Learning, Coursera, Stanford University, https://www.coursera.org/learn/ machine-learning

12. Java Machine Learning Library, http://java-ml.sourceforge.net/
13. Encog Machine Learning Framework, https://www.heatonresearch.com/encog/
14. A Java Library for Technical Analysis, https://github.com/ta4j/ta4j

Chapter 8
Airport Data Analysis Using Common Statistical Methods and Knowledge-Based Techniques

Athanasios T. Ballis and John A. Paravantis

Abstract Airports are the nodal points of the aviation network and a valuable source of information as all airplane movements, together with their associated passenger and cargo carried volumes, are systematically recorded. These data can be used for various types of analyses, depending on the scope of the specific work. This chapter outlines an analysis performed for 28 Greek island airports aiming to reveal key aspects of demand (seasonality and connectivity) and of supply (practical capacity of airports, characteristics of airplanes and air services). The spectrum of issues investigated include: (a) The international passenger connectivity matrix between origin countries and Greek islands, which allowed for the identification of tourist preferences as well as for the detection of missing links (lack of direct flights from certain countries to certain islands). These missing links can be interpreted as market gaps or, put differently, as market opportunities. (b) The monthly, weekly and daily airplane arrival patterns, at airport level, allowing for the classification of the Greek airports according to their air-side operating patterns as well as for the assessment of their practical capacity. (c) The airplane types as well as their itineraries which revealed that three-leg trips can be used in order to implement long haul international flights to islands having short runways.

8.1 Introduction

Airports are the nodal points of the aviation network offering the necessary installation and personnel for flight guidance, landing, takeoff, parking, refueling and maintenance of the aircraft as well as for the processing of passenger and freight

A. T. Ballis (✉)
Department of Transportation Planning and Engineering, National Technical University of Athens, Athens, Greece
e-mail: abal@central.ntua.gr

J. A. Paravantis
Department of International and European Studies, University of Piraeus, Piraeus, Greece
e-mail: jparav@unipi.gr

© Springer Nature Switzerland AG 2019
G. A. Tsihrintzis et al. (eds.), *Machine Learning Paradigms*,
Learning and Analytics in Intelligent Systems 1,
https://doi.org/10.1007/978-3-030-15628-2_8

flows, acting as an interface with all landside related modes and operations. Since the 1970s, air transport has seen major organizational changes caused by its deregulation and liberalization. The joint effects of these changes include the development of airline hubs, the concentration of flight connections by major airlines, the development of airline alliances, as well as the development of low-cost carriers (LCCs) and secondary airports [43, 65]. Airports are major centers of economic activity in regions and urban areas. In the modern economy, they represent nodes of economic activity that contribute significantly to gross domestic product [49]. Furthermore, airport development has catalytic effect on other sectors like domestic and international tourism.

Airports are also a valuable source of information, as all aircraft movements together with their associated passengers and cargo carried volumes, are systematically recorded. The scope of this chapter is to present aspects of the variety of information that may be mined from an airport data set through common statistics but also through knowledge-based mining techniques. An extensive data set, the 2016 database containing detailed information for the 28 airports located in the Greek islands, was used to obtain quantitative information about the demand and supply characteristics, at an airport level and at an airport-system level. Two aspects of the demand side, seasonality (the way the demand is distributed along the year) and connectivity (accessibility to countries being the main sources of tourists to the Greek islands) were investigated. Similarly, three aspects were analyzed from the supply side: the practical capacity of the airports, the operational characteristics of the airplane types taking-off and landing to these airports and the air services (direct and three-leg trips). These issues are analytically presented in the following paragraphs. The remainder of the text is structured as follows. Section 2 hosts the literature review on the subject. Section 3 includes a short description of the case study investigated together with the outcomes of the analysis. Finally, the conclusions are outlined in the fourth section of the chapter.

8.2 Literature Review

International air transport evokes considerable interest of many people: those associated with airlines; airports and communities seeking new air services; users of air transport; air carrier labor; aircraft manufacturers; certain international organizations; people involved in aviation financing, tourism development and trade; people in academia and the communications media; and, at times, members of the general public as well [36]. Key issues that are likely to characterize the 21st century air transportation system include: providing security for air travelers; continuing to improve the environmental consequences of air transportation; dealing with substantial airport capacity shortages; promoting stability in the airline industry; and incorporating new technologies in air traffic control and aircraft design [49]. The literature review of all these subjects is huge, therefore this section focuses only on published works pertinent to issues addressed in this chapter, namely demand seasonality, air connectivity, airport/runway capacity, aircraft performance and aviation and tourism.

Demand seasonality is typical in the aviation world, and the year is separated in three seasons: low, medium and high. Reichard [57] investigated the demand seasonality in air transportation and presented three methods for computing seasonal indices. Shugan and Radas [59] discussed contemporary methodologies for modelling seasonality as well as the implications of those models of seasonality on new product introductions. Snepenger et al. [62] reported on the seasonality of demand faced by tourism businesses in Alaska. Koenig-Lewis and Bischoff [41] performed a state-of-the-art review on seasonality that contains 83 references on the subject. Ćorluka [21] investigated the analysis of seasonality (inbound tourist demand in the region of Adriatic Croatia) and identified differences between the seasonal structures of countries, the nationality of generating markets, and the organization of tourist arrivals. Kraft and Havlíková [43] performed an analysis of seasonality in the offer of flights in Central Europe during 2014, considering the different positions and functions of the airports within the air transport system. More specifically they analyzed the flights offered in ten Central European airports to show the different patterns of spatial and temporal organization of air transport. Vergori [71] investigated the patterns of seasonality and tourism demand forecasting, and analyzed the monthly tourist overnight stays in four European countries (Austria, Finland, Portugal and the Netherlands) for the period from 1990 to 2014. Finally, Duro [25] investigated the seasonality of hotel demand in main Spanish provinces and provided evidence that the domestic market cannot offset the seasonality of the foreign one.

Many publications in scientific journals or professional tourism magazines, especially from less developed countries, have highlighted the strong interrelationship between air connectivity and the successful national or regional tourism growth [23, 44, 50, 68, 69] or the need to develop air connectivity as a necessary tool for the development of tourism [47] or even as an excuse (the lack of sufficient air connectivity) to justify the low level of tourism in certain areas of their countries [17]. Among them, Iñiguez et al. [37] used complex network theory techniques to investigate the implications of air connectivity for tourism. Bannò and Redondi [9] examined the issue of air connectivity (introduction of new routes) and foreign direct investments and provided evidence for the contribution of transport infrastructures to local development. Cetin et al. [19] investigated the impact of direct flights on tourist volume, and focused on the case of Turkish airlines using data for the number of international arrivals to the country before and after air connectivity with certain countries was established. Findings confirmed that direct flights to/from generating regions have a significant impact on the number of arrivals to destinations. Akça [3] compared the connectivity competitiveness for a number of selected airline hubs using computational and sensitivity analysis for the investigation of connectivity measured under different parameters and practical scenarios of real life. Ram et al. [56] dealt with the issue of air transport competitiveness and connectivity in the Caribbean. PwC [55] investigated the air transport and current and potential air connectivity gaps in the CESE (Central, Eastern and South-Eastern Europe) region.

Strong demand seasonality has a negative effect as an airport must provide disproportionally larger installations that are fully used for a few months, being underutilized during the remaining period (in relation to another airport with similar annual traffic that operates under a smooth traffic demand pattern). Airports consist of various subsystems, among them the runways and taxiways subsystem, the aircraft stands, and the terminal building. The airport capacity is defined by the capacity of its weaker subsystem. In a typical airport, runway capacity is definitely the less flexible subsystem: it cannot be increased above its limits as this will jeopardize the safety of airplanes. In contrast, the capacity of a terminal building can be increased above planning levels at the expense of reduced level of service (e.g. delays, overcrowding of lounges, etc.) for the passengers. The relevant issues are discussed in references addressing the various design aspects of airport such as Horonjeff et al. [34] and Ashford et al. [4] as well as in scientific publications.

The interest in the subjects of airport/runway design and airport system planning is not new. Office of Technology Assessment [51] performed an extended analysis for the airport system development in the United States. Trani et al. [67] developed a computer simulation and optimization model to estimate runway design characteristics. The main output of the program was the estimation of the weighted average runway occupancy time for a user defined aircraft population. Widener [72] dealt with the measurement of airport efficiency to minimize airport delays. Tilana [66] confirmed that the overall capacity of an airport is determined by the airfield, particularly the runway system. It was concluded that the lack of adequate airport capacity cannot be tackled by providing additional runway capacity, but also requires innovative ways to use the existing facilities more efficiently. Bubalo [16] worked on the determinants of an airport productivity benchmark.

A significant number of published works concern the issue of airport runway optimization. Runway capacity is influenced by a range of factors such as: runway layout and resulting dependencies, as correlation may exist between runways, e.g. converging and diverging runways, intersecting runways, mixed mode, parallel runways and ground restrictions; regulations such as separation regulations such as wake turbulence, noise and emissions' regulations; local conditions such as visibility and wind direction; fleet mix; taxi and apron systems; and other factors including human factors, system state and performance [22]. Balakrishnan and Chandran [5] investigated the aircraft landing problem for a fixed set of aircraft (the static case) to maximize the runway throughput (equivalently minimizing the landing time of the last aircraft or makespan). Chandran and Balakrishnan [20] introduced an algorithm to compute the trade-off curve between the robustness (reliability of a schedule) and throughput. Bäuerle et al. [11] studied the problem as a special queuing system with the incoming aircraft as customers of different types and separation time between aircraft as the service time. Based on this assumption, they investigated several routing heuristic strategies in respect to the average delay for assigning aircraft to two runways. Potts et al. [54] performed a literature review on airport runway optimization.

Ackert [1] drafted a report allowing for an in-depth understanding of the issues of aircraft payload in relation to flight range. Aircraft operational weights were studied, and their cause and effect relationship on payload-range performance were investigated. In particular, payload-range analysis involved the examination of maximum takeoff weights. The report provided evidence of how multi-range versions of an aircraft type may help an airline achieve both operational flexibility and cost advantages. Martinez-Val et al. [45] had previously used the same material (payload-range diagrams) to show the evolution of jet airliners. Kaplas [39] studied the airport processes at Helsinki Airport in terms of capacity and utilization. He created a model of the airport as a system of processes including the capacity of the runway process. Yutko [73] explored approaches to represent aircraft fuel efficiency performance for the purpose of a commercial aircraft certification standard. He found that most of the available published data usually concern peak performance, while most operations do not normally take place at these maximum levels. Skorupski and Wierzbicki [60] focused on the minimization of runway occupancy time and presented an analysis which showed that the way of braking during the landing roll has an essential impact on runway throughout and thus on airport capacity. Similarly, Kolos-Lakatos and Hansman [42] investigated the same subject of the influence of runway occupancy by analyzing the wake vortex separation requirements on runway throughput. A number of authors, among them Vancroonenburg [70] and Brandt [15], investigated airport/runway capacity issues (aircraft loading and takeoff constraints, flight range, etc.) to optimize cases of air cargo selection and weight balancing. Finally, Schaar and Sherry [58] addressed the issue of benchmarking airport performance efficiency, an issue of academic and practical interest.

The relationship between aviation/airports and tourism has been the subject of many published works. One category addressed the positive impacts of aviation liberalization on tourism [18, 24, 26, 29, 31, 53, 64]. Marrocu and Paci [46] attempted to identify determinants of tourism flows. They stressed the fact that the various tourism destinations are extremely differentiated while, at the same time, consumers are characterized by heterogeneous preferences. Therefore, they confirmed the findings of previous works [52, 61] according to which the diverse features of the leisure products play a key role in determining the flows of different tourists to different destinations. Similar are the results of a questionnaire-based research work [38] that focused on the South Aegean islands in Greece. It was identified that different European nationalities prefer different islands as well as that the main purposes of tourists visiting these islands were "*sun and sea*", "*visiting friends and relatives*", "*discovering landscape and nature*", "*culture and religion*" and "*city break*".

Bieger and Wittmer [12] also worked on the interrelationship between air transport and tourism and identified perspectives and challenges for destinations, airlines and governments. Similarly, Forsyth [30] stressed the benefits of tourism benefits in relation to aviation policy. Khadaroo and Seetanah [40] investigated the role of transport infrastructure in international tourism development. Henderson [33] outlined the case of Indonesia to indicate the connection between transport and tourism destination development. Barros [10] focused on island resorts, and analyzed the travel behavior of tourists holidaying on the island of Madeira, based on their air transport type (scheduled airline flight, low-cost flight or charter flight). Duval [26] performed an extended literature review on air transportation as it is related to tourism, focusing on three salient issues faced by international commercial air transport and their resulting implications on global tourist flows (the wider aeropolitical environment, developments in airline operations, and the issue of carbon pricing on aviation) and confirmed that tourism and air transport are intricately linked.

8.3   Airport Data Analysis

Data analysis is the process of evaluating data using analytical and statistical tools to discover useful information, and help business decision making [63]. Data Analysis requires the obtaining of data (see Sect. 3.1). A critical issue in most data analyses is the scope of the analysis (in Sect. 3.2 the orientation of the current analysis is clearly outlined) that defines the orientation and tools to be used as well as the expected results (starting with Sect. 3.3).

8.3.1   Data Collection and Cleansing

EC Regulation 1358/2003 [27], implementing Regulation 437/2003 of the European Parliament and of the Council on statistical returns in respect of the carriage of passengers, freight and mail by air, mentions three datasets: the Flight Stage dataset (called A1), the On-Flight Origin/Destination dataset (called B1), and the Airport dataset (called C1). Dataset A1 contains periodic flight stage data registered for airport-to-airport routes, and broken down by arrivals/departures, scheduled/non-scheduled, passenger service/all-freight and mail service, airline information, aircraft type and available seats. The values provided concern passengers on board, freight and mail on board, commercial air flights as well as available passenger seats. Dataset B1 contains periodic on flight origin/destination data registered for airport-to-airport routes, and broken down by arrivals/departures, scheduled/non-scheduled, passenger service/all-freight and mail service and airline information. The values provided concern carried passengers and loaded or unloaded freight and mail.

It is noted that the required information in datasets A1 and B1 appear to have many common records (replicated information), yet this is not always true due to the different rules to be followed for the transit and transfer passengers, on-flight and at the airport (see below). Dataset C1 contains periodic (at least annual) airport data registered for declaring airports, and broken down by airline information. The values provided concern total passengers carried, total direct transit passengers, total freight and mail loaded or unloaded, total commercial aircraft movements and aircraft movements. This dataset must contain at least annual data. The data collection process and the data suppliers differ from country to county in the EU family, but usually data are partially provided by airports and (directly or indirectly) partially from airline companies or their handling agents.

Completing the required Data Collection Forms is not always an easy task especially when, in addition to passengers travelling directly to the destination airport, direct transit or transfer passengers are on-board. Direct transit passengers are the passengers who, after a short stop, continue their journey on the same aircraft on a flight having the same flight number as the flight on which they have arrived. These passengers must be counted only once. Transfer or indirect transit passengers are passengers arriving and departing on a different aircraft within 24 h, or on the same aircraft bearing different flight numbers. They are counted twice: once upon arrival and once on departure. Take for example a New York-London-Paris flight that transports passengers traveling from New York to London, passengers from New York to Paris, and passengers from London to Paris. The information filled in the Data Collection Forms in Paris airport depends on whether (a) a different airplane will be used in the London-Paris route, or (b) the leg London-Paris will be performed with the same airplane arriving from New York. In the latter case, A1 and B1 data sets differ significantly in information content. The understanding of the above data structure allows retrieving useful information concerning effective operational forms used by certain airlines that implement triage services (for reasons elsewhere in this chapter). Various details must also be taken into consideration, for instance: passenger baggage is excluded from freight; diplomatic bags are also excluded from freight, infants in arms are included as passengers, etc. There are even more complicated situations: on some flights with intermediate stops, the flight number changes at an airport to designate the change between an inbound and outbound flight. An example is a flight from Barcelona to Hamburg, where the flight continues to Frankfurt before returning to Barcelona. When passengers for an intermediate destination continue their journey on the same aircraft in such circumstances, they should be counted as direct transit passengers. Another complex case is that of passengers who change aircraft due to technical problems, but continue on a flight with the same flight number (they should be counted as direct transit passengers). A special case concerns flights with more than one intermediate airport, like the Rhodes-Kos-Kalymnos-Leros-Astypalaia flight itinerary that connects these five Greek islands all year round.

A case study of airport data in Greece is used as a case study, where various aviation-related statistical information was collected and analyzed, including the 2016 airport database (700,000 records of all national and international aircraft movements in all airports of the country). The analysis was performed in various stages. The first stage was the detection, correction and elaboration of database content. This is due to the fact that the complexity in completing the *Data Collection Forms* (as described in the introduction) may justify erroneous recordings especially in small airports where the relevant information is recorded manually. As a result there are cases where data from the origin airport do not match the data recorded in the destination airport, thus a data cleansing process may be required to identify and somehow correct conflicting records (e.g. the automatic data entry in big airports was considered more accurate than the manual data entry performed in small airports). A data elaboration is also needed to identify and filter non-repeated flights like training flights, aircraft flight tests, fire-fighting flights, occasional visits of Learjets for tourist purposes, etc. Depending on the scope of the analysis these outliers may be excluded, like in the case of origin-destination matrices reflecting the standard connectivity among airports in a region.

8.3.2 Case Study Description and Scope of Current Analysis

Greece is a small country of about 11 million inhabitants, with 15% of them living in its more than 220 inhabited islands. The multi-island character of the country requires an extended network of maritime and airport connections (see Fig. 1). However, only a small percentage of the maritime routes serving the islands are direct. Most islands are served through round trips (e.g. Piraeus, Paros, Naxos, Ios, Santorini) with long travelling times especially for the islands in the last leg of the trip, due to the procedures (ship docking, passenger embarkation/disembarkation, sailing) that take place in all previous ports/islands. For time-sensitive passengers, air transportation is a mandatory option.

Tourism is another reason justifying the need of this extended transport network. Tourism is a major pillar of the Greek economy and the main source of income for many local communities, especially in the islands. Tourism growth depends strongly on the provision of adequate transport infrastructures. This, in combination with the preference of tour operators for direct flights to vacation destinations gave rise to airport development for various Greek areas. Figure 1 presents the structure of the domestic Greek airport network. It contains a main airport hub in Athens (where

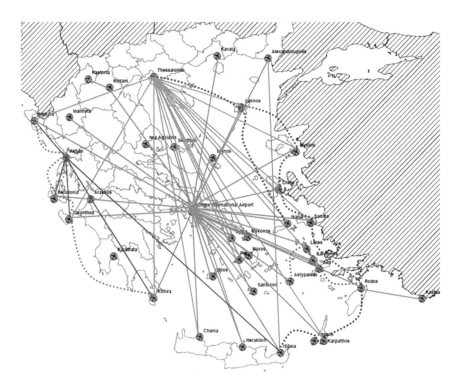

Fig. 1 Structure of the airport system in Greece

the vast majority of other national airports are connected), a secondary hub in Thessaloniki as well as three local hubs in Rhodes, Heraklion and Corfu. The airport of Athens has scheduled air service connections with 50 countries (93 cities) in Western and Eastern Europe, Middle East, USA, Africa and the Rest of Asia. Thessaloniki Airport is also providing international connections to 31 countries (64 cities). Moreover, during the summer period, there are direct charter flights connecting 39 countries (164 cities) to the Greek islands. However, most island airports are small, and due to their strong dependency on national and international tourism, exhibit strong seasonality (see Sect. 3.3) and need to be subsidized.

The economic recession of 2008 had a dramatic effect on Greece, revealing the country's fundamental structural weaknesses in its economy. With more than 300 billion Euros in public debt (gigantic for a country of 10 million natives) the State has assigned the leverage of its private property, part of which is the main transport infrastructure of the country, to the Hellenic Republic Asset Development Fund (HRADF) which proceeded in an international tender procedure for the management and operation of 14 Greek regional airports. This tender, resulted in the declaration of Fraport AG-Slentel Ltd. consortium as the preferred bidder in November of 2014. The 40-year concession agreement signed and ratified by the Parliament in November 2017, concerned the concession of the right to upgrade, maintain, manage and operate two clusters each one having seven regional airports. Cluster A is composed of the airports of Thessaloniki, Corfu, Kefalonia, Aktio, Zakynthos, Kavala and Chania, while Cluster B includes the airports of Rhodes, Samos, Skiathos, Mytilini, Mykonos, Santorini and Kos.

The 23 remaining regional Greek airports (among them the island airports of Astypalaia, Chios, Ikaria, Kalymnos, Karpathos, Kasos, Kastelorizo, Kithira, Leros, Limnos, Milos, Naxos, Paros, Sitia, Skyros and Syros) were left to the State to operate, through the Hellenic Civil Aviation Authority (HCAA airports). The passenger traffic of the above 23 airports accounts for only 2.9% of the total air passenger traffic in Greece. This situation may raise questions about the profitability, potential and necessity of these airports.

The profitability of smaller airports is a major issue for Europe. Under the current market conditions the profitability prospects of commercially run airports also remain highly dependent on the level of throughput, with airports having less than one million passengers per annum typically struggling to cover their operating costs. Consequently, the vast majority of regional airports are subsidized by public authorities on a regular basis (EC 2014/C 99/03 Guidelines on State aid to airports and airlines).

This chapter is part of a wider effort that aims to identify the potential of the above island airports, and to propose projects and initiatives towards their further development. The first part of this effort was devoted in the formation of a multiple regression model that attempted to identify the relationship between the number of tourists in Aegean islands and specific explanatory variables (area, population, accommodation, etc.). The model was used to identify less-developed islands with good potential to increase their tourist product [8]. This chapter constitutes the second part of the above research effort, which describes the analysis of airport data with the aim of identifying certain decisive characteristics of transport demand and supply (concerning the weekly and daily traffic patterns), and missing tourist markets and airplane types used by tour operators in relation to the landing and takeoff capacities of the island airports.

8.3.3 Demand Seasonality

The air passenger traffic in Greece is characterized by strong seasonality mainly due to the tourism demand in the summer period: international tourist flows are concentrated in the period from May to September while domestic tourist flows have a stronger concentration around August. Seasonality affects the major airports of Athens and Thessaloniki, but mostly the peripheral airports, especially the ones that serve tourist destination, like in the islands (see Fig. 2). As the passenger traffic at the island airports is severely reduced during the winter period, making the associated transport services non-profitable, there is a need for State subsidies in order to operate Public Service Obligation services and ensure an adequate level of service.

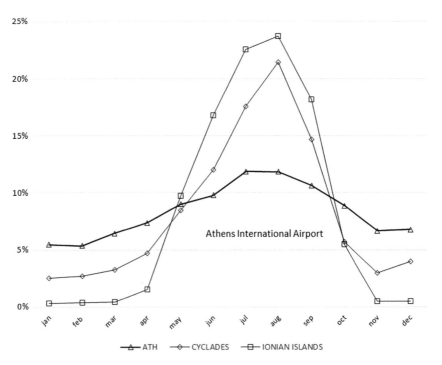

Fig. 2 Demand seasonality patterns of Ionian and Cyclades islands in relation to Athens International Airport (authors' work based on 2016 data of HCAA)

8.3.4 International Passenger Connectivity Matrix

Table 1 presents the Origin-Destination matrix of international passengers arriving in the regions of Ionian Islands, South Aegean (Cyclades and Dodecanese) and North Aegean directly via air. These direct flights are the main transport way that European tourists use to visit the above Greek islands. Less than 10% of the international tourists arriving by air, first fly to Athens for a short stay, and then travel to Greek islands using domestic flights or shipping lines through the port of Piraeus. These tourists have the option of visiting Greek islands that do not have an airport.

The connectivity matrix allows for the identification of tourist preferences in relation to Greek islands and, in another reading, the gaps (missing markets) for certain islands. For instance, an overview of Table 1 reveals that European tourists are not proportionally spread among all islands. There is a strong preference of the British for Corfu (Kerkyra), Rhodes (Rodos) and Zakynthos, while Germans prefer Kos to Zakynthos. Such a variance is understandable and is related to the specific preferences of each nationality (as the relevant literature indicates) as well as to the deals and traditional alliances of tour operators with accommodation owners. Another explanatory reason may be the air distance of the country from Greece in relation to the airplane types used, the runway length of the airport (later on in this chapter). Nevertheless, certain extreme cases like the case of Karpathos where no direct flights from the UK and France are recorded may be interpreted as market gaps or, put differently, as market opportunities.

8.3.5 Weekly and Daily Airport Operating Patterns

Weekly and hourly airport operating patterns give a detailed view of an airport's utilization, peak demand and capacity margins. Although traffic patterns are affected by the yearly seasonality, weekly and hourly airplane traffic patterns are typically stable during one season, for a number of reasons. Traditional carriers operate according to a fixed timetable for each period, while even charter flights bringing tourists to a vacation destination follow a stable weekly pattern for organizational reasons. Also, flights organized by tour operators fly on a specific day every week to a specific airport; upon landing, the "new" vacationers are disembarked and the "old" vacationers (those brought by tour operators the previous week) are embarked. This way the flight is highly utilized on the go and return trips, and the accommodation (e.g. hotels with long-term contracts with the tour operator) is used all days of the week. The airport data analysis of the weekly and hourly patterns of the Greek island airports allowed for their assignment into three classes, namely (1) less busy airports, (2) busy airports, and (3) congested airports. Figure 3 presents one representative case from each of these classes.

Figure 3 also indicates the growth pattern of airport capacity: when airplane traffic is small, it is concentrated on certain preferable days of the week and hours of the day.

Table 1 Origin-Destination matrix of international passengers arriving by air in the Greek islands (Authors' work based on 2016 data of HCAA)

From	To		Ionian Islands				Cyclades		Dodecanese			North Aegean		
Airport name			Corfu	Zakynthos	Kefalonia	Kithira	Santorini	Mykonos	Rhodes	Kos	Karpathos	Samos	Mytilini	Limnos
Code			CFU	ZTH	EFL	KIT	JTR	JMK	RHO	KGS	AOK	SMI	MJT	LXS
Runway (m)			2,373	2,228	2,436	1,461	2,125	1,902	3,305	2,390	2,399	2,044	2,406	3,016
United Kingdom	1,504,284	26.13%	482,484	306,358	157,157		108,484	62,605	365,615			5,018	9,109	7,454
Germany	777,454	13.50%	178,984	9,062			18,884	11,252	329,723	205,630	4,214	16,486	3,219	
Italy	572,778	9.95%	70,742	33,815	23,662		106,615	133,395	115,987	62,816	20,798	4,864	84	
France	217,373	3.78%	62,549	2,494	19		22,968	16,014	91,355	21,966		8		
Poland	338,673	5.88%	101,963	89,060	9,655		2,805	9	88,654	46,527				
Russia	222,191	3.86%	40,911	8,890			195	4	149,353	22,736				102
Netherlands	312,937	5.44%	47,644	55,300	14,993	3,074	7,717	3,202	66,234	75,126	11,871	15,584	11,145	1,047
Sweden	255,656	4.44%	14,646	17,653	6,571		10,310	2,005	155,791	24,247	9,938	14,495		
Israel	209,123	3.63%	11,044	13,024	18		3,425	10,176	131,064	35,407	4,957		8	
Norway	140,389	2.44%	11,528	6,590	1,162		13,481	138	81,933	10,135	4,702	9,582	1,138	
Denmark	150,944	2.62%	13,121	13,211	2,497	743	9,684		80,210	15,459	3,004	9,155	3,860	
Austria	173,677	3.02%	19,627	19,469	6,245		30,399	11,489	50,834	17,194	9,194	6,935	2,291	
Belgium	135,527	2.35%	35,443	8,516	2		5,058	3,207	51,599	29,818		1,884		
Switzerland	153,341	2.66%	21,142	6,882	3		16,774	22,801	43,819	39,161		2,751	8	
Czech Republic	157,056	2.73%	33,869	31,709	4,114		3,974	8	58,241	17,799	3,556	2,679		1,107

(continued)

Table 1 (continued)

From	To		Ionian Islands				Cyclades		Dodecanese			North Aegean		
Airport name			Corfu	Zakynthos	Kefalonia	Kithira	Santorini	Mykonos	Rhodes	Kos	Karpathos	Samos	Mytilini	Limnos
Code			CFU	ZTH	EFL	KIT	JTR	JMK	RHO	KGS	AOK	SMI	MJT	LXS
Runway (m)			2,373	2,228	2,436	1,461	2,125	1,902	3,305	2,390	2,399	2,044	2,406	3,016
Finland	99,869	1.73%	4,272	4,134			4,275		66,446	13,232	2,690	4,820		
Romania	39,716	0.69%	6,943	13,539			3,665	1,080	14,489					
Slovakia	42,256	0.73%	14,326	3,642				3	22,399	1,249	617		20	
Hungary	47,212	0.82%	17,861	12,548	2,490			2	11,741		2,568	2		
Cyprus	12,286	0.21%	980	611			1,766	4,459	3,654	416	22	378		
Lithuania	25,017	0.43%	3,576	870			286		12,157	8,073			55	
Serbia	28,703	0.50%	8,319	4,073	2,691		856	1	10,082	0	559	1,340		782
Ireland	19,369	0.34%	9,479	3,306					3,285	3,299				
Spain	30,615	0.53%	2,320	2	8		11,403	8,552	8,324	6				
Luxembourg	18,917	0.33%	4,490		5				9,214	5,208				
Other countries	71,605	1.24%	14,747	7,592	2,839		6,793	11,442	23,352	816	2,632	849	543	0
Total per island	5,756,968		1,233,010	672,350	234,131	3,817	389,817	301,844	2,045,555	656,320	81,322	96,830	31,480	10,492
Total per region				2,143,308	37.2%		691,661	12.0%		2,783,197	48.3%		138,802	2.4%

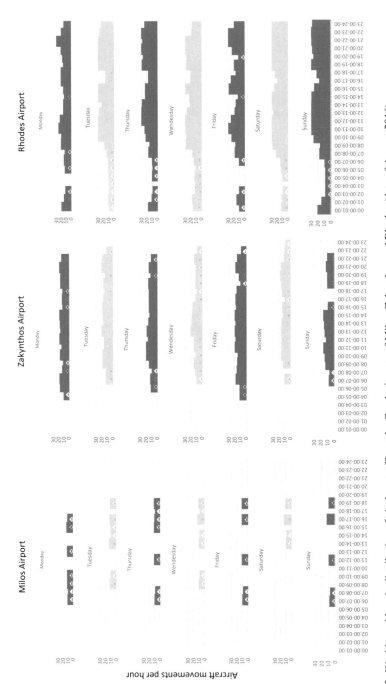

Fig. 3 Weekly and hourly distribution of airplane traffic at the Greek airports of Milos, Zakynthos and Rhodes (data of August 2016)

As traffic increases, the idle periods are gradually populated. When the capacity of the morning and afternoon periods is saturated, the night period may also be used (at least for attractive tourist destinations with high demand) given that the noise disturbance remains at acceptable levels. The term acceptable is interpreted according to the land uses in the surrounding area (e.g. industrial uses are less sensitive than residential ones) or even the tolerance of the local communities to accept some annoyance if that is to be compensated by additional income from tourist activities.

In Europe, night flying restrictions called curfews have been established. According to Greek legislation [32] the quiet hours are defined as follows: From April to September (coinciding to high and medium season) the quiet hours are 15:00–17:30 and 23:00–07:00. For the rest of the year, the quiet hours are between 15:30–17:30 and 22:00–07:30 (coinciding to low season). During the high and low season, the time periods from 23:00–07:00 and 22:00–07:30 are referred to as night, respectively.

The data of hourly patterns of the airplanes can also be used for the calculation of the level of service in the airport terminal facilities. Ballis et al. [6] used this technique for the simulation of airport terminal facilities in the airport of Heraklion on the island of Crete. It was observed that the vast majority of international tourists arrived at the airport terminal building in groups, by using chartered buses, two hours before their flight departure (while the few remaining passengers arrived randomly later on). Knowing the airplane departure plan, a simulation model generated the passenger flows arriving at the terminal building from the landside. These simulated flows were then processed through check in, passport control, X-ray and departure lounge facilities to calculate the level of service offered. The same approach was used for the simulation of the terminal facilities in the Greek airports of Kavala and Alexandroupolis [7].

8.3.6 Airplane Types and Associated Runway Length Requirements

The astonishing growth and evolution of air transport could not have been achieved without the technological advancements that make the airplanes safe and efficient. The new-generation aircraft manufactured by Boeing and Airbus are exponentially more sophisticated than their predecessors. Not only are they more fuel efficient and quieter, but they also improve the aircraft's range and payload capabilities [13]. According to IATA [35], improvements in the materials and aerodynamics of aircraft along with enhancements in engine technology, make modern aircraft more than twice as fuel efficient as they were 40 years ago. Also, the range of modern aircraft has been increased allowing airlines to expand the size of their networks. Furthermore, airlines can maximize their profit on a particular route by using greater payload capability to transport more passengers and cargo on the same airplane.

Table 2 presents the types of aircraft landings and takeoffs to island airports of the Aegean. The flight range and frequency of use (in percentage) of each airplane type in the airports of the Aegean region are also presented. The data show the dominance of Boeing 737-800 and Airbus A320 which account for 55% of all aircraft, with the A320 having a slightly larger share than B737-800. The Airbus A320 family consists of short to medium-range, narrow-body, commercial passenger twin-engine jet airliners manufactured by Airbus. The baseline A320 has given rise to a family of aircraft (A318, A319, A320, the A321 and more recent models like A320neo, etc.) which share a common design, but with passenger capacity ranging from 107 seats on the A318 (typical configuration, 132 seats maximum in economy configuration), to 140–160 seats on the A319neo, 165–194 seats on the A320neo, and 206–244 seats in A321neo [2].

Boeing 737 is a short to medium-range twinjet narrow-body airliner developed and manufactured by Boeing Commercial Airplanes in the United States. The 737 entered airline service in 1968 and has been developed into a family of ten passenger models with capacities from 85 to 215 passengers [48]. The Boeing 737 twin-engine airliner was considered the best-selling jetliner in the world. The first generation of 737s, first flew in 1967 and production was finished in 2000. The design of the next-generation family of 737s began in 1991. The version 737-700 Next-Generation has 149 seats in all-economy configuration, similarly to the Airbus A319. The Boeing 737-800 is a stretched version of the 737-700. The airplane has 189 passenger seats in a one-class layout. It competes with the Airbus A320. All the above aircraft require a considerable length of runway for their takeoff. Therefore it is not possible to operate on islands with runways of ICAO Category 1, 2 and 3, which are served by aircraft with lower capacity and aircraft range such as AT72 and DH8D (that serve mainly ICAO Category 3 airports) as well as the smaller AT43 and DH8A that are able to takeoff from all the island airports of the South Aegean.

The Bombardier Dash 8 (DH8D) is a series of twin-engine, medium-range, turboprop airliners which featured extreme short takeoff and landing performance. The aircraft has been delivered in four series, having from 39 up to 78 passenger seats [14]. The ATR 72 is also a twin-engine turboprop, short-haul regional airliner produced by the aircraft manufacturer ATR. The airplane has a standard seating configuration of 72–78 passenger seats in a single-class arrangement.

Table 2 also shows how different airplane types require different minimum runway lengths. The calculation process is depicted in Fig. 4 and is based on the Payload/Range and Takeoff Field Length Charts, which are unique for each airplane type. It can be used to determine either the payload/range of an aircraft or the runway length required for takeoff at a given weight.

Firstly, the Operating Empty Weight (OEW) of the airplane is calculated. OEW is composed of the total weight of airplane body, the crew, the passengers and their handbags and luggage plus the mail and cargo loaded in the airplane. Then, the weight of the fuel is added. Calculations are carried out according to specific airplane consumption and the flight range (millage), in order to establish the Brake Release Gross Weight. This gives the required takeoff runway length at Sea Level on the Standard Day temperature (59 °F/15 °C) and for a zero slope runway. This runway length must be modified according to the slope, altitude and average temperature of the specific airport to provide the balanced field length required for the takeoff

Table 2 Aircraft movements (landings and takeoffs) in the Greek islands according to aircraft type

Aircraft ICAO code		DH8A	AT43	DH8D	AT72	B463	JS41	RJ85	B712	A319	E190	B733	A321	B737	A320	B734	B752	B738	B763	B753
Max passenger seats		37	48	78	66	112	30	112	110	153	106	149	220	149	179	170	239	189	290	289
Airport name	Runway length																			
Chania	3,348		10	20	6	30		12		484	88	476	1,394	548	5,040	36	18	10,040	236	110
Rhodes	3,305	1,866	462	214	36	2	768	48	24	1,156	276	840	2,818	1,020	11,048	332	1,286	11,616	402	254
Limnos	3,016	24	720	762	240		552			14		16	66	48	118			56		
Skiros	3,002	36	74	322			382					2		14	6					
Heraklion	2,714		2,074	6	108	16	814	292	178	1,462	394	940	4,132	1,046	19,804	902	1,108	12,336	438	474
Corfu	2,373	2	288	702	138		466	62	158	964	106	240	1,262	614	6,300	234	646	7,070	110	8
Kefalonia	2,436		24	832	2		462			226	38	122	402	76	666	16	60	998		
Mytilini	2,406	104	1,036	934	196	88	728	2	98	178	4		194	44	1,774		6	218		
Karpathos	2,399	1,456	176	904	30	2	4			128		28	66	36	232	52	28	594		
Kos	2,390	460	386	1,610	116	14	30		10	486	208	162	804	708	3,210	182	338	5,230	136	68
Zakynthos	2,228		136	1,026			436		72	418	82	336	1,066	870	1,962	382	338	2,972	2	78
Samos	2,044	102	698	1,894	250	52	284		32	168		56	2	252	384	12		654		
Santorini	2,125	4	856	706	10	158	16	118	724	592	62	276	1,790	466	4,380	110	176	2,685		
Sitia	2,074	456	590	2	22		658						2		6					
Mikonos	1,902		364	1,102	66	100	6		456	628	240	122	150	468	4,180	24		1,010		
Skiathos	1,628	26	62	900				4	280	54		106	480	314	538	136	256	480		

(continued)

Table 2 (continued)

Aircraft ICAO code		DH8A	AT43	DH8D	AT72	B463	JS41	RJ85	B712	A319	E190	B733	A321	B737	A320	B734	B752	B738	B763	B753
Max passenger seats		37	48	78	66	112	30	112	110	153	106	149	220	149	179	170	239	189	290	289
Chios	1,511	60	929	2,218	334	226	317			2				82				8		
Kithira	1,461	46	94	530	16		216	30						50	12					
Paros	1,400	1,260	148	692																
Ikaria	1,387	42	506	600	112		8													
Syros	1,080	376	244	56																
Kalymnos	1,015	946	146	52																
Leros	1,012	1,332	34																	
Astypalaia	989	656	90																	
Kassos	983	892	92				2													
Naxos	900	1,034	144																	
Kastelorizo	798	492																		
Milos	795	1,348	278																	
Domestic Aircraft movements	13,020	10,644	15,852	1,593	653	6,146	507	432	562	41	220	3,127	119	20,394	224	3	6,442		1
International Aircraft movements	0	17	232	89	35	3	61	1,600	6,398	1,457	3,502	11,501	6,537	39,266	2,194	4,257	49,525	1,323	991
Total Aircraft movements	13,020	10,661	16,084	1,682	688	6,149	568	2,032	6,960	1,498	3,722	14,628	6,656	59,660	2,418	4,260	55,967	1,324	992

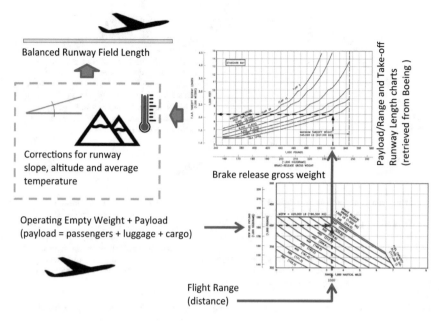

Fig. 4 Variables and process for the calculation of balance takeoff field length

of the airplane. The existence of obstacles along the flight paths around the airport (along with other technicalities) increases the complexity of the calculations. Similar calculations can be performed for airplane landing (the use of the landing charts is similar to that of the takeoff charts) but in general, takeoff is more length-demanding than landing.

In many cases the last (in descending order) populated cells in each column of Table 2 are "suspicious" in the sense that special requirements must be fulfilled so that the airplane can takeoff from these relatively short (in comparison to the usual aircraft's performance) runways. For example, it could be a flight that due to its small payload (fewer passengers) or fuel can takeoff from shorter runways. This could be a plausible explanation to a certain degree, but when the runway is much shorter than anticipated, another explanation should be sought. Following an advanced searching process within the database, it was revealed that there are cases of three-leg trips that may explain the use of certain short runways. The whole transport process is depicted in Fig. 5.

An airplane takes off from Airport A at full load (Operating Empty Weight plus passengers) for the Airport B and C destinations. The runway has sufficient length for the specific airplane type. During its long haul flight, the airplane consumes most of its fuel, thus becomes lighter, and can land in the relatively short runway of Airport B. The passengers for destination B are disembarked, but new passengers (the tourists that have completed their vacation) are embarked, so there is no weight gain. Then, the airplane takes off, but it is lighter as it carries the minimum fuel required to fly

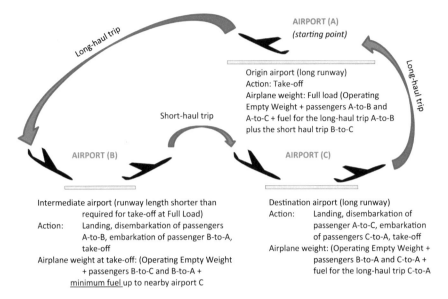

Fig. 5 Three-leg trip allowing for an aircraft to visit an airport with a relatively short runway

up to the nearby Airport C. In this airport, the passengers from Airport A to C are disembarked and the passengers from C to A are embarked. The airplane is refueled with enough kerosene in order to perform its long haul trip to Airport C, but now the weight is not a problem, as the runway length of Airport C is adequate for the takeoff of the airplane at full load. One such flight is, for example, the Amsterdam-Kithira-Kalamata-Amsterdam flight with a Boeing 737, which can takeoff from the 1,461 m runway of the Kithira airport having only the fuel required until its landing at the nearby airport of Kalamata, where it can be refueled and takeoff again at full load from the 2,703 m runway of the airport.

There are certain disadvantages to this technique: the airplane pays the full fee for an additional airport stop, although only a part of its passengers are disembarked at the intermediate airport. Fortunately, transit passengers are not charged at all, but must accept the inconvenience of an intermediate stop. In fact, further research work on this subject is warranted, and it could be an in-depth investigation of three-leg flights with: interviews with managers of airline companies and tour operators; detailed cost analysis; identification of initiatives and measures that make this option more attractive for the tourist market actors, etc. If feasible, this technique could allow for big airplanes to use airports having medium size runways (1,600–1,800 m), therefore some airports could avoid a runway elongation, especially when significant cost, land availability and environmental barriers exist.

8.4   Conclusions

Airport data are a valuable source of information as all aircraft movements, together with their associated passengers and cargo carried volumes, are systematically recorded. These data can be used for various types of analysis, depending on the scope of a specific work.

This chapter outlined an analysis performed using the 2016 data of 28 Greek island airports. With Greece being a small country of about 11 million inhabitants, a 15% of whom live on more than 220 inhabited islands, the multi-island character of the country requires an extended network of maritime and airport connections. Most islands are served through round trips, having long travelling times especially for the islands in the last leg of the trip. For time-sensitive passengers, air transportation is a mandatory option. Tourism is another reason justifying the need of this extended island airport network. Tourism is the main source of income for many island communities and its growth strongly depends on the provision of adequate transport infrastructures. That, in combination with the preference of tour operators for direct flights to vacation destinations, provided justification for airport development in various areas of the country. However, most island airports are small, and due to their strong dependency on national and international tourism suffer from strong seasonality and need to be subsidized.

The analysis attempted to reveal key aspects of demand (seasonality and connectivity) and supply (practical capacity of airports, characteristics of airplanes and air services). The air passenger traffic in Greece is characterized by strong seasonality mainly due to the seasonality of tourism demand in the summer period: international tourist flows are concentrated in the period from May to September, while domestic tourist flows have a stronger concentration around August. Air connectivity is another critical issue: many publications in research journals and professional tourism magazines, especially from less developed countries, have highlighted: (1) the strong interrelationship between air connectivity and successful national or regional tourism growth; (2) the need to develop air connectivity as a necessary tool for the development of tourism; and (3) the lack of sufficient air connectivity as an excuse to justify the low level of tourism in certain areas of their countries. An international passenger connectivity matrix from each tourist source (country) to each Greek island was developed, which allowed for the identification of tourist preferences as well as for the detection of missing links (lack of direct flights from certain countries to certain islands) which may be interpreted as market gaps or regarded as market opportunities.

Furthermore, the monthly, weekly and daily airplane arrival patterns were obtained, at airport level, which allowed for the classification of the Greek airports according to their air-side operating patterns (less busy airports, busy airports, and congested airports that operate even during the night hours). It also demonstrated the way demand evolves: when traffic is small, it is concentrated on certain preferred days of the week and hours of the day. As traffic increases, the idle periods are gradually populated. When the capacity of the morning and afternoon periods is saturated, the night period may also be used (at least for attractive tourist destinations with high demand) given that the noise disturbance remains at acceptable levels (according to the land uses in the surrounding area and the tolerance of the local communities).

Finally the airplane types used and their itineraries were analyzed. The data has shown the dominance of Boeing 737-800 (189 passenger seats in a one-class layout) and Airbus A320 (with a similar passenger seat capacity) as they account for 55% of all aircraft, with the A320 having a slightly larger share than B737-800. These aircraft require a considerable length of runway for their takeoff, and therefore it is not possible to operate on islands with runways of ICAO Category 1, 2 and 3; these are served by aircraft with lower capacity and aircraft range, such as AT72 and DH8D (serving mainly ICAO Category 3 airports) as well as the smaller AT43 and DH8A types (that seem to be able to takeoff from all the island airports of South Aegean).

Further analysis identified that three-leg trips can be used to land and takeoff airplanes from shorter than required runways (since few fuel was left, thus airplane is lighter and requires shorter runway length), which otherwise would not be possible. Additional research is required for this latter finding, as this technique may allow big airplanes to use airports having medium size runways, therefore an airport can avoid its runway elongation when significant cost, land availability and environmental barriers exist.

References

1. S. Ackert, Aircraft payload-range analysis for financiers. Aircraft Monitor, Technical Report, April 2013
2. Airbus, Commercial Aircrafts, www.airbus.com. Accessed 19 July 2018
3. Z. Akça, Comparative analysis with a new hub connectivity measure considering revenue and passenger demand. J. Air Transp. Manag. **67**, 34–35 (2018)
4. N. Ashford, S. Mumayiz, P. Wright, *Airport Engineering: Planning, Design, and Development of 21st-Century Airports*, 4th edn. (Wiley, 2011)
5. H. Balakrishnan, B. Chandran, Scheduling aircraft landings under constrained position shifting, in *AIAA Guidance, Navigation, and Control Conference and Exhibit, Guidance, Navigation, and Control and Co-located Conferences*, https://doi.org/10.2514/6.2006-6320. Accessed 21–24 Aug 2006
6. A. Ballis, A. Stathopoulos, E. Sfakianaki, Sizing of processing and holding air terminal facilities for charter passengers using simulation tools. Int. J. Transp. Manag. **1**(2), 101–113 (2002)
7. A. Ballis, Simulation of airport terminal facilities in the Greek airports of Kavala and Alexandroupolis. Oper. Res. Int. J **2**(3), 391–406 (2002)
8. A. Ballis, J. Paravantis, T. Moschovou, Assessing the tourism potential of the Greek islands of South Aegean, in *9th International Conference on Information, Intelligence, Systems, and Applications (IISA 2018)*, 23–25 July, Zakynthos, Greece (2018)

9. M. Bannò, R. Redondi, Air connectivity and foreign direct investments: economic effects of the introduction of new routes. Eur. Transp. Res. Rev. **6**(4), 355–363 (2014)
10. G.V. Barros, Transportation choice and tourists' behaviour. Tour. Econ. **18**(3), 519–531 (2012)
11. N. Bäuerle, O. Engelhardt-Funke, M. Kolonko, On the waiting time of arriving aircrafts and the capacity or airports with one or two runways. Eur. J. Oper. Res. **177**, 1180–1196 (2007)
12. T. Bieger, A. Wittmer, Air transport and tourism—perspectives and challenges for destinations, airlines and governments. J. Air Transp. Manag. **12**(1), 40–46 (2006)
13. Boeing, Current Market Outlook. Boeing Corporation, 2014–2033 (2015)
14. Bombardier website, Bombardier Celebrates Aviation Milestones: 1,000th Dash 8/Q-Series Turboprop and 400th Global Business Jet, https://www.bombardier.com/en/media/newsList. Accessed 25 July 2018
15. F. Brandt, The Air Cargo Load Planning Problem, Ph.D. Thesis, Karlsruher Institut für Technologie (KIT), http://dx.doi.org/10.5445/IR/1000075507 (2017)
16. B. Bubalo, Determinants of an Airport Productivity Benchmark. Aerlines Magazine. Online Journal on Air Transport for Aviation Business Students and Professionals, October 10, 2010, https://aerlinesmagazine.wordpress.com
17. Business Line, With better air connectivity, North–East rising on tourism map, https://www.thehindubusinessline.com/news/with-better-air-connectivity-north-east-rising-on-tourism-map/article24187253.ece (2018)
18. D. Campisi, R. Costa, P. Mancuso, The effects of low cost airlines growth in Italy. Mod. Econ. **1**, 59–67. https://doi.org/10.4236/me.2010.12006 (2010)
19. G. Cetin, O. Akova, D. Gursoy, F. Kaya, Impact of direct flights on tourist volume: case of Turkish Airlines. J. Tourismol. **2**(2), 36–50 (2016)
20. B. Chandran, H. Balakrishnan, A dynamic programming algorithm for robust runway scheduling, in *Proceedings of the American Control Conference*, New York, NY, USA, 11–13 July 2007
21. G. Ćorluka, Analysis of seasonality-inbound tourist demand in Croatia, in *Proceedings of the International Scientific and Professional Conference "Contemporary Issues in Economy and Technology" CIET 2014*, Split, 19–21 June 2014, vol. 1 (2014)
22. J.C. Delsen, Flexible Arrival and Departure Runway Allocation Using Mixed-Integer Linear Programming A Schiphol Airport Case. MSc Thesis, Delft University of Technology (2016)
23. D. Dimitriou, M.F. Sartzetaki, Air transport connectivity development in tourist regions. Working papers. SIET 2018. ISSN 1973-3208 (2018)
24. F. Dobruszkes, V. Mondou, A. Ghedira, Assessing the impacts of aviation liberalisation on tourism: some methodological considerations derived from the Moroccan and Tunisian cases. J. Transp. Geogr. **50**, 115–127 (2016)
25. J.A. Duro, Seasonality of hotel demand in the main Spanish provinces: measurements and decomposition exercises. Tour. Manag. **52**, 52–63 (2016)
26. D.T. Duval, Critical issues in air transport and tourism. Tour. Geogr. **15**(3), 494–510 (2013)
27. European Commission, Commission Regulation (EC) No 1358/2003 of 31 July 2003 implementing Regulation (EC) No 437/2003 of the European Parliament and of the Council on statistical returns in respect of the carriage of passengers, freight and mail by air and amending Annexes I and II thereto (2007)
28. European Commission, DG TREN, Impact Assessment on Airport Capacity, Efficiency and Safety Framework. Contract for Ex-ante evaluations and Impact Assessments (TREN/A1/46-2005) (2006)
29. P.J. Forsyth, The gains from the liberalisation of air transport. J. Transp. Econ. Policy **32**, 73–92 (1998)
30. P.J. Forsyth, Martin Kunz memorial lecture. Tourism benefits and aviation policy. J. Air Transp. Manag. **12**(1), 3–13 (2006)
31. X. Fu, T.H. Oum, A. Zhang, Air transport liberalization and its impacts on airline competition and air passenger traffic. Transp. J. **49**(4), 24–41 (2010)
32. Government Gazette (Greece). FEK 15/B/12-01-1996 (1996)

33. J. Henderson, Transport and tourism destination development: an Indonesian perspective. Tourism and Hospitality Research **9**(3), 199–208 (2009)
34. R.M. Horonjeff, F.X. McKelvey, W.J. Sproule, S. Young, *Planning and Design of Airports*, 4th edn. (McGraw-Hill, Inc., 1993)
35. IATA, Vision: 2050, IATA (2011)
36. ICAO, Manual on the Regulation of International Air Transport, DOC 9626, Provisional Edition (2016)
37. T. Iñiguez, M. Plumed, M. Lattore Martínez, Ryanair and Spain: air connectivity and tourism from the perspective of complex networks. Tour. Manag. Stud. **10**(1), 46–52 (2014)
38. Institute of Greek Tourism Confederation (INSETE), Research, Analysis and Mapping of the Tourist Environment of the Southern Aegean Region, project funded by the European Social Fund (2015)
39. O. Kaplas, Airport capacity modeling-Case Helsinki Airport, Aalto University, Master Thesis (2014)
40. J. Khadaroo, B. Seetanah, The role of transport infrastructure in international tourism development: a gravity model approach. Tour. Manag. **29**(5), 831–840 (2008)
41. N. Koenig-Lewis, E. Bischoff, Seasonality research: the state of the art. Int. J. Tour. Res. **7**, 201–219 (2005). Published online in Wiley InterScience, www.interscience.wiley.com
42. T. Kolos-Lakatos, R.J. Hansman, The influence of runway occupancy time and wake vortex separation requirements on runway throughput. Report No. ICAT-2013-08 August 2013. MIT International Center for Air Transportation (ICAT) (2013)
43. S. Kraft, D. Havlíková, Anytime? Anywhere? The seasonality of flight offers in Central Europe. Moravian geographical reports. Inst. Geonics **24**(4), 26–37 (2016)
44. MaltaProfile, Air Connectivity is a priority for Tourism sector, https://maltaprofile.info/article/air-connectivity-remains-a-strategic-priority-for-the-tourism-sector. Accessed 15 June 2018 (undated)
45. R. Martinez-Val, J.F. Palacin, E. Perez, The evolution of jet airliners explained through the range equation. Proc. IMechE Part G: J. Aerosp. Eng. **222**(6), 915–919 (2008)
46. E. Marrocu, R. Paci, Different tourists to different destinations. Evidence from spatial interaction models. Tour. Manag. **39**, 71–83 (2013)
47. R. Maslen, A route to success: how global air connectivity drives tourism and economic growth. Connected Visitor Economy Bulletin, February 2016 Edition. Published online by Pacific Asia Travel Association (PATA), https://pata.org/store/wp-content/uploads/2016/02/Ve_February_A4-2.pdf
48. MERC (Miller Engineering & Research Corporation), Boeing 737 & 777 Aircraft, https://www.merc-md.com/portfolio/boeing-737-777/. Accessed 19 July 2018
49. T.E. Nissalke, The air transportation system in the 21st century. Sustain. Built Environ. **2** Oxford, EOLSS Publishing House (2009)
50. OECD/ITF, International Transport Forum. Defining, Measuring and Improving Air Connectivity. Corporate Partnership Board Report (2018)
51. Office of Technology Assessment, Airport System Development (Washington, D. C., U.S. Congress, Office of Technology Assessment, OTA-STI-231, NTIS Order #PB85-127793, August 1984)
52. A. Papatheodorou, Why people travel to different places. Ann. Tour. Res. **28**, 164–179 (2001)
53. A. Papatheodorou, Civil aviation regimes and leisure tourism in Europe. J. Air Transp. Manag. **8**, 381–388 (2002)
54. C.N. Potts, M. Mesgarpour, J. Bennell, Airport Runway Optimization, Ph.D. Thesis, Southampton University, School of Mathematics (2009)
55. PwC, Overview of air transport and current and potential air connectivity gaps in the CESE region, Paper B. Final Report, 4 December 2014, European Commission
56. J. Ram, D. Reeves, R. James, Air transport competitiveness and connectivity in the Caribbean. CDB Working paper No: CDB/WP/18/02 (2002)
57. H. Reichard, Seasonality in Air Transportation, Department of Aeronautics & Astronautics, Cambridge Mass, MIT, Flight Transportation Laboratory, Report R 88-3 (1988)

58. D. Schaar, L. Sherry, Comparison of data envelopment analysis methods used in airport benchmarking, in *Third International Conference on Research in Air Transportation*, Fairfax, VA, 1–4 June 2008
59. S.M. Shugan, S. Radas, Services and seasonal demand, in *Handbook of Services Marketing & Management*, ed. by Teresa A. Swartz and Dawn Iacobucci (1999)
60. J. Skorupski, H. Wierzbicki, Airport capacity increase via the use of braking profiles. Transp. Res. Part C (2016). http://dx.doi.org/10.1016/j.trc.2016.05.016
61. S. Smith, The tourism product. Ann. Tour. Res. **21**, 582–595 (1994)
62. D. Snepenger, B. Houser, M. Snepenger, *Seasonality of Demand* (Montana State University, USA, 1990), pp. 628–630
63. J. Sridhar, What is data analysis and why is it important?, https://www.makeuseof.com/tag/what-is-data-analysis/ (2018)
64. S. Teles, M. Sarmento, Á. Matias, Tourism and strategic competition in the air transport, in *Advances in Tourism Economics. New Developments* (Physica-Verlag, Heidelberg Edition, 2009), pp. 255–272
65. I.B. Thompson, Air transport liberalisation and the development of third level airports in France. J. Transp. Geogr. **10**, 273–285 (2002)
66. L. Tilana, Impact of strategic demand management on runway capacity at Oliver Reginald Tambo international airport. Research Report, University of the Witwatersrand, Johannesburg (2011)
67. A.A. Trani, A.G. Hobeika, B.J. Kim, V. Nunna, C. Zhong, Runway exit designs for capacity improvement demonstrations, Phase II-Computer model development, Final Report, Virginia Polytechnic Institute and State University, Center for Transportation Research, Blacksburg, Virginia (1992)
68. UNWTO Air Connectivity and its Impact on Tourism in Asia and the Pacific, Madrid (2014)
69. UNWTO and ICF International, Enhancing tourism competitiveness through improved air connectivity, in *International Tourism Fair*, Madrid, 20–24 January 2016
70. W. Vancroonenburg, J. Vestichel, K. Tavernier, Berghe G. Vanden, Automatic air cargo selection and weight balancing: a mixed integer programming approach. Transp. Res. Part E **65**, 70–83 (2014)
71. A.S. Vergori, Patterns of seasonality and tourism demand forecasting. Tour. Econ. **23**(5), 1011–1027 (2017)
72. S.D. Widener, Measuring Airport Efficiency with Fixed Asset Utilization to Minimize Airport Delays. Open Access Dissertations. 485, https://scholarlyrepository.miami.edu/oa_dissertations/485 (2010)
73. B. Yutko, Approaches to Representing Aircraft Fuel Efficiency Performance for the Purpose of a Commercial Aircraft Certification Standard. Report No ICAT-2011-05, May 2011, MIT International Center for Air Transportation (ICAT) (2011)

Chapter 9
A Taxonomy and Review of the Network Data Envelopment Analysis Literature

Gregory Koronakos

Abstract Performance measurement deals with ongoing monitoring and evaluation of the operations of the organizations so as to be able to improve their productivity and performance. Thus, the adoption of performance evaluation methods is necessary, which are capable of taking into account all the environmental factors of the organization, identifying the inefficient production processes and suggesting adequate ways to improve them. Such a method is Data Envelopment Analysis (DEA), which is the most popular non-parametric and data driven technique for assessing the efficiency of homogeneous decision making units (DMUs) that use multiple inputs to produce multiple outputs. The DMUs may consist of several sub-processes that interact and perform various operations. DEA has a wide application domain, such as public sector, banks, education, energy systems, transportation, supply chains, countries and so forth. However, the classical DEA models treat the DMU as a "black box", i.e. a single stage production process that transforms some external inputs to final outputs. In such a setting, the internal structure of the DMU is not taken into consideration. Thus, the conventional DEA models fail to mathematically represent the internal characteristics of the DMUs, as well as they fall short to provide precise results and useful information regarding the sources that cause inefficiency. To consider for the internal structure of the DMUs, recent methodological advancements are developed, which extend the standard DEA and constitute a new field, namely the network DEA. The network DEA methods are capable of reflecting accurately the DMUs' internal operations as well as to incorporate their relationships and interdependences. In network DEA, the DMU is considered as a network of interconnected sub-units, with the connections indicating the flow of intermediate products. In this chapter, we describe the underlying notions of network DEA methods and their advantages over the classical DEA ones. We also conduct a critical review of the state-of-the art methods in the field and we provide a thorough categorization of a great volume of network DEA literature in a unified manner. We unveil the relations and the differences of the existing network DEA methods. In addition, we

G. Koronakos (✉)
Department of Informatics, University of Piraeus, 80, Karaoli and Dimitriou,
18534 Piraeus, Greece
e-mail: gkoron@unipi.gr; gregkoron@gmail.com

© Springer Nature Switzerland AG 2019
G. A. Tsihrintzis et al. (eds.), *Machine Learning Paradigms*,
Learning and Analytics in Intelligent Systems 1,
https://doi.org/10.1007/978-3-030-15628-2_9

report their limitations concerning the returns to scale, the inconsistency between the multiplier and the envelopment models as well as the inadequate information that provide for the calculation of efficient projections. The most important network DEA methods do not secure the uniqueness of the efficiency scores, i.e. the same level of overall efficiency is obtained from different combinations of the efficiencies of the sub-processes. Also, the additive efficiency decomposition method provides biased efficiency assessments. Finally, we discuss about the inability of the existing approaches to be universally applied on every type of network structure.

9.1 Introduction

Improving organization's performance requires accurate understanding as well as systematic assessment of its internal structure, which is often a tough task because of organization's complexity. Thus, the performance measurement is a subject of major importance.

Two main approaches, the parametric and the non-parametric, are suggested in the literature for the performance measurement of production units. In the parametric approach, a production function is explicitly assumed so as to describe the relationships among the inputs and the outputs that participate in the production process. However, the production function can be hardly formulated or is completely unknown. On the contrary, the non-parametric approach does not require any a priori specification of the underlying functional form that relates the inputs with the outputs. Data Envelopment Analysis (DEA) is a powerful non-parametric technique that is widely used for evaluating the performance of a set of comparable entities, called decision making units (DMUs), which use multiple inputs to produce multiple outputs. DEA circumvents the problem of specifying an explicit form of the production function by constructing an empirical best practice production frontier. This is accomplished by enveloping the observed data of the DMUs. The linear programming is the underlying mathematical method that enables DEA to determine the efficient production frontier and calculate the efficiency score of each DMU. The efficiencies provided by DEA are relative rather than absolute, because each unit is evaluated relative to the production frontier, i.e. the best practice units. DEA is capable of uncovering the sources of inefficiency and providing prescriptions for improving the inefficient units. DEA takes into account the returns to scale and the orientation of the analysis in calculating efficiency. The CCR (Charnes et al. [21]) and the BCC (Banker et al. [11]) models, under constant returns to scale (CRS) and variable returns to scale (VRS) assumption respectively, have established the foundation for further research in this field. A rapid and continuous growth has been reported since then, both in theoretical and application level. A remarkable body of literature has been developed with a wide range of applications to measure the efficiency in various sectors such as business and finance, public services, education, health care, transportation, agriculture, supply chains, etc.

The DMUs may have a complex structure that includes several interdependent sub-processes with series, parallel or series-parallel arrangement. The traditional

DEA models, however, regard the DMU as a black box, treating them as single stage production processes that transform some external inputs to final outputs. In such a setting, the internal structure and the interactions among the comprised operations of the DMUs are not taken into consideration. Cook and Zhu [38] stressed out that in conventional DEA the DMU is treated as a *black box* and its internal structure and operations are ignored. Kao and Hwang [75] showed that the standard DEA models may deem a DMU overall system efficient even though all their sub-units are inefficient. Conclusively, the standard DEA models fail to adequately capture and mathematically represent the aforementioned characteristics of the DMUs. Also, they fall short to shed light on the sources of inefficiency as well as to provide succinct guidance for the improvement of the inefficient DMUs and sub-units.

On the other hand, network DEA is an extension of conventional DEA developed to take account of the internal structure of DMUs. In network DEA, the DMU is considered as a network of interconnected sub-units, with the connections indicating the flow of intermediate products (commonly called intermediate measures or links). In the literature, these sub-units are also known as stages, divisions, sub-DMUs, sub-systems, sub-processes, processes, procedures, components and functions. Albeit in this chapter we may use these terms interchangeably, we mainly adopt the term "stage" when we refer to the sub-units of the DMUs. The advantage of the network DEA models is their ability to reflect accurately the DMUs' internal operations as well as to incorporate their relationships and interdependences. Therefore, they yield more representative and precise results than the conventional DEA models and provide more information regarding the sources that cause inefficiency. Cook and Seiford [37] included the network DEA models to the methodological developments of DEA and mentioned that these models allow the detailed examination of the inner workings of a production process, which leads to a greater understanding of that process. Indeed, having a full picture of the internal structure of DMUs and examining their sub-units in a coordinated manner will provide further insights for the performance assessment and will assist better the decision making.

Fare and Primont [54] and Charnes et al. [22] were the first studies, to the best of our knowledge, in the field later named network DEA by Fare and Grosskopf [52]. Fare and Primont [54] distinguished the internal structure of multi-plant firms, i.e. firms that own many plants. They defined the firm's technology by constructing first the technology of each plant. Their approach was applied to a selected sample of electric generating firms which consists of nineteen plants in Illinois. Later, the approach of Kao [69] was applied for the efficiency assessment of eight Taiwanese forest districts with 34 working circles. The performance of each working circle was measured based on the technology constructed from all of them. Charnes et al. [22] assumed that the US army recruitment is comprised of two processes, namely the awareness creation and the contract establishment. The work of Fare and Grosskopf [52] is considered pioneering in the field of network DEA. Although the terms "*black box*" and *network technology* had been earlier reported in the studies of Fare [48], Fare and Whittaker [56], Fare and Grosskopf [51] and Fare et al. [49], it was Fare and Grosskopf [52] who first coined the term *network DEA* and provided a consolidated framework of the aforementioned studies for multi-stage processes with various structures.

Network DEA has already attracted the interest of researchers and a significant body of research is devoted to both theory and applications. Kao [74] noticed that the number of publications before 2000 was two or three per year, thereafter though it has rapidly grown. Liu et al. [92], in their citation-based literature survey for DEA for the period 1978–2010, considered the field of network DEA as a subarea which is relatively active in recent years. However, from 2010 onwards there has been a blast of publications on network DEA. Some of these studies explore the properties of the existing methods while others apply them to real world problems. The application field of network DEA as we will also see below is very wide, e.g. supply chains, banking, education, sports just to mention some. The network DEA methods can now be straightforwardly and effectively employed for the performance evaluation of a supply chain and its members which is undoubtedly a rough task. Agrell and Hatami-Marbini [3] provided a thorough review for network DEA methods including studies devoted to supply chain performance analysis. They also remarked that the supply chains are complex multi-stage systems with interrelations, which use multiple inputs to produce multiple outputs. Hence, the network DEA methods can be adequately employed for their performance assessment. Many prominent approaches are developed to deal with the variety of the structures, the interdependencies and the conflicting interests of the sub-units. Most of the network DEA studies are dedicated to the performance assessment of DMUs with a specific internal structure. A DMU may consist of sub-units arranged in series, in parallel or a mixture of these. Cook et al. [36] and Chen et al. [25] provide insights and directions for further research for the two-stage network structures arranged in series. Sotiros et al. [114] introduced the dominance property in network DEA as a minimum requirement that all network DEA models should comply with. The dominance property relies on the rational assumption that higher divisional efficiency scores should lead to higher overall efficiency. Koronakos et al. [80] reformulated some of the basic network DEA methodologies in a common multi-objective programming framework, which differentiate only in the definition of the overall system efficiency and the solution procedure that they adopt. Castelli et al. [17] and Halkos et al. [63] provided comprehensive categorized overviews of models and methods developed for different multi-stage production architectures. Kao [74] provides an excellent review and classification of network DEA methods according to the model they use and the network structure that they examine. Moreover, a collection of network DEA methods is given in Cook and Zhu [38].

The aim of this chapter is to describe the underlying notions of network DEA, to present the state of the art in the field and to review the most significant network DEA methods. Also, our goal is to provide a comprehensive insight and categorization of the network DEA literature in a unified manner. In particular, we present the possible network structures that a DMU may be characterized of, we demonstrate the advantages of the network DEA over the standard DEA, we provide a critical review of the most influential approaches and we discuss their extensions, inherent limitations and shortcomings. In addition, we track the majority of multi-stage DEA applications and we classify them according to the method they utilize. Hence, this chapter presents a complete survey of the network DEA literature.

9.2 DMU's Internal Network Structures and Assessment Paradigms

The DMUs may have various types of internal structures. However, we discern that their production processes may be arranged either in series, in parallel or in series-parallel. The series and the parallel production processes are two distinctive network architectures studied extensively in the literature. In this section, we provide some illustrative examples of network structures which are used as the basis for the development of the most significant network DEA methods. The four types of series two-stage network processes depicted in Fig. 1 are the basis for the development of network DEA theory and applications.

In the Type I two-stage process (Fig. 1a) the first stage uses external inputs (X) to produce the intermediate measures (Z), which are subsequently used as inputs to the second stage which produces the final outputs (Y). In Type I structure, nothing but the external inputs to the first stage enters the system and nothing but the outputs of the second stage leaves the system. This is the elementary network structure that has drawn the attention of most of the research work. Wang et al. [123] and Seiford and Zhu [110] are the first who studied processes of Type I structure.

In the production process of Type II (Fig. 1b) each DMU uses the external inputs (X) in the first stage to yield the intermediate measures (Z), which then are used along with the additional external inputs (L) to the second stage to yield the final outputs (Y), as depicted in Fig. 1b. That is, the second stage uses except from the intermediate measures additional external inputs (L) for exclusive use. Liang et al.

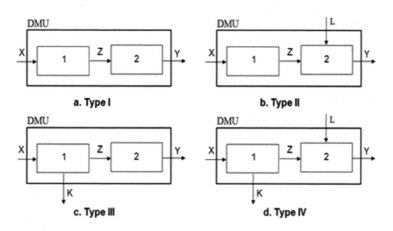

Fig. 1 The four types of series two-stage processes (*Source* Despotis et al. [46])

[86], under game theoretic concepts, studied a supply chain with two stages, the seller and the buyer, where the buyer (second stage) uses extra inputs. Notice that the Type II structure may be varied by assuming that the external inputs (X) can be freely shared between the stages in conjunction with or without the additional inputs (L). Such a variation is considered in Chen et al. [29], where the impact of the Information Technology (IT) on firm performance is examined.

In the production process of Type III (Fig. 1c), the first stage produces some final outputs (K) beyond the intermediate measures (Z), while in the production process of Type IV (Fig. 1d) external inputs and final outputs appear in both stages. The first stage uses the inputs (X) to generate the final outputs (K) and the intermediate measures (Z). The second stage uses the intermediate measures (Z) and the additional external inputs (L) for the production of the final outputs (Y). This type of network structure was first studied in [22, 48, 51, 56].

The four types of network structures portrayed in Fig. 1 can be generalized to series structures with more than two stages. Another basic network structure of the DMUs is a production process whose sub-processes are configured in parallel. Figure 2 depicts the internal structure of a DMU with v parallel processes without interdependencies. Each sub-process transforms the external inputs (X) to final outputs (Y). A characteristic example of units that can be considered as parallel production processes are the academic departments, where teaching and research are two distinct functions. Analogously, a university can be viewed as a DMU and its departments can be regarded as the individual parallel sub-units. A modification of the parallel structure of Fig. 2 may involve shared flows among the stages, i.e. the sub-processes, instead of consuming dedicated inputs, they share common resources (external inputs). Such a case was examined by Fare et al. [49] who assessed the performance of 57 grain farms with one shared input, namely the land. The land is allocated among four different agricultural operations, specifically the crops of corn, soybeans, wheat and the double crop soybeans. Vaz et al. [122], based on Fare et al. [49], studied 78 Portuguese retail stores, each one comprised of five sections, namely groceries, perishables, light bazaar, heavy bazaar and textiles. These sections operate in parallel and share the floor area. In the educational sector, Beasley [13] and Mar Molinero [104] developed nonlinear models to measure teaching and research

Fig. 2 A production process with parallel sub-processes

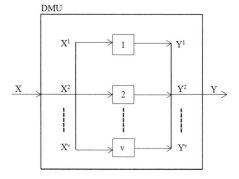

Fig. 3 Mixed network structure with series and parallel sub-processes (*Source* Lewis and Sexton [83])

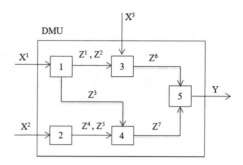

performance as parallel academic operations, in the presence of two shared inputs, namely general and equipment expenditures.

The basic series and parallel configurations are not always sufficient to describe real world situations. Therefore, more complex network structures, mixtures of afore-mentioned ones can be used to represent in detail the relationships among the sub-processes. Figure 3 portrays such a network structure composed of a combination of series and parallel structures. Such a structure was studied by Lewis and Sexton [83] who assessed the performance of 30 teams (DMUs) of the Major League Baseball in the US. They assumed that the operation of each team is represented by a network of five distinct sub-processes. The first and the second correspond to the team's front office operations, while the other three correspond to the team's on-field operations.

9.3 Assessment Paradigms

The performance assessment of DMUs within the network DEA framework is carried out by a variety of approaches, which can be categorized into two assessment paradigms, namely, the *independent assessment paradigm* and the *joint assessment paradigm*. In the *independent assessment paradigm*, the standard DEA models are used to assess the performance of the DMUs and the sub-processes independently. In the *joint assessment paradigm*, the internal structure of the DMUs and the inter-dependencies among the sub-processes are taken into account, also the efficiency assessment of the sub-processes and the whole system is made simultaneously. There are four general approaches that follow the *joint assessment paradigm*: the *efficiency decomposition approach*, the *composition approach*, the *slacks-based measure approach* and the *system-centric approach*.

We give below the notation that will be employed in the current chapter:

$j \in J = \{1, \ldots, n\}$	The index set of the n DMUs.
$j_0 \in J$	Denotes the evaluated DMU.
$\gamma \in \Gamma = \{1, \ldots, \nu\}$	The index set of the ν processes that each DMU$_j$ is composed.
$X_j = (x_{ij}, i = 1, \ldots, m)$	The vector of external inputs used by DMU$_j$.

$Z_j = \left(z_{pj}, p = 1, \ldots, q\right)$ The vector of intermediate measures for DMU$_j$.

$Y_j = \left(y_{sj}, r = 1, \ldots, s\right)$ The vector of final outputs produced by DMU$_j$.

$\eta = (\eta_1, \ldots, \eta_m)$ The vector of weights for the external inputs in the fractional model.

$v = (v_1, \ldots, v_m)$ The vector of weights for the external inputs in the linear model.

$\varphi = \left(\varphi_1, \ldots, \varphi_q\right)$ The vector of weights for the intermediate measures in the fractional model.

$w = \left(w_1, \ldots, w_q\right)$ The vector of weights for the intermediate measures in the linear model.

$\omega = (\omega_1, \ldots, \omega_s)$ The vector of weights for the outputs in the fractional model.

$u = (u_1, \ldots, u_s)$ The vector of weights for the outputs in the linear model.

λ The intensity vector for the first stage.

μ The intensity vector for the second stage.

e_j^o The overall efficiency of DMU$_j$.

e_j^1 The efficiency of the first stage for DMU$_j$.

e_j^2 The efficiency of the second stage for DMU$_j$.

E_j^1 The independent efficiency score of the first stage for DMU$_j$.

E_j^2 The independent efficiency score of the second stage for DMU$_j$.

9.3.1 Independent Assessments

The *independent approach* is an elementary method for the assessment of DMUs that consist of sub-processes. Although the internal structure of the DMUs is recognized, the stage efficiencies and the overall system efficiency are calculated independently. The standard DEA model is used separately in each stage without considering possible conflicts and connections among them. In this approach, the stages are treated as operating independently of one another and are assessed as independent DMUs, hence the impact of each stage to the overall efficiency cannot be measured. Consider the basic input-oriented DEA models under CRS assumption [21], which estimate independently the stage-1, the stage-2 and the overall efficiency for the evaluated unit j_0 with the simple Type-I (Fig. 1a):

Stage 1:

$$E_{j_0}^1 = max \frac{\varphi Z_{j_0}}{\eta X_{j_0}}$$

s.t.

$$\frac{\varphi Z_j}{\eta X_j} \leq 1, \quad j = 1, \ldots, n$$

$$\eta \geq 0, \varphi \geq 0 \tag{1}$$

Stage 2:

$$E_{j_0}^2 = max \frac{\omega Y_{j_0}}{\hat{\varphi} Z_{j_0}}$$

s.t.

$$\frac{\omega Y_j}{\hat{\varphi} Z_j} \leq 1, \quad j = 1, \ldots, n$$

$$\hat{\varphi} \geq 0, \omega \geq 0 \tag{2}$$

Overall:

$$E_{j_0}^o = max \frac{\omega Y_{j_0}}{\eta X_{j_0}}$$

s.t.

$$\frac{\omega Y_j}{\eta X_j} \leq 1, \quad j = 1, \ldots, n$$

$$\eta \geq 0, \omega \geq 0 \tag{3}$$

Notice that the output-oriented variants of the above models can be also used, as well as variable returns to scale assumption. The input-oriented models (1) and (2), under constant returns to scale assumption, yield the independent stage efficiencies while model (3) provides the overall efficiency of the DMU j_0. In model (3) only the external inputs (X) and the final outputs (Y) are used for the assessment of the evaluated unit j_0, whereas in models (1) and (2) for stage-1 and stage-2 respectively, only their individual inputs and outputs are taken into account, i.e. (X)–(Z) for stage-1 and (Z)–(Y) for stage-2. As a result, the fact that the outputs of the first stage are the inputs to the second stage is ignored. Moreover, the overall efficiency is not connected to the individual efficiencies since they are evaluated independently. In effect, the efficiency scores derived by the independent approach are misleading. This has been reported in [75, 76], as a DMU may be overall efficient while the individual stages are not. Such irregular results are attributed to the fact that no coordination between the stages is assumed. Finally, the stage and the overall efficiency scores obtained by the independent approach serve as upper bounds of the stage and system efficiencies respectively. Because of its simplicity the independent approach can be applied to any network structure since the relationships among the stages are not taken into account.

Significant studies that employed the independent approach are, among others, that of Charnes et al. [22], Chilingerian and Sherman [32], Wang et al. [123], Seiford and Zhu [110], Zhu [135], Sexton and Lewis [112] and Lewis and Sexton [83]. In

Charnes et al. [22], the army recruitment was viewed as a two-stage process, namely the awareness creation and the contract establishment. Chilingerian and Sherman [32] modeled the medical service as a two-stage process, where the first stage is under the control of the management and the second stage is controlled by the physician. In stage-1 the management handles the assets of the hospitals and provides with clinical outputs which are used as inputs to the stage-2. In the second stage, the physicians decide how to utilize these inputs so as to provide medical care to the patients. To be more specific, the inputs of stage-1 are nurses, management and support staff, medical supplies, various expenditures, capital and fixed costs. The intermediate measures generated by stage-1 and conveyed to stage-2 include hours of nursing care, counseling services and therapy, volume of diagnostic tests, drugs dispensed and other quantitative indicators about the medical treatment issued. The final outputs of stage-2 are research grants and quantitative indicators for the patients and the trained staff. Wang et al. [123] studied the impact of IT on the performance of 36 banks. They assumed a simple two-stage process of Type I (Fig. 1a) where the first stage represents the *funds collection* and the second the *investment*. Seiford and Zhu [110] studied the performance of the top 55 commercial banks in USA by considering both the operational and the market performance. They modeled the bank operations as a simple two-stage process of Type I, with the stage-1 representing profitability and the stage-2 marketability. Within a similar framework, Zhu [135] evaluated the performance of profitability and marketability of the Fortune 500 companies. Sexton and Lewis [112] evaluated the performance of 30 teams of the Major League Baseball in the USA, by modelling the whole team's operations as a two-stage process of Type I, with stage-1 representing front-office operation and stage-2 representing on-field operation. Lewis and Sexton [83] extended their previous study by modelling the team's operations with the network structure depicted in Fig. 3. In particular, the first two stages correspond to the team's front office operations, which use funds (player salaries) to acquire talent, whereas the rest three stages correspond to the team's on-field operations, which utilize talent to win games.

9.3.2 Joint Assessments

The independent approach neglects the conflicts or connections between the stages. On the contrary, according to the joint assessment paradigm the relationships among the stages and the overall efficiency are taken into account. The efficiency decomposition approach, the composition approach, the slacks-based measure (SBM) approach and the system-centric approach are the characteristic families that follow the joint assessment paradigm. The categorization is based on the perspective of each approach about the relationships between the system (DMU) and the stage efficiencies as well as on the kind of information provided for the performance of the individual stages and the system. The interaction between the sub-processes is taken into account by these approaches, however their difference lies on the way that the overall and the stage efficiencies are derived. In particular, the efficiency decomposition approach measures the system efficiency first and then the stage efficiencies are calculated ex post. On the contrary, in the composition approach the stage efficiencies are calculated first and the overall efficiency is derived a posteriori. In the slacks-based measure approach the stage efficiencies and the overall efficiency are simultaneously obtained from the assessment model, but a different functional form for the overall efficiency is implicitly assumed according to the selected orientation. As system-centric we characterize the network DEA methods that take into account the internal structure of the DMUs and the interdependencies among the stages, but they provide only an overall performance measure without generating the stage efficiencies. In the system-centric methods there is no functional form that connects the overall and the stage efficiencies.

9.3.2.1 Efficiency Decomposition Approach

A major characteristic of the decomposition approach is that, apart from the definition of the efficiency of the individual stages (stage efficiencies), it premises the definition of the overall efficiency of the DMU together with a model to decompose the overall efficiency to the stage efficiencies. The driver of the assessment is the overall efficiency that is optimized first. Then, the efficiency scores of the stages derive as offspring of the overall efficiency of the unit. The two basic decomposition methods dominating the literature on two-stage DEA, i.e. the multiplicative method of Kao and Hwang [75] and the additive method of Chen et al. [26] assume the same definitions of stage efficiencies but they differ substantially in the definition of the overall system efficiency as well as in the way they conceptualize the decomposition of the overall efficiency to the efficiencies of the individual stages. In multiplicative efficiency decomposition, the overall efficiency is defined as a product of the stage efficiencies, whereas in the additive efficiency decomposition, the overall efficiency is defined as a weighted average of the stage efficiencies.

Multiplicative Efficiency Decomposition

The multiplicative efficiency decomposition method is introduced by Kao and Hwang [75] and Liang et al. [87] for the simple two-stage network structure of Type I (Fig. 1a). Specifically, Liang et al. [87] studied the efficiency decomposition of the two-stage process using game theoretic concepts. Under the multiplicative decomposition method, the efficiency of the entire process is decomposed into the product of the efficiencies of the two individual stages. The overall efficiency and the stage efficiencies of the DMUj, under the CRS assumption, are defined as follows:

$$e_j^o = \frac{\omega Y_j}{\eta X_j}, e_j^1 = \frac{\varphi Z_j}{\eta X_j}, e_j^2 = \frac{\omega Y_j}{\hat{\varphi} Z_j} \tag{4}$$

$$e_j^o = \frac{\omega Y_j}{\eta X_j} = \frac{\varphi Z_j}{\eta X_j} \cdot \frac{\omega Y_j}{\hat{\varphi} Z_j} = e_j^1 \cdot e_j^2 \tag{5}$$

In order to link the efficiency assessments of the two stages, it is universally accepted that the values of the intermediate measures (virtual intermediate measures) should be the same for both stages, i.e. the weights associated with the intermediate measures should be the same ($\hat{\varphi} = \varphi$), no matter if these measures are considered as outputs of the first stage or inputs to the second stage. As can be deduced from the decomposition model (5) the overall efficiency is defined as the *square geometric average* of the stage efficiencies. Given the above definitions, the CRS input-oriented model (6) assesses the overall efficiency and the stage efficiencies of the evaluated unit j_0. Notice that the constraints $\omega Y_j / \eta X_j \leq 1$, $j = 1, \ldots, n$ are redundant and thus omitted. Model (6) is a fractional linear program that can be modeled and solved as a linear program (7) by applying the Charnes and Cooper transformation [20] (C-C transformation hereafter). The correspondence of variables is: $v = t\eta$, $u = t\omega$, $w = t\varphi$ where t is a scalar variable such that: $t\eta X_{j_0} = 1$.

$$e_{j_0}^o = max \frac{\omega Y_{j_0}}{\eta X_{j_0}}$$

$s.t.$

$$\frac{\varphi Z_j}{\eta X_j} \leq 1, \quad j = 1, \ldots, n$$

$$\frac{\omega Y_j}{\varphi Z_j} \leq 1, \quad j = 1, \ldots, n$$

$$\eta \geq 0, \varphi \geq 0, \omega \geq 0 \tag{6}$$

$$e_{j_0}^o = max \, uY_{j_0}$$

$s.t.$

$$vX_{j_0} = 1$$

$$wZ_j - vX_j \leq 0, \quad j = 1, \ldots, n$$

$$uY_j - wZ_j \leq 0, \quad j = 1, \ldots, n$$

$$v \geq 0, w \geq 0, u \geq 0 \tag{7}$$

Once an optimal solution (v^*, w^*, u^*) of model (7) is obtained, the overall efficiency and the stage efficiencies are calculated as $e_{j_0}^o = u^* Y_{j_0}$, $e_{j_0}^1 = w^* Z_{j_0}$, $e_{j_0}^2 = e_{j_0}^0 / e_{j_0}^1$.

In parallel, Liang et al. [87] developed the multiplicative decomposition in the light of game theoretic concepts. They characterized the multiplicative decomposition method described above as a cooperative or a centralized game, i.e. they refer to model (7) as centralized. In addition, they presented the case of non-cooperative game between the stages, where preemptive priority is given to one stage like the leader-follower situations in decentralized control systems; this paradigm is also referred to as the Stackelberg game. A DMU may be seen as a supply chain with two parts, consisting for example of a manufacturer and a retailer. In such a setting, the manufacturer acts as a leader whereas the retailer is treated as a follower. Assuming that the first stage is the leader then its performance is computed first by applying the conventional DEA model. The leader (first stage) seeks to maximize its performance without considering the follower (second stage). The performance of the follower (second stage) is calculated subject to the requirement that the leader's efficiency is fixed at its optimal value. Alternatively, if the second stage is assumed to be the leader then its efficiency score is optimized first. The leader-follower modelling approach yields the maximum achievable efficiency score for each stage when it acts as a leader, i.e. it generates the independent efficiency scores.

Notice that the overall efficiency is obtained as the optimal value of the objective function of model (7), the stage-1 efficiency is given by the total virtual intermediate measure, whereas the stage-2 efficiency derives as offspring of the overall and stage-1 efficiencies. A major shortcoming of the multiplicative method is that the decomposition of the overall efficiency to the stage efficiencies is not unique. Indeed, as the term wZ_{j_0} does not appear in either the objective function or in the normalization constraint, its value may vary and still maintain the optimal value of the objective function (i.e. the overall efficiency) and the inequality constraints of model (7). Also, the above deficiency renders the comparison of stage efficiencies among all DMUs lack a common basis. That is why Kao and Hwang [75] and Liang et al. [87] propose solving a pair of linear programs, in a post-optimality phase, to obtain the largest scores for $e_{j_0}^1$ and $e_{j_0}^2$ while maintaining the overall efficiency score obtained by model (7). In particular, they developed a procedure for testing the uniqueness of the efficiency decomposition by maximizing the efficiency of one stage under the constraint that the optimal overall efficiency obtained by (7) is maintained. Then the efficiency of the other stage is calculated from (5). The highest efficiency for the first stage is obtained by model (8) below:

$$e_{j_0}^{1U} = max \ wZ_{j_0}$$

$$s.t.$$

$$vX_{j_0} = 1$$

$$uY_{j_0} = e_{j_0}^o$$

$$wZ_j - vX_j \leq 0, \quad j = 1, \ldots, n$$

$$uY_j - wZ_j \leq 0, \quad j = 1, \ldots, n$$

$$v \geq 0, w \geq 0, u \geq 0 \tag{8}$$

$$e_{j_0}^{2U} = max \, uY_{j_0}$$
$$s.t.$$
$$wZ_{j_0} = 1$$
$$uY_{j_0} - e_{j_0}^o vX_{j_0} = 0$$
$$wZ_j - vX_j \leq 0, \quad j = 1, \ldots, n$$
$$uY_j - wZ_j \leq 0, \quad j = 1, \ldots, n$$
$$v \geq 0, w \geq 0, u \geq 0 \tag{9}$$

Once an optimal solution (v^*, w^*, u^*) of model (8) is obtained then $e_{j_0}^{1U} = w^* Z_{j_0}$ and the efficiency of the second stage is derived by $e_{j_0}^{2L} = e_{j_0}^0 / e_{j_0}^{1U}$. Alternatively, if priority is given to the second stage, then its highest efficiency level is calculated first by model (9). Given an optimal solution (v^*, w^*, u^*) of model (9), the highest efficiency score of stage-2 is $e_{j_0}^{2U} = u^* Y_{j_0}$ and the resulting efficiency of the first stage is $e_{j_0}^{1L} = e_{j_0}^0 / e_{j_0}^{2U}$. If $e_{j_0}^{1U} \neq e_{j_0}^{1L}$ or $e_{j_0}^{2U} \neq e_{j_0}^{2L}$ then the efficiency decomposition is not unique, in other words there are alternative optimal solutions that yield the same level of overall efficiency, i.e. $e_{j_0}^0 = e_{j_0}^1 \cdot e_{j_0}^2 = e_{j_0}^{1U} \cdot e_{j_0}^{2L} = e_{j_0}^{1L} \cdot e_{j_0}^{2U}$.

The purpose of models (8) and (9), beyond checking the uniqueness of the efficiency decomposition, is to provide also alternative solutions in case of non-uniqueness. The argument is that one might wish giving priority to the first or the second stage in the efficiency assessments. Although there is a rationale in this argument, the non-uniqueness of the decomposition is still a problem, especially in the case that no priority is conceived by the management. Notice that the above procedure can be also applied when output orientation is selected using the output-oriented models accordingly.

A major limitation of the multiplicative decomposition method is its inability to be straightforwardly applied under the VRS assumption. This is because the extra free-in-sign variables introduced in the VRS model render it highly non-linear. Kao and Hwang [77] proposed an approach to decompose technical and scale efficiencies of the two-stage process. They derived the scale efficiencies for the two stages assuming an input oriented VRS model for the first stage and an output oriented VRS model for the second stage. Thus, the system efficiency is decomposed into the product of the technical and scale efficiencies of the stages.

Duality

Contrary to the standard DEA context where the multiplier and envelopment DEA models are dual models and equivalent, as also remarked by Chen et al. [27] and Chen et al. [25], such is not necessarily true for the two forms of network DEA models. As they further noted, the duals to the multiplier-based network DEA models may not provide the frontier projections without exerting appropriate modifications to them. The above is also observed for the dual model (10) of the CRS input-oriented model (7). The model (10) does not provide the stage efficiency scores. Also, the usual procedure of adjusting the inputs and outputs by the efficiency scores is not adequate to provide a frontier projection. These irregularities may be attributed to the conflicting nature of the intermediate measures and to the fact that may none DMU be overall efficient, i.e. may none DMU be efficient in both stages. Thus, new techniques are needed for the determination of the efficient frontier of a two-stage process. Chen et al. [27] developed the alternative envelopment model (11), in order to overcome the reported inadequacies and generate the efficient frontier. They replaced the observed levels of intermediate measures by variables and separated the constraints associated with the intermediate measures.

$$\min \theta$$
$$s.t.$$
$$X\lambda \leq \theta X_{j_0}$$
$$Y\mu \geq Y_{j_0}$$
$$Z\lambda - Z\mu \geq 0$$
$$\lambda \geq 0, \mu \geq 0 \qquad\qquad (10)$$

$$\min \tilde{\theta}$$
$$s.t.$$
$$X\lambda \leq \tilde{\theta} X_{j_0}$$
$$Y\mu \geq Y_{j_0}$$
$$Z\lambda \geq \tilde{Z}_{j_0}$$
$$Z\mu \leq \tilde{Z}_{j_0}$$
$$\lambda \geq 0, \mu \geq 0, \tilde{Z}_{j_0} \geq 0 \qquad\qquad (11)$$

Chen et al. [27] showed that model (11) and model (10) yield the same overall efficiency score i.e. $\theta = \tilde{\theta}$. Also, the model (11) provides additionally sufficient information on how to project inefficient DMUs onto the efficient frontier. The projection $(\hat{X}_{j_0}, \hat{Z}_{j_0}, \hat{Y}_{j_0})$ for DMU$_{j_0}$ is derived by the optimal solution of model (11) as $\tilde{\theta}^* X_{j_0}, \tilde{Z}_{j_0}^*, Y_{j_0}$. The findings discussed above are characterized as pitfalls of network DEA models by Chen et al. [25]. They proposed that under network DEA the envelopment models should be used for deriving the frontier projection for inefficient DMUs and the multiplier ones for the determination of the efficiency scores.

Chen et al. [30] and Cook et al. [36] examined the relations and equivalences of the multiplicative decomposition approach with other existing network DEA methods. In particular, they established the equivalence between the studies of Fare and Grosskopf [51], Chen and Zhu [31], Kao and Hwang [75] and Liang et al. [87]. In particular, Chen et al. [30] showed that the model of Chen and Zhu [31] under CRS assumption is equivalent to the output-oriented model of Kao and Hwang [75] and the centralized output oriented model of Liang et al. [87]. Also, Cook et al. [36] illustrated that model (11), the dual of model (7), is equivalent to the models proposed by Fare and

Grosskopf [51]. All these models, under CRS assumption, provide the same overall efficiency score for the two-stage process of Fig. 1a.

Extensions of the Multiplicative Efficiency Decomposition Approach

The multiplicative decomposition method can be readily applied to series multi-stage processes of Type I but not to general network structures because the assumption that the overall efficiency is the product of the stage efficiencies renders the resulting models highly nonlinear. In Kao [70–73] it is shown that the overall efficiency of a DMU with the parallel structure of Fig. 2 is the weighted average of the stage efficiencies, where the weights are derived from the proportions of inputs utilized by each stage. In these studies, the multiplicative decomposition method of Kao and Hwang [75] is modified to be applied to any type of series and series-parallel multi-stage processes. Their modelling approach is based on the common assumption that the weights associated with the intermediate measures are the same. Also, they deal with general series and series-parallel multi-stage processes by transforming the multi-stage process under evaluation. In particular, dummy sub-processes are introduced in the original multi-stage process, which operate in a parallel configuration with the actual sub-processes. By applying this transformation, the overall efficiency of the system is derived as the product of the efficiencies of the sub-systems, where the efficiency of each modified process (sub-system) is a weighted average of the efficiencies of the processes (real and dummies).

Below we give an example of the aforementioned technique applied to the Electricity Service System (Fig. 4) originally discussed in Tone and Tsutsui [117]. The first process (Generation division) generates electric power (Z^1), which then is used to the second process (Transmission division) in order to be sold to large customers as output (Y^2) or to be sent as intermediate measure (Z^2) to the third process (Distribution division) so as to provide electricity to small customers (Y^3).

The series network structure described above can be transformed via the approach introduced in Kao [70] to the network structure of Fig. 5. The squares and circles represent the actual and dummy processes respectively. The modified network structure contains three subsystems arranged in series, where each of them consists of an actual process and a dummy process operating in parallel. The dummy processes are introduced so as to convey the inputs and the outputs dedicated to specific processes throughout the system. The overall efficiency of the DMUj, the stage efficiencies and the efficiencies of the three sub-systems are defined as follows:

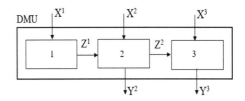

Fig. 4 Electric power generation, transmission and distribution (*Source* Tone and Tsutsui [117])

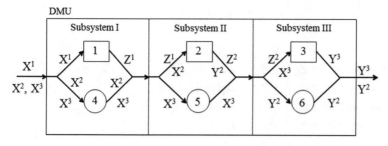

Fig. 5 The transformed network process

$$e_j^o = \frac{u_2 Y_j^2 + u_3 Y_j^3}{v_1 X_j^1 + v_2 X_j^2 + v_3 X_j^3} \qquad e_j^I = \frac{w_1 Z_j^1 + v_2 X_j^2 + v_3 X_j^3}{v_1 X_j^1 + v_2 X_j^2 + v_3 X_j^3},$$

$$e_j^1 = \frac{w_1 Z_j^1}{v_1 X_j^1}, e_j^2 = \frac{w_2 Z_j^2 + u_2 Y_j^2}{w_1 Z_j^1 + v_2 X_j^2}, \quad e_j^{II} = \frac{w_2 Z_j^2 + u_2 Y_j^2 + v_3 X_j^3}{w_1 Z_j^1 + v_2 X_j^2 + v_3 X_j^3}, \tag{12}$$

$$e_j^3 = \frac{u_3 Y_j^3}{w_2 Z_j^2 + v_3 X_j^3} \qquad e_j^{III} = \frac{u_2 Y_j^2 + u_3 Y_j^3}{w_2 Z_j^2 + v_3 X_j^3 + u_2 Y_j^2}$$

As mentioned above, the overall efficiency of a system, whose stages are in parallel, is the weighted average of the stage efficiencies. Thus, in the transformed network (Fig. 5) the efficiency of each sub-system is obtained as the weighted average of the efficiencies of the actual and dummy process. The weights are derived endogenously from the optimization process as the proportions of inputs consumed by each process. Notice that the dummy processes have the same inputs and outputs, therefore their efficiency score is one.

- $e_j^I = t_j^1 e_j^1 + t_j^4 e_j^4 = t_j^1 e_j^1 + \left(1 - t_j^1\right)$ where $t_j^1 = \left(v_1 X_j^1\right) / \left(v_1 X_j^1 + v_2 X_j^2 + v_3 X_j^3\right)$ and $t_j^1 + t_j^4 = 1$

- $e_j^{II} = t_j^2 e_j^2 + t_j^5 e_j^5 = t_j^2 e_j^2 + \left(1 - t_j^2\right)$ where $t_j^2 = \left(w_1 Z_j^1 + v_2 X_j^2\right) / \left(w_1 Z_j^1 + v_2 X_j^2 + v_3 X_j^3\right)$ and $t_j^2 + t_j^5 = 1$

- $e_j^{III} = t_j^3 e_j^3 + t_j^6 e_j^6 = t_j^3 e_j^3 + \left(1 - t_j^3\right)$ where $t_j^3 = \left(w_2 Z_j^2 + v_3 X_j^3\right) / \left(w_2 Z_j^2 + v_3 X_j^3 + u_2 Y_j^2\right)$ and $t_j^3 + t_j^6 = 1$.

From the above mathematical relationships, it follows that the overall (system) efficiency of the DMUj can be calculated as the product of the three sub-systems efficiencies.

$$e_j^o = e_j^I \cdot e_j^{II} \cdot e_j^{III} = \left[t_j^1 e_j^1 + \left(1 - t_j^1\right)\right] \cdot \left[t_j^2 e_j^2 + \left(1 - t_j^2\right)\right] \cdot \left[t_j^3 e_j^3 + \left(1 - t_j^3\right)\right]$$

The resulting model for the performance assessment for DMU j_0 with the above network structure is given as:

$$e_{j_0}^o = max \ u_2 Y_{j_0}^2 + u_3 Y_{j_0}^3$$

$$s.t.$$

$$v_1 X_{j_0}^1 + v_2 X_{j_0}^2 + v_3 X_{j_0}^3 = 1$$

System $\quad u_2 Y_j^2 + u_3 Y_j^3 - v_1 X_j^1 - v_2 X_j^2 - v_3 X_j^3 \leq 0, \quad j = 1, \ldots, n$

1st process $\ w_1 Z_j^1 - v_1 X_j^1 \leq 0, j = 1, \ldots, n$ $\hspace{3cm}$ (13)

2nd process $u_2 Y_j^2 + w_2 Z_j^2 - w_1 Z_j^1 - v_2 X_j^2 \leq 0, \quad j = 1, \ldots, n$

3rdprocess $\ u_3 Y_j^3 - w_2 Z_j^2 - v_3 X_j^3 \leq 0, \quad j = 1, \ldots, n$

$$v_1, v_2, v_3, w_1, w_2, u_1, u_2 \geq 0$$

Once an optimal solution of model (13) is obtained, the overall and the actual stage efficiencies are calculated from the relationships (12). However, the decomposition of the overall efficiency to the stage efficiencies might be not unique [58]. To summarize, the shortcoming of non-unique efficiency scores may occur in the assessment of any type of network structure when the multiplicative decomposition method is applied.

Alternative Multiplicative Efficiency Decomposition Methods

As shown above, a modified version of the multiplicative decomposition approach is proposed in [70, 73] so as to be applicable to any type of series and series-parallel multi-stage processes. Beyond that, alternative methods have been developed whose common characteristic is that the overall efficiency is defined as the product of the stage efficiencies. However, generalizing this assumption to multi-stage networks different to Type I leads to high non-linear models which are difficult to solve. A common solution practice is to use parametric techniques.

Zha and Liang [134] studied a modified two-stage process of Type II (Fig. 1b) where the external inputs are freely allocated between the stages. In the study of Zha and Liang [134], the overall efficiency of the system is derived as the product of the efficiencies of the two individual stages, as in Kao and Hwang [75]. Zha and Liang [134] incorporated game-theory framework and a heuristic procedure so as to overcome the linearization issues raised by the adoption of the multiplicative format of the overall efficiency. In particular, using the concept of Stackelberg (non-cooperative) game, they first computed the lower and upper bounds of the stage efficiencies. Then they incorporated this information into a non-linear cooperative model and by treating the efficiency of one stage as a parameter they succeeded to transform it to a parametric linear program. Their method is illustrated by using the dataset of 30 top U.S. commercial banks which originally studied by Seiford and Zhu [110].

Li et al. [84] studied also a two-stage production process of Type II (Fig. 1b) in the view of cooperative (centralized control) and non-cooperative games (decentralized control). They developed a parametric approach in order to obtain the stage efficiency scores and then the overall efficiency is computed by the product of the stage efficiencies. The evaluation model proposed by Li et al. [84] is non-linear, hence the authors resorted to a heuristic search procedure in order to deal with the non-linearity issues and estimate a global optimal solution that yields the maximum achievable level of overall efficiency. Their approach is demonstrated by evaluating the research and development of 30 Chinese regions.

Additive Efficiency Decomposition

The additive efficiency decomposition method is introduced by Chen et al. [26] for the assessment of the two-stage process of Type I (Fig. 1a) and then is extended by Cook et al. [28] for the evaluation of multi-stage processes of varying structures. Both studies have already received great attention from the research community. In contrast to the multiplicative efficiency decomposition method, the overall efficiency is obtained as a weighted average of the stage efficiencies, where the weights represent the portion of all inputs utilized by each stage. Notably, this aggregation method is used previously in some network DEA studies without, however, being part of a well-established efficiency decomposition framework. For instance, it is first appeared in Beasley [13], who evaluated the efficiency of teaching and research of the UK chemistry and physics departments and viewed them as two processes that operate in parallel and share some resources. The aforementioned aggregation method was also adopted by Cook and Hababou [34], Cook and Green [33] and Jahanshahloo et al. [68], who similarly examined parallel production processes with shared inputs. Amirteimoori and Kordrostami [5] and Amirteimoori and Shafiei [6], aimed to measure the performance of series processes using the aforementioned aggregation method about the overall and the stage efficiencies, however they treated the stages in a non-coordinated manner. In particular, the weights associated with the intermediate measures were different for each stage. In the context of the additive efficiency decomposition method, the overall efficiency and the stage efficiencies, under CRS assumption, of the DMU j are defined as follows:

$$e_j^o = \frac{\omega Y_j + \varphi Z_j}{\eta X_j + \varphi Z_j}$$

$$e_j^1 = \frac{\varphi Z_j}{\eta X_j}, e_j^2 = \frac{\omega Y_j}{\varphi Z_j} \tag{14}$$

$$e_j^o = \frac{\omega Y_j + \varphi Z_j}{\eta X_j + \varphi Z_j} = t_j^1 \frac{\varphi Z_j}{\eta X_j} + t_j^2 \frac{\omega Y_j}{\varphi Z_j}, t_j^1 + t_j^2 = 1$$

$$t_j^1 = \frac{\eta X_j}{\eta X_j + \varphi Z_j}, t_j^2 = \frac{\varphi Z_j}{\eta X_j + \varphi Z_j} \tag{15}$$

The definitions of the stage efficiencies are the same as in the multiplicative method, but the additive method differentiates in the definition of the overall efficiency. In (14) the intermediate measures appear in both terms of the fraction that defines the overall efficiency, meaning that they are considered as inputs and as outputs simultaneously. The decomposition model (15) indicates that the overall efficiency is expressed as a *weighted arithmetic average* of the stage efficiencies. The argument given in Chen et al. [26] for the weights t_j^1 and t_j^2 is that they represent the relative contribution of the two stages to the overall performance of the DMU. The "size" of each stage, as measured by the portion of total resources devoted to each stage, is assumed to reflect their relative contribution to the overall efficiency of the DMU. It is worth to note that as the weights are functions of the optimization variables, they depend on the unit being evaluated and, obviously, they generally differentiate from one unit to another. Thus, the "size" of a stage is not an objective reality, as it is viewed differently from each DMU. Given the above definitions, the input-oriented CRS linear fractional model (16) assesses the overall efficiency of the evaluated unit j_0. By applying the C-C transformation to model (16), its linear equivalent program (17) is obtained:

$$e_{jo}^o = max \frac{\omega Y_{j_o} + \varphi Z_{j_o}}{\eta X_{j_o} + \varphi Z_{j_o}}$$

$$s.t.$$

$$\frac{\varphi Z_j}{\eta X_j} \leq 1, \quad j = 1, \dots, n$$

$$\frac{\omega Y_j}{\varphi Z_j} \leq 1, \quad j = 1, \dots, n$$

$$\eta \geq 0, \varphi \geq 0, \omega \geq 0 \tag{16}$$

$$e_{jo}^o = max\, u Y_{j_o} + w Z_{j_o}$$

$$s.t.$$

$$v X_{j_o} + w Z_{j_o} = 1$$

$$w Z_j - v X_j \leq 0, \quad j = 1, \dots, n$$

$$u Y_j - w Z_j \leq 0, \quad j = 1, \dots, n$$

$$v \geq 0, w \geq 0, u \geq 0 \tag{17}$$

Once an optimal solution (v^*, w^*, u^*) of model (17) is obtained, then the overall efficiency and the stage efficiencies are calculated as follows:

$$e_{jo}^o = u^* Y_{j_o} + w^* Z_{j_o} \qquad e_{jo}^1 = \frac{w^* Z_{j_o}}{v^* X_{j_o}},$$
$$t_{jo}^1 = v^* X_{j_o}, t_{jo}^2 = w^* Z_{j_o}\, e_{jo}^2 = \frac{e_{jo}^o - t_{jo}^1 e_{jo}^1}{t_{jo}^2} = \frac{u^* Y_{j_o}}{w^* Z_{j_o}} \tag{18}$$

The overall efficiency e_{jo}^o is obtained as the optimal value of the objective function, the weight t_{jo}^1 is obtained as the optimal virtual input, the weight t_{jo}^2 is obtained as the optimal virtual intermediate measure and the efficiency of the first stage e_{jo}^1 is given by the ratio of the two weights whereas the efficiency of the second stage e_{jo}^2 is obtained as offspring of $e_{jo}^o, e_{jo}^1, t_{jo}^1, t_{jo}^2$.

A major shortcoming of the additive decomposition method, as demonstrated in [42, 44], is that it biases the efficiency assessments in favor of the second stage against the first one. Indeed, from the relationships (18) and the definition of the endogenous weights we derive that $t_j^2 \leq t_j^1$, since:

$$\frac{t_j^2}{t_j^1} = \frac{uY_j}{wZ_j} = e_j^1 \leq 1$$

The maximum value that t_j^2 can attain is 0.5 and e_j^2 increases (e_j^1 decreases) as t_j^2 decreases. As long as the individual efficiency scores are biased, the overall efficiency score is biased as well. Notice, that this finding is based upon an input-oriented framework, though it is still valid in the output-oriented case. Also, this conclusion can be easily drawn for other types of series multi-stage processes, regardless of the number of stages. Specifically, when the additive decomposition method is applied to multi-stage processes of Type I, under both input and output orientations, then it suffers from biased efficiency assessments. Also, when it is applied to multi-stage processes of Type III (Fig. 1c), then the aforementioned shortcoming is reported only if input orientation is chosen [9]. When the additive method is applied to parallel network structures (Fig. 2) or under the VRS assumption of any type of network structures, then we cannot predetermine the relationship of the weights of the stages.

Similar to the case of the multiplicative efficiency decomposition, the additive decomposition of the overall efficiency to the stage efficiencies is non-unique. Chen et al. [26] developed a procedure, similar to that of Kao and Hwang [75] and Liang et al. [87], so as to derive extreme efficiency decompositions.

The modelling approach adopted by Chen et al. [26] for the additive efficiency decomposition, enables the straightforward assessment of the two-stage process of Type I (Fig. 1a) under variable returns to scale assumption. Although the additive efficiency decomposition can be straightforwardly applied to variable returns to scale assumption, the standard property that the VRS efficiency scores are not less than the CRS efficiency scores does not hold.

Extensions of the Additive Efficiency Decomposition Approach

Cook et al. [39] extended the additive decomposition method of Chen et al. [26] to series, parallel and series-parallel multi-stage processes. Notably, their modelling approach inherits the defects of additive decomposition method in the sense that the overall efficiency decomposition to stage efficiencies is not unique. However, it can be adapted to meet the VRS assumption.

Chen et al. [28] also extended the work of Chen et al. [26] for the efficiency assessment of two-stage production processes with shared resources. Particularly, they assumed a two-stage process as in Fig. 6 where the second stage uses, beyond the intermediate measures (Z), a portion of the external inputs (X). They applied their models to the assessment of the benefits of information technology in banking industry, originally studied by Wang et al. [123].

Alternative Additive Efficiency Decomposition Methods

As alternative additive decomposition methods are considered those in which, the overall efficiency is defined as a weighted arithmetic average of the stage efficiencies and the weights are predetermined and given as parameters instead of being endogenously estimated by the optimization process. However, notice that using this aggregation method for the efficiency assessment of network structures of any form, leads to non-linear models. Parametric techniques are commonly used to handle the non-linearity issues and yield a global optimal solution.

Liang et al. [86] proposed that the operations of a seller-buyer supply chain can be modelled, under both cooperative and non-cooperative concepts, as a two-stage process of Type II (Fig. 1b). They unified the performance assessment models of the two stages, based on the common assumption that the weights of the intermediate measures are the same in both stages. They defined the overall efficiency as the simple arithmetic average of the stage efficiencies.

Chen et al. [29] examined the IT impact on banking industry previously studied by Wang et al. [123] and Chen and Zhu [31], they remarked though that these studies do not fully characterize the IT impact on firm performance. Therefore, they proposed that the external IT-related inputs of stage-1 should be shared with stage-2 (Fig. 6). Similar to Liang et al. [86], they calculate the overall efficiency from the simple arithmetic average of the stage efficiencies.

Liang et al. [88] studied a serial two-stage production process with feedback, as depicted in Fig. 7. In this system, some outputs from the second process are fed back as inputs to the first process, i.e. they have a double role serving both as inputs and outputs. Similar to the aforementioned studies the overall efficiency is derived as the arithmetic average of the stage efficiencies. Liang et al. [88] illustrated their approach by measuring the performance of 50 Chinese universities. In particular, they assumed as inputs to the first stage the *fixed assets*, the *researchers*, the *graduate students*, and the *size* of each university, while they assumed as outputs the numbers of SCI papers, SCI citations, and national awards. These outputs serve as inputs to the second stage, i.e. they are the intermediate measures of the system, in order to attract research funds from the granting agency. The research funding, which is the only output of the second stage, is fed back to the first stage i.e. it serves also as input.

Fig. 6 Two-stage production process with shared resources

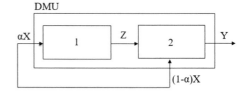

Fig. 7 Two-stage process with feedback (*Source* Liang et al. [88])

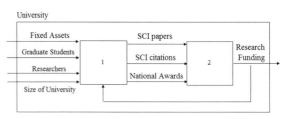

9.3.2.2   The Composition Approach

The composition approach is introduced by Despotis et al. [44] and is based on a reverse perspective on how to obtain and aggregate the stage efficiencies as opposed to the decomposition approach. Unlike the decomposition approach, the composition approach does not require an a priori definition of the overall efficiency and firstly calculates the stage efficiency scores. Estimating the stage efficiencies of multi-stage processes simultaneously can be considered as a multi-objective optimization problem where the efficiency of each stage is treated as a separate objective function and their contradictory nature being taken into account. The modelling framework of the composition methods is based on multi-objective programs that are derived by the composition of the individual assessment models of the stages. The following multi-objective program (MOP) hosts the different composition methods for the efficiency assessment of the evaluated unit j_0.

$$\max e_{j_o}^1$$
$$\max e_{j_o}^2$$
$$s.t.$$
$$e_j^1 \le 1, j = 1, \ldots, n$$
$$e_j^2 \le 1, j = 1, \ldots, n \tag{19}$$

Despotis et al. [44] built a MOP for the performance assessment of the two-stage processes of Type I (Fig. 1a) under CRS assumption, by considering an output-oriented model (20) for stage-1 and an input-oriented model (21) for stage-2. In models (20) and (21) the weights associated with the intermediate measures are assumed the same for both stages, similar to the decomposition methods:

Stage I: Output-oriented

$$min \, \frac{\eta X_{j_0}}{\varphi Z_{j_0}}$$
$$s.t.$$
$$\frac{\eta X_j}{\varphi Z_j} \ge 1, \quad j = 1, \ldots, n$$
$$\eta \ge 0, \varphi \ge 0 \tag{20}$$

Stage II: Input-oriented

$$max \frac{\omega Y_{j_0}}{\varphi Z_{j_0}}$$

s.t.

$$\frac{\omega Y_j}{\varphi Z_j} \leq 1, j = 1, \ldots, n$$

$$\varphi \geq 0, \omega \geq 0 \qquad\qquad\qquad (21)$$

$$min \frac{\eta X_{j_0}}{\varphi Z_{j_0}}$$

s.t.

$$\frac{\eta X_j}{\varphi Z_j} \geq 1, \quad j = 1, \ldots, n$$

$$\frac{\omega Y_j}{\varphi Z_j} \leq 1, \quad j = 1, \ldots, n$$

$$\eta \geq 0, \varphi \geq 0, \omega \geq 0 \qquad\qquad (22)$$

$$max \frac{\omega Y_{j_0}}{\varphi Z_{j_0}}$$

s.t.

$$\frac{\omega Y_j}{\varphi Z_j} \leq 1, \quad j = 1, \ldots, n$$

$$\frac{\eta X_j}{\varphi Z_j} \geq 1, \quad j = 1, \ldots, n$$

$$\eta \geq 0, \varphi \geq 0, \omega \geq 0 \qquad\qquad (23)$$

The models (20) and (21) provide the independent efficiency scores $1/E_{j_0}^1, E_{j_0}^2$ for the first and the second stage respectively. Appending the constraints of model (20) to (21) and vice versa the augmented models (22) and (23) are derived for the first and the second stage respectively. Notice that an optimal solution of model (20) is also optimal in model (22). Indeed, one can always choose small enough values for ω in model (22) to make any optimal solution of model (20) feasible, yet optimal, in model (22). Analogously, an optimal solution of model (21) is also optimal in model (23), as one can choose large enough values for η in model (23) to make any optimal solution of model (21) feasible, yet optimal, in model (23). Models (22) and (23) have common constraints and, thus, can be jointly considered as a bi-objective program (24), which can be formulated and solved as a multi-objective linear program

(MOLP) (25) by applying the C–C transformation. The correspondence of variables is $v = \tau \eta$, $u = \tau \omega$, $w = \tau \varphi$ where τ is a scalar variable such that $\tau \varphi Z_{j_0} = 1$.

$$min \frac{\eta X_{j_0}}{\varphi Z_{j_0}}$$

$$max \frac{\omega Y_{j_0}}{\varphi Z_{j_0}}$$

$s.t.$

$$\frac{\eta X_j}{\varphi Z_j} \geq 1, \quad j = 1, \ldots, n$$

$$\frac{\omega Y_j}{\varphi Z_j} \leq 1, \quad j = 1, \ldots, n$$

$$\eta \geq 0, \varphi \geq 0, \omega \geq 0 \tag{24}$$

$$E_{jo}^1 = min \, v X_{j_0}$$

$$E_{jo}^2 = max \, u Y_{j_0}$$

$s.t.$

$$w Z_{j_0} = 1$$

$$w Z_j - v X_j \leq 0, \quad j = 1, \ldots, n$$

$$u Y_j - w Z_j \leq 0, \quad j = 1, \ldots, n$$

$$v \geq 0, w \geq 0, u \geq 0 \tag{25}$$

Optimizing the first and the second objective function separately one gets the independent efficiency scores of the two stages $\left(1/E_{jo}^1 \leq 1, E_{jo}^2 \leq 1 \right)$. In terms of MOLP, the vector $\left(E_{jo}^1 \geq 1, E_{jo}^2 \leq 1 \right)$ constitutes the ideal point of the bi-objective linear program (25) in the objective functions space. Thus, the efficiencies of the two stages can be obtained by solving the MOLP (25). However, as the ideal point is not generally attainable, solving a MOLP means finding non-dominated feasible solutions in the variable space that are mapped on the Pareto front in the objective functions space, i.e. solutions that they cannot be altered to increase the value of one objective function without decreasing the value of at least one other objective function. As already noticed, a usual approach in solving a MOLP is the scalarizing approach, which transforms the MOLP in a single objective linear program (LP), whose optimal solution is a Pareto optimal (non-dominated) solution of the MOLP. Despotis et al. [44] built a scalarizing function by employing the unweighted Tchebycheff norm (L_∞ norm). By adopting this scalarization method they locate a unique solution on the Pareto front by minimizing the maximum of the deviations $v X_{j_o} - E_{j_o}^1$ and

$E_{j_o}^2 - uY_{j_o}$ of (vX_{j_o}, uY_{j_o}) from the ideal point $\left(E_{j_o}^1, E_{j_o}^2\right)$. This is accomplished by the following min-max model, where δ denotes the largest deviation:

$$min\ \delta$$

$$s.t.$$

$$vX_{j_0} - \delta \leq E_{jo}^1$$

$$uY_{j_0} + \delta \geq E_{jo}^2$$

$$wZ_{j_0} = 1$$

$$wZ_j - vX_j \leq 0, \quad j = 1, \ldots, n$$

$$uY_j - wZ_j \leq 0, \quad j = 1, \ldots, n$$

$$v \geq 0, w \geq 0, u \geq 0, \delta \geq 0 \tag{26}$$

Solving model (26) means searching for a solution where the deviations from the ideal point are equal and minimized. The main advantage of model (26) over the decomposition models (7) and (17) is that it provides a unique point, not necessarily extreme (vertex), on the Pareto front, i.e. unique efficiency scores for the two stages. Also, it provides neutral efficiency scores as opposed to the additive decomposition model (17).

The stage efficiency scores for unit j_0 are derived from the optimal solution (δ^*, v^*, w^*, u^*) of model (26) and then the overall efficiency can be calculated. As noted by Despotis et al. [44] their method grants the flexibility, as the stage efficiencies are assumption-free, to select the aggregation method for the calculation of the overall efficiency a posteriori. Cook et al. [36] noticed that it is reasonable to define the overall efficiency of the two-stage process either as the average (arithmetic mean) of the efficiencies of the two individual stages or as their product. In this line of thought, the overall efficiency of unit j_0 can be aggregated as follows:

$$\hat{e}_{j_o}^o = \frac{1}{2}\left(\hat{e}_{j_o}^1 + \hat{e}_{j_o}^2\right) \text{or}\ \hat{e}_{j_o}^o = \hat{e}_{j_o}^1 \cdot \hat{e}_{j_o}^2 = \frac{1}{v^*X_{j_o}} \cdot u^*Y_{j_o} = \frac{u^*Y_{j_o}}{v^*X_{j_o}}$$

Extensions of the Composition Approach

Despotis et al. [46] extended the composition approach [44] so as to be applicable to the performance evaluation of general series multi-stage processes. They differentiate from Despotis et al. [44] by employing two models with the same orientation, which estimate the stage-1 and the stage-2 efficiencies so as to build the assessment model of their method. In particular, for the performance assessment of the two-stage processes of Type I (Fig. 1a) under CRS assumption, they appended the constraints of model (1) to model (2), and vice versa, without affecting their optimal efficiency scores. Thus, the augmented forms of the input-oriented models (1) and (2) can be written as follows:

$$E_{j_0}^1 = max\frac{\varphi Z_{j_0}}{\eta X_{j_0}}$$

$s.t.$

$$\varphi Z_j - \eta X_j \leq 0, \quad j = 1, \ldots, n$$
$$\omega Y_j - \varphi Z_j \leq 0, \quad j = 1, \ldots, n$$
$$\eta \geq \varepsilon, \varphi \geq \varepsilon, \omega \geq \varepsilon \tag{27}$$

$$E_{j_0}^2 = max\frac{\omega Y_{j_0}}{\varphi Z_{j_0}}$$

$s.t.$

$$\varphi Z_j - \eta X_j \leq 0, \quad j = 1, \ldots, n$$
$$\omega Y_j - \varphi Z_j \leq 0, \quad j = 1, \ldots, n$$
$$\eta \geq \varepsilon, \varphi \geq \varepsilon, \omega \geq \varepsilon \tag{28}$$

In models (27) and (28) the parameter ε is a non-Archimedean infinitesimal. Models (27) and (28) have common constraints, hence they can be jointly considered as a bi-objective program (29). The bi-objective mathematical program (29) or its equivalent model (30) is used for the performance assessment of the two-stage process of Type I (Fig. 1a):

$$max\, e^1 = \frac{\varphi Z_{j_0}}{\eta X_{j_0}}$$
$$max\, e^2 = \frac{\omega Y_{j_0}}{\varphi Z_{j_0}}$$

$s.t.$

$$\varphi Z_j - \eta X_j \leq 0, \quad j = 1, \ldots, n$$
$$\omega Y_j - \varphi Z_j \leq 0, \quad j = 1, \ldots, n$$
$$\eta \geq \varepsilon, \varphi \geq \varepsilon, \omega \geq \varepsilon \tag{29}$$

$$max\, e^1 = wZ_{j_0}$$
$$max\, e^2 = \frac{uY_{j_0}}{wZ_{j_0}}$$

$s.t.$

$$vX_{j_0} = 1$$
$$wZ_j - vX_j \leq 0, \quad j = 1, \ldots, n$$
$$uY_j - wZ_j \leq 0, \quad j = 1, \ldots, n$$
$$v \geq \varepsilon, w \geq \varepsilon, u \geq \varepsilon \tag{30}$$

Model (30) derives from model (29) by applying the C–C transformation with respect to the first objective function, i.e. by multiplying all the terms of the fractional objective functions and the constraints by $t > 0$, such that $t\eta X_{j_0} = 1$ and setting $t\eta = v$, $t\omega = u$, $t\varphi = w$. Notice that in model (30) the second objective function is still in fractional form.

The efficiencies of the two stages can be obtained by solving the bi-objective program (30). Despotis et al. [46] converted the bi-objective program (30) into a single objective program by employing the unweighted Tchebycheff norm (L_∞ norm). The single objective program (31) below, locates a point on the upper-right boundary of the feasible region in the objectives functions space, by minimizing the maximum of the deviations $\left(E_{j_0}^1 - e_{j_0}^1\right)$ and $\left(E_{j_0}^2 - e_{j_0}^2\right)$ of the stage efficiencies $\left(e_{j_0}^1 = wZ_{j_0}, e_{j_0}^2 = uY_{j_0}/wZ_{j_0}\right)$ from the ideal point $(E_{j_0}^1, E_{j_0}^2)$. The ideal efficiency scores $E_{j_0}^1$ and $E_{j_0}^2$ for the two stages are obtained by solving models (27) and (28).

$$\min \delta$$
$$s.t.$$
$$E_{j_0}^1 - wZ_{j_0} \leq \delta$$
$$E_{j_0}^2 - \frac{uY_{j_0}}{wZ_{j_0}} \leq \delta$$
$$vX_{j_0} = 1$$
$$wZ_j - vX_j \leq 0, \quad j = 1, \ldots, n$$
$$uY_j - wZ_j \leq 0, \quad j = 1, \ldots, n$$
$$v \geq 0, w \geq 0, u \geq 0 \tag{31}$$

The largest deviation is denoted by δ in the min-max model (31). Although model (31) is non-linear, it can be easily solved by bisection search [41]. Clearly, $0 \leq \delta \leq 1$. Hence bisection search can be performed in the bounded interval [0, 1] as follows. Let $\underline{\delta}$ be a lower bound of δ for which the constraints of (31) are not consistent (initially $\underline{\delta} = 0$) and $\bar{\delta}$ an upper bound of δ for which the constraints are consistent (initially $\bar{\delta} = 1$). Then the consistency of the constraints is tested for $\delta' = (\underline{\delta} + \bar{\delta})/2$. If they are consistent, δ' will replace $\bar{\delta}$; if they are not, then it will replace $\underline{\delta}$. The bisection continues until both bounds come sufficiently close to each other. At optimality, at least one of the first two constraints in (31) will be binding. Let $(\delta^*, v^*, w^*, u^*)$ be an optimal solution of (31), then the efficiency scores for unit j_0 in the first and the second stage are respectively:

$$e_{j_0}^{1*} = \frac{w^*Z_{j_0}}{v^*X_{j_0}} = w^*Z_{j_0}, e_{j_0}^{2*} = \frac{u^*Y_{j_0}}{w^*Z_{j_0}}$$

Every optimal solution of (31) is *weakly* efficient (*weakly* Pareto optimal) solution for (30). The model (32) below provides a Pareto optimal solution to (30). The model (32) is equivalent to employing lexicographically (in a second phase) the L_1 norm on the set of optimal solutions of (31).

$$\max s_1 + s_2$$

$$s.t.$$

$$E_{j_0}^1 - wZ_{j_0} + s_1 = \delta^*$$

$$\left(E_{j_0}^2 - \delta^*\right)wZ_{j_0} - uY_{j_0} + s_2 w^* Z_{j_0} = 0$$

$$vX_{j_0} = 1$$

$$wZ_j - vX_j \leq 0, \quad j = 1, \ldots, n$$

$$uY_j - wZ_j \leq 0, \quad j = 1, \ldots, n$$

$$v \geq \varepsilon, w \geq \varepsilon, u \geq \varepsilon$$

$$\delta^* \geq s_1 \geq 0, \delta^* \geq s_2 \geq 0 \tag{32}$$

In (32), δ^* is the optimal value of the objective function of (31) and $w^* Z_{j_0}$ is the optimal virtual intermediate measure derived by model (31). Notice here that the term $w^* Z_{j_0}$ is used as an effective substitute of wZ_{j_0} to secure the linearity of the model. In case that $s_2 > 0$ in the optimal solution of (32), the program is solved iteratively by replacing in each iteration the weights w in the coefficient of s_2 with the optimal weights w obtained in the preceding iteration, until the stage efficiencies in two successive iterations remain unchanged [41]. The optimal solution $\left(\hat{s}_1, \hat{s}_2, \hat{v}, \hat{w}, \hat{u}\right)$ of model (32) is a Pareto optimal solution of (30) and the efficiency scores for unit j_0 in the first and the second stage respectively are:

$$\hat{e}_{j_0}^1 = \frac{\hat{w}Z_{j_0}}{\hat{v}X_{j_0}} = \hat{w}Z_{j_0}, \hat{e}_{j_0}^2 = \frac{\hat{u}Y_{j_0}}{\hat{w}Z_{j_0}}$$

Since the optimal solution of (31) is weakly Pareto optimal, in (32), at most one of the two optimal values of the variables $\hat{s}_1$ and $\hat{s}_2$ will be strictly positive. If $\hat{s}_1 = 0$ and $\hat{s}_2 = 0$, then the optimal solution of (31) is Pareto optimal. Once the Pareto optimality of the stage efficiencies is secured, then the overall efficiency of the system is optimized on the set of optimal solutions of model (32), i.e. by incorporating the constraints $e_{j_0}^1 \geq \hat{e}_{j_0}^1$ and $e_{j_0}^2 \geq \hat{e}_{j_0}^2$ into the assessment.

$$\max uY_{j_0}$$

$$s.t.$$

$$wZ_{j_0} \geq \hat{w}Z_{j_0}$$

$$uY_{j_0} - \left(\hat{w}Z_{j_0}\right)wZ_{j_0} \geq 0$$

$$vX_{j_0} = 1$$

$$wZ_j - vX_j \leq 0, \quad j = 1, \ldots, n$$

$$uY_j - wZ_j \leq 0, \quad j = 1, \ldots, n$$

$$v \geq 0, w \geq 0, u \geq 0 \tag{33}$$

Given an optimal solution (u', w', v') of model (33), then the overall efficiency score is calculated as:

$$e_{j_0}^{\prime o} = \frac{u'Y_{j_0}}{v'X_{j_0}} = u'Y_{j_0}, \; e_{j_0}^{\prime 1} = \hat{e}_{j_0}^1, \; e_{j_0}^{\prime 2} = \hat{e}_{j_0}^2$$

The method of Despotis et al. [46] inherits the properties of the composition approach, i.e. it provides unique and unbiased efficiency scores. Despotis et al. [43] applied the method of Despotis et al. [46] to the assessment of the academic research activity. They considered the research activity of an individual staff member as a two-stage process of Type II (Fig. 1b), where the first stage represents the *productivity* of the individual and the second stage represents the impact and the recognition that the research work of the individual has in academia. The inputs in stage-1 are *time in post* and *total salary* since appointment, while the output of the stage-1 is *publications*. The publications made by an individual before his appointment are extra inputs to the stage-2, while the outputs of the stage-2 are the number of *citations* and the *academic achievements* of the individual, such as being member of editorial boards, being invited as keynote speaker in conferences, etc.

Alternative Composition Methods

Despotis et al. [45] developed the "*weak link*" approach in the frame of the composition approach. They introduced a novel definition of the system efficiency in two-stage processes of varying complexity, inspired by the "weak link" notion in supply chains and the maximum-flow/minimum-cut problem in networks [12].

The multiplicative decomposition method estimates the overall system efficiency of the evaluated unit as the squared geometric average of the stage efficiencies. Given that $e_{j_0}^1 \leq 1$ and $e_{j_0}^2 \leq 1$ it is $e_{j_0}^o \leq min\left\{e_{j_0}^1, e_{j_0}^2\right\}$. The latter hold if at least one of the two stages is efficient, i.e. if $e_{j_0}^1 = 1$ and/or $e_{j_0}^2 = 1$. This property declares that the less efficient stage is determinant of the overall system efficiency. This is a natural property that can be easily identified in multi-stage processes, such as in supply chains. In such a context, the less efficient stage is called the "weak link" of the supply chain. In this line of thought, Kao [73] states that "*efficiency decomposition enables decision makers to identify the stages that cause the inefficiency of the system, and to effectively improve the performance of the system*". However, in order to draw safe conclusions about the system efficiency, the identification of the weak link should meet two properties: (a) uniqueness and (b) being supported by a reasonable and meaningful search orientation. As mentioned in Kao and Hwang [75], the stage efficiency scores obtained by the multiplicative method are not unique in general, i.e. different efficiency scores can be obtained for the two stages that maintain the same overall efficiency. Consequently, the weak link might be interchanged between the two stages, depending on the decomposition selected. Thus, as already remarked, the uniqueness property is not met by the model (7).

Fig. 8 An alternative representation of the two-stage process of Type I

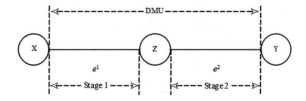

Despotis et al. [45] defined the system efficiency as the minimum of the stage efficiencies:

$$e^o = min\{e^1, e^2\} \tag{34}$$

They provided am alternative representation of the two-stage process of Type I as a *max flow-min cut* analogue. Figure 8 depicts the alternative representation of the basic two-stage process, where the role of nodes and links is interchanged. The link that connects the nodes X and Z represents the first stage of the process, with the nodes X and Z representing the inputs to and the outputs from the first stage. The second link represents the second stage, whose inputs and outputs are Z and Y respectively. At the DMU level, X and Y are the external inputs and outputs respectively and Z represents the intermediate measures linking the two-stages. The labels e^1 and e^2 assigned to the links represent their capacity, that is the efficiencies of the stages. Given the stage efficiencies e^1 and e^2, the system efficiency e^o can be viewed as the maximum flow through the two-stage network and can be estimated as the min-cut of the network, which, in the case of the simple network of Fig. 8, is given by the minimal of the capacities of the two links, i.e. $e^o = min\{e^1, e^2\}$.

The aim of the "weak link" approach is to estimate the capacities (individual efficiency scores) of the two stages in a manner that the minimal capacity (the weak link) and, thus, the overall system efficiency gets the maximum possible value. The mathematical representation of this notion is expressed by the weighted max-min formulation which seeks to maximize the minimum weighted achievement from zero-level efficiency:

$$e^o_{j_o} = max_{v,w,u}\left[min\{q_1 e^1_{j_o}, q_2 e^2_{j_o}\}\right] \tag{35}$$

where $q_1 > 0$ and $q_2 > 0$ are strictly positive parameters (weights). A reasonable pair of values for these parameters is $q_1 = 1/E^1_{j_o}$ and $q_2 = 1/E^2_{j_o}$ [14, 89]. i.e. the driver of the assessment is the ideal stage efficiency scores. This implies that the estimated stage efficiency scores will be proportional to their ideal (independent) counterparts.

The modelling framework of the "weak link" approach is based on the MOP (30) and the max-min scalarization method (35). With respect to the bi-objective mathematical program (30), the search for the individual scores of the two stages is made in two phases: *Phase* I locates a point on the upper-right boundary of the feasible region in the objective functions space of (30) by means of the following max-min model (36), which maximizes the minimal efficiency score, whereas Phase II provides a Pareto optimal solution.

Phase I:

$$
\begin{aligned}
& max \; \theta \\
& s.t. \\
& wZ_{j_0} \geq \theta \; \mathrm{E}^1_{j_0} \\
& \frac{uY_{j_0}}{wZ_{j_0}} \geq \theta \; \mathrm{E}^2_{j_0} \\
& vX_{j_0} = 1 \\
& wZ_j - vX_j \leq 0, \quad j = 1, \ldots, n \\
& uY_j - wZ_j \leq 0, \quad j = 1, \ldots, n \\
& v \geq \varepsilon, w \geq \varepsilon, u \geq \varepsilon, \theta \geq 0
\end{aligned}
\tag{36}
$$

Model (36) is the canonical form of the max-min model (35). The solution of a weighted max-min problem, such as model (36), is weakly Pareto optimal. At optimality, at least one of the first two constraints in (36) will be binding. Although model (36) is non-linear, it can be solved by bisection search in terms of θ in the bounded interval $[0, 1]$, since $0 \leq \theta \leq 1$ [44]. Let $\underline{\theta}$ be a lower bound of θ for which the constraints of (36) are consistent (initially $\underline{\theta} = 0$) and $\bar{\theta}$ an upper bound of θ for which the constraints are not consistent (initially $\bar{\theta} = 1 + \varepsilon$), where ε is a very small positive number. Then the consistency of the constraints is tested for $\theta' = (\underline{\theta} + \bar{\theta})/2$. If they are consistent, θ' will replace $\underline{\theta}$; if they are not it will replace $\bar{\theta}$. The bisection search continues until both bounds come sufficiently close to each other.

Let $(\theta^*, v^*, w^*, u^*)$ be an optimal solution of (36), which is *weak Pareto* for the MOP (30), then the stage efficiencies are calculated as:

$$
e^{1*}_{j_0} = \frac{w^*Z_{j_0}}{v^*X_{j_0}} = w^*Z_{j_0}, e^{2*}_{j_0} = \frac{u^*Y_{j_0}}{w^*Z_{j_0}}
$$

As the *weak Pareto* is a weaker property than *Pareto optimality*, it is not unlikely that the solution of (36) is Pareto optimal. However, Despotis et al. [45] proposed, for the Phase II, the model (37) that provides a Pareto optimal solution to (30). The model (37) is equivalent to employing lexicographically the L_1 norm on the set of optimal solutions of (36).

Phase II:

$$max \ s_1 + s_2$$
$$s.t.$$
$$wZ_{j_0} - s_1 = e_{j_0}^{1*}$$
$$uY_{j_0} - s_2 w^* Z_{j_0} - e_{j_0}^{2*} w Z_{j_0} = 0$$
$$vX_{j_0} = 1$$
$$wZ_j - vX_j \le 0, \quad j = 1, \ldots, n$$
$$uY_j - wZ_j \le 0, \quad j = 1, \ldots, n$$
$$0 \le s_1 \le E_{j_0}^1, 0 \le s_2 \le E_{j_0}^2$$
$$v \ge \varepsilon, w \ge \varepsilon, u \ge \varepsilon \tag{37}$$

As long as the solution obtained in Phase I is weakly Pareto, in Phase II at most one of the two optimal values of the variables $\hat{s}_1$ and $\hat{s}_2$ will be strictly positive (i.e. $\hat{s}_1 \hat{s}_2 = 0$). If $\hat{s}_1 = 0$ and $\hat{s}_2 = 0$, then the Phase I solution is Pareto optimal.

In model (37), the second constraint originates from its original non-linear form $\left(uY_{j_0}/wZ_{j_0} \right) - s_2 = e_{j_0}^{2*}$ or $uY_{j_0} - s_2 wZ_{j_0} - e_{j_0}^{2*} wZ_{j_0} = 0$, where the virtual measure wZ_{j_0} in the second term is replaced by $s_1 + e_{j_0}^{1*} = s_1 + w^* Z_{j_0}$, as per the first constraint, to get $uY_{j_0} - s_1 s_2 - s_2 w^* Z_{j_0} - e_{j_0}^{2*} wZ_{j_0} = 0$. At optimality, it is $s_1 s_2 = 0$, because at least one of the two variables will be zero. Therefore, the non-linear term $s_1 s_2$ can be omitted without altering the optimal solution, to get the linear form $uY_{j_0} - s_2 w^* Z_{j_0} - e_{j_0}^{2*} wZ_{j_0} = 0$. Once the optimal solution $\left(\hat{s}_1, \hat{s}_2, \hat{v}, w, \hat{u} \right)$ of (37) is obtained, the individual stage efficiency scores for unit j_0 as well as the overall efficiency of the system, according to the definition (34), are respectively:

$$\hat{e}_{j_0}^1 = \frac{\hat{w} Z_{j_0}}{\hat{v} X_{j_0}} = \hat{w} Z_{j_0}, \hat{e}_{j_0}^2 = \frac{\hat{u} Y_{j_0}}{\hat{w} Z_{j_0}}, \hat{e}_{j_0}^o = min \left\{ \hat{e}_{j_0}^1, \hat{e}_{j_0}^2 \right\}$$

Illustration

We utilize the synthetic data originally presented in Despotis et al. [46] so as to provide a graphical illustration of the similarities and the dissimilarities of the efficiency decomposition and composition methodologies. The data set comprises of 30 DMUs with two inputs (X1, X2), two intermediate measures (Z1, Z2) and two outputs (Y1, Y2).

Figure 9 depicts the Pareto front (curve AD) and the optimal efficiency scores obtained for a specific DMU, namely the DMU 18, by employing the decomposition and the composition approaches. The coordinates of the point E(0.5046, 1) are the independent efficiency scores of the two stages. This is the ideal point in the multi-objective programming terminology. The points A(0.2378,1) and D(0.5046, 0.4910) are the extreme points on the upper-right boundary of the feasible set in the objective functions space.

Fig. 9 The Pareto front for
unit 18 in the (e^1, e^2) space

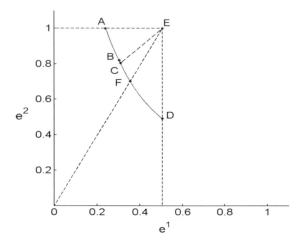

The points B(0.3021, 0.8201) and D represent the two extreme efficiency decompositions derived by the multiplicative efficiency decomposition method, i.e. from models (9) and (8) respectively. The segment BD of the Pareto front depicts alternative efficiency decompositions, all maintaining the same optimal level of overall efficiency $e^O = 0.2477$. Among them, D is located as an optimal solution by the multiplicative decomposition model (7). Point D depicts also the stage efficiency scores derived by the additive decomposition model (17). The additive efficiency decomposition for the DMU 18 is unique.

In the frame of the composition approach, the min-max method proposed by Despotis et al. [46], i.e. models (31) and (32), yields the stage efficiency scores depicted by the unique Pareto optimal point C(0.3082, 0.8037). The point C is derived by the intersection of the Pareto front AD with a ray from the ideal point E with direction $(-1, -1)$. The point F(0.3536, 0.7007) depicts the unique Pareto optimal point obtained by model (37)-same as from model (36). The point F derived by the "*weak link*" approach is formed by the intersection of the dotted line OE with the Pareto front. The overall system efficiency of unit 18 is $\hat{e}^o_{18} = min\{0.3536, 0.7007\} = 0.3536$.

The modelling techniques used in the min-max method and the "weak-link" approach resemble structurally, since they achieve their optimal solutions by employing similar scalarizing functions. However, they differ in the reference points and the search direction employed for locating the Pareto optimal solutions. The min-max method locates a point on the Pareto front AD by reaching down from the ideal point towards the feasible set along a ray with direction $(-1, -1)$. On the contrary, the max-min optimization technique, used in the "weak-link" approach, locates a point on the Pareto front AD by pushing out from the zero-level efficiency point $(0, 0)$ towards the ideal point E. If the components of the point E are equal $(E^1_{j_0} = E^2_{j_0})$, then the two methods locate the same point on the Pareto front. This holds because, in such a case, the ray from the origin to the point E coincides with the search direction of the un-weighted min-max method.

9.3.2.3 Slacks-Based Measure Approach

Tone and Tsutsui [117] introduced the network slacks-based measure (NSBM), based on the SBM and the weighted SBM proposed by Tone [116] and Tsutsui and Goto [121] respectively. This approach assesses simultaneously the overall and the stage efficiencies of the evaluated units. The method is built in the envelopment form based on a generalized production possibility set that describes the relationships of the multi-stage processes. In particular, it is assumed that a DMU consists of ν sub-processes ($\gamma = 1, ..., \nu$), where each sub-process consumes external inputs X^γ to produce some outputs Y^γ (the superscript γ denotes the sub-process). The sub-processes are connected and interact via the intermediate measures $Z^{(\gamma,\psi)}$, where the superscripts γ and ψ ($\gamma \neq \psi$) represent the source sub-process and the recipient sub-process respectively. The generalized production possibility set $\{X^\gamma, Z^{(\gamma,\psi)}, Y^\gamma\}$ under VRS assumption is defined as:

$$
\begin{aligned}
&X^\gamma h^\gamma \leq X^\gamma, \quad \gamma = 1, \ldots, \nu \\
&Z^{(\gamma,\psi)} h^\gamma = Z^{(\gamma,\psi)}, \quad \forall (\gamma, \psi) \ (as\ outputs\ from\ \gamma) \\
&Z^{(\gamma,\psi)} h^\psi = Z^{(\gamma,\psi)}, \quad \forall (\gamma, \psi) \ (as\ inputs\ to\ \psi) \\
&\Upsilon^\gamma h^\gamma \geq Y^\gamma, \quad \gamma = 1, \ldots, \nu \\
&h^\gamma \geq 0, \quad \forall (\gamma) \\
&e h^\gamma = 1, \quad \forall (\gamma)
\end{aligned}
\tag{38}
$$

Notice that the intensity vector h^γ is specific to each sub-process γ ($\gamma = 1, ..., \nu$). The above VRS production possibility set can be also used under CRS assumption by removing the last set of convexity constraints ($h^\gamma = 1$). Tone and Tsutsui [117] proposed two options for representing the constraints corresponding to the intermediate measures:

(a) The "free" link case, where the linking flows are freely determined, i.e. $Z^{(\gamma,\psi)} h^\gamma = Z^{(\gamma,\psi)} h^\psi, \forall (\gamma, \psi)$. In this case, the intermediate measures that link the stages are tested in the light of the other DMUs. Hence, the intermediate measures may increase or decrease in order to preserve the continuity of being simultaneously outputs of one stage and inputs to some other.

(b) The "fixed" link case, where the intermediate measures are kept unchanged on their initial levels, i.e. $Z^{(\gamma,\psi)} h^\gamma = Z_{j_o}^{(\gamma,\psi)}, Z^{(\gamma,\psi)} h^\psi = Z_{j_o}^{(\gamma,\psi)}, \forall (\gamma, \psi)$. Notice that the subscript j_o denotes the DMU under evaluation. Tone and Tsutsui [117] remarked that if all the intermediate measures are fixed to their original levels then the analysis to follow will treat the stages separately similar to the independent assessment.

By incorporating the input and output slacks and one of the above options for the constraints of the intermediate measures, the DMUj_o under evaluation is expressed as follows:

$$
\begin{aligned}
X^\gamma h^\gamma + s^{\gamma-} &= X_{j_o}^\gamma, & \gamma &= 1, \ldots, \upsilon \\
Y^\gamma h^\gamma - s^{\gamma+} &= Y_{j_o}^\gamma, & \gamma &= 1, \ldots, \upsilon \\
e h^\gamma &= 1, & \gamma &= 1, \ldots, \upsilon \\
h^\gamma &\geq 0, s^{\gamma-} \geq 0, s^{\gamma+} \geq 0, & \gamma &= 1, \ldots, \upsilon
\end{aligned}
\tag{39a}
$$

$$
Z^{(\gamma,\psi)} h^\gamma - Z^{(\gamma,\psi)} h^\psi = 0, \quad \forall(\gamma, \psi) \quad \textit{(free link)} \tag{39b}
$$

or

$$
\begin{aligned}
Z^{(\gamma,\psi)} h^\gamma &= Z_{j_o}^{(\gamma,\psi)}, & \forall(\gamma, \psi) \\
Z^{(\gamma,\psi)} h^\psi &= Z_{j_o}^{(\gamma,\psi)}, & \forall(\gamma, \psi)
\end{aligned}
\quad \textit{(fixed link)} \tag{39c}
$$

Tone and Tsutsui [117], similar to the conventional SBM, proposed three different efficiency measures based on the orientation, they formed the input, the output and the non-oriented situation. When input orientation is selected then the overall efficiency of the DMUj_o is derived as a weighted arithmetic mean of the slacks-based measures of the individual stages, i.e. $e_{j_o}^o = \sum_{\gamma=1}^{\upsilon} w^\gamma \cdot e_{j_o}^\gamma$, with $\sum_{\gamma=1}^{\upsilon} w^\gamma = 1$ and $w^\gamma \geq 0$. The weights w^γ are predefined by the analyst and represent the importance of each stage. The input-oriented NSBM model for the efficiency assessment of the DMUj_o is as follows:

$$
e_{j_0}^o = \min \sum_{\gamma=1}^{\upsilon} w^\gamma \left[1 - \frac{1}{m_\gamma} \left(\sum_{i=1}^{m_\delta} \frac{s_i^{\gamma-}}{x_{i_{j_o}}^\gamma} \right) \right]
$$
subject to (39a), (39b) *or* (39c) $\tag{40}$

In model (40), the number of inputs consumed by each stage γ is denoted by m_γ, also the *free* or the *fixed* link case can be used to represent the constraints corresponding to the intermediate measures. When an optimal solution of model (40) is obtained then the overall efficiency can be directly obtained from its objective function and the stage efficiencies are calculated using the optimal input slacks $s^{\gamma-*}$ as follows:

$$
e_{j_0}^\gamma = 1 - \frac{1}{m_\gamma} \left(\sum_{i=1}^{m_\gamma} \frac{s_i^{\gamma-*}}{x_{i_{j_o}}^\gamma} \right), \gamma = 1, \ldots, \upsilon \tag{41}
$$

When output orientation is selected, then the following NSBM model is used for the performance assessment of the DMUj_o:

$$\frac{1}{\theta_{jo}^o} = max \sum_{\gamma=1}^{\nu} w^{\gamma} \left[1 + \frac{1}{s_{\gamma}} \left(\sum_{r=1}^{s_{\gamma}} \frac{s_r^{\gamma+}}{y_{rjo}^{\gamma}} \right) \right]$$

subject to (39a), (39b) or (39c)   (42)

The output oriented overall efficiency for DMUj_o is derived from the optimal value of the objective function of model (42). The authors in order to confine the efficiency scores into the range [0, 1], they expressed the output-oriented stage efficiency scores using the optimal output slacks $s^{\gamma+*}$ as:

$$\theta_{jo}^{\gamma} = \frac{1}{1 + \frac{1}{s_{\gamma}} \left(\sum_{r=1}^{s_{\gamma}} \frac{s_r^{\gamma+*}}{y_{rjo}^{\gamma}} \right)}, \gamma = 1, \ldots, \nu \qquad (43)$$

As can be deduced the NSBM output oriented overall efficiency is the weighted harmonic mean of the stage efficiency scores:

$$\frac{1}{\theta_{jo}^o} = \sum_{\gamma=1}^{\nu} \frac{w^{\gamma}}{\theta_{jo}^{\gamma}} \qquad (44)$$

In case non-orientation is selected, i.e. when both input and output slacks are taken into consideration in the assessment, then the non-oriented NSBM model is expressed as:

$$\zeta_{jo}^o = min \frac{\sum_{\gamma=1}^{\nu} w^{\gamma} \left[1 - \frac{1}{m_{\gamma}} \left(\sum_{i=1}^{m_{\gamma}} \frac{s_i^{\gamma-}}{x_{ijo}^{\gamma}} \right) \right]}{\sum_{\gamma=1}^{\nu} w^{\gamma} \left[1 + \frac{1}{s_{\gamma}} \left(\sum_{r=1}^{s_{\gamma}} \frac{s_r^{\gamma+}}{y_{rjo}^{\gamma}} \right) \right]}$$

subject to (39a), (39b) or (39c)   (45)

Given the optimal solution of model (45), then the non-oriented overall efficiency is straightforwadly derived from the objective function while the non-oriented stage efficiency scores are calculated as follows:

$$\zeta_{jo}^{\gamma} = min \frac{1 - \frac{1}{m_{\gamma}} \left(\sum_{i=1}^{m_{\gamma}} \frac{s_i^{\gamma-*}}{x_{ijo}^{\gamma}} \right)}{1 + \frac{1}{s_{\gamma}} \left(\sum_{r=1}^{s_{\gamma}} \frac{s_r^{\gamma+*}}{y_{rjo}^{\gamma}} \right)}, \gamma = 1, \ldots, \nu \qquad (46)$$

Once an optimal solution $(h^{\gamma*}, s^{\gamma-*}, s^{\gamma+*})$ of models (40), (42) or (45) is obtained, then the projections of the inputs and outputs onto the efficient frontier can be calculated as follows:

$$X_{j_o}^{\gamma*} = X_{j_o}^{\gamma} - s^{\gamma-*} = X^{\gamma}h^{\gamma*}, \gamma = 1, \ldots, \nu$$
$$Y_{j_o}^{\gamma*} = Y_{j_o}^{\gamma} + s^{\gamma+*} = \Upsilon^{\gamma}h^{\gamma*}, \gamma = 1, \ldots, \nu \qquad (47)$$

If the *fixed link* case is used in the assessment, then the intermediate measures will remain unchanged to their initial levels. Otherwise, if the *free link* case is selected, then the projections of the intermediate measures are computed as follows:

$$Z_{j_o}^{(\gamma,\psi)*} = Z^{(\gamma,\psi)}h^{\gamma*}, \ \forall(\gamma,\psi) \qquad (48)$$

From the above we conclude that in the non-oriented case the relationship between the overall efficiency and the stage efficiencies cannot be defined explicitly. Tone and Tsutsui [117] noticed that alternative forms of the overall efficiency could be used in the non-oriented case. For instance, Lu et al. [101] modified the non-oriented NSBM (45) by deriving the non-oriented overall efficiency as the simple arithmetic mean of the non-oriented stage efficiencies.

Notice that the above models are given under VRS assumption, however the CRS models can be also formed regardless the orientation by removing the corresponding convexity constraints. The experimentation of Tone and Tsutsui [117] revealed that under CRS assumption and employing the *free link* case the NSBM may deem inefficient all the DMUs under evaluation in each individual stage. This finding contradicts with the characteristics of traditional DEA models where at least one DMU is deemed efficient so as to construct the efficient frontier. On the other hand, the authors proved that under VRS assumption there is always at least an efficient DMU in each sub-process. As they further pointed out, this also holds when the *fixed link* case is utilized under CRS assumption.

Fukuyama and Mirdehghan [58] showed, by providing adequate examples, that the NSBM of Tone and Tsutsui [117] fails to identify the efficiency status of DMUs because the slacks concerning the intermediate measures are not considered in the definitions of the efficiencies. They proposed a revised PPS and a two-phase approach which identifies sufficiently the efficiency status under the *fixed link* case only. Mirdehghan and Fukuyama [106] developed another two-phase approach by incorporating the notions of mathematical dominance, which deals effectively with the *free link* case also.

Chen et al. [25] noticed that the network DEA methods that are developed on the basis of the production possibility set, such as the slacks-based method [117] should be re-examined with respect to the definition of the stage efficiencies. Especially, they discovered that the NSBM of Tone and Tsutsui [117] provides only the overall efficiency when it is applied for the performance assessment of the two-stage processes of Type I (Fig. 1a). Chen et al. [25] argued that since the intermediate measures are the only outputs from stage-1 and the only inputs to stage-2, then neither the input-oriented NSBM for stage-2 nor the output-oriented NSBM for stage-1 can be formed. This relates, as noted above, with the absence of the slacks associated

Table 1 Applicability of NSBM

Network structure	Input oriented	Output oriented	Non-oriented
Series-Type I	–	–	–
Series-Type II	✓	–	–
Series-Type	–	✓	–
Series-Type	✓	✓	✓
Generalized Series	✓	✓	✓
Parallel	✓	✓	✓
Series-Parallel (Mixed)	✓	✓	✓

with the intermediate measures in the definitions of the efficiencies. They regarded this finding as a pitfall and they concluded that the NSBM models can only yield the overall efficiency of the Type I two-stage process. In Table 1 we demonstrate the applicability of the NSBM on various types of network structures.

9.3.2.4   System-Centric Approach

The network DEA methods that are characterized as system-centric, they do not provide the stage efficiencies but only the overall efficiency of the DMU under evaluation. Most of these methods are modelled in the envelopment form which is based on the unification of the production possibility sets of the individual stages. Kao [74] referred to such methods as "*system distance measure*" methods, where an input or output oriented distance measure model is employed to measure the overall efficiency of each DMU.

Notice that most system-centric methods originate from the pioneer work of Fare [48], who studied DMUs with the structure of Type IV (Fig. 1d) and combined the production technology of the two stages to derive the entire-expanded technology of the DMU. The proposed model however yields only the overall efficiency of the DMU. Fare and Whittaker [56] employed the approach of Fare [48] for the performance assessment of 137 dairy farms in USA. Fare and Grosskopf [51] studied the same network structure and they followed the same practice to formulate the system technology. They built upon Fare [48] to construct Malmquist productivity indices [18, 50] to draw efficiency comparisons between periods. Fare and Grosskopf [52], as mentioned above, unified the methods introduced in [48, 49, 51, 56] to a generalized framework for modelling various types of network structures.

9.4   Classification of Network DEA Studies

In this section we provide a thorough classification of network DEA studies involving theoretical developments and applications. They are basically categorized according to the assessment paradigm they follow.

Table 2 presents the studies that are based on independent assessments. For each study we provide the reference, the network structure of the DMUs, the number of stages and the returns to scale assumed to form the production possibility set (PPS). We also indicate the studies that provide theoretical developments or these that consists of applications and we give a short description of the application field.

In Table 3 the fifth column (*Model*) indicates whether the study is seminal or modification and extension of an existing one.

The following directed graphs depict the starting points and the advancements of the *multiplicative* and the *additive* efficiency decomposition methods. Each node represents one or more studies that constitute a milestone on each efficiency decomposition approach. The edges indicate relationship between studies, i.e. the direction of each edge points from the study used as theoretical basis to the study that extend this basis. By employing this representation method, we highlight the development of the efficiency decomposition approaches and the knowledge flow paths. Notice that the colors on each node indicate the type of network structures that examined in each study (Fig. 10).

Below we provide the schematic representation of the evolution of the additive efficiency decomposition method (Fig. 11).

The applications of the multiplicative efficiency decomposition method are presented in Table 4. The fifth column of Table 4 provides the model used in each study.

Table 5 presents the applications of the additive efficiency decomposition method. Similar to Table 4, in Table 5 the fifth column reports the model used in each study.

The following table reports the studies that are based on the slacks-based measure approach. The fifth column of Table 6 indicates whether the study is seminal, extension or application of an existing one.

Table 7 summarizes the studies that are characterized as System-centric. The fifth column (*Model*) of Table 7 indicates whether the study is seminal, extension or application of an existing one.

Table 2 Studies based on independent assessments

References	Network structure	No of stages	PPS—returns to scale	Theoretic	Application	Application field
Fare and Primont [54]	Parallel	$v \geq 2$	VRS	✓	✓	US coal-fired steam electric generating plants
Charnes et al. [22]	Series-Type IV	2	CRS	✓	✓	US Military
Chilingerian and Sherman [32]	Series-Type I	2	CRS	✓	✓	Medical services
Fare et al. [53]	Parallel	$v \geq 2$	CRS	✓		–
Fare and Primont [55]	Parallel	$v \geq 2$	VRS	✓		–
Wang et al. [123]	Series-Type I	2	VRS	✓	✓	IT on banks
Kao [69]	Parallel	$v \geq 2$	VRS		✓	Taiwanese forests
Seiford and Zhu [110]	Series-Type I	2	CRS/VRS	✓	✓	US commercial banks
Soteriou and Zenios [113]	Mixed	3	VRS	✓	✓	Branches of a Cyprus bank
Zhu [135]	Series-Type I	2	CRS/VRS		✓	Fortune 500 companies
Keh and Chu [78]	Series-Type I	2	VRS		✓	Grocery stores
Sexton and Lewis [112]	Series-Type I	2	VRS	✓	✓	Teams of USA Major League Baseball

(continued)

Table 2 (continued)

References	Network structure	No of stages	PPS—returns to scale	Theoretic	Application	Application field
Luo [103]	Series-Type I	2	CRS/VRS		✓	US large banks
Lewis and Sexton [83]	Mixed	5	VRS	✓	✓	Teams of USA Major League Baseball
Abad et al. [1]	Series-Type II	2	VRS		✓	Stocks in the Spanish manufacturing industry
Keh et al. [79]	Series-Type I	2	VRS		✓	Asia–Pacific hotels
Lu [100]	Series-Type I	2	CRS/VRS		✓	Taiwanese IC-design firms
Lo and Lu [96]	Series-Type I	2	VRS		✓	Taiwanese financial holding companies
Lo [95]	Series-Type I	2	VRS		✓	US S&P 500 firms
Tsolas [119]	Series-Type I	2	VRS		✓	Greek commercial banks
Tsolas [120]	Series-Type I	2	VRS		✓	Greek construction firms
Adler et al. [2]	Mixed	3	VRS	✓	✓	European airports

Table 3 Seminal efficiency decomposition approaches, modifications and extensions

References	Network structure	No of stages	PPS—returns to scale	Model	Application field
Beasley [13]	Parallel	2	CRS	Seminal Study	UK Chemistry and Physics departments
Mar Molinero [104]	Parallel	2	CRS	Modification of Beasley [13]	UK Chemistry and Physics departments
Tsai and Mar Molinero [118]	Parallel	5	VRS	Extension of Mar Molinero [104]	National Health Service trusts in England
Cook et al. [35]	Parallel	2	CRS	Modification of Beasley [13]	Branches of a Canadian bank
Cook and Hababou [34]	Parallel	2	VRS	Extension of Cook et al. [35]	Branches of a Canadian bank
Cook and Green [33]	Parallel	4	CRS	Extension of Cook et al. [34, 35]	Manufacturing plants in steel industry
Jahanshahloo et al. [68]	Parallel	3	CRS	Extension of Cook et al. [35]	Branches of an Iranian bank
Amirteimoori and Kordrostami [5]	Generalized Series	$v \geq 2$	CRS	Extension of Beasley [13] and Cook et al. [35]	Illustrative data
Amirteimoori and Shafiei [6]	Generalized series/Series-Type IV	$v \geq 2$	CRS	Extension of Beasley [13] and Cook et al. [35]	Illustrative data
Chen et al. [29]	Series-Type II	2	CRS	Extension of Tsai and Mar Molinero [118]	IT on banks
Liang et al. [86]	Series-Type II	2	CRS	Extension of Tsai and Mar Molinero [118]	Illustrative data on Supply Chains
Kao and Hwang [75]	Series-Type I	2	CRS	Seminal Study	Non-life Insurance Companies in Taiwan
Liang et al. [87]	Series-Type I	2	CRS	Seminal Study	IT on banks/US commercial banks
Chen et al. [26]	Series-Type I	2	CRS/VRS	Extension of Beasley [13], Amirteimoori and Kordrostami [5] and Amirteimoori and Shafiei [6]	Non-life Insurance Companies in Taiwan

(continued)

Table 3 (continued)

References	Network structure	No of stages	PPS—returns to scale	Model	Application field
Kao [70]	Generalized series/Parallel/ Mixed	$v \geq 2$	CRS	Extension of Beasley [13], Cook et al. [35], Amirteimoori and Kordrostami [5], Amirteimoori and Shafiei [6] and Kao and Hwang [75]	Non-life Insurance Companies in Taiwan/Illustrative data
Kao [71]	Parallel	$v \geq 2$	CRS	Extension of Beasley [13] and Cook et al. [35]	Taiwanese forests
Chen et al. [27]	Series-Type I	2	CRS	Extension of Kao and Hwang [75]	Non-life Insurance Companies in Taiwan
Cook et al. [39]	Generalized series/Mixed	$v > 2$	CRS	Extension of Chen et al. [26]	Electric power companies/Illustrative data
Chen et al. [28]	Series-Type II	2	VRS	Extension of Chen et al. [26]	IT on banks
Zha and Liang [134]	Series—Type II	2	CRS	Extension of Kao and Hwang [75]	US commercial banks
Kao and Hwang [76]	Generalized series/Parallel/ Mixed	$v \geq 2$	CRS	Unification of Kao [70, 71]	IT on banks/Illustrative data
Kao and Hwang [77]	Series-Type I	2	VRS	Modification of Kao and Hwang [75]	–
Liang et al. [88]	Series with feedback	2	CRS	Extension of Chen et al. [29] and Liang et al. [86]	Chinese universities
Li et al. [84]	Series-Type II	2	CRS	Extension of Kao and Hwang [75]	Regional R&D in China
Kao [73]	Generalized Series/Parallel/ Mixed	$v \geq 2$	CRS	Extension of Kao [70, 71] and Kao and Hwang [76]	Electric power companies/Illustrative data
Li et al. [85]	Series-Type I	2	VRS	Extension of Kao and Hwang [75]	Nations in 2012 London summer Olympic Games
An et al. [8]	Series-Type I	2	CRS	Extension of Kao and Hwang [75]	Non-life Insurance Companies in Taiwan

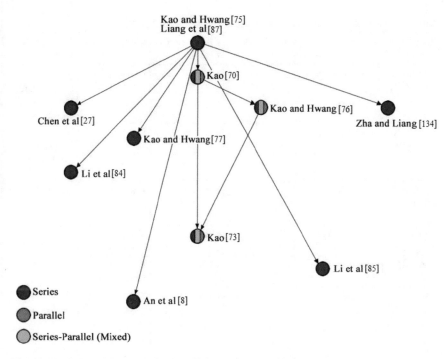

Fig. 10   Evolution of the multiplicative efficiency decomposition method

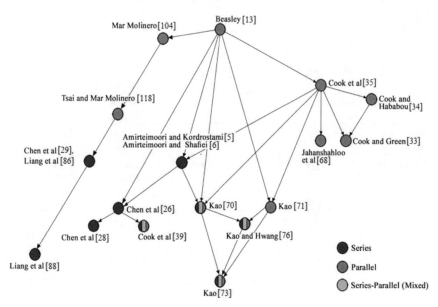

Fig. 11   Evolution of the additive efficiency decomposition method

Table 4 Applications of the multiplicative efficiency decomposition method

References	Network structure	No of stages	PPS—returns to scale	Model	Application field
Liu and Wang [94]	Series-Type I	2	CRS	Kao and Hwang [75]	Printed circuit board industry in Taiwan
Guan and Chen [59]	Series-Type II	2	CRS	Kao [70]	High-tech innovations in Chinese provinces
Hsieh and Lin [67]	Mixed	4	CRS	Kao [70]	International hotels in Taiwan
Cao and Yang [15]	Series-Type I	2	CRS	Kao and Hwang [75]	Internet companies
Zhu [136]	Series-Type I	2	CRS	Kao and Hwang [75]	Airlines
Lee and Johnson [81]	Series-Type I	3	CRS	Kao [70]	Firms of semiconductor manufacturing industry
Lee and Johnson [82]	Mixed	4	VRS	Kao [70]	US airlines
Chen et al. [24]	Series-Type I	2	CRS	Kao and Hwang [75]	Automotive industry
Limaei [90]	Series-Type I	2	CRS	Kao and Hwang [75]	Iranian forests
Wanke [125]	Series-Type I	2	CRS	Kao and Hwang [75]	Brazilian ports
Wanke and Barros [126]	Series-Type I	2	CRS	Kao and Hwang [75]	Brazilian banks
Wanke et al. [127]	Series-Type I	2	CRS	Kao and Hwang [75]	Australian public schools
Carayannis et al. [16]	Series-Type	2	VRS	Kao and Hwang [75]	National and regional innovation systems of European countries

Table 5 Applications of the additive efficiency decomposition method

References	Network structure	No of stages	PPS—returns to scale	Model	Application field
Diez-Ticio and Mancebon [47]	Parallel	2	VRS	Tsai and Mar Molinero [118]	Spanish police service
Yu [131]	Parallel	2	VRS	Tsai and Mar Molinero [118]	Taiwan's bus transit system
Yu and Fan [133]	Mixed	3	CRS	Mar Molinero [104]	Taiwan's bus transit system
Liu [93]	Series-Type I	2	CRS	Chen et al. [26]	Taiwanese financial holding companies
Guan and Chen [60]	Series-Type IV	2	CRS/VRS	Chen et al. [26]	Innovation activities of OECD countries
Premachandra et al. [107]	Series-Type II	2	VRS	Chen et al. [26]	US mutual funds
Lu et al. [102]	Series-Type I	2	VRS	Chen et al. [26]	US airlines
Kao [72]	Parallel	2	CRS/VRS	Beasley [13], Kao [71] and Kao and Hwang [76]	UK Chemistry and Physics departments
Rogge and Jaeger [109]	Parallel	6	CRS	Mar Molinero [104]	Solid waste in municipalities
Da Cruz et al. [40]	Parallel	2	CRS	Rogge and Jaeger [109]	Water utilities
Amirteimoori [7]	Series-Type IV	2	VRS	Chen et al. [26]	Car distribution and service
Wang et al. [124]	Series-Type I	2	VRS	Chen et al. [26]	Chinese commercial banks
Yang et al. [130]	Series-Type I	2	VRS	Chen et al. [26]	National Basketball Association (NBA) teams

(continued)

Table 5 (continued)

References	Network structure	No of stages	PPS—returns to scale	Model	Application field
Toloo et al. [115]	Series-Type II	2	CRS	Chen et al. [29]	IT on banks/UK Chemistry and Physics departments
Halkos et al. [64]	Series-Type I	2	VRS	Chen et al. [26]	Secondary education in 65 countries
Halkos et al. [65]	Series-Type I	2	VRS	Chen et al. [26]	Sustainability of European regions
Halkos et al. [66]	Series-Type I	2	VRS	Chen et al. [26]	Sustainable development of countries with advanced economies
Guo et al. [61]	Generalized Series/Mixed	$v \geq 2$	CRS	Chen et al. [26]	Automotive industry/Synthetic data from Despotis et al. [46]
Guo et al. [62]	Series-Type I/Type II	2	CRS	Liang et al. [86]	Non-life Insurance Companies in Taiwan/Regional R&D in China

Table 6 Studies based on the slacks-based measure approach

References	Network structure	No of stages	PPS—returns to scale	Model	Application field
Tone and Tsutsui [117]	Generalized series/mixed	$v > 2$	CRS/VRS	Seminal Study	Electric power companies
Avkiran [10]	Mixed	3	VRS	Tone and Tsutsui [117]	UAE domestic commercial banks

(continued)

Table 6 (continued)

References	Network structure	No of stages	PPS—returns to scale	Model	Application field
Yu [132]	Mixed	3	CRS	Tone and Tsutsui [117]	Domestic airports of Taiwan
Fukuyama and Weber [57]	Series-Type I	2	CRS	Extension of Tone and Tsutsui [117]	Japanese banks
Matthews [105]	Generalized Series	3	VRS	Tone and Tsutsui [117]	Chinese banks
Lin and Chiu [91]	Mixed	4	VRS	Tone and Tsutsui [117]	Taiwanese domestic banks
Akther et al. [4]	Series-Type I	2	CRS	Fukuyama and Weber [57]	Bangladeshi banks
Lu et al. [101]	Series-Type I	2	VRS	Tone and Tsutsui [117]	National innovation system among countries
Lozano and Gutierrez [97]	Series-Type II	2	VRS	Tone and Tsutsui [117]	European airlines
Chang et al. [19]	Series-Type IV	2	VRS	Tone and Tsutsui [117]	International cruise lines

Table 7 Studies based on the system-centric approach

References	Network structure	No of stages	PPS—returns to scale	Model	Application field
Fare [48]	Series-Type IV	2	CRS	Seminal Study	–
Fare and Whittaker [56]	Series-Type IV	2	VRS	Fare [48]	US dairy farms
Fare and Grosskopf [51]	Series-Type IV	2	CRS	Extension of Fare [48]	–

(continued)

Table 7 (continued)

References	Network structure	No of stages	PPS—returns to scale	Model	Application field
Fare et al. [49]	Parallel	4	CRS	Seminal Study	US grain farms
Lothgren and Tambour [99]	Series-Type IV	2	CRS	Fare and Grosskopf [51]	Swedish pharmacies
Fare and Grosskopf [52]	Generalized series/Parallel/ Mixed	v ≥ 2	CRS	Unification of Fare [48], Fare et al. [49], Fare and Grosskopf [51] and Fare and Whittaker [56]	–
Prieto and Zofio [108]	Mixed	4	CRS	Fare and Grosskopf [52]	OECD countries
Sheth et al. [111]	Series-Type I	2	VRS	Fare and Grosskopf [52]	Bus routes in Virginia State of USA
Vaz et al. [122]	Parallel	5	VRS	Fare et al. [49]	Portuguese retail stores
Yang et al. [129]	Series	2	CRS	Seminal Study	Supply Chains/Branches of China Construction Bank
Chen and Yan [23]	Series/Mixed	2/3	CRS	Seminal Study	Illustrative data on Supply Chains
Lozano et al. [98]	Series-Type IV	2	VRS	Fare and Grosskopf [52]	Spanish airports
Wu et al. [128]	Series-Type IV	2	CRS	Seminal Study	Industrial production and pollution treatment of Chinese regions

9.5   Conclusion

In this chapter we provided a detailed survey of the network DEA literature. We also demonstrated the usefulness of network DEA and its advantages over the standard DEA for the assessment of multi-stage processes. We presented in detail the most important network DEA methods and we discussed their extensions and modifications. In addition, we reported most of the studies that apply the existing network DEA methods to real word problems. The network DEA studies were classified according to the model developed or used. We illustrated that the additive and the multiplicative decomposition methods provide non-unique stage efficiency scores. We showed that the additive efficiency decomposition method also suffers from biased efficiency results emerging from the endogenous weights. In addition, we demonstrated that the composition approach is capable of overcoming these drawbacks. Finally, we discussed about the inability of the existing approaches to be universally applied on every type of network structure.

A subject for future research is the extension of the composition approach to general network structures involving series and parallel processes. A universal network DEA method that could be applied to every type of network structure would be advantageous. Also, future studies could be focused on the revision of the system-centric methods, so as to yield the stage efficiency scores except from the overall efficiency. Moreover, a topic that should be revisited is the returns to scale in network DEA models. Another issue that is worth investigating in network DEA is the perfect mapping between the multiplier and the envelopment models of each approach, in order to provide efficient projections. Undoubtedly, further development of the network DEA methods is needed so as to widen the application field, to aid the decision makers to address the increasing complexity of the organizations and improve their performance.

References

1. C. Abad, S.A. Thoreb, J. Laffarga, Fundamental analysis of stocks by two-stage DEA. Manag. Decis. Econ. **25**, 231–241 (2004)
2. N. Adler, V. Liebert, E. Yazhemsky, Benchmarking airports from a managerial perspective. Omega Int. J. Manag. Sci. **41**, 442–458 (2013)
3. P.J. Agrell, A. Hatami-Marbini, Frontier-based performance analysis models for supply chain management: state of the art and research directions. Comput. Ind. Eng. **66**(3), 567–583 (2013)
4. S. Akther, H. Fukuyama, W.L. Weber, Estimating two-stage network slacks-based inefficiency: an application to Bangladesh banking. Omega Int. J. Manag. Sci. **41**, 88–96 (2013)
5. A. Amirteimoori, S. Kordrostami, DEA-like models for multi-component performance measurement. Appl. Math. Comput. **163**, 735–743 (2005)
6. A. Amirteimoori, M. Shafiei, Measuring the efficiency of interdependent decision making sub-units in DEA. Appl. Math. Comput. **173**, 847–855 (2006)
7. A. Amirteimoori, A DEA two-stage decision processes with shared resources. Cent. Eur. J. Oper. Res. **21**, 141–151 (2013)

8. Q. An, H. Yan, J. Wu, L. Liang, Internal resource waste and centralization degree in two-stage systems: an efficiency analysis. Omega Int. J. Manag. Sci. **61**, 89–99 (2016)
9. S. Ang, C.M. Chen, Pitfalls of decomposition weights in the additive multi-stage DEA model. Omega Int. J. Manag. Sci. **58**, 139–153 (2016)
10. N. Avkiran, Opening the black box of efficiency analysis: an illustration with UAE banks. Omega Int. J. Manag. Sci. **37**, 930–941 (2009)
11. R.D. Banker, A. Charnes, W.W. Cooper, Some models for estimating technical and scale inefficiencies in data envelopment analysis. Manag. Sci. **30**, 1078–1092 (1984)
12. M.S. Bazaraa, J.J. Jarvis, H.D. Sherali, *Linear programming and network flows* (John Wiley & Sons Inc, Hoboken, New Jersey, 2011)
13. J.E. Beasley, Determining teaching and research efficiencies. J. Oper. Res. Soc. **46**(4), 441–452 (1995)
14. J. Buchanan, L. Gardiner, A comparison of two reference point methods in multiple objective mathematical programming. Eur. J. Oper. Res. **149**(1), 17–34 (2003)
15. X. Cao, F. Yang, Measuring the performance of Internet companies using a two-stage data envelopment analysis model. Enterp. Inf. Syst. **5**(2), 207–217 (2011)
16. E.G. Carayannis, E. Grigoroudis, Y. Goletsis, A multilevel and multistage efficiency evaluation of innovation systems: a multiobjective DEA approach. Expert Syst. Appl. **62**, 63–80 (2016)
17. L. Castelli, R. Pesenti, W. Ukovich, A classification of DEA models when the internal structure of the decision making units is considered. Ann. Oper. Res. **173**(1), 207–235 (2010)
18. D.W. Caves, L.R. Christensen, W.E. Diewert, The economic theory of index numbers and the measurement of input, output, and productivity. Econ. **50**(6), 1393–1414 (1982)
19. Y.T. Chang, S. Lee, H.K. Park, Efficiency analysis of major cruise lines. Tour. Manag. **58**, 78–88 (2017)
20. A. Charnes, W.W. Cooper, Programming with linear fractional functional. Nav. Res. Logist. **9**, 181–185 (1962)
21. A. Charnes, W.W. Cooper, E. Rhodes, Measuring the efficiency of decision making units. Eur. J. Oper. Res. **2**(6), 429–444 (1978)
22. A. Charnes, W.W. Cooper, B. Golany, R. Halek, G. Klopp, E. Schmitz, D. Thomas, Two phase data envelopment analysis approach to policy evaluation and management of army recruiting activities: tradeoffs between joint services and army advertising. Research Report CCS no. 532, Center for Cybernetic Studies, The University of Texas, Austin, Texas (1986)
23. C. Chen, H. Yan, Network DEA model for supply chain performance evaluation. Eur. J. Oper. Res. **213**(1), 147–155 (2011)
24. C. Chen, J. Zhu, J.Y. Yu, H. Noori, A new methodology for evaluating sustainable product design performance with two-stage network data envelopment analysis. Eur. J. Oper. Res. **221**(2), 348–359 (2012)
25. Y. Chen, W.D. Cook, C. Kao, J. Zhu, Network DEA pitfalls: divisional efficiency and frontier projection under general network structures. Eur. J. Oper. Res. **226**, 507–515 (2013)
26. Y. Chen, W.D. Cook, N. Li, J. Zhu, Additive efficiency decomposition in two-stage DEA. Eur. J. Oper. Res. **196**, 1170–1176 (2009)
27. Y. Chen, W.D. Cook, J. Zhu, Deriving the DEA frontier for two-stage DEA processes. Eur. J. Oper. Res. **202**, 138–142 (2010)
28. Y. Chen, J. Du, H.D. Sherman, J. Zhu, DEA model with shared resources and efficiency decomposition. Eur. J. Oper. Res. **207**, 339–349 (2010)
29. Y. Chen, L. Liang, F. Yang, J. Zhu, Evaluation of information technology investment: a data envelopment analysis approach. Comput. Oper. Res. **33**, 1368–1379 (2006)
30. Y. Chen, L. Liang, J. Zhu, Equivalence in two-stage DEA approaches. Eur. J. Oper. Res. **193**, 600–604 (2009)
31. Y. Chen, J. Zhu, Measuring information technology's indirect impact on firm performance. Inform. Tech. Manag. **5**, 9–22 (2004)
32. J.A. Chilingerian, H.D. Sherman, Managing physician efficiency and effectiveness in providing hospital services. Health Serv. Manag. Res. **3**(1), 3–15 (1990)

33. W.D. Cook, R.H. Green, Multicomponent efficiency measurement and core business identification in multiplant firms: a DEA model. Eur. J. Oper. Res. **157**, 540–551 (2004)
34. W.D. Cook, M. Hababou, Sales performance measurement in bank branches. Omega Int. J. Manag. Sci. **29**, 299–307 (2001)
35. W.D. Cook, M. Hababou, H.J.H. Tuenter, Multicomponent efficiency measurement and shared inputs in data envelopment analysis: an application to sales and service performance in bank branches. J. Prod. Anal. **14**, 209–224 (2000)
36. W.D. Cook, L. Liang, J. Zhu, Measuring performance of two-stage network structures by DEA: a review and future perspective. Omega Int. J. Manag. Sci. **38**(6), 423–430 (2010)
37. W.D. Cook, L.M. Seiford, Data envelopment analysis (DEA)—thirty years on. Eur. J. Oper. Res. **192**(1), 1–17 (2009)
38. W.D. Cook, J. Zhu, *Data envelopment analysis—a handbook of modeling internal structure and network* (Springer, New York, 2014)
39. W.D. Cook, J. Zhu, G. Bi, F. Yang, Network DEA: additive efficiency decomposition. Eur. J. Oper. Res. **207**(2), 1122–1129 (2010)
40. N.F. Da Cruz, P. Carvalho, R.C. Marques, Disentangling the cost efficiency of jointly provided water and wastewater services. Util. Pol. **24**, 70–77 (2013)
41. D.K. Despotis, Fractional minmax goal programming: a unified approach to priority estimation and preference analysis in MCDM. J. Oper. Res. Soc. **47**, 989–999 (1996)
42. D.K. Despotis, G. Koronakos, Efficiency assessment in two-stage processes: a novel network DEA approach. Procedia Comput. Sci. **31**, 299–307 (2014)
43. D.K. Despotis, G. Koronakos, S. Sotiros, A multi-objective programming approach to network DEA with an application to the assessment of the academic research activity. Procedia Comput. Sci. **55**, 370–379 (2015)
44. D.K. Despotis, G. Koronakos, D. Sotiros, Composition versus decomposition in two-stage network DEA: a reverse approach. J. Prod. Anal. **45**(1), 71–87 (2016)
45. D.K. Despotis, G. Koronakos, D. Sotiros, The "weak-link" approach to network DEA for two-stage processes. Eur. J. Oper. Res. **254**(2), 481–492 (2016)
46. D.K. Despotis, D. Sotiros, G. Koronakos, A network DEA approach for series multi-stage processes. Omega Int. J. Manag. Sci. **61**, 35–48 (2016)
47. A. Diez-Ticio, M.J. Mancebon, The efficiency of the Spanish police service: an application of the multiactivity DEA model. Appl. Econ. **34**, 351–362 (2002)
48. R. Fare, Measuring farrell efficiency for a firm with intermediate inputs. Acad. Econ. Paper. **19**, 829–840 (1991)
49. R. Fare, R. Grabowski, S. Grosskopf, S. Kraft, Efficiency of a fixed but allocatable input: a non-parametric approach. Econ. Lett. **56**, 187–193 (1997)
50. R. Fare, S. Grosskopf, Malmquist productivity indexes and fisher ideal indexes. Econ. J. **102**(410), 158–160 (1992)
51. R. Fare, S. Grosskopf, Productivity and intermediate products: a frontier approach. Econ. Lett. **50**, 65–70 (1996)
52. R. Fare, S. Grosskopf, Network DEA. Soc. Econ. Plann. Sci. **34**(1), 35–49 (2000)
53. R. Fare, S. Grosskopf, S.K. Li, Linear programming models for firm and industry performance. Scand. J. Econ. **94**(4), 599–608 (1992)
54. R. Fare, D. Primont, Efficiency measures for multiplant firms. Oper. Res. Lett. **3**, 257–260 (1984)
55. R. Fare, D. Primont, Measuring the efficiency of multiunit banking: an activity analysis approach. J. Bank. Finance **17**, 539–544 (1993)
56. R. Fare, G. Whittaker, An intermediate input model of dairy production using complex survey data. J. Agr. Econ. **46**(2), 201–223 (1995)
57. H. Fukuyama, W.L. Weber, A slacks-based inefficiency measure for a two-stage system with bad outputs. Omega Int. J. Manag. Sci. **38**, 398–409 (2010)
58. H. Fukuyama, S.M. Mirdehghan, Identifying the efficiency status in network DEA. Eur. J. Oper. Res. **220**, 85–92 (2012)

59. J.C. Guan, K.H. Chen, Measuring the innovation production process: a cross-region empirical study of China's high-tech innovations. Technovation **30**, 348–358 (2010)
60. J.C. Guan, K.H. Chen, Modeling the relative efficiency of national innovation systems. Res. Pol. **41**, 102–115 (2012)
61. C. Guo, F. Wei, T. Ding, L. Zhang, L. Liang, Multistage network DEA: Decomposition and aggregation weights of component performance. Comput. Ind. Eng. **113**, 64–74 (2017)
62. C. Guo, R.A. Shureshjani, A.A. Foroughi, J. Zhu, Decomposition weights and overall efficiency in two-stage additive network DEA. Eur. J. Oper. Res. **257**(3), 896–906 (2017)
63. G.E. Halkos, N.G. Tzeremes, S.A. Kourtzidis, A unified classification of two-stage DEA models. Surv. Oper. Res. Manag. Sci. **19**, 1–16 (2014)
64. G.E. Halkos, N.G. Tzeremes, S.A. Kourtzidis, Weight assurance region in two-stage additive efficiency decomposition DEA model: an application to school data. J. Oper. Res. Soc. **66**(4), 696–704 (2015)
65. G.E. Halkos, N.G. Tzeremes, S.A. Kourtzidis, Regional sustainability efficiency index in Europe: an additive two-stage DEA approach. Oper. Res. **15**(1), 1–23 (2015)
66. G.E. Halkos, N.G. Tzeremes, S.A. Kourtzidis, Measuring sustainability efficiency using a two-stage data envelopment analysis approach. J. Ind. Ecol. **20**(5), 1159–1175 (2016)
67. L.F. Hsieh, L.H. Lin, A performance evaluation model for international tourist hotels in Taiwan: an application of the relational network DEA. Int. J. Hospit. Manag. **29**, 14–24 (2010)
68. G.R. Jahanshahloo, A.R. Amirteimoori, S. Kordrostami, Multi-component performance, progress and regress measurement and shared inputs and outputs in DEA for panel data: an application in commercial bank branches. Appl. Math. Comput. **151**, 1–16 (2004)
69. C. Kao, Measuring the efficiency of forest districts with multiple working circles. J. Oper. Res. Soc. **49**(6), 583–590 (1998)
70. C. Kao, Efficiency decomposition in network data envelopment analysis: a relational model. Eur. J. Oper. Res. **192**(3), 949–962 (2009)
71. C. Kao, Efficiency measurement for parallel production systems. Eur. J. Oper. Res. **196**, 1107–1112 (2009)
72. C. Kao, Efficiency decomposition for parallel production systems. J. Oper. Res. Soc. **63**, 64–71 (2012)
73. C. Kao, Efficiency decomposition for general multi-stage systems in data envelopment analysis. Eur. J. Oper. Res. **232**(1), 117–124 (2014)
74. C. Kao, Network data envelopment analysis: a review. Eur. J. Oper. Res. **239**(1), 1–16 (2014)
75. C. Kao, S.-N. Hwang, Efficiency decomposition in two-stage data envelopment analysis: an application to non-life insurance companies in Taiwan. Eur. J. Oper. Res. **185**, 418–429 (2008)
76. C. Kao, S.-N. Hwang, Efficiency measurement for network systems: IT impact on firm performance. Decis. Support Syst. **48**, 437–446 (2010)
77. C. Kao, S.-N. Hwang, Decomposition of technical and scale efficiencies in two-stage production systems. Eur. J. Oper. Res. **211**, 515–519 (2011)
78. H.T. Keh, S. Chu, Retail productivity and scale economies at the firm level: a DEA approach. Omega Int. J. Manag. Sci. **31**(2), 75–82 (2003)
79. H.T. Keh, S. Chu, J. Xu, Efficiency, effectiveness and productivity of marketing in services. Eur. J. Oper. Res. **170**(1), 265–276 (2006)
80. G. Koronakos, D. Sotiros, D.K. Despotis, Reformulation of network data envelopment analysis models using a common modelling framework. Eur. J. Oper. Res. (2018). https://doi.org/10.1016/j.ejor.2018.04.004
81. C.Y. Lee, A.L. Johnson, A decomposition of productivity change in the semiconductor manufacturing industry. Int. J. Prod. Res. **49**(16), 4761–4785 (2011)
82. C.Y. Lee, A.L. Johnson, Two-dimensional efficiency decomposition to measure the demand effect in productivity analysis. Eur. J. Oper. Res. **216**, 584–593 (2012)
83. H.F. Lewis, T.R. Sexton, Network DEA: efficiency analysis of organizations with complex internal structure. Comput. Oper. Res. **31**, 1365–1410 (2004)

84. Y. Li, Y. Chen, L. Liang, J. Xie, DEA models for extended two-stage network structures. Omega Int. J. Manag. Sci. **40**(5), 611–618 (2012)
85. Y. Li, X. Lei, Q. Dai, L. Liang, Performance evaluation of participating nations at the 2012 London Summer Olympics by a two-stage data envelopment analysis. Eur. J. Oper. Res. **243**(3), 964–973 (2015)
86. L. Liang, F. Yang, W.D. Cook, J. Zhu, DEA models for supply chain efficiency evaluation. Ann. Oper. Res. **145**(1), 35–49 (2006)
87. L. Liang, W.D. Cook, J. Zhu, DEA models for two-stage processes: game approach and efficiency decomposition. Nav. Res. Logist. **55**, 643–653 (2008)
88. L. Liang, Z.Q. Li, W.D. Cook, J. Zhu, Data envelopment analysis efficiency in two-stage networks with feedback. IIE Trans. **43**, 309–322 (2011)
89. M. Lightner, S. Director, Multiple criterion optimization for the design of electronic circuits. IEEE Trans. Circuits Syst. **28**(3), 169–179 (1981)
90. S.M. Limaei, Efficiency of Iranian forest industry based on DEA models. J. Forest. Res. **24**, 759–765 (2013)
91. T.Y. Lin, S.H. Chiu, Using independent component analysis and network DEA to improve bank performance evaluation. Econ. Model. **32**, 608–616 (2013)
92. J.S. Liu, L.Y.Y. Lu, W.M. Lu, B.J.Y. Lin, Data envelopment analysis 1978–2010: a citation-based literature survey. Omega Int. J. Manag. Sci. **41**(1), 3–15 (2013)
93. S.T. Liu, Performance measurement of Taiwan financial holding companies: an additive efficiency decomposition approach. Expert Syst. Appl. **38**, 5674–5679 (2011)
94. S.T. Liu, R.T. Wang, Efficiency measures of PCB manufacturing firms using relational two-stage data envelopment analysis. Expert Syst. Appl. **36**, 4935–4939 (2009)
95. S.F. Lo, Performance evaluation for sustainable business: a profitability and marketability framework. Corp. Soc. Responsib. Environ. Manag. **17**, 311–319 (2010)
96. S.F. Lo, W.M. Lu, An integrated performance evaluation of financial holding companies in Taiwan. Eur. J. Oper. Res. **198**, 341–350 (2009)
97. S. Lozano, E. Gutierrez, A slacks-based network DEA efficiency analysis of European airlines. Transp. Plan. Technol. **37**(7), 623–637 (2014)
98. S. Lozano, E. Gutierrez, P. Moreno, Network DEA approach to airports performance assessment considering undesirable outputs. Appl. Math. Model. **37**, 1665–1676 (2013)
99. M. Lothgren, M. Tambour, Productivity and customer satisfaction in Swedish pharmacies: a DEA network model. Eur. J. Oper. Res. **115**, 449–458 (1999)
100. W.C. Lu, The evolution of R&D efficiency and marketability: evidence from Taiwan's IC-design industry. Asian J. Technol. Innov. **17**, 1–26 (2009)
101. W.M. Lu, Q.L. Kweh, C.L. Huang, Intellectual capital and national innovation systems performance. Knowl. Base. Syst. **71**, 201–210 (2014)
102. W.M. Lu, W.K. Wang, S.W. Hung, E.T. Lu, The effects of corporate governance on airline performance: production and marketing efficiency perspectives. Transp. Res. Part E Logist. Transp. Rev. **48**(2), 529–544 (2012)
103. X.M. Luo, Evaluating the profitability and marketability efficiency of large banks: an application of data envelopment analysis. J. Bus. Res. **56**, 627–635 (2003)
104. C. Mar Molinero, On the joint determination of efficiencies in a data envelopment analysis context. J. Oper. Res. Soc. **47**(10), 1273–1279 (1996)
105. K. Matthews, Risk management and managerial efficiency in Chinese banks: a network DEA framework. Omega Int. J. Manag. Sci. **41**(2), 207–215 (2013)
106. S.M. Mirdehghan, H. Fukuyama, Pareto-Koopmans efficiency and network DEA. Omega Int. J. Manag. Sci. **61**, 78–88 (2016)
107. I.M. Premachandra, J. Zhu, J. Watson, D.U.A. Galagedera, Best performing US mutual fund families from 1993 to 2008: evidence from a novel two-stage DEA model for efficiency decomposition. J. Bank. Financ. **36**, 3302–3317 (2012)
108. A.M. Prieto, J.L. Zofío, Network DEA efficiency in input-output models: with an application to OECD countries. Eur. J. Oper. Res. **178**, 292–304 (2007)

109. N. Rogge, S. Jaeger, Evaluating the efficiency of municipalities in collecting and processing municipal solid waste: a shared input DEA-model. Waste Manag. **32**, 1968–1978 (2012)
110. L.M. Seiford, J. Zhu, Profitability and marketability of the top 55 US commercial banks. Manag. Sci. **45**(9), 1270–1288 (1999)
111. C. Sheth, K. Triantis, D. Teodorovic, Performance evaluation of bus routes: a provider and passenger perspective. Transp. Res. Part E Logist. Transp. Rev. **43**(4), 453–478 (2007)
112. T.R. Sexton, H.F. Lewis, Two-stage DEA: an application to major league baseball. J. Prod. Anal. **19**(2–3), 227–249 (2003)
113. A. Soteriou, S.A. Zenios, Operations, quality, and profitability in the provision of banking services. Manag. Sci. **45**, 1221–1238 (1999)
114. D. Sotiros, G. Koronakos, D.K. Despotis, Dominance at the divisional efficiencies level in network DEA: the case of two-stage processes. Omega Int. J. Manag. Sci. (2018). https://doi.org/10.1016/j.omega.2018.06.007
115. M. Toloo, A. Emrouznejad, P. Moreno, A linear relational DEA model to evaluate two-stage processes with shared inputs. J. Comput. Appl. Math. **1**, 1–17 (2015)
116. K. Tone, A slacks-based measure of efficiency in data envelopment analysis. Eur. J. Oper. Res. **130**(3), 498–509 (2001)
117. K. Tone, M. Tsutsui, Network DEA: a slacks-based measure approach. Eur. J. Oper. Res. **197**, 243–252 (2009)
118. P.F. Tsai, C. Mar Molinero, A variable returns to scale data envelopment analysis model for the joint determination of efficiencies with an example of the UK health service. Eur. J. Oper. Res. **141**, 21–38 (2002)
119. I.E. Tsolas, Relative profitability and stock market performance of listed commercial banks on the Athens Exchange: a non-parametric approach. IMA J. Manag. Math. **22**, 323–342 (2011)
120. I.E. Tsolas, Modeling profitability and stock market performance of listed construction firms on the Athens Exchange: two-stage DEA approach. J. Construct. Eng. Manag. **139**, 111–119 (2013)
121. M. Tsutsui, M. Goto, A multi-division efficiency evaluation of US electric power companies using a weighted slacks-based measure. Socio. Econ. Plan. Sci. **43**(3), 201–208 (2009)
122. C.B. Vaz, A.S. Camanho, R.C. Guimaraes, The assessment of retailing efficiency using network data envelopment analysis. Ann. Oper. Res. **173**, 5–24 (2010)
123. C.H. Wang, R. Gopal, S. Zionts, Use of data envelopment analysis in assessing information technology impact on firm performance. Ann. Oper. Res. **73**, 191–213 (1997)
124. K. Wang, W. Huang, J. Wu, Y.N. Liu, Efficiency measures of the Chinese commercial banking system using an additive two-stage DEA. Omega Int. J. Manag. Sci. **44**, 5–20 (2014)
125. P.F. Wanke, Physical infrastructure and shipment consolidation efficiency drivers in Brazilian ports: a two-stage network-DEA approach. Transp. Policy **29**, 145–153 (2013)
126. P. Wanke, C. Barros, Two-stage DEA: an application to major Brazilian banks. Expert Syst. Appl. **41**, 2337–2344 (2014)
127. P. Wanke, V. Blackburn, C.P. Barros, Cost and learning efficiency drivers in Australian schools: a two-stage network DEA approach. Appl. Econ. **48**(38), 3577–3604 (2016)
128. J. Wu, P. Yin, J. Sun, J. Chu, L. Liang, Evaluating the environmental efficiency of a two-stage system with undesired outputs by a DEA approach: an interest preference perspective. Eur. J. Oper. Res. **254**(3), 1047–1062 (2016)
129. F. Yang, D. Wu, L. Liang, G. Bi, D.D. Wu, Supply chain DEA: production possibility set and performance evaluation model. Ann. Oper. Res. **185**(1), 195–211 (2011)
130. C.H. Yang, H.Y. Lin, C.P. Chen, Measuring the efficiency of NBA teams: additive efficiency decomposition in two-stage DEA. Ann. Oper. Res. **217**(1), 565–589 (2014)
131. M.M. Yu, Measuring the efficiency and return to scale status of multimode bus transit—evidence from Taiwan's bus system. Appl. Econ. Lett. **15**, 647–653 (2008)
132. M.M. Yu, Assessment of airport performance using the SBM-NDEA model. Omega Int. J. Manag. Sci. **38**(6), 440–452 (2010)
133. M.M. Yu, C.K. Fan, Measuring the performance of multimode bus transit: a mixed structure network DEA model. Transp. Res. Part E Logist. Transp. Rev. **45**, 501–515 (2009)

134. Y. Zha, L. Liang, Two-stage cooperation model with input freely distributed among the stages. Eur. J. Oper. Res. **205**, 332–338 (2010)
135. J. Zhu, Multi-factor performance measure model with an application to Fortune 500 companies. Eur. J. Oper. Res. **123**(1), 105–124 (2000)
136. J. Zhu, Airlines performance via two-stage network DEA approach. J. Cent. **4**(2), 260–269 (2011)

Part IV
Learning and Analytics in Intelligent Safety and Emergency Response Systems

Chapter 10
Applying Advanced Data Analytics and Machine Learning to Enhance the Safety Control of Dams

João Rico, José Barateiro, Juan Mata, António Antunes and Elsa Cardoso

Abstract The protection of critical engineering infrastructures is vital to today's society, not only to ensure the maintenance of their services (e.g., water supply, energy production, transport), but also to avoid large-scale disasters. Therefore, technical and financial efforts are being continuously made to improve the safety control of large civil engineering structures like dams, bridges and nuclear facilities. This control is based on the measurement of physical quantities that characterize the structural behavior, such as displacements, strains and stresses. The analysis of monitoring data and its evaluation against physical and mathematical models is the strongest tool to assess the safety of the structural behavior. Commonly, dam specialists use multiple linear regression models to analyze the dam response, which is a well-known approach among dam engineers since the 1950s decade. Nowadays, the data acquisition paradigm is changing from a manual process, where measurements were taken with low frequency (e.g., on a weekly basis), to a fully automated process that allows much higher frequencies. This new paradigm escalates the potential of data analytics on top of monitoring data, but, on the other hand, increases data quality issues related to anomalies in the acquisition process. This chapter presents the full data lifecycle in the safety control of large-scale civil engineering infrastructures (focused on dams), from the data acquisition process, data processing and storage,

J. Rico · J. Barateiro (✉) · J. Mata · A. Antunes
Laboratório Nacional de Engenharia Civil, Av. Brasil, 101, Lisbon, Portugal
e-mail: jbarateiro@lnec.pt

J. Rico
e-mail: jmrico@lnec.pt

J. Mata
e-mail: jmata@lnec.pt

A. Antunes
e-mail: aantunes@lnec.pt

J. Barateiro · E. Cardoso
INESC-ID, R. Alves Redol, 9, Lisbon, Portugal

E. Cardoso
ISCTE-Instituto Universitário de Lisboa, Av. Forças Armadas, Lisbon, Portugal
e-mail: earc@iscte-iul.pt

© Springer Nature Switzerland AG 2019   315
G. A. Tsihrintzis et al. (eds.), *Machine Learning Paradigms*,
Learning and Analytics in Intelligent Systems 1,
https://doi.org/10.1007/978-3-030-15628-2_10

data quality and outlier detection, and data analysis. A strong focus is made on the use of machine learning techniques for data analysis, where the common multiple linear regression analysis is compared with deep learning strategies, namely recurrent neural networks. Demonstration scenarios are presented based on data obtained from monitoring systems of concrete dams under operation in Portugal.

10.1   Introduction

Dam safety is a continuous requirement due to the potential risk in terms of environmental, social and economical disasters. In the International Commission on Large Dams' bulletin number 138 [24], it is referred that the assurance of the safety of a dam or any other retaining structure requires "a series of concomitant, well directed, and reasonably organized activities. The activities must: (i) be complementary in a chain of successive actions leading to an assurance of safety, (ii) contain redundancies to a certain extent so as to provide guarantees that go beyond operational risks".

Continuous dam safety control must be done at various levels, and must include an individual assessment (dam body, its foundation, appurtenant works, adjacent slopes, and downstream zones) and as a whole, in the various areas of dam safety: environmental, structural and hydraulic/operational. Environmental safety, as the name suggests, is related to the environmental impacts originated by the dam, both in terms of maintenance of ecological flows, with direct influence on the fauna and flora existing upstream and downstream, and in terms of the control of the characteristics and quality of the reservoir water and soil. Hydraulic/operational safety is related to the exploitation and operation of hydraulic devices, as well as the implementation of early warning and alert systems for emergency situations. Structural safety can be understood as the dam's capacity to satisfy the structural design requirements, avoiding accidents and incidents during the dam's life. Structural safety includes all activities, decisions and interventions necessary to ensure the adequate structural performance of the dam.

Structural safety control is based on making decisions during the different phases of a dam's life through safety control activities. Thus, the main aim of structural safety control is the multiple assessment of the expected dam behavior based on models and on the measurements of parameters that characterize the dam's behavior and its condition. The main concern is the assessment of the real and actual dam behavior, under exploitation conditions, in order to early detect possible malfunctions.

To be effective, dam safety control must be considered as an ongoing process. The assessment of the dam's structural behavior and condition through the use of monitoring systems is a continuous improvement process based on three activities: monitoring, data analysis and interpretation of the dam's behavior, and dam safety assessment and decision-making, as represented in Fig. 10.1.

Monitoring considers the observation of a phenomenon or event, involving visual inspections and taking measurements to quantify in order to better describe it. The purpose of the analysis and interpretation activities is to provide the necessary back-

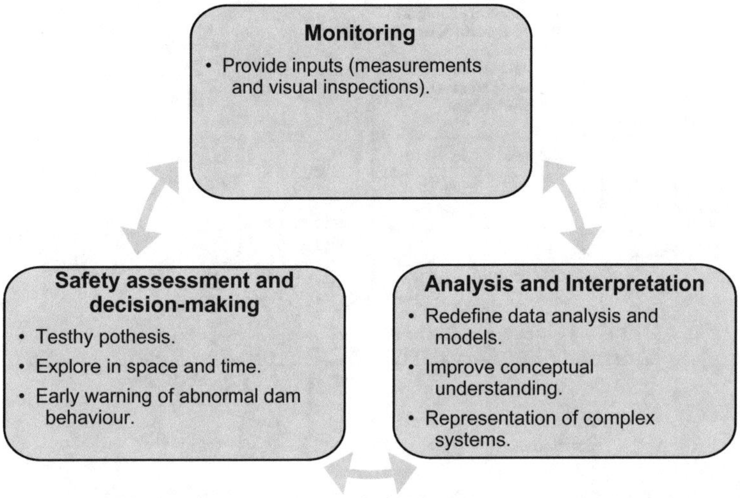

Fig. 10.1 Main activities of structural dam safety control

ground about dam behavior for a better definition of the requirements (data selection, type of models, etc.), to enhance the conceptual understanding and to represent the dam's behavior through models. Once a model is constructed, the assessment of the dam's condition is based on test hypotheses and scenarios using monitoring data, and the prediction of the structural behaviour in space and time.

The procedure of providing information for the assessment of the dam's structural condition and behavior is itself a form of critical thinking and analysis in order to reduce the degree of uncertainty about the structural behavior. Our ideas about the actual dam behavior, or even the validity of the hypothesis and models, are put into question when new evidence that do not match the existing hypothesis are found.

In the design phase of a dam, the intention is to create a structural form which, together with the foundation and the environment, will most economically: (i) perform its function satisfactorily without appreciable deterioration during normal scenarios expected to occur in the dam's life (ensuring performance and dam safety conditions) and, (ii) will not fail catastrophically during the most unlikely (but possible) extreme hazard scenarios which may occur (ensuring dam safety condition).

During the dam's life, the performance and dam safety conditions are reassessed for the same scenarios, and for other scenarios "suggested" by the observed behavior through the analysis of relevant parameters (such as self weight, water level and temperature variations, among others), as represented in Fig. 10.2. Typically, these parameters will describe: the loads or operating conditions to which the system may be subjected, the materials from which the structure is constructed, the materials upon which the structure is to be founded, and the structural response of the dam.

The assessment of the structural dam behavior and dam condition must be performed for each dam independently, even for dams of the same type, because of sev-

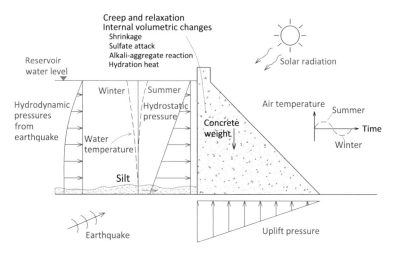

Fig. 10.2 Some parameters analyzed for the assessment of dam safety and performance

eral aspects, such as heterogeneities in the dam's foundation and in the surrounding areas of the dam, or different loads (as a consequence of environmental or operational conditions).

Models used for dam safety control should incorporate all information available. Note, however, that in the point of view of the design, the values for the structural properties and the loads are assigned and the dam safety condition is assessed based on the responses obtained through numerical models, whereas, in the point of view of the safety control, and when there is a large amount of observations of loads and responses, the structural properties that characterize the model can be determined and compared with the values of these structural properties effectively observed.

During the dam's life phases, the models used are updated to take into account the observed dam behavior through the monitoring systems. This is the case of quantitative interpretation models, whose parameters can be updated based on the measured dam response over time.

Present and past observations of loads and responses provide a continuous representation of the dam state, creating an updated model that accurately represents the real behavior of each dam. Such data is the most powerful tool to support decision making by dam specialists, since:

- it supports informed decision making as a data-driven decision support system, where dam specialists can analyze the actual behavior of each dam;
- it provides historical data about relations between actions/responses, making it possible to create predictions about the future dam responses, allowing a model-driven decision support system.

As a consequence, dam safety specialists create behavior prediction models to represent the expected behavior of each dam. The predicted values are then compared with the real response (values observed by the monitoring system), aiming to identify deviations from the real behavior to the expected one (considered as the normal behavior). Identified deviations between the predicted behavior and the observed behavior can mean: (i) structural anomaly; (ii) structural adaptation to new conditions, which means that the prediction model is outdated; (iii) inadequate prediction. Indeed, it is of most importance to adequately identify deviations that represent any kind of structural anomaly, reducing as much as possible any deviations related to outdated or inadequate prediction models.

Traditionally, dam specialists use Multiple Linear Regression models [40] to predict the expected behavior of each dam. Current advances in the machine learning field, specially on the class of deep learning models, can be seen as a breakthrough for dam safety if they manage to better predict the dam behavior in time. Based on this challenge, this chapter proposes the use of Recurrent Neural Networks to model the behavior of dams, represented by the sensor data generated by the monitoring system installed at each dam.

The remainder of this chapter is organized as follows:

- Section 10.2 presents the data lifecycle in the safety control of concrete dams, explaining how the data is captured; what are the main processing and data storage capabilities that are required for dam safety data; how data quality and outlier detection can be achieved is this field; how quantitative models can be used in dam safety and, finally, what is the current state of the art in the usage of machine learning techniques in the safety control of concrete dams;
- Section 10.3 surveys the use of deep learning for sensor data prediction with a specific focus on deep learning in time series;
- Section 10.4 presents the research method followed in this chapter to design the proposal of the usage of deep learning strategies to predict the structural dam behaviour;
- Section 10.5 details the deep learning methods used to predict the structural dam behaviour;
- Section 10.6 presents the evaluation of the methods proposed on Sect. 10.5, using a real case study of the Alto Lindoso dam;
- Section 10.7 details the main conclusions of this chapter.

10.2   The Data Lifecycle in the Safety Control of Concrete Dams

To understand the implications of dam safety data we need to consider its lifecycle. Figure 10.3 shows the dam safety engineering oversight proposed by Ljunggren et al. [38], where the data management lifecycle includes: (i) raw data collection; (ii) processing and data storage; (iii) data analysis.

Based on this structure, this section analyses raw data collection in Sect. 10.2.1, processing and data storage capabilities in Sect. 10.2.2, data quality and outlier detection in Sect. 10.2.3, analysis of the structural response with quantitative interpretation

Fig. 10.3 Dam safety engineering oversight. Retrieved from [38]

models in Sect. 10.2.4 and, finally, Sect. 10.2.5 surveys how machine learning techniques are currently being used in dam safety to support the analysis process.

10.2.1 Raw Data Collection

Visual inspections,[1] tests[2] and measurements provided by monitoring systems are the methods used to maintain an updated knowledge required to exercise safety control.

Dam safety control begins with the preparation of a monitoring plan, in principle before the start of construction. The monitoring plan must pay attention to the hypotheses and critical aspects considered in the project, taking into account the assessment of potential risks, and the definition of the necessary resources to guarantee the safety control and the dam functionality over time, and the timely detection of any abnormal phenomena. The monitoring plan is established for the entire life

---

[1] Inspections are either of a routine nature, or may follow unusual occurrences, such as earthquakes or large floods.

[2] Laboratory and in situ tests, and long term monitoring are used to measure changes in structural properties, actions, and their effects and consequences.

of the dam, however, it must be understood as being dynamic and must be revised and updated, if necessary.

During construction, the good quality of materials and construction processes should be ensured. During the first filling of the reservoir, the dam behavior must be followed with particular attention, not only because it is in this period that a potential risk is created with the formation of the reservoir (it is the first load test), but also because experience has shown it to be a critical period of the dam safety (the dam is subjected to loads that was never subjected). During operation, monitoring focuses on supporting the analysis and interpretation of the dam behavior. At this point in the life of the dam, a significant body of information has most likely been developed about the dam and dam site. The information and data collected during previous phases of the dam can be used to identify the dam safety issues of current concern [46].

The assessment of dam condition, through the use of the information provided by the monitoring system, is achieved by having an up-to-date knowledge of the dam so that anomalous behavior is detected in sufficient time to allow appropriate intervention to correct the situation or to avoid serious consequences.

A monitoring system, defined in the monitoring plan, is designed according to the possible accident and incident scenarios, taking into account aspects related to: (i) the dam safety and functionality, (ii) characterization of the dam behavior (actions, structural properties and effects), (iii) accuracy of the instrument concomitant with the expected range of the physical quantities that will be measured, (iv) reliability and redundancy, (v) access to the dam (some dams have no access during the winter, for example).

Monitoring systems must be adequate and reliable. The insurance of good performance and the functionality of the monitoring system throughout all of the dam's life phases are main requirements. For these reasons, the use of instruments that can easily be replaced or repaired without compromising the continuity of the monitoring process (both time continuity and in compatibility of the measurements) is recommended. The redundancy of measurements must be considered to avoid possible wrong conclusions based on a possible malfunction of a single instrument. Besides the economic aspects, the instrumentation must take into account practical aspects. For example, the installation of high precision instruments with short range, may not be acceptable.

The variables to be measured, the general information about the devices to be installed, and the procedures to be followed in the installation and maintenance are also presented in the monitoring plan. The methods used for dam monitoring can be classified by their purpose [22]:

- Characterization of the structural properties: in situ and laboratory tests of samples of the materials used in the dam construction, forced vibration tests, and tests under fast loads or permanent loads over time in cells installed in the dam.
- Monitoring of the actions: the observation of the sequence and techniques used in the dam construction, water levels, air and water temperatures, earthquakes (including induced earthquakes by the filling of the reservoir), through the use of

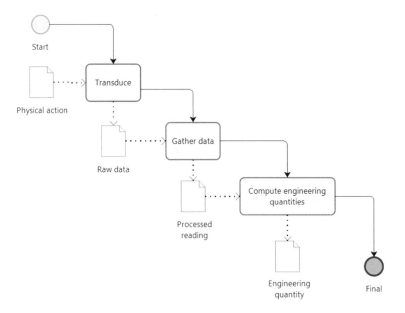

Fig. 10.4 Data transformation workflow for an electrical instrument

limnimetric scales, thermohygrographs, maximum and minimum thermometers and seismographs, among others.
- Monitoring of the direct effects of the actions: the pressures in the pores and cracks, seepage and leakage, and concrete temperatures, through the use of piezometers, drains, weirs and thermometers.
- Monitoring of the indirect mechanical effects: absolute and relative displacements, joint movements, stresses and strains, through survey methods, and through the use of direct and inverted pendulums, rod extensometers, jointmeters, stress meters and strain gauges.

Instrument measurements are manually collected by human operators, using specific measuring instruments, or automatically collected by data acquisition units connected to a network of sensors.

10.2.2 Processing and Data Storage

After the acquisition process, the data is transformed into engineering quantities (e.g. relative displacements, seepage) by specific algorithms that use a set of calibration constants. In fact, the term reading does not correspond to raw data, since a reading is also a transformation from the raw data.

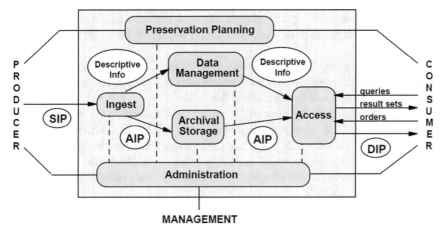

Fig. 10.5 OAIS functional model. Retrieved from [25]

Figure 10.4 illustrates a typical data transformation workflow for an electrical instrument (e.g., a Carlson Extensometer). Instruments (transducers[3]) convert a physical action (e.g., displacement) into an electrical signal (raw data produced as a voltage in mV), which is then converted by a gathering instrument (or by the sensor) into processed readings (e.g., resistance, relation of resistances). Finally, the readings are converted into engineering quantities (e.g., extension), which is the information used to assess the structural behavior of the dam. The dam safety monitoring information includes, essentially, instrument properties, calibration constants, readings and engineering quantities to quantify the physical actions and the response of the dam.

Since dams can be in-service for several decades or even a century, past data collected during the construction and exploitation phases is critical to support the assessment of current structural safety. As a consequence, the data lifecycle in the safety control of concrete dams must be seen as a long-term cycle where data must be preserved. Data management of such long cycles is usually known as digital preservation, where current efforts are often built upon the Open Archival Information System (OAIS) reference model [25], which addresses fundamental issues surrounding trust and provides the basis of a certification standard for digital repositories [26].

The OAIS model provides a high level model designed to support static processes and static information types for longterm preservation. Figure 10.5 shows the high-level functional entities of OAIS in relation to its contextual environment, which is comprised of producers, consumers, and management. Content enters the archive through the Ingest function in the form of a Submission Information Package (SIP). It is processed and passed onto Archival Storage as an Archival Information Package (AIP). For access, a Dissemination Information Package (DIP) is created upon the request of a Consumer. These primary functions are managed, supported and controlled by Preservation Planning, Data Management, and Administration. The

---

[3] A transducer is a device that converts any type of energy into another.

OAIS also defines a conceptual information model describing the structure of the information packages handled within the archive during ingest, archival storage, and access.

The OAIS focus on the "inner walls" of an archive, ensuring that data captured during the dam entire lifecycle is accessible for analysis and reuse, dealing with the data storage requirement for dam safety data.

Due to the properties and value of dam safety data, specially the one generated by sensors that can not be recovered or repeated in case of loss, dam owners and national authorities must use long-term repositories that ensure adequate management of dam safety data and long term preservation to support access to data during the entire data lifecycle in the safety control of dams, which can encompass several decades or even a century of observations.

10.2.3 Data Quality Assessment and Outlier Detection

Data quality is focal when talking about dam safety engineering. As seen in Fig. 10.3, dam's data quality is impacted by every single process in the cycle, however, in the reverse direction, data quality directly impacts data analysis and dam safety assessment. The effects of bad data quality will only be observed in the final stages of the cycle, with the greater risk of not being detected at all and misinforming dam safety experts.

To correctly analyze the behavior of a dam, measurements of physical quantities, collected by the dam monitoring system, should be representative of the dam real behavior, i.e., the measurement result (collected value) and the correspondent measurand (real physical value) should have the same value. The real value is always unknown, however there are ways to determine if the measured value is wrong (e.g., domain limits). If a measurements does not correspond to the real physical value of the quantity measured, the dataset contains errors (dam behavior interpretation is distorted) and therefore is not of quality [41].

With the advances of technology, automated systems are utilized to monitor dams, providing an increase of information that beforehand had to be collected manually and with less frequency. However, with this increased amount of information, the potential for measurement errors also increases. Automated measurements can be compared with the manual ones (manual measurements) and use them as reference elements to assess the quality of stored data [42], as seen in Fig. 10.6. Estimating the Probability Density Functions (PDF) can be useful to characterize both the manual and the automated measurements (similar PDF are expected if the sensors are paired).

Regarding gross errors, such as outliers, they can be caused by various factors. Mechanical and instrumental errors (like sensor failure), IT errors (overrides in databases or data corruption), human errors (manual entries) or even deviations in the system behavior (if the outlier is caused by this, the information gathered by detection is very important to the creation of knowledge) [20]. In order to detect outliers, several techniques can be applied from simple 2D scatter-plots or Whiskers-Box-

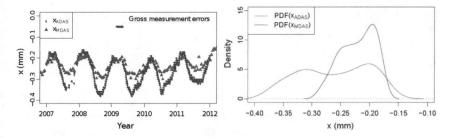

Fig. 10.6 Left: manual data aquisition system (MDAS, in red) versus automatic data acquisition systems (ADAS, in blue) measurements; Right: MDAS (in red) versus ADAS (in blue) PDF comparison. Retrieved from [42]

plot to Machine Learning (ML) algorithms. However, some techniques are faster and more accurate than others, and a manual outlier detection process does not suffice in an Automatic Data Acquisition Systems (ADAS) scenario, where acquisition frequency is higher and therefore a larger amount of data is collected and needs to be pre-processed before it is available to dam safety experts.

Multiple Linear Regressions (MLR) and other predictive algorithms can be used in outlier detection. After predicting a set of values, incoming measurements can be labeled as outliers if they are found outside of a certain distance from the expected value limits (defined by a boundary based on standard deviations) [63]. Other ML algorithms, like Density-Based Spatial Clustering of Applications with Noise (DBSCAN), an unsupervised clustering algorithm, can also be useful in this task [2].

10.2.4   Data Analysis and Dam Safety Assessment Based on Quantitative Interpretation Models

Quantitative interpretation models for the prediction of the structural response of concrete dams, typically used by dam engineers, are based on the estimation of parameters, and on several simplifying assumptions concerning the behavior of materials, such as:

  (i) The analyzed effects refer to a period in the life of a concrete dam for which there are no relevant structural changes.
 (ii) The effects of the normal structural behavior for normal operating conditions can be represented by two parts: a part of elastic nature (reversible and instantaneous, resulting from the variations of the hydrostatic pressure and the temperature) and another part of the inelastic nature (irreversible) such as a time function.
(iii) The effects of the hydrostatic pressure, temperature, and time changes can be evaluated separately.

Table 10.1 HST and HTT models—quantification of the thermal effect through multiple linear regression

Model	Advantages	Disadvantages
HST	• Simple	• Thermal effect estimated without knowledge of the air, water or dam body temperatures evolution
	• Thermal effect obtained through sinusoidal functions	• The prediction approximates the maximum measured values by default and the minimum measured values by excess
HTT	• More accurate quantification of the thermal effect through the knowledge of the embedded thermometers in the dam body	• Thermometers in the body that present a phase offset similar to the structural response under analysis is required
		• Difficulty in the selection of the thermometers and its adequate functioning along time

Quantitative interpretation models are typically obtained through multiple linear regression (MLR). If the hypotheses that support the MLR models are true, the separation of effects is valid, which is advantageous to quantify the contribution that a particular action has on the structural response. The main actions are the hydrostatic pressure variation caused by the variation of the water in the reservoir, the temperature changes in the dam body that results from the air and water temperature variations, and other phenomena not reversible in time (mainly due to changes in the properties of the materials). HST (Hydrostatic, Seasonal, Time) models are statistical models widely used because they consider that the thermal effect considered as the sum of sinusoidal functions with an annual period, similar to variations of air and water temperatures [40]. However, the effect of the real annual wave of the air temperature variation does not follow a shape similar to a sinusoidal function (the winter and summer periods are not well represented by sinusoidal waves, and, as a consequence, extreme values on the structural response due to the temperature effect are not accurate). To overcome this drawback, some models use recorded temperatures, also known as HTT (Hydrostatic, Thermal, Time) models that better represent the thermal effect on dam behavior (instead of the seasonal function like HST models). However, the choice of the thermometers (usually devices embedded in the dam body) to be considered or even the use of air temperature measurements also have its difficulties.

Table 10.1 shows the main advantages and disadvantages of multiple linear regression models (HST and HTT models) regarding the quantification of the temperature effect. In both types of models the air temperature is not considered because the structural response presents, in average, a phase offset and a amplitude change when compared with the annual air temperatures variation.

Structural safety control activities are based on the interpretation of the observed behaviour. Models are used to support the interpretation of the observed structural behaviour along time. The portion not explained by the model, called residues ($\varepsilon = \delta_{measured} - \delta_{modelprediction}$), is obtained through the difference between the observed behaviour and the model prediction. The interpretation of the residues (and of their evolution) is equally important since they are related to the measurement uncertainty and the portion of the structural response that could not be explained by the used model.

Formulation by Separation of the Reversible and Irreversible Effects in HST Models

HST models consist in approximating the shape of the deterministic indicators through simple functions which are easier to manipulate [63]. It is considered that the effects (such as horizontal displacements at the crest of the dam) associated with a limited time period at a specific point can be approximated by

$$\delta_{HST} = \delta_H + \delta_S + \delta_T + k \tag{10.1}$$

where δ_{HST} is the observed structural response; δ_H is the portion of the structural response due to the elastic effect of hydrostatic pressure; δ_S is the elastic portion of the structural response due to the effect of temperature depending on the thermal conditions represented by seasonal terms; and δ_T is the portion of the structural response due to the effect function of time considered irreversible.

The separation of effects requires the consideration of a constant k due to the fact that the structural response, measured on the reference date, has a value different from zero.

The portion of the structural response due to the effect of hydrostatic load, δ_H, is usually represented by polynomials depending on the height of the water in the reservoir h:

$$\delta_H (h) = \beta_1 h + \beta_2 h^2 + \beta_3 h^3 + \beta_4 h^4. \tag{10.2}$$

The portion of the structural response due to the effect of the temperature changes can be considered as a proportional function of the environmental temperature changes, with a phase shift, depending on the depth into the section. The portion of the structural response due to temperature changes is considered instantaneous with respect to the temperature field in the dam body, but it is deferred with respect to the measured air and water temperatures.

Very simple models, like these HST models, usually do not use temperature measurements because it is assumed that the annual thermal effect $\delta_S (d)$ can be represented by the sum of sinusoidal functions with a one-year period. Thus, the effect of temperature variations is defined by a linear combination of sinusoidal functions, which only depend on the day of the year:

$$\delta_S (d) = \beta_5 \sin(d) + \beta_6 \cos(d) + \beta_7 \sin^2(d) + \beta_8 \cos^2(d) \tag{10.3}$$

where $d = \frac{2\pi \cdot j}{365}$ and j represents the number of days between the beginning of the year (January 1) until the date of observation ($0 \leq j \leq 365$).

To represent the time effects, δ_t, it is usual to consider the functions presented in Eq. 10.4, where t is the number of days since the beginning of the analysis.

$$\delta_T (t) = \beta_9 t + \beta_{10} e^{-t} \tag{10.4}$$

10.2.5 Data Analysis and Dam Safety Assessment Based on Machine Learning Models

Machine learning techniques are used to aid in the interpretation of the observed behavior and the structural safety control of dams. Predicted values, obtained through the use of machine learning models, are compared to the real value of sensor readings, in order to detect any major deviation of response (which can point to possible structural damage). Studies in the field of research mainly focus on obtaining a meaningful set of predictors (e.g., water level, air and water temperature, etc), and a good machine learning method capable of correctly predict the response variable (e.g., radial displacements, crack openings, etc).

Tatin et al. [64] presented the HST-Grad model, a hybrid of the HST model with the inclusion of air and water thermal variation. The model can be seen as a Multiple Linear Regression (MLR) where the effects of lag predictors (temperature gradients) were enough to slightly increase the performance of the previous models for radial displacement prediction in French dams. HST models can also be used to individually quantify a specific effect in the dam response due to one of the main actions (e.g., effect of hydrostatic pressure in crest displacement shown by De Sortis and Paoliani [9]). In Tatin et al. [65], the structure is seen as a group of horizontal layers that allow the observation of the effect of thermal effect of both water and air temperature throughout the dam.

Several adaptations from the HST model can be found. The utilization of measured concrete temperatures instead of the seasonal temperature variation (the HTT model, as seen in Perner and Obernhuber [50]); the inclusion of an Error Correction Model (ECM) for increased precision in time-series, proposed by Li et al. [33]; the use of a hybrid with a genetic algorithm, increasing robustness and predictive power of MLR shown in Stojanovic et al. [60]; the Hydrostatic Seasonal State (HSS) model, proposed by Li et al. [34], that represents time-effect deformation as a state equation proved able to provide a better fitting to radial deformations than the HST model; and the EFR (EFfet Retard—Delayed effect) model, utilized by Guo et al. [14] to predict pore water pressure by taking in account the delayed hydrostatic effect, are some examples.

Feed-forward Neural Networks (FNN) proved to be an powerful tool in assessing concrete dam behavior, as shown by Mata [40]. The FNN model's prediction for horizontal displacements, using water level and seasonal temperatures variations, obtained better results when compared to MLR models. FNNs models can also be used to predict piezometric water levels in piezometers. Ranković et al. [54] compare FFN with MLR (both trained with three values of water level from previous days of each measurement) and concludes that FNN provide better prediction of the target variable. Tayfur et al. [66] obtained improved performance, on average, using FNN when compared with a Finite Element Model (FEM), however this was not the case in every analyzed dataset. Kang et al. [29] utilizes Extreme Learning Machine, a type of FNN with a single hidden nodes layer, obtaining better results of displacements prediction when compared to MLR, backpropagation-trained NNs, and Stepwise Regression.

Modification to Support Vector Machines (SVM) were used by Cheng and Zheng [5] in order to create a model able to simulate the non linear mapping between environmental and latent variables. The LS-SVM (Least squares SVM) model presented good performance when predicting uplift pressure and horizontal displacements. Ranković et al. [53] use Support Vector Regression (SVR), an application of SVM for function estimation, to predict tangential displacements. To predict displacements, Su et al. [61] combines SVM with other methods, such as wavelet analysis to resolve some problems identified with SVM (e.g, kernel function and parameter optimization). SVM can also be useful to detect failure, categorized damage states and predict local responses, as can be seen in the FEM-SVM based hybrid methodology presented by Hariri-Ardebili and Pourkamali-Anaraki [15].

Principal Component Analysis (PCA) is used by Yu et al. [71] to provide data reduction, noise filtering and multivariate analysis and monitoring. PCA is followed by a HST model to predict crack size opening. Mata et al. [43] makes use of PCA to obtain a HTT model, where the goal of PCA was to select which thermometers had more impacting/correlation to the response variable, and therefore should be used as predictors. This was also done by Prakash et al. [51] when obtaining a Hydrostatic, Seasonal, Temperature, and Time (HSTT) model. PCA was utilized as a data reduction tool before using HSTT to predict the dam responses (displacement and strain).

Bui et al. [4] presents SONFIS (Swarm Optimized Neural Fuzzy Inference System) as a promising tool for modeling horizontal displacements. When compared to other algorithms, like SVM and Random Forest (RF), SONFIS outperformed the presented benchmarks by using the PSO (Particle Swarm Optimization) algorithm to optimize parameters for the neural fuzzy inference system (water level, air temperature and time are used as predictors).

Other regressions besides the HST model can be found in Jung et al. [27], where a Robust Regression Analysis is used with PCA to predict piezometric readings in time-series data with periodic or dominant variations, and in Xu et al. [68], which combined a genetic algorithm (for predictor variables selection) and a Partial Least Squares (PLS) regression to predict crack opening in a Chinese dam. Random Forest Regression (RFR) is used by Dai et al. [8] to predict horizontal displacements and in Li et al. [35] to create a uplift pressure model, obtaining better results than SVM. Li et al. [35] also studies 18 different predictors, stating that the model can be used to extract correlations and rules between the variables (the influence of rainfall was considered the smallest when compared to the other factors). Boosted Regression Trees (BRT) are used in Salazar et al. [56] to predict radial displacements and leakage flows. Different predictors were explored, and relative influence of each predictor was obtained for each target variable. BRT can also be used in a anomaly detection scenario as seen in Salazar et al. [58].

Most of the authors use HST models or adapt it to obtain good predictive models. However, the diversity of ML algorithms available is increasing daily, and researchers are using them to create better models. PCA, SVM, and NN are some algorithms already used by the community (Salehi and Burgueño [59] presents a survey of ML techniques used in structural engineering) that help increasing prediction performance, and therefore, increasing safety control mechanisms accuracy in concrete dams. In Salazar et al. [55], RF, BRT, NN, SVM and Multivariate Adaptive Regression Splines (MARS) are utilized to predict radial and tangential displacements and leakage (in a total of 14 different datasets). An extensive comparison between several machine learning techniques applied to dam behavior prediction is presented in Salazar et al. [57].

10.3 Data Analysis and Data Prediction Using Deep Learning Models—An Overview

Deep learning methods [31] are a class of machine learning methods that learn multiple layers, or levels, of representations by composing simple non-linear modules at one layer into more abstract representations one layer above. What is more, this representation learning is done mostly automatically, without a strong dependence of a human doing manual feature engineering requiring time and expert domain knowledge—the features are learned from the input data by a general-purpose learning mechanism. Of course the domain specialist is still fundamental to interpret and validate the quality of the model within their knowledge expertise. In recent years, deep learning has achieved state of the art or highly competitive results in fields such as image recognition [30], natural language understanding [7], drug discovery [39], recommendation systems [67], and board and video games playing [45].

Recurrent Neural Networks (RNN) are a class of deep learning models which have proved effective at solving tasks involving input and/or output sequences [12]. They process an input sequence one element at a time (possibly a vector) and maintain an internal "state vector" that contains information about the sequence of previous inputs. In principle, a RNN can map from the entire history of the previous inputs to each output, effectively allowing it to simultaneously capture dependencies on multiple timescales. RNNs have recently seen a surge in application including in

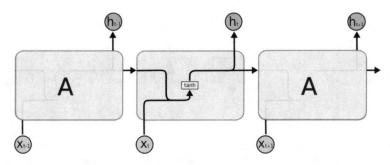

Fig. 10.7 Graphical representation of the recurrence relations defining a RNN. In this diagram, the non-linear activation function σ is represented as the hyperbolic tangent, a common choice. Retrieved from [48]

such diverse fields as machine translation [62] and natural language processing [44], urban mobility and traffic prediction [72], speech recognition [13], clinical diagnosis [37] and DNA sequencing [52].

Given an input sequence $(x_1, x_2, \ldots, x_t)$ where each $x_t \in \mathbb{R}^p$, p is the number of input features and t is the number of timesteps, a RNN is defined by the following recurrent relation:

$$h_t = \sigma(Wx_t + Uh_{t-1} + b), \tag{10.5}$$

where $W \in \mathbb{R}^{n \times p}$, $U \in \mathbb{R}^{n \times n}$, $b \in \mathbb{R}^n$ are matrices representing the hidden state whose values will be learned during training, and n is the dimension or size of the RNN cell. The n-dimensional hidden state vectors at time t and $t - 1$ are denoted by h_t and h_{t-1} respectively, and σ is a non-linear activation function such as the hyperbolic tangent, the logistic sigmoid or the rectified linear unit. Figure 10.7 shows a graphical representation of the recurrent relation above, and Fig. 10.8 depicts two representations of an RNN, namely (a) a compact (or folded) cell in which outputs are connected back as inputs, and (b) an unrolled series of cells explicitly showing each time step. This re-feeding of outputs as inputs and the sharing of the cell's parameters across layers (or timesteps) are the main differences from the classical feedforward neural network.

The most common architectures of RNNs are sketched in Fig. 10.9 in an unrolled representation. For a timeseries prediction problem one will typically use the sequence vector to single vector architecture for the one-step prediction (b) or, in the case of multi-step prediction, one of the two sequence-to-sequence architectures (d) and (e).

Despite its representational power, training RNNs has been considered difficult [3]. When unrolled in time, RNNs resemble a very deep FFN with as many layers as timesteps. The naive application of backpropagation leads to the problem of exploding and vanishing gradients. In 1997, Hochreiter and Schmidhuber introduced the Long Short-Term Memory (LSTM) network as a solution to this problem

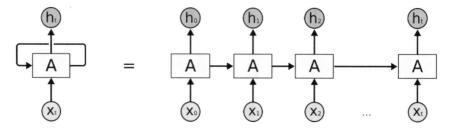

Fig. 10.8 Two representations of the same RNN model. Left: A compact (or folded) representation, in which the parameter sharing between different timesteps is accentuated. Right: An unrolled representation of an RNN putting in evidence the relationship between different elements of the input sequence, the internal state of the network and the output. Retrieved from [48]

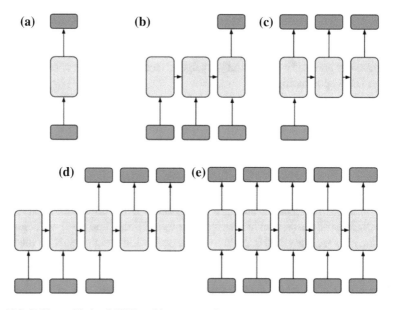

Fig. 10.9 Different kinds of RNN architectures. **a** One-to-one: the traditional feedforward neural network; **b** Many-to-one: sequence prediction or classification, includes one-step time series prediction tasks, such as, for example, financial or weather forecasting, product recommendations, anomaly detection, sentiment analysis or DNA sequence classification; **c** One-to-many: sequence generation, including for example image labeling; **d** and **e** Many-to-many: sequence to sequence learning of which some example applications would be sentence translations, multi-step time series prediction, program execution, text summarization, and text and music generation. Retrieved from [36]

by adding to the vanilla RNNs a memory cell and a gating mechanism that regulates the information flow [19]. This gating mechanism (which includes an input, forget and output gate) is responsible for managing which information persists, for how long, and when to be read from the memory cell. Because the cell state is updated with an addition operation (and not a sigmoidal transformation as in vanilla RNNs) it

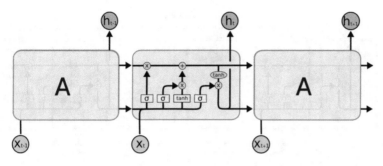

Fig. 10.10 Graphical representation of the recurrence relations defining a LSTM cell. Retrieved from [48]

does not suffer from the vanishing gradient problem. Figure 10.10 shows a depiction of this mechanism, which can be formalized in the following equations:

$$f_t = \sigma(W_f x_t + U_f h_{t-1} + b_f) \tag{10.6}$$

$$i_t = \sigma(W_i x_t + U_i h_{t-1} + b_i) \tag{10.7}$$

$$\tilde{c}_t = \tanh(W_c x_t + U_c h_{t-1} + b_c) \tag{10.8}$$

$$c_t = i_t \odot \tilde{c}_t + f_t \odot c_{t-1} \tag{10.9}$$

$$o_t = \sigma(W_o x_t + U_o h_{t-1} + b_o) \tag{10.10}$$

$$h_t = o_t \odot \tanh(c_t) \tag{10.11}$$

where, $f_t, i_t, \tilde{c}_t, c_t, o_t$ and h_t, represent the outputs of forget gate, input gate, candidate state, cell state, output gate and the final cell output, respectively, $\odot$ represents the element-wise Hadamard product, and h_t can be used as the final output.

There are several extensions, or deep learning alternatives, to the vanilla RNN and LSTM architectures and numerous related techniques which we do not pursue in this work, but which have already shown to be promising approaches to effectively model timeseries prediction tasks. These include attention mechanisms [69], convolutional neural networks [70], stochastic regularization techniques (such as dropout [18]), grid LSTMs [28], multimodal learning [47] and probabilistic methods [11].

10.4 Adopted Problem Solving Process—The Design Science Research Methodology

The Design Science Research Methodology (DSR) is a problem solving process that focuses on the relevance of creating and evaluating different artifacts to meet and solve relevant objectives and problems [16]. As proposed by Peffers et al. [49], the design science research methodology encompasses six steps triggered by possible research entry points, as represented in Fig. 10.11.

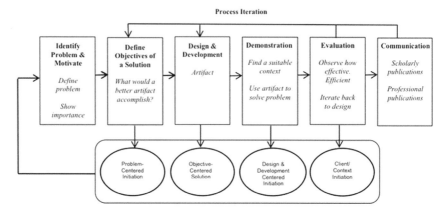

Fig. 10.11 Design science research methodology. Retrieved from [49]

This chapter follows a DSR methodology and is driven by the relevance of dam structural safety, being a recognized problem, not only due to the value produced by these critical infrastructures, but also to the potential catastrophic consequences in the case of structural failures. As such, the objective of this research is to improve the support to the decision making process performed by dam safety specialists, through better prediction models for data captured by sensors installed on dams. To accomplish this objective, methods based on prediction models that use recurrent neural networks (namely Long Short Term-Memory networks) are presented on Sect. 10.5, demonstrated and evaluated on a real case study provided by the Alto Lindoso dam on Sect. 10.6. Note that the evaluation of the proposed method is performed based on the results of the multiple linear regression, which is the traditional method used by dam specialists and, as a consequence, is used as the baseline for this study.

10.5   Proposed Methodology—Adding Value to the Interpretation of the Monitored Dam Behaviour Through the Use of Deep Learning Models

HST models based on MLR methods are widely used and allow a big picture of the structural dam behaviour, but a portion of the measured behaviour is not explained. To overcome this drawback, the model proposed aims to take advantage of the knowledge of:

- the multiple linear regression models (HST models in this case study), commonly used, through the use of the predicted values obtained by the model.
- the evolution of the air temperature, represented by mean ($T_{air,mean}$), the 10-quantile ($T_{air,10Q}$) and the 90-quantile ($T_{air,90Q}$) of the air temperature. It is expected that the LSTM neural network model will be able to identify the structural

response pattern (phase offset and amplitude change) due to the air temperature effect, which the MLR model is not able to.

- the reservoir water level evolution along time. For example, an increase of the reservoir water level tend to increase the uplift pressures in the dam foundation, and indirectly, the evolution, along time, of the observed displacement measured in the dam body.

Based on the referred before, two strategies to take advantage of property of LSTM models in processing sequential data could be adopted: (i) the use of a LSTM to model the structural behaviour or (ii) the use of a LSTM model to model the part of the structural response that is not explained by the MLR model.

1. The use of LSTM to model the structural behaviour. In this case, the main inputs of the LSTM model are the predicted values obtained from the MLR model δ_{MLR}, to represent the overall pattern regarding to the hydrostatic, thermal and irreversible effects; $T_{air,mean}$, $T_{air,10Q}$, $T_{air,90Q}$ and h^4 to represent the "missing component" related reversible effect; and t or e^{-t} to represent the "missing component" related to the irreversible effects along time. The output is the observed structural behaviour, δ.
2. The use of a LSTM model to model the part of the structural response that is not explained by the MLR. In this case, the main inputs of the LSTM model are $T_{air,mean}$, $T_{air,10Q}$ and $T_{air,90Q}$ and h^4. The output is the observed structural behaviour not explained by the MLR model, $\delta - \delta_{MLR}$, Eq. 10.12. Besides the information related to a possible anomalous phenomena, the residuals $\varepsilon_{MLR} = \delta - \delta_{MLR}$ contain information related to errors (measurements and model) and other unknown effects.

The last strategy was adopted because the innovation based on the proposal is clear for both academy and future stakeholders and final users. Knowledge developed along years is used (taking into account that worldwide countries have different levels of knowledge and practices regarding the dam safety activities) and improved based on the use of LSTM models.[4]

In this way, we model the part of the structural response not explained by the MLR,

$$\delta = \delta_{MLR} + \varepsilon_{MLR} \qquad (10.12)$$

The unexplained pattern ($\varepsilon_{MLR} = \delta - \delta_{MLR}$) of the structural behaviour (obtained from the MLR model, Eq. 10.12) will be explained by the LSTM model taking advantage of the knowledge of the MLR models (designed as $\delta_{LSTM|MLR}$), as follows

$$\delta - \delta_{MLR} = \delta_{LSTM|MLR} + \varepsilon_{LSTM|MLR} \qquad (10.13)$$

---

[4]New developments must take into account that dam safety is a continuous requirement due to the potential risk in terms of environmental, social and economical disasters.

Finally, the predicted value for the structural behaviour is obtained by summing both δ_{MLR} and $\delta_{LSTM|MLR}$, being the $\varepsilon_{LSTM|MLR}$ the new residuals, as in

$$\delta = \delta_{MLR} + \delta_{LSTM|MLR} + \varepsilon_{LSTM|MLR}. \tag{10.14}$$

The model we propose is a single layer LSTM that takes as input a sequence of N x_t input vectors from x_{T-N} to x_T to predict the value of $\delta - \delta_{MLR}$ at time T. Each input vector x_t has as components the values of $T_{air,mean}$, $T_{air,10Q}$, $T_{air,90Q}$ and h^4 at time t. In general, this model can be easily expanded by adding other input features as new components of x_t.

The hyperparameters of the LSTM are chosen through a validation set approach. These hyperparameters include the size of LSTM cell, the number of timesteps of each input sequence, the loss function, the optimization method and the batch size. To carry out this approach one splits the available dataset into three disjoint sets: the training, the validation and the test set. Holding out the test set, one trains the LSTM model on the training set optimizing the loss function for different combinations of the hyperparameters. The error of each of the resulting models is evaluated on the validation set, and one which achieves the best compromise between a low value of the validation error and computational cost (for example, more timesteps and a larger value of the size of the LSTM cell correspond to longer training and inference times).

10.6 Demonstration and Evaluation—Assessment and Interpretation of the Monitored Structural Behaviour of a Concrete Dam During Its Operation Phase

10.6.1 The Case Study—The Alto Lindoso Dam

The Alto Lindoso dam, depicted in Fig. 10.12, is a double curvature concrete dam whose construction finished in 1992 in a symmetrical valley of the Lima river, in the north of Portugal. The dam is 110 m high, the crest elevation is 339.0 m, and the total crest length is 297 m. The thickness of the central block is 4 m at the crest and 21 m at the base. There are three internal horizontal inspection galleries (GV1, GV2 and GV3) across the dam and a drainage gallery (GGD) close to the foundation [10]. The dam is founded in a good quality granitic rock mass, but with some heterogeneity. The rock mass deformability was characterised through mechanical "in situ" and laboratory tests, and geophysical tests for the determination of propagation velocities of longitudinal waves were performed, before and after the foundation treatment [10].

The dam body was built between April 1987 and July 1990. The injection of contraction joints was carried out between March and May 1991. The reservoir first filling was initiated in January 1992, with the reservoir water elevation at 234 m, and

Fig. 10.12 Alto Lindoso dam

the retention water level with the reservoir water elevation at 338.0 m was achieved in April 1994.

In 2008, an analysis of the structural dam behaviour, carried out by the Portuguese National Laboratory for Civil Engineering, concluded that the Alto Lindoso dam presented, globally, satisfactory structural dam behaviour [32].

In accordance with best technical practices, the monitoring system of the Alto Lindoso dam aims at the evaluation of the loads; the characterisation of the geological, thermal and hydraulic properties of the materials; and the evaluation of the structural response.

The monitoring system of the Alto Lindoso dam consists of several devices which make it possible to measure quantities such as: concrete and air temperatures, reservoir water level, seepage and leakage, displacements in the dam and in its foundation, joint movements, strains and stresses in the concrete, and pressures, among others.

The system used for the measurement of the reservoir water level comprises a high precision pressure meter with a quartz pressure cell, which provides a record of the water height over time, and a level scale. The air temperature and humidity are measured in an automated weather station placed on the right bank, approximately 50 m apart from the dam crest.

The concrete temperature is measured in 70 electrical resistance thermometers distributed across the dam thickness in 16 sections of several blocks. The location of the thermometers was defined taking into account the remaining electrical resistance devices (strain gauges, embedded jointmeters and stress gauges) that also allow for the measurement of the concrete temperature.

Displacements are measured using an integrated system that includes five pendulums, 18 rod extensometers (Fig. 10.13) and geodetic observations. The relative movements between blocks are measured by superficial and embedded jointmeters.

The deformation of the concrete is measured with electrical strain gauges arranged in groups, distributed in radial sections, allowing the determination of the stress state through the knowledge of the deformation state and of the deformability law of the

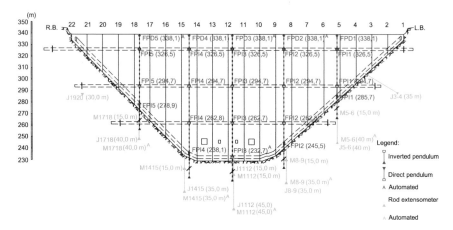

Fig. 10.13 Pendulum and rod extensometer distribution in the Alto Lindoso dam

concrete. Stress gauges, which allow for the direct measurement of normal stress components, were also placed.

The quantities of drained and infiltrated water are measured individually, in drains of the drainage system installed in the dam foundation and in weirs that differentiate the total quantity of water that flows in the drainage gallery in several zones of the dam. The drainage system comprises a set of 52 drains, distributed over the drainage gallery with two drains per block, except for the central blocks 11/12–13/14 and block 18/19 where four drains were executed. All the water extracted from drains and leakages is collected in three weirs. Weir named Bica 1 collects the water from blocks 1/2–9/10 on the left bank, while weir named Bica 2 collects water from blocks 14/15–21/22, on the right bank. Finally, weir named Bica 3 receives all the water that flows in the drainage gallery.

The measurement of the uplift pressure in the foundation is performed by a piezometric network that comprises 23 piezometers. The pressures within the concrete are observed by two groups of three pressure gauges embedded in the concrete in two sections (at levels 310 m and 236 m) in the central block (block 11–12).

In the recent past, an automated data acquisition system was installed but it is still in a testing phase. ADAS includes the measurement of horizontal displacement along pendulums (telecoordinometers), relative displacements in the foundation (rod extensometers), relative movements between blocks (superficial jointmeters), discharges (in weirs) and the uplift pressure (piezometers). Figure 10.14 illustrates the location of the ADAS devices of the Alto Lindoso dam. Manual measurement is also possible in these places.

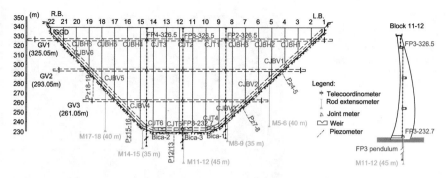

Fig. 10.14 Location of ADAS devices in the Alto Lindoso dam

10.6.2 The Dataset—Horizontal Displacements Measured by the Pendulum Method

As referred before, horizontal displacements in the Alto Lindoso dam, with reference to a vertical, are measured by the pendulum method. Pendulums are installed in shafts either built during construction or drilled after. The pendulum method is based on the position of a steel wire through a vertical line that crosses the dam body. One extremity of the wire is fixed and defines one of two possible variants for this method, direct pendulum or inverted pendulum [23].

In the direct pendulum, one end of the wire is fixed on a high point of the dam, whereas on the opposite end, a weight of approximately 600 N strains the wire. In this case, displacements obtained in the various access points to the wire are relative to the fixed high point.

In the inverted pendulum, one end of the wire is fixed in a deep zone of the dam foundation, out of the zone affected by the main actions. The other end of the wire its connected to a float. In this case, absolute displacements are obtained.

As a consequence of the geometry of the dam, the horizontal displacements are obtained by a combination of direct and indirect pendulums.

In specific points near the pendulum, measuring tables are installed, fixed to the dam, to support the sensitive reading devices which allow for the measurement of the data raw (distance of the wire to the dam) according to the radial and tangential components, Figs. 10.15 and 10.16. The most sensitive reading devices are optical instruments (e.g. coordinometers or electro-optical coordinometers). Reading instruments accuracy may be greater than 0.1 mm [21] (for example, the coordinometers used in Portugal presents resolution equal to 0.01 mm), Fig. 10.17. In recent years, automated systems (such as telecoordinometers with an accuracy equal to 0.01 mm) have been adopted to measure displacements by the pendulum method [23], Fig. 10.15.

Figure 10.18 provides an overview of the dataset used to evaluate the LSTM models for dam safety monitoring and prediction. This dataset provides data back to the year 1992 to the present. The first grid represents the daily evolution of water height

Fig. 10.15 Pendulum, measuring table and automated system for horizontal displacement measurements

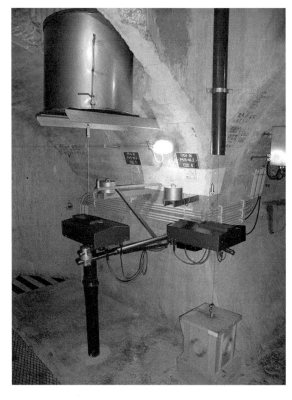

Fig. 10.16 Measuring table

Fig. 10.17 Coordinometer

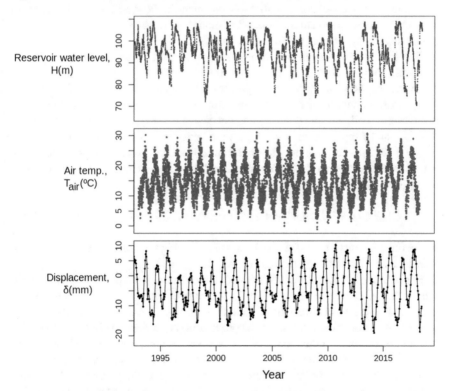

Fig. 10.18 Evolution of the reservoir water height, air temperature (daily average) and horizontal radial displacement at the FP3-326.5m for the training, validation and test periods

(in meters), while the second grid represent the average of the daily air temperature evolution in (°C) and the third grid represents the horizontal radial displacement measured, $\delta_{measured}$, at the FP3-326.5m, near the crest arch at the block 11–12, through the pendulum method (in millimeters).

Note that reservoir water height and air temperature represent the main actions to the dam and can be directly or indirectly used as input features to the model, while the horizontal radial displacement is the final target for prediction by the proposed model.

10.6.3 Main Results and Discussion

In this section we present, evaluate and comment on the LSTM|MLR model put forward in Sect. 10.5 and designed to predict $\delta - \delta_{MLR}$, that is, the difference between the values of δ, the horizontal radial displacement in the crest of the Alto Lindoso dam and the predicted values of the MLR model, at timestep t. We will refer to this model interchangeably as LSTM|MLR or LSTM.

The LSTM model evaluated in this section is built on top of MLR model, as described above in Sect. 10.5. The inputs of the MLR model are: the reservoir water height standardized, h, to the fourth power (h^4) to represent the effect of the hydrostatic pressure; and the effect of the temperature was considered through the implementation of both the sine ($sin(d)$) and the cosine ($cos(d)$) of annual period, where $d = \frac{2\pi \cdot j}{365}$ and j represent the day of the year, between 01 January and 31 December ($0 \leq j \leq 365$, the time effect did not seem to have a significant importance in the period examined by this study. The MLR model can be represented in equation form in the following way:

$$\delta_{MLR} = -3.73 \times h^4 - 4.22 \times sin(d) - 2.75 \times cos(d) - 4.30, \quad (10.15)$$

where the notation of Eqs. 10.1–10.3 was followed. Figure 10.19 shows the measured displacements δ and the predictions of the MLR model.

For both the MLR and the LSTM model, the dataset was split into three disjoint sets: training, validation and test set. In the case of the MLR model, there are no hyperparameters to choose with a validation approach, and we have denoted the union of the validation and test sets as prediction set, as depicted in Fig. 10.19. The test set corresponds to the period of the last year of data, and the validation set to period of two years before the test set. Accordingly, the prediction set corresponds to the last three years of data. Figure 10.20 shows the residues of the MLR model (the differences between the measured value of the displacement and the prediction of the MLR model) which are the quantity to be learned by the LSTM model.

The inputs to the LSTM model are: (i) the time series related to hydrostatic effect, based on h^4, and (ii) the temperature effect through the use of the average, the 10th and the 90th percentile of the air temperature recorded during the time period of 15 days immediately before each time-step.

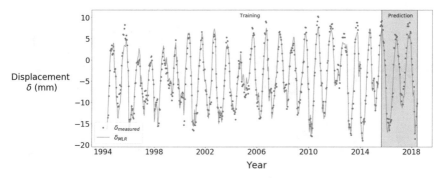

Fig. 10.19 Measured displacements ($\delta_{measured}$) and MLR model (δ_{MLR}) fit on the training set (1994 to 2015) and evaluated of the prediction set (2015–2018)

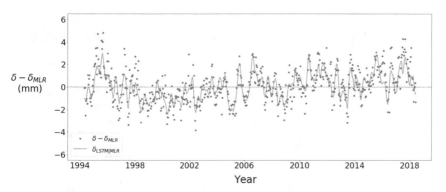

Fig. 10.20 Differences between the measured displacements and the MLR predictions ($\delta_{measured} - \delta_{MLR}$) and LSTM|MLR model fit to these differences (in orange)

As mentioned in Sect. 10.5, the hyperparameters for the LSTM model were chosen via a validation set approach. Each model consisted of an LSTM cell of dimension 32. The training was carried using the Python deep learning library Keras [6] with Tensorflow [1] as a backend, a batch size of 4, 'Rmsprop' [17] as an optimizer and Mean Squared Error (MSE) as a loss function. The models receive a sequence of 30 time-steps of the inputs (corresponding to 15 months of data). The MLR model was trained on the dataset with the last 3 years of data held out, and used as a prediction set. For the LSTM model, the last year of data was used test set. The data corresponding to the period of two years before the last year was used as validation set to choose the hyperparameters.

Figure 10.20 shows the fit of the LSTM model to the residues of the MLR, and in Fig. 10.21 the predicted values, $\delta_{MLR} + \delta_{LSTM|MLR}$, are depicted. These predictions are shown in greater detail for last three years of data, corresponding to the validation and the test set, in Fig. 10.22. Collectively these figures illustrate the adequateness of the strategy proposed since the LSTM|MLR can in fact learn the pattern of the structural behaviour prevailing in the residues of the MLR model (Fig. 10.20) and

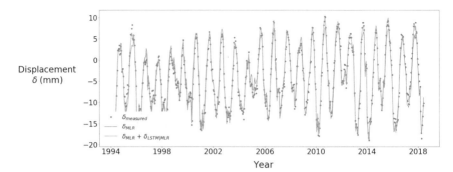

Fig. 10.21 Measured displacements, MLR model predictions and MLR + LSTM|MLR model predictions

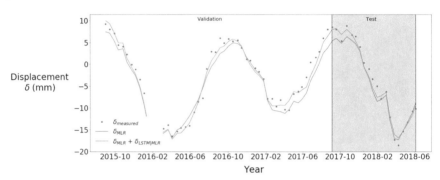

Fig. 10.22 Measured displacements, MLR model predictions and MLR + LSTM|MLR model predictions in the Validation and Test sets

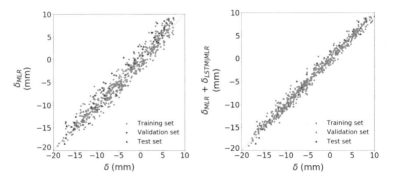

Fig. 10.23 Left: Predicted values δ_{MLR} versus measured displacements δ, Right: Predicted values $\delta_{MLR} + \delta_{LSTM|MLR}$ versus measured displacements δ

Table 10.2 Performance parameters obtained from the MLR model and from the new MLR+LSTM|MLR model

	Model	MSE (mm²)	MAE (mm)	SD (mm)	MaxAE (mm)
Training set	δ_{MLR}	2.08	1.15	0.87	4.81
	$\delta_{MLR} + \delta_{LSTM|MLR}$	0.72	0.66	0.54	3.10
	Gain (%)	65	43	39	35
Validation set	δ_{MLR}	3.48	1.53	1.07	4.27
	$\delta_{MLR} + \delta_{LSTM|MLR}$	1.66	1.03	0.77	3.12
	Gain (%)	52	33	28	27
Test set	δ_{MLR}	4.29	1.67	1.23	4.26
	$\delta_{MLR} + \delta_{LSTM|MLR}$	1.36	0.97	0.66	2.57
	Gain (%)	68	42	46	40

provide through the MLR+LSTM|MLR model better predictions of the relevant quantity, namely the displacements, than the baseline MLR model (Fig. 10.21).

We have plotted the predicted values of the MLR model and of the MLR + LSMT|MLR model against the measured values in Fig. 10.23. Again, comparing the two plots it is clear that the MLR+LSTM|MLR model is a model which provides a better prediction, since the cloud of points depicted is denser and closer to the identity line.

Table 10.2 presents the main performance parameters such as the mean squared error (MSE), the mean absolute error (MAE), the standard deviation of the errors (SD), and the maximum absolute error for each of the models for the training dataset (period between April 1994 and August 2015), the validation set (period between August 2015 and June 2017) the test dataset (period between July 2017 and June 2018). The last row of Table 10.2 shows the gain between the two models in the different error metrics. The MLR+LSTM|MLR model has a gain in every metric ranging from 27% (for the MaxAE of the validation set) to 68% (for the MSE of the test set).

The results presented in this section, demonstrate the added value of the LSTM|MLR model for the monitoring of dam safety. This model is capable of learning the pattern of residues obtained from the MLR model by capturing the non-linearities and the sequential, long-term effects. The remaining error of the MLR+LSTM|MLR model corresponds to measurement errors and other smaller errors not explained by this model. It is also expected that with more data, such as higher frequency of data collection or longer observation time, these results would improve since the LSTM model would be better able to capture the non-linearities and long-term dependencies. The gain of the MLR+LSTM|MLR model is particularly significant since monitoring of dam safety is focused on the analysis of the residues.

10.7 Final Remarks

Current societies strongly depend on civil engineering infrastructures that support basic services, such as water supply, energy production and transport. As a consequence, structural safety problems can produce catastrophic consequences, especially if we consider critical infrastructures such as large dams, bridges or nuclear facilities. In order to manage structural safety risks, large civil engineering structures are continuously monitored by several sensors to provide accurate models of the current state of each structure.

This chapter analyzed the main activities of the safety control of concrete dams, explaining the common data lifecycle since the dam construction until its decommissioning, and surveying the current state of the art with regard to the use of machine learning techniques to aid the decision making process during the structural safety control of concrete dams.

Taking advantage of advances in deep learning, namely in the field of time-series prediction, this chapter proposed prediction methods based on recurrent neural networks with gains approximately near to 30% regarding to maximum absolute error. The design of these methods followed the design science research methodology and were evaluated in the real case study of Alto Lindoso dam located in the North of Portugal. The evaluation results showed an improvement when compared to traditional baseline method based on multiple linear regressions.

The contribution of the deep learning based approach to the analysis and interpretation of the monitoring data for dam safety control is not limited to the proposed methodologies. It is considered that the benefits of deep learning over traditional machine learning approaches illustrate the potential improvement that these models can provide to the analysis of the monitoring data and to the interpretation of structural dam behavior. They clearly illustrate the benefits in the knowledge extraction from the information embedded in the monitoring data.

This chapter demonstrated how recurrent neural networks can be used to support decision making of dam specialists for concrete dam safety control. The improved results are promising and motivate detailed exploitation (e.g., additional input features, distinct target predictions) to create improved models for significant monitoring systems, improving the capability to anticipate structural safety problems in large civil engineering structures.

References

1. M. Abadi, P. Barham, J. Chen, Z. Chen, A. Davis, J. Dean, M. Devin, S. Ghemawat, G. Irving, M. Isard et al., Tensorflow: a system for large-scale machine learning. OSDI **16**, 265–283 (2016)
2. A.L. Antunes, E. Cardoso, J. Barateiro, Adding value to sensor data of civil engineering structures: automatic outlier detection, in *ML-ISAPR 2018: 1st Workshop on Machine Learning,*

Intelligent Systems and Statistical Analysis for Pattern Recognition in Real-life Scenarios (2018)

3. Y. Bengio, P. Simard, P. Frasconi, Learning long-term dependencies with gradient descent is difficult. Lang. Resour. Eval. **5**(2), 157–166 (1994)

4. K.T.T. Bui, D.T. Bui, J. Zou, C. Van Doan, I. Revhaug, A novel hybrid artificial intelligent approach based on neural fuzzy inference model and particle swarm optimization for horizontal displacement modeling of hydropower dam. Neural Comput. Appl. **29**(12), 1495–1506 (2018)

5. L. Cheng, D. Zheng, Two online dam safety monitoring models based on the process of extracting environmental effect. Adv. Eng. Softw. **57**, 48–56 (2013)

6. F. Chollet et al., Keras (2015)

7. R. Collobert, J. Weston, L. Bottou, M. Karlen, K. Kavukcuoglu, P. Kuksa, Natural language processing (almost) from scratch. J. Mach. Learn. Res. 12(Aug), 2493–2537 (2011)

8. B. Dai, C. Gu, E. Zhao, X. Qin, Statistical model optimized random forest regression model for concrete dam deformation monitoring. Struct. Control Health Monit. **25**(6), e2170 (2018)

9. A. De Sortis, P. Paoliani, Statistical analysis and structural identification in concrete dam monitoring. Eng. Struct. **29**(1), 110–120 (2007)

10. EDP, Design of Alto Lindoso dam (in Portuguese) Technical report EDP - Energias de Portugal, Oporto (1983)

11. Y. Gal, Z. Ghahramani, Dropout as a Bayesian approximation: representing model uncertainty in deep learning, in *International Conference on Machine Learning* (2016), pp. 1050–1059

12. A. Graves, Supervised sequence labelling, in *Supervised Sequence Labelling with Recurrent Neural Networks* (Springer, Berlin, 2012), pp. 5–13

13. A. Graves, A.r. Mohamed, G. Hinton, Speech recognition with deep recurrent neural networks, in *2013 IEEE International Conference on Acoustics, Speech And Signal Processing (ICASSP)* (IEEE, 2013), pp. 6645–6649

14. X. Guo, J. Baroth, D. Dias, A. Simon, An analytical model for the monitoring of pore water pressure inside embankment dams. Eng. Struct. **160**, 356–365 (2018)

15. M.A. Hariri-Ardebili, F. Pourkamali-Anaraki, Support vector machine based reliability analysis of concrete dams. Soil Dyn. Earthq. Eng. **104**, 276–295 (2018)

16. A.R. Hevner, S.T. March, J. Park, S. Ram, Design science in information systems research. MIS Q **28**(1), 75–105 (2004), http://dl.acm.org/citation.cfm?id=2017212.2017217

17. G. Hinton, N. Srivastava, K. Swersky, Neural networks for machine learning lecture 6a overview of mini-batch gradient descent. Coursera (2012a)

18. G.E. Hinton, N. Srivastava, A. Krizhevsky, I. Sutskever, R.R. Salakhutdinov, Improving neural networks by preventing co-adaptation of feature detectors (2012b). arXiv:12070580

19. S. Hochreiter, J. Schmidhuber, Long short-term memory. Neural Comput. **9**(8), 1735–1780 (1997)

20. V. Hodge, J. Austin, A survey of outlier detection methodologies. Artif. Intell. Rev. **22**(2), 85–126 (2004)

21. ICOLD, General considerations on instrumentation for concrete dams. Bulletin number 23. International Commission on Large Dams, Paris (1972)

22. ICOLD, Ageing of dams and appurtenant works. Review and recommendations, in *Bulletin Number 93. International Commission on Large Dams, Paris* (1994)

23. ICOLD, Automated dam monitoring systems - guidelines and case histories, in *Bulletin Number 118. International Commission on Large Dams, Paris* (2000)

24. ICOLD, Surveillance: basic elements in a "Dam Safety" process, in *Bulletin Number 138. International Commission on Large Dams, Paris* (2009)

25. ISO 14721:2012, *Space Data and Information Transfer Systems - Open Archival Information System (OAIS) - Reference Model. Standard, International Organization for Standardization, Geneva, CH* (2012)

26. ISO 16363:2012, *Space Data and Information Transfer Systems - Audit and Certification of Trustworthy Digital Repositories. Standard, International Organization for Standardization, Geneva, CH* (2012)

27. I.S. Jung, M. Berges, J. Garrett, J.C. Kelly, Interpreting the dynamics of embankment dams through a time-series analysis of piezometer data using a non-parametric spectral estimation method, in *Computing in Civil Engineering - Proceedings of the 2013 ASCE International Workshop on Computing in Civil Engineering* (2013), pp. 25–32

28. N. Kalchbrenner, I. Danihelka, A. Graves, Grid long short-term memory (2015). arXiv:150701526

29. F. Kang, J. Liu, J. Li, S. Li, Concrete dam deformation prediction model for health monitoring based on extreme learning machine. Struct. Control Health Monit. **24**(10), e1997 (2017)

30. A. Krizhevsky, I. Sutskever, G.E. Hinton, Imagenet classification with deep convolutional neural networks, in *Advances in Neural Information Processing Systems* (2012), pp. 1097–1105

31. Y. LeCun, Y. Bengio, G. Hinton, Deep learning. Nature **521**(7553), 436 (2015)

32. N. Leitão, Alto Lindoso dam. Behaviour analysis report (in Portuguese). Technical report, Portuguese National Laboratory for Civil Engineering, Lisbon (2009)

33. F. Li, Z. Wang, G. Liu, Towards an error correction model for dam monitoring data analysis based on cointegration theory. Struct. Saf. **43**, 12–20 (2013)

34. F. Li, Z. Wang, G. Liu, C. Fu, J. Wang, Hydrostatic seasonal state model for monitoring data analysis of concrete dams. Struct. Infrastruct. Eng. **11**(12), 1616–1631 (2015)

35. X. Li, H. Su, J. Hu, The prediction model of dam uplift pressure based on random forest, in *IOP Conference Series: Materials Science and Engineering, IOP Publishing*, vol. 229 (2017), p. 012025

36. Z.C. Lipton, J. Berkowitz, C. Elkan, A critical review of recurrent neural networks for sequence learning (2015). arXiv:150600019

37. Z.C. Lipton, D.C. Kale, C. Elkan, R. Wetzel, Learning to diagnose with LSTM recurrent neural networks (2015). arXiv:151103677

38. M. Ljunggren L. Tim, P. Campbell, Is your dam as safe as your data suggest, in *NZSOLD/ANCOLD Conference*, vol. 1 (2013)

39. J. Ma, R.P. Sheridan, A. Liaw, G.E. Dahl, V. Svetnik, Deep neural nets as a method for quantitative structure-activity relationships. J. Chem. Inf. Model. **55**(2), 263–274 (2015)

40. J. Mata, Interpretation of concrete dam behaviour with artificial neural network and multiple linear regression models. Eng. Struct. **33**(3), 903–910 (2011). https://doi.org/10.1016/j.engstruct.2010.12.011, http://www.sciencedirect.com/science/article/pii/S0141029610004839

41. J. Mata, Structural safety control of concrete dams aided by automated monitoring systems. Ph.D. thesis, Instituto Superior Técnico - Universidade de Lisboa, Lisbon (2013)

42. J. Mata, T. de A. Castro, Assessment of stored automated measurements in concrete dams. Dam World 2015, Portugal (2015)

43. J. Mata, A. Tavares de Castro, J. Sá da Costa, Constructing statistical models for arch dam deformation. Struct. Control Health Monit. **21**(3), 423–437 (2014)

44. T. Mikolov, M. Karafiát, L. Burget, J. Černockỳ, S. Khudanpur, Recurrent neural network based language model, in *Eleventh Annual Conference of the International Speech Communication Association* (2010)

45. V. Mnih, K. Kavukcuoglu, D. Silver, A.A. Rusu, J. Veness, M.G. Bellemare, A. Graves, M. Riedmiller, A.K. Fidjeland, G. Ostrovski et al., Human-level control through deep reinforcement learning. Nature **518**(7540), 529 (2015)

46. B. Myers, J. Stateler, Why include instrumentation in dam monitoring programs? Technical report, U.S. Society on Dams - Committee on monitoring of dams and their foundations, United States of America (2008)

47. J. Ngiam, A. Khosla, M. Kim, J. Nam, H. Lee, A.Y. Ng, Multimodal deep learning, in *Proceedings of the 28th International Conference on Machine Learning (ICML-11)* (2011), pp. 689–696

48. C. Olah, Understanding LSTM networks. GITHUB blog, posted on August 27 2015 (2015)

49. K. Peffers, T. Tuunanen, M. Rothenberger, S. Chatterjee, A design science research methodology for information systems research. J. Manag. Inf. Syst. **24**(3), 45–77 (2007). https://doi.org/10.2753/MIS0742-1222240302

50. F. Perner, P. Obernhuber, Analysis of arch dam deformations. Front. Arch. Civ. Eng. China **4**(1), 102–108 (2010)
51. G. Prakash, A. Sadhu, S. Narasimhan, J.M. Brehe, Initial service life data towards structural health monitoring of a concrete arch dam. Struct. Control Health Monit. **25**(1), e2036 (2018)
52. D. Quang, X. Xie, Danq: a hybrid convolutional and recurrent deep neural network for quantifying the function of dna sequences. Nucleic Acids Res. **44**(11), e107–e107 (2016)
53. V. Ranković, N. Grujović, D. Divac, N. Milivojević, Development of support vector regression identification model for prediction of dam structural behaviour. Struct. Saf. **48**, 33–39 (2014)
54. V. Ranković, A. Novaković, N. Grujović, D. Divac, N. Milivojević, Predicting piezometric water level in dams via artificial neural networks. Neural Comput. Appl. **24**(5), 1115–1121 (2014)
55. F. Salazar, M. Toledo, E. Oñate, R. Morán, An empirical comparison of machine learning techniques for dam behaviour modelling. Struct. Saf. **56**, 9–17 (2015)
56. F. Salazar, M.Á. Toledo, E. Oñate, B. Suárez, Interpretation of dam deformation and leakage with boosted regression trees. Eng. Struct. **119**, 230–251 (2016)
57. F. Salazar, R. Morán, M.Á. Toledo, E. Oñate, Data-based models for the prediction of dam behaviour: a review and some methodological considerations. Arch. Comput. Methods Eng. **24**(1), 1–21 (2017)
58. F. Salazar, M.Á. Toledo, J.M. González, E. Oñate, Early detection of anomalies in dam performance: a methodology based on boosted regression trees. Struct. Control Health Monit. **24**(11), e2012 (2017)
59. H. Salehi, R. Burgueño, Emerging artificial intelligence methods in structural engineering. Eng. Struct. **171**, 170–189 (2018)
60. B. Stojanovic, M. Milivojevic, M. Ivanovic, N. Milivojevic, D. Divac, Adaptive system for dam behavior modeling based on linear regression and genetic algorithms. Adv. Eng. Softw. **65**, 182–190 (2013)
61. H. Su, X. Li, B. Yang, Z. Wen, Wavelet support vector machine-based prediction model of dam deformation. Mech. Syst. Signal Process. **110**, 412–427 (2018)
62. I. Sutskever, O. Vinyals, Q.V. Le, Sequence to sequence learning with neural networks, in *Advances in Neural Information Processing Systems* (2014), pp. 3104–3112
63. Swiss Committee on Dams, Methods of analysis for the prediction and the verification of dam behaviour, in *21st Congress of the International Commission on Large Dams, Montreal, Switzerland* (2003)
64. M. Tatin, M. Briffaut, F. Dufour, A. Simon, J.P. Fabre, Thermal displacements of concrete dams: accounting for water temperature in statistical models. Eng. Struct. **91**, 26–39 (2015)
65. M. Tatin, M. Briffaut, F. Dufour, A. Simon, J.P. Fabre, Statistical modelling of thermal displacements for concrete dams: influence of water temperature profile and dam thickness profile. Eng. Struct. **165**, 63–75 (2018)
66. G. Tayfur, D. Swiatek, A. Wita, V.P. Singh, Case study: finite element method and artificial neural network models for flow through jeziorsko earthfill dam in Poland. J. Hydraul. Eng. **131**(6), 431–440 (2005)
67. H. Wang, N. Wang, D.Y. Yeung, Collaborative deep learning for recommender systems, in *Proceedings of the 21th ACM SIGKDD International Conference on Knowledge Discovery and Data Mining, ACM* (2015), pp 1235–1244
68. C. Xu, D. Yue, C. Deng, Hybrid GA/SIMPLE as alternative regression model in dam deformation analysis. Eng. Appl. Artif. Intell. **25**(3), 468–475 (2012). https://doi.org/10.1016/j.engappai.2011.09.020, http://www.sciencedirect.com/science/article/pii/S0952197611001734
69. K. Xu, J. Ba, R. Kiros, K. Cho, A. Courville, R. Salakhudinov, R. Zemel, Y. Bengio, Show, attend and tell: neural image caption generation with visual attention, in *International Conference on Machine Learning* (2015), pp. 2048–2057
70. J. Yang, M.N. Nguyen, P.P. San, X. Li, S. Krishnaswamy, Deep convolutional neural networks on multichannel time series for human activity recognition. IJCAI **15**, 3995–4001 (2015)

71. H. Yu, Z. Wu, T. Bao, L. Zhang, Multivariate analysis in dam monitoring data with PCA. Sci. China Technol. Sci. **53**(4), 1088–1097 (2010)
72. J. Zhang, Y. Zheng, D. Qi, Deep spatio-temporal residual networks for citywide crowd flows prediction, in *AAAI* (2017), pp. 1655–1661

Chapter 11
Analytics and Evolving Landscape of Machine Learning for Emergency Response

Minsung Hong and Rajendra Akerkar

Abstract The advances in information technology have had a profound impact on emergency management by making unprecedented volumes of data available to the decision makers. This has resulted in new challenges related to the effective management of large volumes of data. In this regard, the role of machine learning in mass emergency and humanitarian crises is constantly evolving and gaining traction. As a branch of artificial intelligence, machine learning technologies have the huge advantages of self-learning, self-organization, and self-adaptation, along with simpleness, generality and robustness. Although these technologies do not perfectly solve issues in emergency management. They have greatly improved the capability and effectiveness of emergency management. In this paper, we review the use of machine learning techniques to support the decision-making processes for the emergency management and discuss their challenges. Additionally, we discuss the challenges and opportunities of the machine learning approaches and intelligent data analysis to distinct phases of emergency management. Based on the literature review, we observe a trend to move from narrow in scope, problem-specific applications of machine learning to solutions that address a wider spectrum of problems, such as situational awareness and real-time threat assessment using diverse streams of data. This chapter also focuses on crowd-sourcing approaches with machine learning to achieve better understanding and decision support during an emergency.

Keywords Emergency management · Crisis analytics · Data analysis ·
Data mining · Machine learning · Decision making · Situational awareness ·
Real-time assessment · Deep learning · Data streams

M. Hong · R. Akerkar (✉)
Vestlandsforsking, Box 163, 6851 Sogndal, Norway
e-mail: rak@vestforsk.no

M. Hong
e-mail: msh@vestforsk.no

© Springer Nature Switzerland AG 2019
G. A. Tsihrintzis et al. (eds.), *Machine Learning Paradigms*,
Learning and Analytics in Intelligent Systems 1,
https://doi.org/10.1007/978-3-030-15628-2_11

11.1 Introduction

In the contemporary society, a variety of emergencies take place more and more frequently. A considerable number of emergency incidents have threatened to human life, environmental protection, social stability and even political relationship of all countries around the world [29]. In this regard, sociologists have been working to define emergency for decades. There is a broad consensus that emergencies are social phenomena, characterized by a disruption of routine and of social structure, norms, and/or values. It implies that the severity of an emergency is more related to the extent of the disruption of social life as aspects of governments, business and individuals, than the measurable magnitude of the hazard [25]. Therefore, the negative effects of emergencies emphasize the need to improve the emergency management capability and strengthen the security worldwide.

11.1.1 Emergency Management

Unlike other, more structured disciplines, emergency management has expanded and contracted in response to events, congressional desires and leadership styles [41]. Some representative definitions in the literature are as follows:

- According to definition of the Federal Emergency Management Agency (FEMA) in USA, the process of emergency management consists of preparing for, mitigating, responding to and recovering from an emergency when a disaster arises [2].
- More modern emergency management involves processes to apply modern technologies and management methods to effectively and efficiently monitoring, response to, control and process events, by integrating various social resources and analysing scientifically the cause [28].
- Another definition for emergency management is "a discipline that deals with risk and risk avoidance." Risk represents a wide range of issues and the range of situations that might possibly involve emergency management or the emergency management system is vast. This supports the premise that emergency management is essential to the security of everyone's daily lives and should be integrated into daily decisions and not just called on during times of disasters [41].

In essence, emergency management is a complex and multifaceted task that involves a variety of activities, of managers and stakeholders, when emergency is not only arising but also before and after of emergency, so as to prevent the occurrence of unexpected events, to reduce the social damages and to mitigate the impacts. Based on the definitions of the emergency management, the evolution of an emergency can be distinguished as three stages, namely pre-emergency, in-emergency and post-emergency, as shown in Fig. 11.1 [3]. Chen et al. described emergency management as a '4R' process, namely reduction, readiness, response and recovery. Reduction is referred to the pre-emergency phase, readiness and response belong to the

Fig. 11.1 The lifecycle of
emergency management

in-emergency phase, and recovery is referred to the post-emergency phase. In each
phase, the outcome of decision-making impacts substantially the evolution of events
and the effectiveness of emergency management [28].

As aforementioned, emergency management is a multifaceted process to prevent,
reduce, respond to and recover from the impact of the emergency on the society.
Because of the scale of events, emergency response requires the participation and
cooperation of multiple organizations (e.g., government, public and private). This
emphasizes the need for efficient and effective decision support systems, as it is
practically impossible for a human decision maker to understand and manage the
complexity of the situation. Instead, problems such as situational awareness [144]
and building a common operating picture, shared among multiple actors who often
have only partial view of the situation, are becoming some of the most urgent needs
of emergency management [140].

However, the emergency data used to these decision support systems arises from
the problems of delivering repetitive information and information overload [4].
Therefore, to improve the capability and effectiveness of emergency management,
machine learning techniques have been utilised.

11.1.2 Machine Learning

Emergency management is concerned not only with predicting the course and consequences of emergencies, but also mitigating those undesired consequences. This
process is undoubtedly a challenging task by the unprecedented volumes of data (e.g.,

forecast, news, web pages, data of social network service, and sensing data) and the pressure of time [140]. Machine learning techniques have been proven to successfully support the decision making processes in managing many complex problems. In that sense, emergency management is no exception; however, it presents various challenge to machine learning techniques for the emergency response. In this section, we briefly introduce category of machine learning.

Machine learning has progressed dramatically over the past two decades, from laboratory curiosity to a practical technology in widespread commercial use [58]. Machine learning and data mining often use the same methods and overlap significantly, but while machine learning focuses on prediction, data mining concentrates on the discovery of (previously) unknown properties in the data. According to depending on whether there is labeled instance which is consists of label and data, Machine learning are typically classified into three categories as follows:

– **Supervised learning**: Supervised machine learning makes predictions about future instances using externally supplied instances that consist of values and a label. Its goal is to build a concise model of the distribution of class labels, and then a classifier based on the model is used to assign class labels to the testing instances [65].
– **Unsupervised learning**: Unsupervised learning is inferring directly the properties of this probability density without the help of externally provided instances providing correct label or degree-of-error for each observation [43]. There are representative algorithms like as Apriori algorithm, K-means and so on.
– **Reinforcement learning**: Reinforcement learning deduces labels of instances with a dynamic environment. There are two main strategies. The first is to search in the space of behaviours in order to find one that performs well in the environment, such as genetic algorithms, and the second is to use statistical techniques and dynamic programming methods to estimate the utility of taking actions in states of the world [60].

11.1.3 Scope and Organizations

This chapter focuses on the application of machine learning techniques to support the decision-making processes for the emergency management. We start with the data-driven methodologies within the frameworks of machine learning and their roles and challenges in supporting different phases of emergency management. We then discuss the characteristics of disaster data akin to 5 Vs of big data and summarize various applications cases of big data analysis. Next, with respect to emphasizing the advance of the social media, we focus on reviewing the crowdsourcing approaches with machine learning in emergency management, and issues of the approaches are discussed in terms of the data analysis. Finally, some examples of the tweet related to emergency are discussed to more deeply contemplate the challenges and opportunities.

Existing survey papers for machine learning techniques in emergency management have reviewed according to categories of the machine learning, or have considered only some part of emergency tasks. Whereas, in this chapter, we review the approaches of machine learning, have been discussed from 2010 to current, along each task of emergencies. Therefore, we believe that readers can easily find topics related to their interests and compare with existing approaches to little more concretely grasp potentiality of their methods.

11.2 Applications of Machine Learning in Emergency Response

As discussed earlier, in the present-day emergency management, the immediate and accurate decision making more relies on the capability of data analysis and processing. Therefore, there is an urgent need to enhance the machine learning functionality of emergency management, such as, to develop scalable and real-time algorithms for time-sensitive decisions, to integrate structured, unstructured and semi-structured data [29]. In this section, we attempt to introduce the tasks of machine learning in each phase of emergency management and review the challenges and benefits of various machine learning techniques for the emergency management.

11.2.1 Machine Learning Techniques for Emergency Management Cycles

Successful emergency management requires a variety of tasks based on various technologies of machine learning within across the board three phases mentioned in Sect. 11.1.1. Figure 11.2 shows the tasks related to machine learning for each phase of the emergency management as follows: (1) predicting the occurrence of potential events and discovering the early warning signs; (2) during the emergency, detecting the events occurred and tracking change of the incidents, and recognizing situations of people, supply, and so on; (3) evaluating the loss caused by incidents and the execution of response, and simultaneously adjusting volunteer efforts based on crowdsourcing to recover from an emergency.

– **Event Prediction**: It is forecasting emergency using technology and interpretation methods and can be achieved by extracting features or pattern. Although there is no prediction method with perfect accuracy, early detection of natural disasters reduces hazards in nearby locations [87].
– **Warning Systems**: To detect impending emergency can give that information to people at risk and enable those in danger to make decisions and early take action [123]. These systems have improved drastically in recent years but they are not perfect yet.

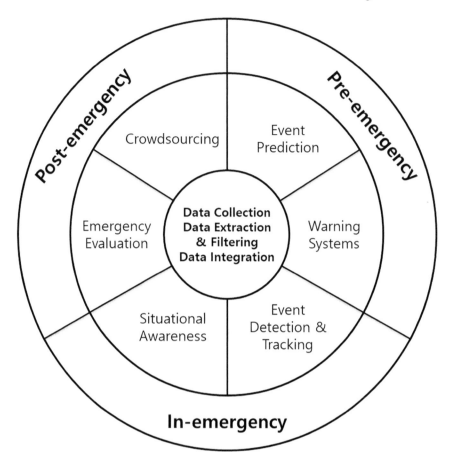

Fig. 11.2 Tasks related to machine learning in emergency management

- **Event Detection and Tracking**: Most systems based on machine learning during
  crises start with detecting and tracking events. The events are mainly associated
  with a specific time and location [21]. However, due to the online nature of collected
  data, events may or may not be necessarily associated with physical locations.
- **Situational Awareness**: It provides more deep recognition of events in emergency
  using social media data related to specific information (e.g., caution, advice, dona-
  tions, casualties and damage) and smart-phones which typically mount various
  sensors such as camera, GPS and accelerometer [140].
- **Emergency Evaluation**: It is one of critical and complex tasks in emergency
  management [29]. In post-emergency, the activity outcomes (e.g., loss of resources,
  recoverability, performance and social influence) for current emergency should be
  measured to suppress the deterioration of next emergency.
- **Crowdsourcing**: This task is a sourcing model in which organizations use predominantly advanced Internet technologies to harness the efforts of a virtual crowd

to perform specific organizational tasks [117]. It may allow to immediately collecting the statuses and requirements of people after an emergency and analysed and categorized the data collected to support relief operations.

Similarly, there are various tasks for emergencies, and machine learning techniques have been applied to each task to improve the effective and efficient emergency management. In next subsections, the approaches of machine learning will be reviewed for each task.

11.2.2 Event Prediction

Many researchers used Neural Networks techniques for predicting emergency. Shah and Ghazali proposed Improved Artificial Bee Colony (IABC) algorithm to improving the training process of Multilayer Perceptron (MLP) in order to overcome local minimal and slow convergence of ordinary backpropagation [118]. The IACB-MLP has showed that it is outperforms than the conventional backpropagation for forecasting earthquake magnitude with time series data in California. Also, to predict magnitudes of earthquake and the impending event following the occurrence of pre-seismic signals, Moustra et al. evaluated the performances of Artificial Neural Networks (ANNs) with various types of input data from the region of Greece [82]. In their study, a feed-forward MLP type Neural Network was implemented using the backpropagation learning algorithm for training. A feed-forward backpropagation algorithm has been also applied into development of a time-dependent surrogate model of storm surge [63]. As experimental results, storm surge was predicted by the 92 trained networks for approaching hurricane climatological and track parameters in a few seconds. Other approach based on ANN has been studied to forecast probabilities of occurrence and re-occurrences of earthquake in the region of Chile by Reyes et al., and they used input values (e.g., the b-value, the Bath's law and the Omori-Utsu's law) which are strongly correlated with seismicity [109]. The occurrences have been judged by threshold values which are adjusted for obtaining as few as false positives as possible. In addition, a combination an ANN and Genetic Algorithm (GA) has been proposed to predict 1-day-ahead Monsoon flood by Sahay and Scrivastava [111]. Four wavelet transform-genetic algorithm-neural network models (WAGANN) have been developed and evaluated for forecasting flows in two Indian rivers such as the Kosi and Gandak. In their experiments, WAGANN models predicted relatively reasonable estimates for the extreme flows and showed little bias for under-prediction or over-prediction.

A variety of clustering methods have been also applied to the prediction task for emergency management. An approach for the prediction of the seasonal tropical cyclone activity over the western North Pacific has been developed to provide useful probabilistic information on the seasonal characteristics of the tropical cyclone tracks and vulnerable areas [62]. In a developed model, the fuzzy c-means clustering has been used to forecast tropical cyclone tracks and density over the entire basin.

From experiments, seven patterns were founded to draw a map of the seasonal track density of tropical cyclone. Moreover, the k-means clustering technique has been combined with the statistical regression techniques for the inducement the weather phenomenon in forecasting the cloudburst [94]. The approach clusters atmospheric pressure according to areas of strong relative humidity for discovering weather patterns. To predict wildfire risk using weather data, Context-Based Fire Risk (CBFR) model has been developed based on clustering and ensemble learning techniques by considering the inherent challenge arising due to the temporal dynamicity of weather data [115]. These two machine learning techniques are used to anomaly detection. A particle swarm optimization algorithm-based clustering method with abnormally high-dimensional data has been also proposed to forecast earthquake [142]. A model analyses relationships between earthquake precursor data and earthquake magnitude, and an average distance between clusters is set as the evaluation function of the particle swarm optimization clustering algorithm. Experimental results indicate that this model can effectively and validly predict the earthquake magnitude in accordance with the earthquake precursor data than k-means algorithm model. Additionally, an ant-colony clustering algorithm has been introduced in earthquake prediction by Shao et al. [119]. Measure parameters include spatial entropy, mean-fit and un-similar for clustering analysis. As their experiments, it showed that their algorithm could achieve better results than the traditional k-means algorithm to forecast of earthquake like the swarm optimization algorithm-based clustering method.

Decision trees, have often fast and accurate performance in machine learning, have been combined with other techniques to predict emergencies. To predict disaster before it has occurrence, there are many studies which combine the decision tree with various machine learning techniques such as regression [145], hidden Markov model [125], association rule learning [34, 79] and fuzzy logic and particle swarm optimization [106]. In particular, decision tree techniques have been also applied into prediction of surroundings (e.g., flood susceptible areas [127] and landslide susceptible areas [30]) for emergency situation.

11.2.3 Warning Systems

Large magnitude emergency such as earthquake and flood kill and injure tens to thousands of people, inflicting lasting societal and economic disasters. Early warning could provide seconds to minutes of warning, allowing people to move to safe zones and prepare activities like automated slowdown and shutdown of transit and other machinery [64].

For early detection and warning of emergency in environments with wireless sensor networks (WSNs), Bahrepour et al. have tried to consolidate a general decision tree with the reputation-based voting method [12, 13]. In their works, early event warning of emergency are fulfilled with distributed event detection. As experimental results with wild and residential fire datasets, it was showed that their approach not only achieves a high detection rate but also has a low computational overhead and time complexity. In addition, a Random Forest (RF) based decision tree was applied into analysis of the potential factors affecting the satellite signal to announce the flood by Revilla-Romero et al. [108]. They investigated various satellite data for 322 rivers in Africa, Asia, Europe, North America and South America. Their experiments shown that the mean discharge, climatic region, land cover and upstream catchment area are dominant variables to determine good or poor vulnerable of surrounding areas.

ANN techniques have been applied into warning emergency with various perspectives. Kong et al. used a smartphone-based seismic network which consists of smart device contains accelerometers [64]. The ANN was used to separate data collected from personal smartphone sensors into activities of the earthquake and human to warn earthquake. Additionally, to set an early warning threshold level of dam, a continuous monitoring of long-term static deformation based on three ANN approaches (i.e., the static neural network, the dynamic neural network and the auto-associate network) was proposed [61].

Krzhizhanovskaya et al. developed a flood early warning system to monitor sensor networks installed in flood defenses (e.g., dikes, dams and embankments), detect abnormalities in sensor signals, calculate dike failure probability, and simulate possible scenarios of dike breaching and flood propagation [66, 98]. In the warning system, k-means clustering and Neural Clouds (NC) based classification has role to detect abnormality of sensor parameters in critical pre-failure conditions. Social media data that contains social concerns of people was used for early warning system by Avvenuti et al. [11]. Their system applies classification techniques provided by Weka to distinguish Twitter messages into "useful" and "not useful", and several machine learning techniques are utilized to temporal and spatial analysis of messages. There is also another study considers social media data. Fersini et al. implemented a decision support system using machine learning and natural language processing to effectively detect and warn the earthquake [37]. On a real Twitter dataset, their system has shown outperformed results to identify messages related to the earthquakes and critical tremors.

11.2.4   Event Detection and Tracking

The Support vector machine (SVM) is a supervised learning model that defines a kernel function able to transform the data to a high dimensional feature space when the data can be separated by linear models. SVM is designed for binary classification in nature, but can also solve the multi-class classification problems through one-against-one or one-against-all strategy. SVM was found be effective in emergency rescue evacuation support system for detecting and tracking a sudden incident [45, 81]. Pohl et al. took advantage of the Self-Organizing Map (SOM) and the Agglomerative Clustering (AC) for sub-event detection that operate on Flickr and YouTube data [102]. As they mention, multimedia data may be of particular importance to detect and track emergency event. Therefore, Vector Space Model (VSM) was also utilized to represent and annotate the media data before applying clustering techniques. Their experiments showed that social multimedia data in the context of emergency is worth using for detecting sub-events. Similarly a method of cross-media analytic based on the clustering, sentiment analysis and keyword extraction was introduced to detect and track emergency events [134]. Moreover, the semantic expansion and sentiment analysis were adopted to quantify public sentiment time series.

Song et al. developed a model of human behaviour that takes into account several factors have been founded through empirical analysis between human mobility and emergency to detect and monitor human emergency behaviour and their mobility during large-scale emergency [122]. For the model, they used Hidden Markov Model (HMM) to model dependency between human behaviours in emergency. For the Great East Japan Earthquake and the Fukushima nuclear accident, the efficiency of the behaviour model was evaluated. HMM for speech recognition technique was applied into detecting earthquakes and tracking volcano activities [17]. To fit the model parameters in to earthquake detection, Beyreuther et al. introduced state clustering into their model to refine the intrinsically assumed time dependency. As experiments in during around four months, their earthquake detector of single station HMM showed that it can achieve similar detection rates as a common trigger in combination with coincidence sums over two stations. Akin to this, an approach, which used audio data, proposed to identify anthropogenic disasters by Ye et al. [137]. In their approach, acoustic events are detected and learned using the dictionary learning and the spherical k-means clustering. And detected events then are classified into specific sounds (e.g., screaming, shouting, gun shout and explosion) by a clustering technique based on the hierarchical regularized logistic regression model. Experimental results with an audio dataset showed the effectiveness of the proposed hazard sound recognition method. Singh et al. investigated Twitter posts in a flood and proposed an algorithm to identify victims asking for help. To categorize the posts into high or low priority tweets, the SVM, the gradient boosting and the RF are applied. Furthermore, inferring user locations using the Markov model uses historical locations of users. In their experiments, the proposed algorithm worked with its classification accuracy of 81% and location prediction accuracy of 87% [121]. In addition, Caragea et al. developed an enhanced messaging for the emergency response sector, as a reusable information technology infrastructure, to detect and track emergency [24]. They focused on correct classifying messages during disasters by using the SVM classifier. Furthermore, the bag-of-words (BoW) approach, the feature abstraction, the feature selection and the Latent Dirichlet Allocation (LDA) were applied into feature representation as inputs for learning a classifier. Besides, various techniques of machine learning have been used for analysis of social media data to detect and track emergencies [112, 113, 128].

11.2.5 Situational Awareness

The recent advances of mobile devices, that are capable of wireless communications and have sufficient computing power, have caught attention of researchers and practitioners. The devices can serve as automated sensors (typically equipped with GPS, motion sensors, etc.) and are capable of relatively high quality imaging and video recording. Therefore, they can be used to enhance situational awareness by gleaning various information to produce accurate results [132]. Furthermore, the advance of social networking services (e.g., Twitter, Instagram, Flicker and Facebook) allows people post their needs and gather the timely-relevant information. Tweets as one of these messages were also investigated to detect possible seismic events, to compare and contrast the people behaviour during emergency and to extract useful information using several extraction techniques [4]. Like this, social media is also a potential source for situational awareness in emergency management. Recent disasters, such as the Hurricane Sandy of 2012, the Typhoons Haiyan or Hagpuit in 2013–2014, or the Nepal earthquake in 2015 have shown that information provided by eye-witnesses through social networking services can greatly improve situational awareness.

Alam et al., for situational awareness during cyclone emergency, proposed a social media image processing pipeline, which includes a noise filter and a damage assessment classifier [5, 6, 88]. The Convolutional Neural Network (CNN) applied into filtering out irrelevant image content, and the perceptual hashing technique was employed for image de-duplication. Additionally, CNN technique was also used for situational awareness during heavy rainfall by Li et al. [74]. They focused on using social remote sensing data for emergency response. The classification results obtained for the central parts of Wuhan and Shenzhen demonstrated the effectiveness of the CNN method considered for monitoring the heavy rainfall event that happened in both cities.

Shen et al. proposed a method to retrieve events based on event-specific hashtags preliminary to collected for situational awareness of emergencies [120]. In their experiments, the SVN showed best performance for extracting and classifying hashtags from data in Twitter. Then, the hashtags were used to collect relevant messages from not only Twitter but also other social media platforms. The SVM was also introduced to extract features and classify texts [104]. Raginia et al. proposed a hybrid method for segregating and classifying the texts obtained from people who are at risk in the affected region for situational awareness of emergency. The results showed that the text classification algorithm can help the emergency responders to locate the people at risk in real-time.

For understanding situations in disaster response, Li et al. proposed a domain adaptation approach, which learns classifiers from un-labeled target data, in addition to unlabeled data [73]. A Naive Bayes (NB) classifier and an iterative self-training strategy were adopted for tweet classification. Their experiment results showed that the domain adaptation classifiers are better as comparing with the supervised learning using only labeled data. Ramchurn et al. proposed an emergency manage-ment system called HAC-ER for situational awareness from large streams of reports posted by members of the public and trusted organizations [105]. They combined the independent Bayesian classifier combination with the Gaussian process to remove errors and to predict locations of events in affected areas by emergencies. Additionally, Imran et al. presented human-annotated Twitter corpora collected during 19 different crises and compared supervised learning techniques such as the SVM, the NB and the RF in terms of the utility of annotations [55]. Given the complexity of the multiclass classification of short messages, it was indicated that all three classifiers have decedent results. To discover important topics from Twitter and provide useful information of situation awareness during emergency, Yin et al. developed an online incremental clustering algorithm that automatically groups similar tweets into topic clusters [138]. They also adapted optimization techniques (i.e., burst detection, text classification, online clustering and geotagging) to deal with real-time, high-volume text streams. It includes an early indicator identification of unexpected incidents, an impact exploration of events and an incidents evolution monitoring.

In general, to classify social media data may be tedious and time consuming task, since the collected data are not in the form of a labeled data. Therefore, Pandey and Natarajan utilized the semi-supervised machine learning approach to avoid the classification process and concurrently obtain useful information in situational awareness [95]. In addition, they also introduced an interactive map to grasp the vulnerable areas during an emergency. A reinforcement learning technique was also introduced to map dynamic situations in emergency. Sadhu et al. proposed a Multi-Agent Reinforcement Learning (MARL) framework implemented as a mobile application and a back-end server [110]. Via both simulations and real experiments, an evaluation of the framework in terms of effectiveness in tracking random dynamicity of the environment was performed.

11.2.6   Emergency Evaluation

Trekin at el. applied CNN into developing a method of change detection on remote sensing imagery to improve time efficiency of assessment of damaged buildings in disaster affected area [129]. Also, a deep learning-based framework for rapid regional tsunami damage recognition using post-event on synthetic radar imagery was proposed [15]. They applied the SqueezeNet network (as a CNN type) architecture into a selection algorithm, and a recognition algorithm with a modified wide residual network was developed to classify the damaged regions. Via experiments on Tohoku earthquake in 2011 and tsunami area, it was showed that the proposed framework is fast in model training and prediction calculations. The potential of CNN features was also explored for an online classification of satellite image to detect structural damages by Vetrivel et al. [131]. A feature extraction and a classification process are carried out at an object level, where the objects are obtained by over-segmentation of satellite images. The proposed framework outperformed a batch classifier with lesser time and memory requirements. As other usage case of CNN for natural emergency evaluation, analysis of images posted on social media platforms using the CNN was proposed [89]. Experimental results indicated that the domain-specific fine-tuning of deep CNN outperforms Bag-of-Visual-Words (BoVW). In addition, high classification accuracy under both event-specific and cross-event test settings demonstrated that their approach can effectively adapt deep-CNN features to identify the severity of destruction from social media images taken after a disaster strikes. Additionally, Attari et al. introduced a Nazr-CNN, a deep learning pipeline for an object detection and fine-grained classification in aerial images for assessing and monitoring damage [10]. In their work, a hidden layer of a CNN was used to encode the popular BoVW of the segments generated from the first component in order to help discriminate between different levels of damage. Moreover, backpropagation neural network as a kind of multilayer feed-forward network was used in evaluation of city emergency management system for disaster event by Jiang and Li [57].

Cervone et al. proposed a methodology that leverages data harvested from social media for collecting remote-sensing imagery during disasters [26, 27]. The images are then fused with multiple sources for the damage assessment of transportation infrastructure. In this method, DT was used to classify entire scenes acquired. They also evaluated the proposed methodology with considering Colorado floods in 2013 [27]. Zhang and his colleagues proposed a machine learning framework to assess post-earthquake structural safety [143]. In this framework, Classification and Regression Tree (CART) and RF were implemented to map damage patterns to classified structural safety states. For assessment of sensitive area for landslide at the Pauri Garhwal in India, the RF and CART were also compared with Logistic Model Trees (LMT) and Best First Decision Trees (BFDT) [99]. The results showed that a RF model has the highest predictive capability followed by the LMT, BFDT and CART models, respectively. It was showed that although all four methods have shown good results, the performance of the RF method was the best for landslide spatial prediction.

An assessment model based on RF was adopted to evaluate regional flood hazard [133]. The risk assessment method was implemented in Dongjiang River Basin, China. In addition, SVM technique was used for risk assessment as a comparison, as well as an analysis of index importance degree. The spatial distributions of the RF and the SVM-based assessment maps showed a similar correlation coefficient, was indicated that the classification capacity of the two methods is similar in the majority of cases. Joshi et al. introduced a methodology for detection of damage post disasters by examining the textural features from high resolution aerial imagery [59]. The proposed technique considered DT, NB, SVM, RF, Voting Classifier and Adaptive Booster, and were compared to identify damaged regions from aerial images using only pre-event images as the input. As a result, the RF-based classifier comparatively had higher accuracy than other classifiers. Yoon and Jeong applied Cubist and RF techniques into assessment what vulnerability indicators are statistically associated with disaster damage in Korea and found twelve indicators to evaluate vulnerability of 230 local communities to disasters [139].

Zanini et al. proposed a procedure based on Fuzzy Logic (FL) for the evaluation of interactions between existing buildings and urban roadway networks after a seismic event [141]. The methodology was applied to the Municipality of Conegliano in Italy in the potential seismic damage scenario. Their experiments showed it is able to evaluate the network link functionality reductions caused by building damages, through the estimation of the residual road width, without the necessity of carrying out expensive and detailed surveys on the analysed area. Izadi at el. proposed a neuro-fuzzy approach based on GA and SVM for the semi-automatic detection and assessment of damaged roads in urban areas using pre-event vector map and both pre and post-earthquake QuickBird images [56]. Experimental results showed the efficiency and accuracy of the Neuro-Fuzzy systems for road damage assessment. Resch et al. introduced an approach based on analyse social media posts to assess footprints and the damage caused by natural disasters through combining LDA for semantic information extraction with spatial and temporal analysis for hot spot detection [107]. Furthermore, they provided a damage map that indicates where significant losses have occurred. Their experiments showed that earthquake footprints can be reliably and accurately identified in our use case. Nadi and Edrisi introduced a Markov decision process as a multi-agent assessment and a response system with reinforcement learning designed to ensure the integration of emergency response and relief assessment operations [84]. Experiments indicated that the use of the proposed approach in assessing network conditions and true demand during search and rescue operations can decrease death tolls.

11.2.7 Crowdsourcing

Volunteers provide information and resources to the affected people and this process has been facilitated by social media in recent years [87]. In this regard, for some years now, both researchers and practitioners in the areas of emergency management have been exploring the role of crowdsourcing in collecting, processing and sharing information [101]. Although there are various roles of crowdsourcing in tasks of emergency management, in this section, we focus on its usages in tasks for post-emergency, especially related relief activities. The others will be investigated and discussed in Sect. 11.5. First we start with reviewing studies focused on crowdsourcing for post-emergency without considering machine learning techniques.

Landwehr and Carley have reviewed how social media is used in disaster by individuals, first responders and disaster researchers. They have also introduced a variety of software tools that can be used by analysts to work with social media and have discussed of several different directions in which some of the research on social media usage in disaster is currently heading [69]. As aiming at efficiently harnessing crowdsourcing in remote assistance in real-time, Yang et al. designed and developed a crowdsourcing disaster support platform [136]. They considered three unique features as follows: selecting and notifying individual requests, providing collaborative working functionalities, improving answer credibility through "crowd voting." In addition, Dubey et al. attempted to develop a theoretical framework which can assist relief activities using valuable information derived using comprehensive crowdsourcing framework in environments with Internet of Things [35]. They have conducted extensive review of articles published in reputable journals, magazines and blogs by eminent practitioners and policy makers. Murali et al. proposed a multi-platform model to deal with disasters and support relief activities while handling the needs of victims, volunteers and government agencies [83]. Further, they used various techniques, ranging from Natural Language Processing (NLP) to crowdsourcing, for ensuring robustness and scalability of solution.

Various techniques of machine learning have been also combined with crowd-sourcing to support relief activities in post-emergency. Most of the current systems allow volunteers to directly provide input to them [86]. Hence, many social media posts in the aftermath of disasters might contain useful information. There are several tries to extract the information from social media through a variety of machine learning techniques such as LDA and DT [86], RF and NB classifier [103]. In order to detect potential incidents implicated by victims negative emotions in the post-disaster situation, Bai et al. introduced a structured framework including three phases [14]. NB, RF, SVM and KNN techniques were compared each other in terms of to classify emergency-related messages; as a result, the RF and the SVM were outperformed than the others. Harris et al. proposed an approach of the post-phase situational awareness of an earthquake hit area to the rescue task [42]. To ensure the credibility of the crowdsourced data, their system considers the K-means clustering technique and maps coordinates of the calamity area through a short messaging service. In addition, experimentation was carried out to evaluate the time taken to notify via SMS. Imran et al. also applied a clustering technique to the classification of crisis-related messages in microblog streams [52]. Social media messages are clustered together with the textual similarity, and human curators annotated larger clusters first to train the classifier. Liang at el. introduced a semi-supervised learning based cognitive framework to support emergency management through mapping crowdsourced data. The framework first divides the satellite or aerial image into patches leveraging a graph-based clustering approach [76]. The KNN classifier is then used to provide labels for a few patches. With over 50 participants working on three different tasks, their experiments showed that the crowdsourced variant performs well producing noise-tolerate flood maps.

11.3   Analysis of Emergency Data

Information exchange during and after the disaster periods can greatly reduce the losses caused by the disaster. This is because it allows people to make better use of the available resources and provides a channel through which reports on casualties and losses in each affected area can be delivered expeditiously [75]. Furthermore, the success of a disaster relief and response process is largely dependent on timely and accurate information regarding the status of the disaster, the surrounding environment and the affected people [87]. Therefore, understanding, analytic and utilization of data collected in emergency are vital. In this section, we describe data characteristics generated during disaster with considering the 5 Vs of big data and various application cases of Big Data Analysis (BDA) are looked over.

11.3.1   Big Data in Emergency Management

The role of data in emergency management has been evolving. Nowadays, scientists are facing one of the biggest challenges of managing large volumes of data generated at times of disasters. As a huge amount of emergency-related data is getting generated, traditional data storage and processing systems are facing challenges in fulfilling performance, scalability and availability [39]. Therefore, analytic methods to manage and process data in emergency management are particularly challenged due to the combination of its unique characteristics as follows:

– **Rapid increase of data** by many numbers of producers and consumers
– **The timely sensitivity** of detect and response
– **Combination of static and dynamic data** (e.g., maps and crowd emotion) [76, 116]
– **Heterogeneous formats**, ranging from raw data (e.g., sensors) to structured data (e.g., metadata) and unstructured data (e.g., multimedia) [101]
– **Various levels of trustworthiness** of the data sources [47, 126]
– **Possibility of extracting valuable information** like crowdsourcing, generated by people who are actually at the emergency scene for near real-time [75].

These characteristics similar to "5 Vs (i.e., volume, variety, velocity, veracity and value)" of big data. Besides, within the present-day emergency management systems, the immediate and accurate decision-making more and more relies on the capability of data analysis and processing especially in the face of big data [29]. Therefore, there is an urgent need to enhance the BDA technologies of emergency management, such as, to develop scalable and real-time algorithms for time-sensitive decisions, to integrate structured, unstructured and semi-structured data, to deal with the imprecise and uncertain information, to extract dynamic patterns and outline the evolution of these patterns, to work in distributed environment and to present the multi-scale, multilevel and multi-dimensional patterns through various visualization approaches [8].

BDA was often defined as holistic process to process and analyse the 5 Vs in order to create actionable insights for sustained competitive advantages [36]. In this regard, Mehrotra et al. suggested that BDA can aid to create the next generation of emergency management technologies as it has the potential to mitigate the effects of disasters by enabling access to critical information in real-time [80]. Also, they emphasized that "accurate and timely analysis and assessment of the situation can empower decision makers during a crisis to make more informed decisions, take appropriate actions and better manage the response process and associated risks." Thus, it is essential to reconsider how data on disasters should be properly and efficiently produced, organized, stored and analysed [44]. Here, we briefly discuss the BDA for emergency management including data collection, information extraction, data filtering and data integration.

11.3.2 Data Collection

Data collection is the process of inserting to a system data, which is coming from multiple heterogeneous sources. Scalability is an important issue in data collection, since the flow of information may be very high during the time of a critical event [49]. Traditionally, it was done in the form of paper reports or questionnaires. The development of IT allowed for use of word processors, spreadsheets and forms to enter data directly into the databases [140]. Moreover, sensing technologies are also currently undergoing rapid advances and are leading to might use of them significantly increases performance of situational awareness [19]. As another increasingly important source, social media is a new way of communication in the course of disasters. A major difference between social media and traditional sources is the possibility of receiving feedback from the affected people [87]. Additionally, for some years now, both researchers and practitioners in the areas of disaster and emergency management have been exploring the role of crowdsourcing in collecting, processing and sharing information across organizations and affected populations [101].

11.3.3 Information Extraction and Filtering

The goal of information extraction is to automatically extract structured information, i.e., categorized and contextually and semantically well-defined data from a certain domain, from unstructured machine-readable documents [49]. Whereas, according to Wikipedia, data filtering is that removes redundant or unwanted information from an data stream using (semi) automated or computerized methods prior to provide it. Its main goal is the management of the data overload and increment of the semantic information. If all the disaster information were presented to the users, it would cause an overwhelming workload. Therefore, the disaster information should be filtered based on the specific purposes of the users [75]. Especially, in emergency management, the utilization of social media data can be extended for various tasks such as warning [11], detection and tracking [102, 112], situational awareness [6, 86], assessment [27, 107], etc. Therefore, it is more emphasized that the data extraction and filtering became the core of emergency management together with effects of social media data. However, the social media data contain texts, images, videos, tags and so on. Therefore, information is extracted from heterogeneous sources such as social media and monitoring devices. Typically, the disaster information from different sources varies greatly in structure or format. To support further analysis and processing, a specification of common format should be required for disaster information integration. Then, the integrated disaster information in this format can be organized and stored for further processing [75].

11.3.4   Data Integration

Emergency services sometimes should deal with the massive amount of data arriving through multiple channels such as existing records, sensors, satellite networks or social media [1]. Therefore, one of the biggest challenges in emergency management is to develop a data integration protocol. Data integration is the process of combining data residing at different sources and providing the user with a unified view of these data [72]. This process includes the following tasks: (1) to convert contents of different formats into a standard format; (2) to verify the credibility of various crowdsourcing data sources and attempt to leverage it to produce useful information for disaster decision-making; (3) to map images or texts with their corresponding geolocations to better capture the current situation; and (4) to process and analyse the data from different sources [75].

11.3.5   Applications for Data Analysis in Emergency

In this section, we introduce several applications and platforms comprehensively deal with data analysis for emergency managements. **Ushahidi** as the first large-scale crowdsourcing system developed to report Kenyan post-election violence in 2008 and since then has been applied into many major disasters such as Hurricane Sandy and Haiti Earthquake [91]. The Ushahidi is an open source and free systems which can either be deployed on external servers or on its hosting system CrowdMap.

This system collects emergency-related data from several sources, web, Twitter, RSS feeds, emails, SMS and so on. Collected data is then visualized on the map. Last, The Ushahidi allows users to filter information based on several types (e.g. supplies or shelter).

Artificial Intelligence for Disaster Response (AIDR) is a free software platform which can be either run as a web application or created [53]. In the AIDR, tweets are collected according to pre-selected set of keywords. Prior labeled tweets will be used as the training set of a classifier which labels collected tweets based on the keywords. In the training process, n-grams of tweets are used as features and therefore the classifier is retrained for every new category and disaster.

TweetTracker consists of tracking, analysing and understanding tweets related to a specific topic [67]. In the process of tracking the status of and event, data includes keywords, location and users can be collected using a set of criteria from Twitter, Facebook, YouTube, VK and Instagram. Fluctuations in the total number of post or frequency of posts with specific words can be analysed for different time periods. Moreover, keywords, hashtags, links, images and videos with their frequencies are available to understand tweets by the user. For instance, to better understand the geographic distribution of posts, geo-tagged tweets are presented on a map.

DisasterMapper is a CyberGIS framework to ingest and archive massive amounts of social media data [50]. In this framework, to manage massive social media data, Apache Hive is used as the scalable storage solution. Furthermore, Hadoop platform is used as a scalable distributed computing environment to process social media data. Mahout is leveraged to support big data analytic. It can automatically synthesize multi-sourced data, such as social media and socioeconomic data, to track disaster events, to produce maps and to perform spatial and statistical analysis for emergency management.

IDDSS-Sensor is software for Geographic Information System (GIS) implemented to provide the functions of standard-based access, as well as on-the-fly harmonization, integration and usage of multi-agency sensor information [7]. The software has three layers, namely, the storage, service and presentation layer. In storage layer, PostgreSQL is used as an open-source object relational database of integration data models. For service layer, 52ºNorth and GeoServer was employed as SOS implementation and as spatial data service for serving static spatial data. As GUI of the system, third layer was developed by JavaScript and the CESIUM, ExtJS3 and IDDSS were used for displaying the sensor data and visual indicators.

Table 11.1 lists approaches and applications in emergency management in terms of target types of emergency, belonged tasks, used technique of machine learning and data analysis.

Table 11.1 Approaches with machine learning for emergency management

Title	Emergency	Task	Machine learning technique	Data	Source	Data analysis
Prediction of Earthquake Magnitude by an Improved ABC-MLP [118]	Earthquake	Prediction	– Artificial bee colony – Multilayer perceptron – Backpropagation	– Earthquake parameter magnitude	– California earthquake data (2011)[a]	– Labeling
Artificial neural networks for earthquake prediction using time series magnitude data or Seismic Electric Signals [82]	Earthquake	Prediction	– Artificial Neural Network	– Time series magnitude data – Seismic electric signals	– VAN team in Greece Athens earthquake data (1981) – Seismological Institute National Observatory of Athens (SINOA)	– Labeling
A time-dependent surrogate model for storm surge prediction based on an artificial neural network using high-fidelity synthetic hurricane modeling [63]	Hurricane	Prediction	– Artificial neural network – Backpropagation	– Historical hurricane parameters	– Atlantic hurricane database, HURDAT2[b]	– Labeling
Neural Networks to Predict Earthquakes in Chile [109]	Earthquake	Prediction	– Artificial neural network	– Earthquake magnitude	– Chile's National Seismological Service	– Labeling
Predicting Monsoon Floods in Rivers Embedding Wavelet Transform, Genetic Algorithm and Neural Network [111]	Flood	Prediction	– Artificial neural network – Genetic algorithm	– Flow data	– Monsoon flow (2001–07)	– Labeling
Track-pattern-based model for seasonal prediction of tropical cyclone activity in the western North Pacific [62]	Tropical cyclone	Prediction	– Fuzzy c-means clustering	– Tropical cyclone locations	– Regional Specialized Meteorological Centers Tokyo-Typhoon Center	– Labeling
The Application of Ant-Colony Clustering Algorithm to Earthquake Prediction [119]	Earthquake	Prediction	– Ant colony clustering algorithm	– Earthquake magnitude		– Labeling – Anomaly detection
Dynamic and robust wildfire risk prediction system: an unsupervised approach [115]	Wildfire	Prediction	– Clustering algorithms	– Weather data	– Fire & Rescue NSW from the Bureau of Meteorology (BoM)	– Labeling – Anomaly detection

(continued)

Table 11.1 (continued)

Title	Emergency	Task	Machine learning technique	Data	Source	Data analysis
The data mining technology of particle swarm optimization algorithm in earthquake prediction [142]	Earthquake	Prediction	– Particle swarm optimization clustering algorithm	– Earthquake magnitude		– Labeling – Anomaly detection
Spatial prediction of flood susceptible areas using rule based decision tree and a novel ensemble bivariate and multivariate statistical models in GIS [127]	Flood	Prediction	– Rule-based decision tree – Logistic regression statistical methods	– Rainfall data – Spatial database	– Landslide-hazard map [90]	– Labeling – Anomaly detection – Mapping flood susceptibilities of areas
Clustering technique to interpret Numerical Weather Prediction output products for forecast of Cloudburst [94]	Cloudburst	Prediction	– K-means clustering	– Flow pattern data – Temperature – Atmospheric pressure level	– European Center for Medium-range Weather Forecasting	– Labeling
GIS-based landslide susceptibility modelling: a comparative assessment of kernel logistic regression, Naïve-Bayes tree, and alternating decision tree models [30]	Landslide	Prediction	– Kernal logistic regression – Naive Bayes tree – Alternating decision tree	– Landslide inventory map – Landslide conditioning factors – Landsat 8 OLI images – Precipitation data	– Self-collection – NASA[c] – U.S Geological Survey (USGS)[d]	– Collecting landslide inventory map – Labeling – To manually ingrate early reports, aerial photographs and GPSs
Application of decision trees to the analysis of soil radon data for earthquake prediction [145]	Earthquake	Prediction	– Decision tree	– Soil radon data	– Soil radon data [146]	– Labeling
A new data mining model for hurricane intensity prediction [125]	Hurricane	Prediction	– Feature weight learning – Genetic algorithm – Extensible Markov model	– Historical hurricane data	– Atlantic tropical cyclones (1982–2003) from National Center for Atmospheric Research NCAR	– Labeling
Computational Intelligence Techniques for Predicting Earthquakes [79]	Earthquake	Prediction	– Association rules – Decision tree	– Location and magnitude of earthquakes	– Spanishs Geographical Institute (SGI)	– Labeling

(continued)

Table 11.1 (continued)

Title	Emergency	Task	Machine learning technique	Data	Source	Data analysis
Novel method for hurricane trajectory prediction based on data mining [34]	Hurricane	Prediction	– Association rules	– Hurricanes motion characteristics	– Atlantic weather Hurricane/Tropical Data from 1900 to 2008[e]	– Labeling
Spatial Data Mining for Prediction of natural events and emergency management based on fuzzy logic using hybrid PSO [106]	Natural emergency	Prediction	– Fuzzy logic – Particle swarm optimization – K-means clustering – Naïve Bayes classifier	– Water level modifications data – Spatial-temporal data – Text data	– Bangladesh Country Almanac (BCA) dataset	– Labeling and detecting centroid – Filtering inconsistent and unnecessary data – Integrating spatial-temporal data and text
Use of wireless sensor networks for distributed event detection in emergency management applications[13]	Fire	Warning	– Decision tree	– Residential fire data (i.e., temperature, ionization, photoelectric and CO)	NIST[f]	– Labeling
Earthquake emergency management by social sensing [11]	Earthquake	Warning, Evaluation	– Bayesian statistics – Peak-detection algorithms – Corrected conditional entropy – Classification algorithms	– Social media data with keywords – multimedia data (i.e., video and photo)	– Using Streaming API of Twitter	– Collecting social media data – Labeling – Removing noise of different meaning and past event
Evaluation of the satellite-based Global Flood Detection System for measuring river discharge: influence of local factors [108]	Flood	Warning Detection and tracking, Evaluation	– Random forest	– Surface water extent and floodplains data from satellite – River width – Presence of floodplains flooded forest and wetlands – Flood extent – Land cover – Leaf area index – Climatic areas – Presence of river ice – Dam location	– Global Flood Detection System, Global Runoff Data Centre and South African Water Affairs[g] (1998–2010), – SRTM Water Body Database and the HydroSHEDS [135] – Global Lakes and Wetlands Database level 3 [70] – Global Flood Hazard Map [96] – Global Land Cover 2009 [20] – SPOT-VGT[h] – Koppen-Geiger climate map of the world [97] – Circum-Arctic map [22] – Global Reservoir and Dam [71]	– Labeling – Integrating data related to flood

(continued)

Table 11.1 (continued)

Title	Emergency	Task	Machine learning technique	Data	Source	Data analysis
Flood early warning system: design, implementation and computational modules [66] Flood early warning system: sensors and internet [98]	Flood	Warning	– K-means clustering	– Sensor stream data – Digital terrain model data	– Actueel Hoogtebestand Nederland[i]	– Collecting sensor data in real-time – Labeling and detecting anomaly – Integrating map and sensor data
MyShake: A smartphone seismic network for earthquake early warning and beyond [64]	Earthquake	Warning	– Artificial neural network	– Acceleration and GPS data from android phone – Earthquake data	– Self-collection, Northern and Southern California Earthquake Data Centers[j] – Center for Engineering Strong Motion Data[k] – National Research Institute for Earth Science and Disaster Prevention for Japanese[l]	– Collecting smartphone sensor – Labeling – Filtering frequency band – Integrating locations and acceleration according to time-stamps
Monitoring of long-term static deformation data of Fei-Tsui arch dam using artificial neural network-based approaches [61]	Flood	Warning	– Artificial neural network (i.e., static neural network, dynamic neural network and auto-associate network)	– Residual deformation in Fei-Tsui arch dam (i.e., static deformation, water level and temperature distribution)	– Long-term static deformation data of the Fei-Tsui arch dam (Taiwan)	– Collecting static deformation, water level and temperature distribution – Labeling
Earthquake management: a decision support system based on natural language processing [37]	Earthquake	Warning	– Bayesian model averaging – Voting mechanism – Classifiers (i.e., decision tree, support vector machine, naive Bayes, logistic regression and K-near neighbors)	– Social media data with keywords	– Using Streaming API of Twitter	– Collecting social media data – Labeling

(continued)

Table 11.1 (continued)

Title	Emergency	Task	Machine learning technique	Data	Source	Data analysis
Development of ERESS in panic-type disasters: Disaster recognition algorithm by buffering-SVM [81] Disaster detection by statistics and SVM for emergency rescue evacuation support system [45]	Natural emergency	Detection and tracking	– Support vector machine	– Sensor data from smart phone		– Collecting sensor data from smart phone – Extracting feature and labeling
Social media for crisis management: clustering approaches for sub-event detection [102]	Emergency	Detection and tracking	– Self-organizing map – Agglomerative Clustering	– Social multimedia data (i.e., video and photo)	– Using APIs of Flicker[m] and Youtube[n] for 2011 Mississippi Flood, Oslo Bombing, UK Riots and Hurricane Irene	– Collecting social multimedia data – Labeling – Integrating spatial-temporal data, images and videos
Classifying text messages for the Haiti earthquake [24]	Earthquake	Detection and tracking	– Support vector machine – Bag-of-words – Feature abstraction – Feature selection – Latent Dirichlet allocation	– Text messages in phone, e-mail, Twitter and web	– Ushahidi-Haiti[o]	– Labeling – Selecting feature to remove irrelevant and redundant feature
Prediction of human emergency behaviour and their mobility following large-scale disaster [122]	Natural emergency	Detection and tracking	– Hidden Markov model – Bayesian filtering – Maximum entropy inverse	– Human mobility database (GPS records) – Disaster intensity data (i.e., seismic scale and damage level) – Disaster reporting data	– The Great East Japan Earthquake and Fukushima nuclear from Japan – Government statistical reports – Government declarations – Japanese Cabinet Secretarial[p] – News reports [46]	– Collecting GPSs of human – Labeling – Filtering GPSs – Integrating map and human movements
Event classification and location prediction from tweets during disasters [121]	Emergency	Detection and tracking	– Markov chain – Support vector machine – Gradient boosting – Random forest	– Social media data with geotags according to keywords	– Using Streaming API of Twitter	– Collecting social media data – Labeling – Filtering noise and redundancy

(continued)

Table 11.1 (continued)

Title	Emergency	Task	Machine learning technique	Data	Source	Data analysis
Constructing a Hidden Markov Model based earthquake detector: application to induced seismicity [17]	Earthquake	Detection and tracking	– Hidden Markov model	– Geothermal data for geothermal reservoir – Speech recordings for volcano	– Geothermal plant in the municipality of Unterhaching in Germany (2008–2010) – Mt. Merapi volcano 1998	– Labeling – Filtering noise on sound recording
Earthquake shakes Twitter users: real-time event detection by social sensors [112] Tweet analysis for real-time event detection and earthquake reporting system development [113]	Earthquake	Prediction, Warning, Detection and tracking	– Support vector machine – Kalman filtering – Particle filtering	– Tweet feeds – GPS and registered location of user	– Using Streaming API of Twitter	– Collecting social media data – Labeling – Mapping emergency event and tweet feeds
Leveraging cross-media analytics to detect events and mine opinions for emergency management [134]	Natural emergency	Detection and tracking	– Clustering techniques	– Microblog data	– Online news: Tecent rolling news[q] and Sina rolling news[r] – Online forum: New Commentary column of the Tianya forum[s]	– Crawling microblog from online news and forums – Extracting relevance keywords – Filtering document by sentiment analysis – Integrating cross-media data
Audio Data Mining for Anthropogenic Disaster Identification: An Automatic Taxonomy Approach [137]	Man-made Emergency	Detection and tracking	– K-means dictionary learning – Clustering technique based on hierarchical regularized logistic regression model	– Audio clips from four channel	– BBC Sound Effects Library – Urban-Sound8K datasets [114] – Sound Classification [100] – Sound effects from internet sources[t]	– Labeling
Social Media: New Perspectives to Improve Remote Sensing for Emergency Response [74]	Natural emergency	Situational awareness	– Convolutional neural network	– Social media data of Weibo with keywords (in Chinese)	– Using Sina Weibo open platform API for Weibo data in Wuhan (2016) and Shenzhen (2014)	– Collecting social media data – Labeling

(continued)

Table 11.1 (continued)

Title	Emergency	Task	Machine learning technique	Data	Source	Data analysis
Disaster response aided by tweet classification with a domain adaptation approach [73]	Natural emergency	Situational awareness	– Naïve Bayes classifier – Iterative self-training strategy	– Social media data with hashtags	– CrisisLexT6 dataset [92]	– Labeling
Using social media to enhance emergency situation awareness [138]	Natural emergency	Situational awareness	– Incremental clustering algorithm	– Social media data	– Using Streaming API of Twitter (2010)	– Collecting social media data – Labeling – Filtering stop words
How social media can contribute during disaster events? Case study of Chennai floods 2015 [95]	Natural emergency	Situational awareness	– Semi-supervised classification	– Social media data with hashtags and geotags according to keywords – Web Map service and satellite data	– Using Streaming API of Twitter (2015)	– Collecting social media data – Labeling – Integrating map and tweet feeds
Post earthquake disaster awareness to emergency task force using crowdsourced data [42]	Earthquake	Situational awareness, Crowdsourcing	– K-means clustering	– Social media data with spatial-temporal data	– Using Streaming API of Twitter	– Collecting social media data – Labeling – Mapping affected areas by disaster
Mining crisis information: A strategic approach for detection of people at risk through social media analysis [104]	Natural emergency	Situational awareness	– Support vector machine	– Social media data with hashtags and geotags	– Using Streaming API of Twitter – Followthehashtag[a] – Test dataset [93]	– Collecting social media data – Labeling

(continued)

Table 11.1 (continued)

Title	Emergency	Task	Machine learning technique	Data	Source	Data analysis
Image4Act: Online Social Media Image Processing for Disaster Response [5]	Natural emergency	Situational awareness, Crowdsourcing	– Convolutional neural networks – Perceptual Hashing	– Social media data (i.e., image and geotags) according to keywords	– Using Streaming API of Twitter	– Collecting social media data – Labeling – Filtering irrelevant and duplicate images
Information retrieval of a disaster event from cross-platform social media [120]	Natural emergency	Situational awareness	– Support vector machine	– Social media data with hashtags	– Using Streaming API of Twitter – English language disaster datasets [51, 92, 93]	– Collecting social media data – Labeling – Filtering noise – Integrating cross-social media data
Twitter as a Lifeline: Human-annotated Twitter Corpora for NLP of Crisis-related Messages [55]	Natural emergency	Situational awareness	– Naive Bayes – Support vector machine – Random forest	– Social media data	– Using Streaming API of Twitter for 19 different crises (2013–2015)	– Collecting social media data – Labeling – Filtering stop words, URLs and user-mentions
Argus: Smartphone-enabled human cooperation via multi-agent reinforcement learning for disaster situational awareness [110]	Natural emergency	Situational awareness	– Multi-agent reinforcement Learning (i.e., modified version of distributed Q-learning [78])	– Multimedia data (i.e., audio and video)	– Using a mobile application implemented by them-self	– Collecting audio and video from user smartphone – Labeling – Generating 3D map
The evaluation of city emergency management system results based on BP neural network [57]	Natural emergency	Evaluation	– Backpropagation neural network	– Results evaluation of City emergency management system	– Using a survey data for evaluation of city emergency management system	– Survey evaluation data – Labeling

(continued)

Table 11.1 (continued)

Title	Emergency	Task	Machine learning technique	Data	Source	Data analysis
Damage assessment from social media imagery data during disasters [89]	Natural emergency	Evaluation	– Convolutional neural network	– Social media data with hashtags and keywords – Google images data	– Using AIDR [52] Typhoon Ruby/Hagupit (2014), Nepal Earthquake (2015), Ecuador Earthquake (2016) and Hurricane Matthew (2016) – Google search	– Collecting images from web and social media – Labeling
Nazr-CNN: Fine-Grained Classification of UAV Imagery for Damage Assessment [10]	Natural emergency	Evaluation	– Nazr-convolutional neural network	– Unmanned Aerial Vehicles (UAV) imagery	– Cyclone Pam in Vanuatu (2015) by the UAViators[v]	– Collecting UAV imagery – Labeling – Mapping emergency damages
Automatic Image Filtering on Social Networks Using Deep Learning and Perceptual Hashing During Crises [88]	Natural emergency	Evaluation	– Convolutional Neural Networks – Perceptual hashing	– Images from Twitter	– Using AIDR [51] for Typhoon Ruby (2014), Nepal Earthquake (2015), Ecuador Earthquake (2016) and Hurricane Matthew (2016)	– Collecting social media data – Labeling – Filtering duplicates and relevance
A machine learning framework for assessing post-earthquake structural safety [143]	Earthquake	Evaluation	– Regression tree – Random forest	– Response data of engineering demand	– Nonlinear response history analysis	– Labeling
Processing Social Media Images by Combining Human and Machine Computing during Crises [6]	Natural emergency	Evaluation	– Convolutional neural networks – Perceptual hashing – Transfer learning	– Images from Twitter, Flickr, Instagram and so on.	– Using Streaming API of Twitter for Typhoon Ruby (2014), Nepal Earthquake (2015), Ecuador Earthquake (2016) and Hurricane Matthew (2016)	– Collecting social media data – Labeling – Filtering de-duplication and relevancy – Integrating cross-socia l media data

(continued)

Table 11.1 (continued)

Title	Emergency	Task	Machine learning technique	Data	Source	Data analysis
Using Twitter for tasking remote-sensing data collection and damage assessment: 2013 Boulder flood case study [26]	Natural emergency	Evaluation	– Decision tree	– Social media data in Twitter – UAV imagery	– Using Streaming API of Twitter – Falcon UAV	– Collecting UAV imagery and tweet feeds – Labeling – Mapping emergency damages
Damage identification and assessment using image processing on post-disaster satellite imagery [59]	Natural emergency	Evaluation	– Decision tree – Naive Bayes – Support vector machine – Random forest – Voting classifier – Adaptive booster	– Aerial images	– Using GeoEye1 for the Christchurch earthquake (2011) and Japan earthquake and tsunami (2011)	– Collecting images – Labeling – Filtering noise – Mapping emergency damages
Application and comparison of decision tree-based machine learning methods in landslide susceptibility assessment at Pauri Garhwal Area, Uttarakhand, India [99]	Landslide	Evaluation	– Random forest – Logistic model tree – Best first decision trees – Regression tree	– Landslide and Non-landslide Data	– Interpretation of satellite and Google Earth images	– Collecting images – Labeling – Mapping emergency damages
A New Neuro-Fuzzy Approach for Post-earthquake Road Damage Assessment Using GA and SVM Classification from QuickBird Satellite Images [56]	Earthquake	Evaluation	– Neurofuzzy inference – Support vector machine – Genetic algorithm	– QuickBrid (high-resolution satellite images)	– Bam in Iran (2003)	– Collecting images – Labeling – Mapping damages
Combining machine-learning topic models and spatio-temporal analysis of social media data for disaster footprint and damage assessment [107]	Earthquake	Evaluation	– Latent Dirichlet allocation	– Social media data – Earthquake footprints (Experiment)	– Using Streaming API of Twitter Napa earthquake (2014) – Official earthquake footprint by USGS	– Collecting social media data for specific time and place – Labeling – Filtering stop words, Unique words, stemming, etc. – Mapping emergency damages

(continued)

Table 11.1 (continued)

Title	Emergency	Task	Machine learning technique	Data	Source	Data analysis
Satellite imagery analysis for operational damage assessment in Emergency situations [129]	Natural emergency	Evaluation	– Convolutional neural network	– Two aerospace images for before and after emergency	– Two areas of Ventura and Santa Rosa counties(2017) from Digital-globe within Open Data Program[w]	– Labeling – Mapping emergency damages
A framework of rapid regional tsunami damage recognition from post-event TerraSAR-X imagery using deep neural networks [15]	Tsunami	Evaluation	– Convolutional neural network	– TerraSAR-X data (i.e., high-resolution synthetic aperture radar)	– The Pacific coast of the Miyagi prefecture (2011) from the Pasco Corporation	– Collecting data in the StripMap mode – Labeling – Mapping damages
TweetTracker: An Analysis Tool for Humanitarian and Disaster Relief [67]	Natural emergency	Tracking, Situational awareness, Crowdsourcing	– Various machine learning techniques	– Various social data	– Twitter, Facebook, YouTube, VK and Instagram	– Collecting social media data – Labeling – Filtering noise – Mapping emergency information
Emergency-relief coordination on social media: Automatically matching resource requests and offers [103]	Natural emergency	Crowdsourcing	– Random forest – Naive Bayes multinomial	– Social media data with keywords and hashtags	– Using Streaming API of Twitter	– Collecting social media data for specific time and place – Labeling
Finding requests in social media for disaster relief [86]	Natural emergency	Crowdsourcing	– Latent Dirichlet allocation – Decision tree	– Social media data – Request data (Experiment)	– Using Streaming API of Twitter – Request data [103]	– Collecting social media data for specific time and place – Labeling – Removing duplicate tweets, re-tweets, etc. – Mapping request tweet feeds

Table 11.1 (continued)

Title	Emergency	Task	Machine learning technique	Data	Source	Data analysis
A Weibo-based approach to disaster informatics: incidents monitor in post-disaster situation via Weibo text negative sentiment analysis [14]	Natural emergency	Crowd-sourcing	– Random forest – Naive Bayes – Support vector machine – K-nearest neighbor	– Social media data	– Using Sina Weibo open platform API for Yushu earthquake (2010), Beijing rainstorm (2012) and Yuyao flood (2013)	– Collecting social media data – Labeling – Filtering messages of positive and neutral emotion
Ushahidi, or testimony: Web 2.0 tools for crowd-sourcing crisis information [91]	Natural emergency	Warning, Tracking, Crowdsourcing	– Various machine learning techniques	– Web, Twitter, RSS feeds, emails, SMS, etc.	– Using Streaming API of Twitter	– Collecting social media data – Labeling – Filtering noise – Mapping tweet feeds
DisasterMapper: A CyberGIS framework for emergency management using social media data [50]	Natural emergency	Crowdsourcing	– Various machine learning techniques by Mahout	– Social media data	– Using crawler based on Twitter4j[x]	– Collecting social media data – Labeling – Mapping tweet feeds
Identifying valuable information from twitter during natural disasters [130]	Natural emergency	Situational awareness, Crowdsourcing	– Naive Bayes classifiers	– Social media data	– Using Streaming API of Twitter	– Collecting social media data – Labeling – Removing unnecessary data such as URLs, slang, etc.
Data mining Twitter during the UK floods: Investigating the potential use of social media in emergency management [124]	Flood	Situational awareness, Crowdsourcing	– Naive Bayes classifiers	– Social media data	– Using Streaming API of Twitter	– Collecting social media data – Labeling – Filtering noise
Crowdsourcing Incident Information for Disaster Response Using Twitter [68]	Natural emergency	Crowdsourcing	– Naive Bayes classifiers – Support vector machine	– Social media data with specific keywords	– Twitter dataset of Hurricane Sandy [147]	– Labeling – Mapping damages

(continued)

Table 11.1 (continued)

Title	Emergency	Task	Machine learning technique	Data	Source	Data analysis
#ChennaiFloods: Leveraging Human and Machine Learning for Crisis Mapping during Disasters Using Social Media [9]	Flood	Crowdsourcing	– Smooth support vector machine	– Social media data with specific hashtags	– Using Streaming API of Twitter for a hashtag '#ChennaiFloods'	– Collecting social media data – Labeling – Filtering stop words, stemming, etc. – Mapping tweet feeds

[a] http://www.data.scec.org/
[b] http://www.nhc.noaa.gov/data/#hurdat
[c] http://reverb.echo.nasa.gov/reverb/
[d] http://landsat.usgs.gov/landsat8.php,http://www.sxmb.gov.cn/index.php
[e] http://weather.unisys.com/hurricane/
[f] http://smokealarm.nist.gov/
[g] http://www.dwa.gov.za/
[h] http://wdc.dlr.de/
[i] http://www.ahn.nl/
[j] http://www.ncedc.org/
[k] https://www.strongmotioncenter.org/
[l] http://www.bosai.go.jp/e/
[m] http://flickrj.sourceforge.net/
[n] https://developers.google.com/gdata/
[o] http://haiti.ushahidi.com
[P] http://www.cas.go.jp/
[q] http://roll.news.qq.com/
[r] http://roll.news.sina.com.cn/news/gnxw/gdxw1/index.shtml
[s] http://bbs.tianya.cn/list-news-1.shtml
[t] http://sound.natix.org/
[u] http://www.followthehashtag.com/features/Twitter-historical-data-recover/
[v] http://uaviators.org/
[w] https://www.digitalglobe.com/opendata
[x] http://twitter4j.org/en/index.html

11.4 Challenges and Opportunities of Machine Learning in Response

Now we discuss aforementioned approaches to draw challenges and opportunities of machine learning techniques in emergency management. **Information and Knowledge Learning**. For this category, the following challenges are raised as further research issues: (1) Support sustainable annotation and classification on heterogeneous, streaming data from the multisource. To achieve this, effective and efficient learning algorithms must be studied considering the context of the emergency and the stakeholders belong to each task. (2) Facilitate real-time analysis and discover information across multiple streams through developing high-speed and flexible techniques. (3) Construct customized information extraction methods that can learn by the integration between domain experts and existing system. **Integration with Geographic Information System**. GISs supports to integrate, store, share and display geographically referenced information. Users (e.g., stakeholders or decision maker) can understand overall situations and discover insight through GIS for emergency. In this regard, integration between a GIS and other components is an important research. There are worth further investigation as follows: (1) automate or crowdsource the linkage construction between information/data and geo-map in real-time. For achieve this, as a prior work, automatic location extraction techniques from disaster data obtained or collected are essential for the real-time processing. (2) Intelligent alerts and location broadcasting when people enter a dangerous area. In this case, sensor-based or vision-based approaches can be integrated as the anomaly detection techniques. **Emergency Data Analysis**. As aforementioned, data in emergency are in general generated from various sources and are heterogeneous in nature. Therefore, effective and efficient methods for data analysis in emergency management should consider discovering the inter-dependencies of data and extracting useful information and knowledge. An additional challenge is to integrate data with great diversity which may be cased from heterogeneous sources with different levels of redundancy, accuracy and uncertainty or may be due to different characteristics of data (e.g., structured/unstructured, real-time streams/static data). In this regards, some interesting research directions include: (1) a unified method for each specific algorithms to collect, extract, filter for heterogeneous, multisource disaster data; and (2) building an analysis method capable of real-time processing.

In addition, we discuss issues in crowdsourcing with the machine learning for emergency management in terms of data analysis such as data collection, information extraction, data filtering and data integration.

11.4.1 Data Collection

Scalability issues. Large crises often generate an explosion of social media activity. In case of Twitter, although each message contains 140 characters, is around 4KB by considering the metadata. Furthermore, the significant amount of storage space may be required by attached multimedia objects such as images and videos. Data velocity is a more challenging issue, especially with considering frequent occurring drastic variations. The largest peak of tweets during a natural hazard was measured as 16,000 tweets per minute.[1] Finally, redundancy is in general cited as a scalability challenge. Repeated (e.g., retweets) messages are common in time-sensitive social media, even un-trusted tweet such as rumours and spam might gain more concern due to simply repeating more.

Content issues. Even although microtexts are brief and informal, to analysis this type of text is difficult work due to complexity with technological, cross-lingual and cross-cultural factors. This causes severe challenges to computational methods and can lead to poor and misleading results. Additionally, the texts are also highly heterogeneous with multiple sources and varying levels of quality. Quality itself is important question, encompassing many attributes including objectivity, clarity, timeliness, conciseness and so on.

11.4.2 Information Extraction

Inadequate spatial information. As spatial and temporal information, are two components of an event, most systems encounter challenges to determine geographical information of social media that lack GPS information. In this case, additional information (e.g., geo-tag and locations in user profile) can be used.

Combining manual and automatic annotation. In a supervised learning setting, data labeled through manual works is necessary to training a model, but it may be costly to obtain. This is particularly problematic in emergency that attract concerns of a multilingual population, or for tasks that require domain knowledge, related to affected region or characteristics of emergency events. Also, labeled data are not always reliable and may not be available at the time of the emergency. In this case, a hybrid approach that combines human and automatic annotation can be used. An active learning with the selection of items to be labeled by humans can be applied to improve classification accuracy as new labels are received.

---

[1]During Hurricane Sandy in 2012: http://www.cbsnews.com/news/social-media-a-news-source-and-toolduring-superstorm-sandy/.

11.4.3 Data Filtering

Mundane events. People post specific events as well as daily life on social media sites. These data as noise, which creates more challenges for event detection methods to overcome, should be separated of real-life big events like emergencies.

Rumours, spam and social bots. Filtering social media data in crowdsourcing is a necessary process before data usage in any stages of a disaster. Data which is overwhelmed with unwanted content (e.g., rumours, spam and content created by social bots) does not show real opinion of the crowd. To overcome challenge by rumours, methods have been proposed to automatically detect tweets by using their specific behaviours such as the difference of diffusion process of rumours and normal posts, the number of users and the depth related the diffusion, etc. In case of spam, characteristics of spammers can be used, such as posting numerous messages by one account and the few numbers of reciprocal connections. Specifically for bot, three major methods have been proposed through manual annotation, using the suspension mechanism of social media sites, or creating lure bots.

11.4.4 Data Integration

Describing the events. Creating descriptions or labeling for a detected event is in general a challenge task. Although major keywords that are frequently posted during the event are presented as a description, it does not constitute a grammatically well-formed. In addition, with other useful data such as maps, images and video, even this issue will be more complex, but it may be practical and helpful to understand event on the whole.

Domain adaptation. Simply reusing an existing classifier trained on past data does not perform well in practice, as it yields a significant loss in accuracy even when emergencies have a lot of common elements. In machine learning, domain adaptation is a series of methods adapting it to continue to fulfil well on a dataset with different characteristics. Furthermore, this techniques may help to integrate a variety of data from different domain to supplement weaknesses each other. For instance, briefly, sensor-based approach that provides more detail situation can be integrated with crowdsourcing-based methods, which support to detect event in broad area, for fast and accurate event detection for time-sensitive tasks in emergency.

From these challenges, several interest topics are able to be raised as opportunities such as:

– Annotation: What is the best way to collect and aggregate labels for unlabeled data from the crowd? How can we do the annotation in the most cost-efficient manner? What is the most effective way to collect probabilistic data from the crowd? How can we collect data requiring global domain knowledge via crowdsourcing?

- Time-sensitive and complex tasks: How can we design crowdsourcing systems to handle (near) real-time or time-sensitive tasks? How can we deal with work dependencies requiring more complexity?
- Data collection for specific domains: How can machine learning researchers apply the crowdsourcing principles for different domains where privacy and original characteristics are at play?
- Reliability, efficiency and scalability of a system: How can we deal with sparse, noisy and large number of label classes such as tagging images for Deep Learning based computer vision algorithms? How can we efficiently apply a series of useful methods (e.g., optimal budget allocation, label aggregation algorithms and active learning) into crowdsourcing disciplines?

11.5   Crowdsourcing in Emergency Management

In this section, combination crowdsourcing and machine learning for various tasks in emergencies are reviewed, and challenges and opportunities are raised in terms of data analysis. Moreover, we more deeply contemplate, with some examples (tweets in Twitter) for challenges and opportunities discussed.

Volunteering is part of how community reacts to emergencies [31] and this process has been facilitated by social media in recent years. Volunteers provide information and resources to the affected people [87]. In this regard, for some years now, both researchers and practitioners, in the field of emergency management, have been exploring crowdsourcing for collecting, processing and sharing information between stakeholders. Jeff Howe in 2006 fir sly defined the term as "the act of taking a job traditionally performed by a designated agent and outsourcing it to a generally large group of people in the form of an open call [48]". Since Howes definition, an extended range of crowdsourcing researches have been carried out from a number of fields such as computer sciences, management, information systems and so on. Additionally, as a problem-solving method [32], crowdsourcing has caught the attention of emerging paradigms such as collective intelligence, human computation, or social computing [101]. With respect to this, Liu has analysed the distinct skills and expertise of different crowds typically involved in emergency management: (1) affected-populations, (2) social medias and (3) digital volunteer communities [77]. In this framework, affected populations generate local, timely and direct experiential information. And social media make available unexpected and fortuitous experience. Finally, digital volunteers offer their capabilities for processing and managing emergency data. Also, Poblet et al. introduced four different crowd's roles such as sensors, social computers, reporters and micro-taskers. It considered four data types (i.e., raw data, unstructured data, semi-structured data and structured data) and two involvement types (i.e., active or passive) [101]. There have been recently efforts toward crowdsourcing such tasks in emergency management but it is still challenging. Social media posts come at a fast pace and immense volume. Moreover, it is challenging to collect all the posts which are related to a disaster due to the restrictions by social media services. The collected data contains daily information and is only in part insightful information. Another issue is malicious content such as spam and rumours which can cause panic and stress, especially when produced in large scale using bots [87]. In this regard, one of areas to potentially resolve these challenges is machine learning which has been applied to a variety of tasks in emergency management as mentioned in Sect. 11.2.

11.5.1 Crowdsourcing with Machine Learning for Emergency Management

Here, hybrid approach, crowdsourcing with machine learning, for various tasks in emergencies are reviewed, with exception of approaches for relief task, since the approaches are already reviewed in Sect. 11.2.7.

Pandey and Natarajan proposed a prototype solution to provide situation awareness during and post disaster event using semi-supervised machine learning technique based on SVM and creating interactive open street maps for crowdsourcing the user data providing threat and relief information [95]. Their model was evaluated with the data from Chennai flood 2015. Truong et al. developed a Bayesian approach to the classification of tweets during Hurricane Sandy in order to distinguish "informational" from "conversational" tweets [130]. They designed an effective set of features and used them as input to NB classifiers. The NB classifier was also introduced to reduce noise in crowdsourced data related to emergency management [124]. Their approach was assessed with the flood data of Cumbria 2015. As similarly, Imran et al. proposed automatic methods based on NB classifier for extracting information from microblog posts. They also focused on extracting valuable "information nuggets" relevant to disaster response [54]. Kurkcu et al. likewise used NB classifier with TF-IDF to identify keywords from tweets related to emergency [68]. Experimental result indicated that crowdsourced data refined could provide detailed location information of a specific incident according to its intensity and duration. Anbalagan and Valliyammai proposed a system that performs disaster tweet collection based on trending disaster hash tags [9]. They compared naive-Bayesian and Smooth Support Vector Machine (SSVM) classifications on collected tweets to identify the severity of the emergency. As a result, the SSVM outperformed the naive-Bayesian. Also, emergency geographic map was generated for the affected area through location to interpolation cluster proximity. Balena et al. evaluated supervised classification methods (AdaBoost, NB, RF, SVM and Neural Networks) to compare their effectiveness and potential for classifying message requests asking for/offering to help in emergency [16]. From their experiments, the RF and the Neural Network had better performance than the others. Nagy et al. introduced an evaluation of approaches to accurately and precisely identify crowd sentiment in social media data (Tweets) in emergency situation [85]. SVM was used to classify the lexicons linked to the seed.

Their technique performed better than Bayesian networks alone, and the combination with Bayesian networks improved the sentiment detection.

11.5.2 Example: Crowdsourcing and Machine Learning for Tracking Emergency

In this section, we more deeply contemplate, with several examples (tweet feeds in Twitter) for challenges and opportunities discussed in the previous section. Table 11.2 shows the tweet examples which will be discussed in this section.

Discovering location of social media data. Geographical coordinate (known as geotagging) attached in message is useful for a number of tasks in disaster response [106]. For instance, it allows to search or verify information about a local event, by filtering the messages corresponding to a particular affected region; further geotagging can also be used for higher-level tasks, such as predict transmission of infectious disease. Unfortunately, only 2% of emergency-related messages includes GPS coordinates in practice, further, large portion of these messages may be made through the social bot [23]. For example, a place of the tweet E2 is able to be speculated from its content about Tohoku earthquake, while E1 doesn't contain explicit coordinates. In this regard, from user profiles, some information such as a home location (freeform text), preferred language and time zone may be considered for determining the location of the tweet [113].

Table 11.2 Tweet examples related to emergency

No	Tweet
E1	[11-03-2011T 05:49:01] BIG EARTHQUAKE!!!
E2	[11-03-2011T 05:50:00] Massive quake in Tokyo
E3	[29-10-2012T 03:16:51] #havetoacemysatexam this made me laugh, no idea why but omfg haha http://bit.ly/S6fbBt
E4	[15-11-2012T 13:03:15] ATTENTION ANYONE LOOKING FOR A JOB: FEMA needs assistance for South Jersey! $1000.00/per l (904)797-5338!
E5	[15-11-2012T 05:50:13] #Sandy RUMOR CONTROL: The rumor that @fema is offering $300 cash cards for food is FALSE. http://www.fema.gov/hurricane-sandy-rumor-control
E6	[12-05-2015T 05:50:45] Lets not forget the people in #NepalEarthquake let us all say a prayer for the people in Nepal MrAlMubarak
E7	[11-03-2011T 05:48:54] Huge earthquake in TK we are affected!
E8	[11-03-2011T 05:48:54] Earthquake!
E9	[11-03-2011T 05:48:08] Earthquake vertical shake!
E10	[11-03-2011T 05:48:14] Earthquake!!!!
E11	[11-03-2011T 05:52:23] I can see the tsunami coming!!!!

Trustworthiness of crowdsourced data. Crowdsourced data reflects opinion of crowd; which sometimes contains more than the credible data. On the other hand, within crowdsourcing using social media service, rumour, spam and bots can be critical issues. Among the 8 million tweets according to Boston Marathon Bombing in 2013, around 29% were revealed to be rumours and 51% to be general opinions and comments [40]. This insight shows the perils of using keyword or hashtag-based topic definition. Additionally, as an example of spam (E3[2]), many tweets which generated in areas close to the occurrence of Hurricane Sandy had been related to promoting the material for Scholastic Assessment Test (SAT) together with *#HaveToAceMySA-TExam* tag [87]. In case of E4,[3] it may be confirmed by other tweets which notice whether or not a rumor is true as E5.[4] For retrieving the relevant tweets, using the expert knowledge to compose high-precision queries has been emphasized as one possible solution [38].

Integration between different disciplines. Images and videos in tweets may open new opportunities to deeply understand situation and event. Information source such as website URLs, photos and videos in tweets related to emergency has been found to be around 18% [40]. Once images from the photos and videos are obtained, vision-based analysis can be used to find sources of the multimedia [18]. E6 is one example related to Nepal earthquake in 2015 and contains one image for broken apartments. The photo in tweet E6[5] can be searched by Google Image Search to link with one YouTube video[6] which contains a "Nepal" as a title.

Real-time analysis within sudden events that rapid increase data. In emergency such as Haiti earthquake in 2010 and Fukushima nuclear disaster in 2011, crowd power played an important role in emergency response [136]. However, it still difficult and complex to timely analyse the social data generated in emergencies because of a sharp and sudden rise in the number of tweets immediately. We note that the E1-2 and E7-11 were obtained from [33]. For instance, the E7-10 are tweets that have been generated in Tokyo after the Tohoku earthquake happened at the epicenter. In the E7-10 tweets are emerging rapidly with in seconds (even in milliseconds). Further E11 is the first tweet about a tsunami by an eyewitness at 6 min after the earthquake occurred; it indicates emergency may be dramatically changed from moment to moment. Therefore, it is essential to develop efficient and effective analysis algorithms that will be working in real-time; moreover, these algorithms should continuously adapt to existing situations and more correctly predict the next situation.

---

[2]https://twitter.com/NevaKey9/status/262860388083843072.

[3]https://twitter.com/OccupySandyNJ/status/269183900113326082.

[4]https://twitter.com/FEMASandy/status/269180520993259520.

[5]https://twitter.com/saflaher/status/598247877295644674.

[6]https://www.youtube.com/watch?v=aNZiLYEr6to.

11.6   Conclusions

Some emergencies occur more frequently and are seriously affecting human life, society, environmental protection and even relationship between countries in the world. Therefore, stakeholders have been taking efforts to construct or study the emergency response for preventing and reducing damages, while researchers have been applying various techniques of machine learning to improve effect and efficiency of the emergency management. Although it was not perfect, many researches have proven successfully supporting tasks in the emergency phases.

The purpose of this chapter is to discuss a hybrid crowdsourcing and real-time machine learning approaches to rapidly process large volumes of data for emergency response in a time-sensitive manner. We then reviewed the application and the approach of machine learning techniques to support the emergency management for the each task, and the challenges and the opportunities were proposed. We described characteristics of data being generated during disaster and discussed that the data characteristics akin to 5 Vs of big data. In addition, various applications cases of big data analysis were looked over. Moreover, we focused on crowdsourcing with machine learning in emergency management, and their challenges and opportunities were discussed in terms of the data analysis. Finally, example of the tweet feeds related to emergency was discussed to more deeply describe the challenges and opportunities.

Acknowledgements The work is funded from the Research Council of Norway (RCN) and the Norwegian Centre for International Cooperation in Education (SiU) grant through INTPART programme.

References

1. R. Agarwal, V. Dhar, Editorial–big data, data science, and analytics: the opportunity and challenge for IS research. Inf. Syst. Res. **25**(3), 443–448 (2014)
2. F.E.M. Agency, Federal response plan (FRP). Technical Report (Federal Emergency Management Agency, 1999)
3. R. Akerkar, Processing big data for emergency management, in *Smart Technologies for Emergency Response and Disaster Management* (2017), p. 144
4. S. Akter, S.F. Wamba, Big data and disaster management: a systematic review and agenda for future research. Ann. Oper. Res., 1–21 (2017)
5. F. Alam, M. Imran, F. Ofli, Image4act: online social media image processing for disaster response, in *Proceedings of the 2017 IEEE/ACM International Conference on Advances in Social Networks Analysis and Mining 2017* (ACM, 2017), pp. 601–604
6. F. Alam, F. Ofli, M. Imran, Processing social media images by combining human and machine computing during crises. Int. J. Hum. Comput. Interact. **34**(4), 311–327 (2018)
7. F. Alamdar, M. Kalantari, A. Rajabifard, Towards multi-agency sensor information integration for disaster management. Comput. Environ. Urban Syst. **56**, 68–85 (2016)
8. A. Amaye, K. Neville, A. Pope, Bigpromises: using organisational mindfulness to integrate big data in emergency management decision making. J. Decis. Syst. **25**(sup1), 76–84 (2016)

9. B. Anbalagan, C. Valliyammai, # chennaifloods: leveraging human and machine learning for crisis mapping during disasters using social media, in *Proceedings of the IEEE 23rd International Conference on High Performance Computing Workshops (HiPCW)* (IEEE, 2016), pp. 50–59

10. N. Attari, F. Ofli, M. Awad, J. Lucas, S. Chawla, Nazr-cnn: fine-grained classification of UAV imagery for damage assessment, in *Proceedings of the 2017 IEEE International Conference on Data Science and Advanced Analytics, DSAA* (IEEE, 2017), pp. 50–59

11. M. Avvenuti, S. Cresci, P.N. Mariantonietta, A. Marchetti, M. Tesconi, Earthquake emergency management by social sensing, in *Proceedings of the 2014 IEEE International Conference on Pervasive Computing and Communications Workshops (PERCOM Workshops)* (IEEE, 2014), pp. 587–592

12. M. Bahrepour, N. Meratnia, M. Poel, Z. Taghikhaki, P.J. Havinga, Distributed event detection in wireless sensor networks for disaster management, in *Proceedings of the 2nd International Conference on Intelligent Networking and Collaborative Systems (INCOS)* (IEEE, 2010), pp. 507–512

13. M. Bahrepour, N. Meratnia, M. Poel, Z. Taghikhaki, P.J. Havinga, Use of wireless sensor networks for distributed event detection in disaster management applications. Int. J. Space-Based Situat. Comput. **2**(1), 58–69 (2012)

14. H. Bai, G. Yu, A weibo-based approach to disaster informatics: incidents monitor in post-disaster situation via weibo text negative sentiment analysis. Nat. Hazards **83**(2), 1177–1196 (2016)

15. Y. Bai, C. Gao, S. Singh, M. Koch, B. Adriano, E. Mas, S. Koshimura, A framework of rapid regional tsunami damage recognition from post-event terrasar-x imagery using deep neural networks. IEEE Geosci. Remote Sens. Lett. **15**(1), 43–47 (2018)

16. P. Balena, N. Amoroso, C.D. Lucia, Integrating supervised classification in social participation systems for disaster response. a pilot study, in *International Conference on Computational Science and Its Applications* (Springer, 2017), pp. 675–686

17. M. Beyreuther, C. Hammer, J. Wassermann, M. Ohrnberger, T. Megies, Constructing a hidden markov model based earthquake detector: application to induced seismicity. Geophys. J. Int. **189**(1), 602–610 (2012)

18. M. Bica, L. Palen, C. Bopp, Visual representations of disaster, in *Proceedings of the 20th ACM Conference on Computer-Supported Cooperative Work and Social Computing (CSCW 2017)* (2017), pp. 1262–1276

19. P. Boccardo, F.G. Tonolo, Remote sensing role in emergency mapping for disaster response. in *Engineering Geology for Society and Territory*, vol. 5 (Springer, 2015), pp. 17–24

20. S. Bontemps, P. Defourny, E.V. Bogaert, O. Arino, V. Kalogirou, J.R. Perez, GLOBCOVER 2009-Products description and validation report (2011)

21. T. Brants, F. Chen, A. Farahat, A system for new event detection. in *Proceedings of the 26th Annual International ACM SIGIR Conference on Research and Development in Informaion Retrieval* (ACM, 2003), pp. 330–337

22. J. Brown, O.F Ferrians Jr., J. Heginbottom, E. Melnikov, Circum-Arctic map of permafrost and ground-ice conditions. US Geological Survey Reston (1997)

23. S.H. Burton, K.W. Tanner, C.G. Giraud-Carrier, J.H. West, M.D. Barnes, "Right time, right place" health communication on twitter: value and accuracy of location information. J. Med. Internet Res. **14**(6), (2012)

24. C. Caragea, N. McNeese, A. Jaiswal, G. Traylor, H.W. Kim, P. Mitra, D. Wu, A.H. Tapia, L. Giles, B.J. Jansen, et al., Classifying text messages for the haiti earthquake. in *Proceedings of the 8th International Conference on Information Systems for Crisis Response and Management (ISCRAM2011)* (Citeseer, 2011)

25. C. Castillo, *Big Crisis Data: Social Media in Disasters and Time-Critical Situations* (Cambridge University Press, 2016)

26. G. Cervone, E. Sava, Q. Huang, E. Schnebele, J. Harrison, N. Waters, Using twitter for tasking remote-sensing data collection and damage assessment: 2013 boulder flood case study. Int. J. Remote Sens. **37**(1), 100–124 (2016)

27. G. Cervone, E. Schnebele, N. Waters, M. Moccaldi, R. Sicignano, Using social media and satellite data for damage assessment in urban areas during emergencies, in *Seeing Cities Through Big Data* (Springer, 2017), pp. 443–457
28. A. Chen, N. Chen, H. Ni, et al., *Modern Emergency Management Theory and Method* (2009)
29. N. Chen, L. Wenjing, B. Ruizhen, A. Chen, Application of computational intelligence technologies in emergency management: a literature review. Artif. Intell. Rev., 1–38 (2017)
30. W. Chen, X. Xie, J. Peng, J. Wang, Z. Duan, H. Hong, Gis-based landslide susceptibility modelling: a comparative assessment of kernel logistic regression, naïve-bayes tree, and alternating decision tree models. Geomat. Nat. Hazards Risk **8**(2), 950–973 (2017)
31. R. Dimitroff, L. Schmidt, T. Bond, Organizational behavior and disaster. Proj. Manag. J. **36**(1), 28–38 (2005)
32. A. Doan, R. Ramakrishnan, A.Y. Halevy, Crowdsourcing systems on the world-wide web. Commun. ACM **54**(4), 86–96 (2011)
33. S. Doan, B.K.H. Vo, N. Collier, An analysis of twitter messages in the 2011 Tohoku earthquake, in *International Conference on Electronic Healthcare* (Springer, 2011), pp. 58–66
34. X. Dong, D. Pi, Novel method for hurricane trajectory prediction based on data mining. Nat. Hazards Earth Syst. Sci. **13**(12), 3211–3220 (2013)
35. R. Dubey, Z. Luo, M. Xu, S.F. Wamba, Developing an integration framework for crowdsourcing and internet of things with applications for disaster response, in *Proceedings of the IEEE International Conference on Data Science and Data Intensive Systems (DSDIS)* (IEEE, 2015), pp. 520–524
36. B. Fahimnia, J. Sarkis, H. Davarzani, Green supply chain management: a review and bibliometric analysis. Int. J. Prod. Econ. **162**, 101–114 (2015)
37. E. Fersini, E. Messina, F.A. Pozzi, Earthquake management: a decision support system based on natural language processing. J. Ambient Intell. Humaniz. Comput. **8**(1), 37–45 (2017)
38. A. Ghenai, Y. Mejova, Catching Zika fever: application of crowdsourcing and machine learning for tracking health misinformation on twitter. arXiv:1707.03778 (2017)
39. K. Grolinger, E. Mezghani, M. Capretz, E. Exposito, Knowledge as a service framework for collaborative data management in cloud environments-disaster domain, in *Managing Big Data in Cloud Computing Environments*, (2016), pp. 183–209
40. A. Gupta, H. Lamba, P. Kumaraguru, $1.00 per rt# bostonmarathon# prayforboston: analyzing fake content on twitter, in *eCrime Researchers Summit (eCRS), 2013* (IEEE, 2013), pp. 1–12
41. G.D. Haddow, J.A. Bullock, D.P. Coppola, *Introduction to Emergency Management* (Butterworth-Heinemann, 2017)
42. P. Harris, J. Anitha, Post earthquake disaster awareness to emergency task force using crowdsourced data, in *Proceedings of the IEEE International Conference on Industrial and Information Systems (ICIIS)* (IEEE, 2017), pp. 1–6
43. T. Hastie, R. Tibshirani, J. Friedman, Unsupervised learning, in *The Elements of Statistical Learning* (Springer, 2009), pp. 485–585
44. B.T. Hazen, C.A. Boone, J.D. Ezell, L.A. Jones-Farmer, Data quality for data science, predictive analytics, and big data in supply chain management: an introduction to the problem and suggestions for research and applications. Int. J. Prod. Econ. **154**, 72–80 (2014)
45. H. Higuchi, J. Fujimura, T. Nakamura, K. Kogo, K. Tsudaka, T. Wada, H. Okada, K. Ohtsuki, Disaster detection by statistics and svm for emergency rescue evacuation support system, in *Proceedings of the 43th International Conference on Parallel Processing Workshops (ICCPW)* (IEEE, 2014), pp. 349–354
46. R.C. Hoetzlein, Visual communication in times of crisis: the fukushima nuclear accident. Leonardo **45**(2), 113–118 (2012)
47. C. Howard, D. Jones, S. Reece, A. Waldock, Learning to trust the crowd: validating crowdsources for improved situational awareness in disaster response. Procedia Eng. **159**, 141–147 (2016)
48. J. Howe, The rise of crowdsourcing. Wired Mag. **14**(6), 1–4 (2006)
49. V. Hristidis, S.C. Chen, T. Li, S. Luis, Y. Deng, Survey of data management and analysis in disaster situations. J. Syst. Softw. **83**(10), 1701–1714 (2010)

50. Q. Huang, G. Cervone, D. Jing, C. Chang, Disastermapper: a cybergis framework for disaster management using social media data, in *Proceedings of the 4th International ACM SIGSPATIAL Workshop on Analytics for Big Geospatial Data, BigSpatial@SIGSPATIAL 2015*, ed. by V. Chandola, R.R. Vatsavai (ACM, 2015), pp. 1–6

51. M. Imran, AIDR: artificial intelligence for disaster. Ph.D. thesis, Qatar Computing Research Institute (2014)

52. M. Imran, C. Castillo, J. Lucas, P. Meier, J. Rogstadius, Coordinating human and machine intelligence to classify microblog communications in crises, in *ISCRAM* (2014)

53. M. Imran, C. Castillo, J. Lucas, P. Meier, S. Vieweg, AIDR: artificial intelligence for disaster response, in *Proceedings of the 23rd International Conference on World Wide Web* (ACM, 2014), pp. 159–162

54. M. Imran, S. Elbassuoni, C. Castillo, F. Diaz, P. Meier, Extracting information nuggets from disaster-related messages in social media, in *Iscram* (2013)

55. M. Imran, P. Mitra, C. Castillo, Twitter as a lifeline: human-annotated twitter corpora for NLP of crisis-related messages, in *Proceedings of the Tenth International Conference on Language Resources and Evaluation LREC 2016* (2016)

56. M. Izadi, A. Mohammadzadeh, A. Haghighattalab, A new neuro-fuzzy approach for post-earthquake road damage assessment using GA and SVM classification from quickbird satellite images. J. Indian Soc. Remote Sens. **45**(6), 965–977 (2017)

57. D.M. Jiang, Z.B. Li, The evaluation of city emergency management system results based on BP neural network, in *Applied Mechanics and Materials*, vol. 263 (Trans Tech Publication, 2013), pp. 3288–3291

58. M.I. Jordan, T.M. Mitchell, Machine learning: trends, perspectives, and prospects. Science **349**(6245), 255–260 (2015)

59. A.R. Joshi, I. Tarte, S. Suresh, S.G. Koolagudi, Damage identification and assessment using image processing on post-disaster satellite imagery, in *IEEE Transaction on Global Humanitarian Technology Conference (GHTC)* (IEEE 2017), pp. 1–7

60. L.P. Kaelbling, M.L. Littman, A.W. Moore, Reinforcement learning: a survey. J. Artif. Intell. Res. **4**, 237–285 (1996)

61. C.Y. Kao, C.H. Loh, Monitoring of long-term static deformation data of Fei-Tsui arch dam using artificial neural network-based approaches. Struct. Control Health Monit. **20**(3), 282–303 (2013)

62. H.S. Kim, C.H. Ho, J.H. Kim, P.S. Chu, Track-pattern-based model for seasonal prediction of tropical cyclone activity in the western North Pacific. J. Clim. **25**(13), 4660–4678 (2012)

63. S.W. Kim, J.A. Melby, N.C. Nadal-Caraballo, J. Ratcliff, A time-dependent surrogate model for storm surge prediction based on an artificial neural network using high-fidelity synthetic hurricane modeling. Nat. Hazards **76**(1), 565–585 (2015)

64. Q. Kong, R.M. Allen, L. Schreier, Y.W. Kwon, Myshake: a smartphone seismic network for earthquake early warning and beyond. Sci. Adv. **2**(2), e1501055 (2016)

65. S.B. Kotsiantis, I. Zaharakis, P. Pintelas, Supervised machine learning: a review of classification techniques. Emerg. Artif. Intell. Appl. Comput. Eng. **160**, 3–24 (2007)

66. V.V. Krzhizhanovskaya, G. Shirshov, N. Melnikova, R.G. Belleman, F. Rusadi, B. Broekhuijsen, B. Gouldby, J. Lhomme, B. Balis, M. Bubak et al., Flood early warning system: design, implementation and computational modules. Procedia Comput. Sci. **4**, 106–115 (2011)

67. S. Kumar, G. Barbier, M.A. Abbasi, H. Liu, Tweettracker: an analysis tool for humanitarian and disaster relief, in *Proceedings of the Fifth International Conference on Weblogs and Social Media*, ed. by L.A. Adamic, R.A. Baeza-Yates, S. Counts (The AAAI Press, Barcelona, Catalonia, Spain, 2011), 17–21 July 2011

68. A. Kurkcu, F. Zuo, J. Gao, E.F. Morgul, K. Ozbay, Crowdsourcing incident information for disaster response using twitter, in *Proceedings of the 65th Annual Meeting of Transportation Research Board* (2017)

69. P.M. Landwehr, K.M. Carley, Social media in disaster relief, in *Data Mining and Knowledge Discovery for Big Data* (Springer, 2014), pp. 225–257

70. B. Lehner, P. Döll, Development and validation of a global database of lakes, reservoirs and wetlands. J. Hydrol. **296**(1–4), 1–22 (2004)
71. B. Lehner, C.R. Liermann, C. Revenga, C. Vörösmarty, B. Fekete, P. Crouzet, P. Döll, M. Endejan, K. Frenken, J. Magome et al., High-resolution mapping of the world's reservoirs and dams for sustainable river-flow management. Front. Ecol. Environ. **9**(9), 494–502 (2011)
72. M. Lenzerini, Data integration: a theoretical perspective, in *Proceedings of the Twenty-first ACM SIGACT-SIGMOD-SIGART Symposium on Principles of Database Systems*, ed. by L. Popa, S. Abiteboul, P.G. Kolaitis (ACM, 2002), pp. 233–246
73. H. Li, D. Caragea, C. Caragea, N. Herndon, Disaster response aided by tweet classification with a domain adaptation approach. J. Contingencies Crisis Manag. **26**(1), 16–27 (2018)
74. J. Li, Z. He, J. Plaza, S. Li, J. Chen, H. Wu, Y. Wang, Y. Liu, Social media: new perspectives to improve remote sensing for emergency response. Proc. IEEE **105**(10), 1900–1912 (2017)
75. T. Li, N. Xie, C. Zeng, W. Zhou, L. Zheng, Y. Jiang, Y. Yang, H. Ha, W. Xue, Y. Huang, S. Chen, J.K. Navlakha, S.S. Iyengar, Data-driven techniques in disaster information management. ACM Comput. Surv. **50**(1), 1 (2017)
76. J. Liang, P. Jacobs, S. Parthasarathy, Human-guided flood mapping: from experts to the crowd (2018)
77. S.B. Liu, Crisis crowdsourcing framework: designing strategic configurations of crowdsourcing for the emergency management domain. Comput. Support. Coop. Work (CSCW) **23**(4–6), 389–443 (2014)
78. C. Mariano, E. Morales, A new distributed reinforcement learning algorithm for multiple objective optimization problems, in *Advances in Artificial Intelligence* (Springer, 2000), pp. 290–299
79. F. Martínez-Álvarez, A.T. Lora, A. Morales-Esteban, J.C. Riquelme, Computational intelligence techniques for predicting earthquakes, in *Hybrid Artificial Intelligent Systems—6th International Conference, HAIS 2011*, ed. by E. Corchado, M. Kurzynski, M. Wozniak, Wroclaw, Poland, May 23–25, 2011, Proceedings, Part II, Lecture Notes in Computer Science, vol. 6679 (Springer, 2011), pp. 287–294
80. S. Mehrotra, X. Qiu, Z. Cao, A. Tate, Technological challenges in emergency response. IEEE Intell. Syst. **28**(4), 5–8 (2013)
81. K. Mori, T. Nakamura, J. Fujimura, K. Tsudaka, T. Wada, H. Okada, K. Ohtsuki, Development of ERESS in panic-type disasters: disaster recognition algorithm by buffering-SVM, in *Proceedings of the 13th International Conference on ITS Telecommunications (ITST)* (IEEE, 2013), pp. 337–343
82. M. Moustra, M. Avraamides, C. Christodoulou, Artificial neural networks for earthquake prediction using time series magnitude data or seismic electric signals. Expert Syst. Appl. **38**(12), 15032–15039 (2011)
83. S. Murali, V. Krishnapriya, A. Thomas, *Crowdsourcing for disaster relief: a multi-platform model, in Distributed Computing* (Electrical Circuits and Robotics (DISCOVER), IEEE (IEEE, VLSI, 2016), pp. 264–268
84. A. Nadi, A. Edrisi, Adaptive multi-agent relief assessment and emergency response. Int. J. Disaster Risk Reduct. **24**, 12–23 (2017)
85. A. Nagy, J. Stamberger, Crowd sentiment detection during disasters and crises, in *Proceedings of the 9th International ISCRAM Conference* (2012) pp. 1–9
86. T.H. Nazer, F. Morstatter, H. Dani, H. Liu, Finding requests in social media for disaster relief, in *Proceedings of the IEEE/ACM International Conference on Advances in Social Networks Analysis and Mining (ASONAM)* (IEEE, 2016), pp. 1410–1413
87. T.H. Nazer, G. Xue, Y. Ji, H. Liu, Intelligent disaster response via social media analysis a survey. ACM SIGKDD Explor. Newsl. **19**(1), 46–59 (2017)
88. D.T. Nguyen, F. Alam, F. Ofli, M. Imran, Automatic image filtering on social networks using deep learning and perceptual hashing during crises. Comput. Res. Repository (2017). arXiv:1704.02602
89. D.T. Nguyen, F. Ofli, M. Imran, P. Mitra, Damage assessment from social media imagery data during disasters. in *Proceedings of the 2017 IEEE/ACM International Conference on Advances in Social Networks Analysis and Mining 2017* (ACM, 2017) pp. 569–576

90. G.C. Ohlmacher, J.C. Davis, Using multiple logistic regression and gis technology to predict landslide hazard in northeast kansas, usa. Eng. Geol. **69**(3–4), 331–343 (2003)
91. O. Okolloh, Ushahidi, or testimony: Web 2.0 tools for crowdsourcing crisis information. Participatory Learn. Action **59**(1), 65–70 (2009)
92. A. Olteanu, C. Castillo, F. Diaz, S. Vieweg, Crisislex: a lexicon for collecting and filtering microblogged communications in crises, in *Proceedings of the 8th International AAAI Conference on Weblogs and Social Media (ICWSM)* (2014)
93. A. Olteanu, S. Vieweg, C. Castillo, What to expect when the unexpected happens: social media communications across crises, in *Proceedings of the 18th ACM Conference on Computer Supported Cooperative Work & Social Computing* (ACM, 2015), pp. 994–1009
94. K. Pabreja, Clustering technique to interpret numerical weather prediction output products for forecast of cloudburst. Int. J. Comput. Sci. Inf. Technol. (IJCSIT) **3**(1), 2996–2999 (2012)
95. N. Pandey, S. Natarajan, How social media can contribute during disaster events? case study of chennai floods 2015, in *Proceedings of the International Conference on Advances in Computing, Communications and Informatics (ICACCI)* (IEEE, 2016), pp. 1352–1356
96. F. Pappenberger, P. Matgen, K.J. Beven, J.B. Henry, L. Pfister et al., Influence of uncertain boundary conditions and model structure on flood inundation predictions. Adv. Water Resour. **29**(10), 1430–1449 (2006)
97. M.C. Peel, B.L. Finlayson, T.A. McMahon, Updated world map of the köppen-geiger climate classification. Hydrol. Earth Syst. Sci. Discuss. **4**(2), 439–473 (2007)
98. B. Pengel, V. Krzhizhanovskaya, N. Melnikova, G. Shirshov, A. Koelewijn, A. Pyayt, I. Mokhov et al., Flood early warning system: sensors and internet. IAHS Red Book **357**, 445–453 (2013)
99. B.T. Pham, K. Khosravi, I. Prakash, Application and comparison of decision tree-based machine learning methods in landside susceptibility assessment at pauri garhwal area, uttarakhand, india. Environ. Process. **4**(3), 711–730 (2017)
100. K.J. Piczak, ESC: dataset for environmental sound classification, in *Proceedings of the 23rd ACM International Conference on Multimedia* (ACM, 2015), pp. 1015–1018
101. M. Poblet, E. García-Cuesta, P. Casanovas, Crowdsourcing roles, methods and tools for data-intensive disaster management. Inf. Syst. Front., 1–17 (2017)
102. D. Pohl, A. Bouchachia, H. Hellwagner, Social media for crisis management: clustering approaches for sub-event detection. Multimed. Tools Appl. **74**(11), 3901–3932 (2015)
103. H. Purohit, C. Castillo, F. Diaz, A. Sheth, P. Meier, Emergency-relief coordination on social media: automatically matching resource requests and offers. First Monday **19**(1) (2013)
104. J.R. Ragini, P.R. Anand, V. Bhaskar, Mining crisis information: a strategic approach for detection of people at risk through social media analysis. Int. J. Disaster Risk Reduct. **27**, 556–566 (2018)
105. S.D. Ramchurn, T.D. Huynh, Y. Ikuno, J. Flann, F. Wu, L. Moreau, N.R. Jennings, J.E. Fischer, W. Jiang, T. Rodden et al., HAC-ER: a disaster response system based on human-agent collectives, in *Proceedings of the 2015 International Conference on Autonomous Agents and Multiagent Systems, International Foundation for Autonomous Agents and Multiagent Systems* (2015), pp. 533–541
106. K. Ravikumar, A.R. Kannan, Spatial data mining for prediction of natural events and disaster management based on fuzzy logic using hybrid PSO (2018)
107. B. Resch, F. Usländer, C. Havas, Combining machine-learning topic models and spatiotemporal analysis of social media data for disaster footprint and damage assessment. Cartogr. Geogr. Inf. Sci., 1–15 (2017)
108. B. Revilla-Romero, J. Thielen, P. Salamon, T.D. Groeve, G. Brakenridge, Evaluation of the satellite-based global flood detection system for measuring river discharge: influence of local factors. Hydrol. Earth Syst. Sci. **18**(11), 4467 (2014)
109. J. Reyes, A. Morales-Esteban, F. Martínez-Álvarez, Neural networks to predict earthquakes in chile. Appl. Soft Comput. **13**(2), 1314–1328 (2013)
110. V. Sadhu, G. Salles-Loustau, D. Pompili, S. Zonouz, V. Sritapan, Argus: Smartphone-enabled human cooperation via multi-agent reinforcement learning for disaster situational awareness,

in *Proceedings of the IEEE International Conference on Autonomic Computing (ICAC)* (IEEE, 2016), pp. 251–256

111. R.R. Sahay, A. Srivastava, Predicting monsoon floods in rivers embedding wavelet transform, genetic algorithm and neural network. Water Resour. Manag. **28**(2), 301–317 (2014)

112. T. Sakaki, M. Okazaki, Y. Matsuo, Earthquake shakes twitter users: real-time event detection by social sensors. in *Proceedings of the 19th International Conference on World Wide Web* (ACM, 2010), pp. 851–860

113. T. Sakaki, M. Okazaki, Y. Matsuo, Tweet analysis for real-time event detection and earthquake reporting system development. IEEE Trans. Knowl. Data Eng. **25**(4), 919–931 (2013)

114. J. Salamon, C. Jacoby, J.P. Bello, A dataset and taxonomy for urban sound research, in *Proceedings of the 22nd ACM International Conference on Multimedia* (ACM, 2014), pp. 1041–1044

115. M. Salehi, L.I. Rusu, T. Lynar, A. Phan, Dynamic and robust wildfire risk prediction system: an unsupervised approach, in *Proceedings of the 22nd ACM SIGKDD International Conference on Knowledge Discovery and Data Mining* (ACM, 2016), pp. 245–254

116. A. Salfinger, S. Girtelschmid, B. Pröll, W. Retschitzegger, W. Schwinger, *Crowd-sensing meets situation awareness: a research roadmap for crisis management, in Proceedings of the 48th Hawaii International Conference on System Sciences, HICSS* (Hawaii, USA, Kauai, 2015), pp. 153–162

117. G.D. Saxton, O. Oh, R. Kishore, Rules of crowdsourcing: models, issues, and systems of control. Inf. Syst. Manag. **30**(1), 2–20 (2013)

118. H. Shah, R. Ghazali, Prediction of earthquake magnitude by an improved ABC-MLP, in *Developments in E-systems Engineering (DeSE)* (IEEE, 2011), pp. 312–317

119. X. Shao, X. Li, L. Li, X. Hu, The application of ant-colony clustering algorithm to earthquake prediction, in *Advances in Electronic Engineering, Communication and Management*, vol. 2 (Springer, 2012), pp. 145–150

120. S. Shen, N. Murzintcev, C. Song, C. Cheng, Information retrieval of a disaster event from cross-platform social media. Inf. Discov. Deliv. **45**(4), 220–226 (2017)

121. J.P. Singh, Y.K. Dwivedi, N.P. Rana, A. Kumar, K.K. Kapoor, Event classification and location prediction from tweets during disasters. Ann. Oper. Res., 1–21 (2017)

122. X. Song, Q. Zhang, Y. Sekimoto, R. Shibasaki, Prediction of human emergency behavior and their mobility following large-scale disaster, in *Proceedings of the 20th ACM SIGKDD International Conference on Knowledge Discovery and Data Mining* (ACM, 2014), pp. 5–14

123. J.H. Sorensen, Hazard warning systems: review of 20 years of progress. Nat. Hazards Rev. **1**(2), 119–125 (2000)

124. T. Spielhofer, R. Greenlaw, D. Markham, A. Hahne, Data mining twitter during the UK floods: investigating the potential use of social media in emergency management, in *Proceedings of the 3rd International Conference on Information and Communication Technologies for Disaster Management (ICT-DM)* (IEEE, 2016), pp. 1–6

125. Y. Su, S. Chelluboina, M. Hahsler, M.H. Dunham, A new data mining model for hurricane intensity prediction, in *Proceedings of the 2010 IEEE International Conference on Data Mining Workshops (ICDMW)* (IEEE, 2010), pp. 98–105

126. A.H. Tapia, K. Bajpai, B.J. Jansen, J. Yen, L. Giles, Seeking the trustworthy tweet: can microblogged data fit the information needs of disaster response and humanitarian relief organizations, in *Proceedings of the 8th International ISCRAM Conference* (2011), pp. 1–10

127. M.S. Tehrany, B. Pradhan, M.N. Jebur, Spatial prediction of flood susceptible areas using rule based decision tree (DT) and a novel ensemble bivariate and multivariate statistical models in gis. J. Hydrol. **504**, 69–79 (2013)

128. H.N. Teodorescu, Using analytics and social media for monitoring and mitigation of social disasters. Procedia Eng. **107**, 325–334 (2015)

129. A. Trekin, G. Novikov, G. Potapov, V. Ignatiev, E. Burnaev, Satellite imagery analysis for operational damage assessment in emergency situations (2018). arXiv:1803.00397

130. B. Truong, C. Caragea, A. Squicciarini, A.H. Tapia, Identifying valuable information from twitter during natural disasters. Proc. Assoc. Inf. Sci. Technol. **51**(1), 1–4 (2014)

131. A. Vetrivel, N. Kerle, M. Gerke, F. Nex, G. Vosselman, Towards automated satellite image segmentation and classification for assessing disaster damage using data-specific features with incremental learning (2016)
132. S. Vieweg, A.L. Hughes, K. Starbird, L. Palen, Microblogging during two natural hazards events: what twitter may contribute to situational awareness, in *Proceedings of the 28th International Conference on Human Factors in Computing Systems, CHI*, ed. by E.D. Mynatt, D. Schoner, G. Fitzpatrick, S.E. Hudson, W.K. Edwards, T. Rodden (ACM, Atlanta, Georgia, 2010), pp. 1079–1088
133. Z. Wang, C. Lai, X. Chen, B. Yang, S. Zhao, X. Bai, Flood hazard risk assessment model based on random forest. J. Hydrol **527**, 1130–1141 (2015)
134. W. Xu, L. Liu, W. Shang, Leveraging cross-media analytics to detect events and mine opinions for emergency management. Online Inf. Rev. **41**(4), 487–506 (2017)
135. D. Yamazaki, F. O'Loughlin, M.A. Trigg, Z.F. Miller, T.M. Pavelsky, P.D. Bates, Development of the global width database for large rivers. Water Resour. Res. **50**(4), 3467–3480 (2014)
136. D. Yang, D. Zhang, K. Frank, P. Robertson, E. Jennings, M. Roddy, M. Lichtenstern, Providing real-time assistance in disaster relief by leveraging crowdsourcing power. Pers. Ubiquitous Comput. **18**(8), 2025–2034 (2014)
137. J. Ye, T. Kobayashi, X. Wang, H. Tsuda, M. Masahiro, *Audio data mining for anthropogenic disaster identification: an automatic taxonomy approach* (IEEE Trans. Emerg. Top, Comput, 2017)
138. J. Yin, A. Lampert, M. Cameron, B. Robinson, R. Power, Using social media to enhance emergency situation awareness. IEEE Intell. Syst. **27**(6), 52–59 (2012)
139. D.K. Yoon, S. Jeong, Assessment of community vulnerability to natural disasters in Korea by using gis and machine learning techniques, in *Quantitative Regional Economic and Environmental Analysis for Sustainability in Korea* (Springer, 2016), pp. 123–140
140. A.T. Zagorecki, E.J. David, J. Ristvej, Data mining and machine learning in the context of disaster and crisis management. Int. J. Emerg. Manag. **9**(4), 351–365 (2013)
141. M.A. Zanini, F. Faleschini, P. Zampieri, C. Pellegrino, G. Gecchele, M. Gastaldi, R. Rossi, Post-quake urban road network functionality assessment for seismic emergency management in historical centres. Struct. Infrastruct. Eng. **13**(9), 1117–1129 (2017)
142. X.Y. Zhang, X. Li, X. Lin, The data mining technology of particle swarm optimization algorithm in earthquake prediction, in *Advanced Materials Research*, vol. 989 (Trans Tech Publication, 2014), pp. 1570–1573
143. Y. Zhang, H.V. Burton, H. Sun, M. Shokrabadi, A machine learning framework for assessing post-earthquake structural safety. Struct. Saf. **72**, 1–16 (2018)
144. L. Zheng, C. Shen, L. Tang, T. Li, S. Luis, S. Chen, Applying data mining techniques to address disaster information management challenges on mobile devices, in *Proceedings of the 17th ACM SIGKDD International Conference on Knowledge Discovery and Data Mining*, ed. by C. Apté, J. Ghosh, P. Smyth (ACM, 2011), pp. 283–291
145. B. Zmazek, L. Todorovski, S. Džeroski, J. Vaupotič, I. Kobal, Application of decision trees to the analysis of soil radon data for earthquake prediction. Appl. Radiat. Isot. **58**(6), 697–706 (2003)
146. B. Zmazek, M. Živčić, J. Vaupotič, M. Bidovec, M. Poljak, I. Kobal, Soil radon monitoring in the Krško Basin. Slovenia. Appl. Radiat. Isot. **56**(4), 649–657 (2002)
147. A. Zubiaga, H. Ji, Tweet, but verify: epistemic study of information verification on twitter. Soc. Netw. Anal. Min. **4**(1), 163 (2014)

Part V
Learning and Analytics in Intelligent
Social Media

Chapter 12
Social Media Analytics, Types and Methodology

Paraskevas Koukaras and Christos Tjortjis

Abstract The rapid growth of Social Media Networks (SMN) initiated a new era for data analytics. We use various data mining and machine learning algorithms to analyze different types of data generated within these complex networks, attempting to produce usable knowledge. When engaging in *descriptive* analytics, we utilize data aggregation and mining techniques to provide an insight into the past or present, describing patterns, trends, incidents etc. and try to answer the question "What is happening or What has happened". *Diagnostic* analytics come with a pack of techniques that act as tracking/monitoring tools aiming to understand "Why something is happening or Why it happened". *Predictive* analytics come with a variety of forecasting techniques and statistical models, which combined, produce insights for the future, hopefully answering "What could happen". *Prescriptive* analytics, utilize simulation and optimization methodologies and techniques to generate a helping/support mechanism, answering the question "What should we do". In order to perform any type of analysis, we first need to *identify* the correct sources of information. Then, we need APIs to initialize *data extraction*. Once data are available, *cleaning and preprocessing* are performed, which involve dealing with noise, outliers, missing values, duplicate data and aggregation, discretization, feature selection, feature extraction, sampling. The next step involves *analysis,* depending on the Social Media Analytics (SMA) task, the choice of techniques and methodologies varies (e.g. similarity, clustering, classification, link prediction, ranking, recommendation, information fusion). Finally, it comes to human judgment to meaningfully interpret and draw valuable knowledge from the output of the *analysis* step. This chapter discusses these concepts elaborating on and categorizing various mining tasks (supervised and unsupervised) while presenting the required process and its steps to analyze data retrieved from the Social Media (SM) ecosystem.

P. Koukaras · C. Tjortjis (✉)
School of Science & Technology, International Hellenic University,
14th km Thessaloniki, 57001 Moudania, Thermi, Greece
e-mail: c.tjortjis@ihu.edu.gr

P. Koukaras
e-mail: p.koukaras@ihu.edu.gr

© Springer Nature Switzerland AG 2019
G. A. Tsihrintzis et al. (eds.), *Machine Learning Paradigms*,
Learning and Analytics in Intelligent Systems 1,
https://doi.org/10.1007/978-3-030-15628-2_12

Keywords Social media networks · Social media analytics · Social media · Data mining · Machine learning · Supervised/unsupervised learning

12.1 Social Networks and Analytics

Social Media Networks (SMN) can be considered a mixture of data technologies with traditional Machine Learning (ML) and Data mining (DM) algorithms, creating new challenges for the study area of social networks and Social Media (SM) [1]. Such problems have to do with data processing, data storage or data representation. Our efforts on this field mainly focus on DM for extracting patterns, analyzing behaviors, tracking data, visualization and more.

Most of the recent research, deals with social networks as homogeneous networks, meaning that there is no differentiation between nodes and links. This in not reflected on the real networks, where multi-typed components are constantly interacting with others. The new wave of research focuses on *Heterogeneous Information Networks (HINs)* which seem to be more promising in representing multi-typed and interconnected data without losing their semantic meaning after a data analysis [2].

Heterogeneous Networks are characterized by their richer structure and semantic information compared with Homogeneous ones. Having richer data means that more opportunities arise after analyzing these networks. On the other hand, richer data come with more handling issues. There are many data mining tasks, prominent ones are discussed in this chapter.

In such networks the interacting entities or components are bound to interconnected networks forming a structure of multilayer networks [2]. Our aim is to project the best way possible, real systems or networks, highly populated by multi-typed components [3] which might be bio networks, other systems or simple social activities.

To achieve that, we need to perform information network analysis, which recently has gained high attention from the research community, whilst preserving the structural generality of the networks [4]. The concepts of link mining and analysis, social network analysis, hypertext and web mining, network science and graph mining [5] come into play enabling us to effectively perform this analysis.

The nodes and links in heterogeneous networks are extremely rich in information compared with the homogeneous ones, so mining tasks such as *clustering* [6], *classification* [7] *and similarity search* [8] should be addressed reforming or updating currently available implementations. Advanced tasks deal with mining patterns from link associations, making assumptions regarding the type of objects and links, as mentioned in [4].

The concept of *HINs* was first proposed in 2009 [9], followed by the *meta path* concept two years later [10]. Since then, researchers have endeavored to rapidly develop information retrieval techniques. Implicationally, the research world is being drawn to study heterogeneous networks which is becoming an even more interesting topic.

Depending on one's objectives, Social Media Analytics (SMA) can have four different forms, namely, *descriptive* analytics, *diagnostic* analytics [11], *predictive* analytics, and *prescriptive* analytics [12, 13]. These concepts are explained in the following paragraphs.

12.1.1 Descriptive Analytics

It answers the question: "what happened and/or what is happening?" [14]. It is the first step of analysis utilized especially in business analytics. This type of analytics perform data gathering and description of SM information in the form of reports, visualization and clustering in order to comprehend a potential problem or opportunity. It usually mines historical data to find reasons for past failure or success. It is considered a post-mortem analysis; for example, SM user comment analysis [15]. Comment analysis helps us grasp sentiments and discover trends within a Social Network, by using clustering on topics. Descriptive analytics is one of the most frequently used type of SMA [14].

12.1.2 Diagnostic Analytics

It aims to answer the question "why something happened?". For example, while descriptive analytics provide information about a SM marketing campaign success by measuring followers, mentions, posts, page views, etc., diagnostic analytics transform all these data into a singular view that is easier to understand and evaluate its performance compared to previous ones. Inferential statistics, behavioral analytics, correlations & retrospective analysis and outcome, all are all types of diagnostic analytics.

12.1.3 Predictive Analytics

It aims to answer the question "what will happen and/or why will it happen?" [16]. It involves a diversity of methods and techniques from DM, predictive modeling and machine learning, aiming at analyzing current or historical data in order to predict possible outcomes for future or arbitrary speaking "unknown events". For example, an "intention" of buying a specific product in a SMN can be mined to predict the actual purchase. In business perspectives, it has wide usage in sales prediction numbers based on site visiting statistics. It is also used to exploit patterns in transactional data to identify business opportunities or mitigate risks. It essentially uses models to capture relationships associated with a particular set of conditions to aid on a probable decision making by utilizing technical approaches to provide a score or probability of something happening or not. This type of analytics is also used in other science fields such as actuarial science, marketing, financial services, retail, travel, healthcare and more.

12.1.4 Prescriptive Analytics

It is considered the ultimate frontier of analytic capabilities. It usually utilizes a variety of fields of mathematical and computational sciences to make suggestions supporting decision making while taking into consideration the output form descriptive and predictive analytics [17]. This type of analytics can be considered proactive in nature. It does invoke methods for predicting outcomes, yet it makes suggestions for best possible actions/options given specific situations/scenarios. It still has no wide usage in SMA since its application is yet quite ambiguous. For example, given a specific scenario of buyers' pattern behavior regarding specific products, is it possible to improve the actual product offering? Prescriptive analytics include optimization and simulation modeling, multi-criteria decision modeling, expert systems and group support systems.

12.2 Introduction to Social Network Mining

Studies around social networks and SM are connected with Big Data, as techniques from Big Data are used in order to examine the dynamics of such vast networks while they expand and evolve [19]. The work presented in [20, 21] provide some introductory concepts on SM, including historical elements that are useful to better understand the general context of this work.

SM is a relatively new field of studies with lots of prospects on working new wonders. We could easily allege that this rising trend allows us to merge theoretical concepts into practical solutions that could shake the whole connection of the virtual world to the real.

SM data are quirky, unlike conventional data usually handled by classic data mining methodologies and can only be handled using innovative data mining techniques. We should deliver knowledge that comes from user-generated data (content) which is nearly always overwhelmed by social relations. Thus, we are in great need to develop through extensive study, techniques that would help us to achieve that task in DM.

As a definition for *social media mining* we consider: "*the process of representing, analyzing and extracting actionable patterns from social media data*" [22]. To do that, we use new algorithms or modified ones which come from a blend of various methodologies and principles of other science disciplines such as statistics, mathematics, computer science, sociology, machine learning etc. The data we handle are enormous in terms of size, which means that we need individuals who explicitly focus on these kinds of studies.

SM mining scientists need to have the necessary skills to preprocess, measure, model and find patterns in SM data, visualizing them in a meaningful manner. This means a blend of well-established computational and social theory background along with analytical and coding skills are essential for succeeding in these high-demanding tasks.

12.3 Data Structure

There are several challenges that emerge when dealing with complex networks and big data, such as heterogeneity, spurious correlations, incidental endogeneities, noise accumulation [23]. Specifically, data challenges deal with the four (4) "V's" (Volume, Velocity, Veracity, Variety) [24]. Volume and Velocity have to do with storage and timeliness due to the vast amounts of data generation on complex networks, while Veracity and Variety deal with heterogeneity and reliability of data because we have multi-typed data and variety of data sources.

Since we most often deal with graphs and HINs in SM, we handle three (3) types of data [25].

12.3.1 Structured Data

Structured data focus on solving the issue of efficient handling of any kind of data. They aim on taking a form that optimizes specific operation implementation, as well as reducing the complexity of performing tasks. This is the reason why many types of structured data exist to match our needs.

To effectively handle structured data, we use data models that uniquely define how our data will be manipulated (stored, accessed etc.). Examples are numeric, alphabetic, date time, address and more, which normally are accompanied by restrictions regarding their length.

Their advantages usually stem from their ease to be queried and analyzed. On the other hand, nowadays data are vast and most often cannot be squeezed to follow an organized structure. Database tables are most commonly organized using entity-relation models (Fig. 1).

12.3.2 Semi-structured Data

Semi-structured data is the type of data that keeps half of the characteristics of both other types. Although they can be considered structured data, they do not necessarily conform with a strict structure model.

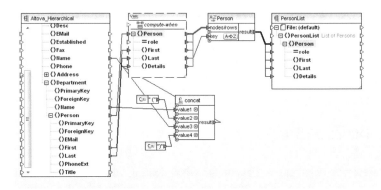

Fig. 1 Example of structured data [26]

Most often these data contain tags or other markers easing the process of identifying specific attributes on them. Also, they might include metadata characterizing the data that they are bound to. For example, word documents, may be labeled (metadata) showing the editor or date time of the last modification. XML as well as other markup languages are the ones most commonly used for handling semi-structured data.

Characteristics:

a. XML format data
b. object→attribute→object
c. relation→connections among attributes.

12.3.3 Unstructured Data

These are data that cannot be classified or conform to a formal form or structure. They are mostly text, which might consist of dates, numbers and so on. They are characterized by ambiguity and irregular ways of data structuring. Unstructured data examples are the ones that might have recognizable entities and extractable relations like photos, videos, pdf files, webpages, word documents etc.

12.4 Data Quality

Before proceeding with the task of presenting data mining algorithms, we must make sure that the data are at a certain level of quality that fits our needs. To do so, the standard requirements are categorized [22] as.

12.4.1 Noise

Noise highly affects the performance of data mining algorithms as it could be executed having incorrect or non-representing data [27]. Noise is inevitable in real world applications, but methods to mitigate its effect can be used [28]. There are two (2) types of noise: (a) implicit errors which are introduced by tools, such as input from different types of sensors or monitoring tools, and (b) random errors introduced during batching processes or individuals responsible for data input, e.g. digitization of documents [29].

12.4.2 Outliers

Outliers represent data instances that are different from others in the same dataset. These instances may have a more significant impact on results by distorting them. For example, in a dataset of people and assets, few very wealthy people who have significantly more assets than the average person will influence a relevant query. So, sometimes we need to either remove outliers or just keep them, always deciding to depend on what is more appropriate for our data mining task. Performance of methods [30] commonly used, varies from case to case, as outlier detection can be a challenging task [31].

12.4.3 Missing Values

Having such values in our dataset is very common. We usually fill in forms, forgetting to input all the fields, or we create accounts without registering all requested information. We should find ways to handle these missing values [32]. To mitigate this issue, we have three main options to choose from.

1. Exclude the missing attributes from the data mining algorithm's functionality (ignore missing values).
2. Completely remove dataset instances that have missing values (instance removal).
3. Populate missing value instances, possibly with the most common ones in the dataset or by a special value such as N/A, or the most probable value (missing value replacement).

12.4.4 Duplicate Data

We face this phenomenon when our dataset is populated by instances having the same feature values. Thus, the problem of detecting and eliminating duplicate data is very disconcerting and is certainly one of the biggest problems in the field of data cleaning and data quality [33]. Once again, depending on what are our mining tasks and our data warehouse, instances can be removed or kept. For instance, duplicate tweets or posts on SM are instances of the duplicate data problem.

12.5 Data Preprocessing

After having verified that the quality of data is adequately matching our needs, the next process is data preprocessing. Standard preprocessing tasks are described below:

12.5.1 Aggregation

We perform the task of aggregation when we need to combine features with multiple instances into one. An example could be, storing a photo in Facebook. We could store the image's width and height or simply store the image dimension as one entry (width_height). This could aid preserving storage capacity, but also provide resistance to data noise.

12.5.2 Discretization

With this process, we simply decide the ranges of continuous features that will be transformed to discrete ones. For example, a feature like height that characterizes a person can be split into discrete values such as tall, normal, short. In addition, it is very important to decide the actual range band of each discrete value with special attention, as it will highly affect the mining task.

12.5.3 Feature Selection

Depending on the mining algorithm we use and whether it takes advantage of the feature selection process, we can boost the performance of the data mining task. Nonetheless, some features in our datasets may not be used, because they are irrelevant with the task at hand. For example, when we want to predict if a person resides in Greece, a feature "name" would be of minor importance or no importance at all for our prediction task. Also, feature selection might require huge amounts of computational power (affecting the speed of our predictions) if we are to run tests on all features on a dataset; thus, the selection should always be made carefully.

12.5.4   Feature Extraction

It is the procedure of the conversion of already available features to a new set of features. This conversion happens on data and new features are extracted to be used for the mining task. The aim of this process is to surpass the performance of an algorithm that uses the default features. If this does not happen, feature extraction is not performed correctly, or it is not needed. An example could be a dataset that contains features about vehicles if we convert the vehicle's speed and time of travel into a new one called distance (distance = speed * time).

12.5.5   Sampling

A very useful and common method for saving time on knowledge extraction from a dataset. It is already mentioned that time and computational power are very important for any mining task and this is mainly due to the huge datasets we deal with. Furthermore, we discuss about SMN where multimedia datasets exist along with large processing streams that accompany them. Sampling comes as a universal tackler for this problem, where a smaller random (in terms of ideally manageable) sample from the whole dataset is selected and the mining task is performed on that. Thus, the whole trick on effective sampling is the method that utilizes the random sampling selection. It is of imperative need that the sample we concluded after the process, realistically (or closely) represents the whole dataset. For sampling, we distinguish three (3) techniques:

- *Random Sampling*. This is the default sampling method where instances in a dataset all have the same probability of being selected for the sample. There is a uniform selection of instances.
- *With or Without Replacement*. In sampling with replacement each instance from the initial dataset can be selected many times and can be added to our new sample dataset. The process without replacement is exactly the opposite of sampling with replacement, meaning that, in case an instance is selected once it cannot be added again to the sample dataset.
- *Stratified Sampling*. This is a two-step method that first uses *bins*, which are partitioned pieces of the initial dataset and then performs random sampling on a fixed number of instances of these bins. It should be noted that this method is proposed for datasets that are not uniformly distributed.

When talking about complex network mining, most of our data have a network form, meaning that they have multiple subsets of nodes and edges. The sampling techniques described above can be used for selecting the appropriate nodes and edges for our task. Also, we can use seed nodes [22] which are small sets of nodes to perform sampling on:

- their connected components,
- the directly connected nodes and edges on them,
- their neighboring nodes and edges.

When preprocessing is complete, we can proceed to the actual mining using algorithms suitable for our goal.

12.6   Network Modeling

Trying to model complex information networks might become a very difficult task. When we handle real world data we find ourselves trying to understand a chaotic mix of physical and abstract data entities which are connected forming huge interconnected multilayer networks.

We must structure all this raw data and form interactions between interconnected multiple types forming the so called semi-structured HINs [34]. Real world applications handle information that deal with big data, social network, medical, e-commerce, engineering datasets which can be modelled using HIN techniques, enabling us to squish out useful information more effectively.

New principles regarding modeling already started to develop rapidly, studying and implementing very powerful algorithms specialized in farming interconnected data such as clustering and classification, meta-path-based similarity search and mining relations. These concepts are being studied and presented with experimental results in [34] where *RankClus, NetClus, GnetMine, RankClass, PathSim* and more are explained.

Next, the three (3) best known models (*random graphs, the small world model* and *the preferential attachment*) for generating networks from real world ones will be discussed, along with real world network characteristics. A rather abstract point will be made on these, since our aim is to cover multiple concepts concerning complex information networks and SM mining.

12.6.1   Real World Networks

Every network has certain characteristics. During the task of designing a network model we must take into consideration these characteristics. Accuracy is very much dependent on how well we can mimic real world situations with artificial ones. Since we deal with inter-connected networks, our first task is to find attributes and draw from them statistics, common amongst most of other interconnected networks.

The measures we can use are *degree distribution, clustering coefficient* and *average path length* [22]. Degree distribution shows the way node degrees are scattered within a network. Clustering coefficient gives the transitivity measure of each of the networks we examine. Finally, the average path length (shortest path) shows the average distance among pairs of nodes [35].

12.6.2   Random Graphs

When modelling with random graphs we assume that all links (edges) that connect nodes (individuals) generate random friendships. However, this assumption does not necessarily mirror real-world situations, e.g. people are not always characterized by friendly relations. Thus, using this technique we take a leap of faith hoping that the relations random graphs form, will generate networks that match the real ones. We find that they exhibit a Poisson degree distribution, a small clustering coefficient and a realistic average path [22].

12.6.3   Small World Model

The small world model comes as an upgrade, fixing some of the random graph model issues regarding the efficient real-world representation. The small-world model was proposed in 1997 [36] as an improvement to suffering of clustering coefficient that comes with the random graphs. This phenomenon (small world) was further discussed [37].

Most individuals in SM have a limited number of connections, such as close friends, family members etc. This model assumes that the number of connections is equal for all members, meaning that all individuals have the same number of neighbors. Still, this seems unrealistic although it helps to model more precisely the clustering coefficient that is found in real world networks. However, even that is not precise enough, since the small-world model is found to produce a degree distribution almost the same as the Poisson degree distribution being found in random graphs [22].

12.6.4   Preferential Attachment Model

The preferential attachment model comes as the most promising one of the models discussed in network modelling literature and it was proposed in 1999 [38]. Simply stated, this model assumes that new nodes that are added to networks have the tendency to connect to already existing nodes because others are already connected to them. Statistically speaking the higher the node's degree, the more probable that new nodes will be connected to that node [22]. This kind of network forming, is called an aristocrat network which is found to be more realistic in average path lengths, but still the clustering coefficient is smaller, an unrealistic characteristic considering real-world networks.

12.7 Network Schemas

12.7.1 Multi-relational Network with Single Typed Objects

The basic characteristic of this type of schema, is that the object type is only one, but the relations that it can enfold are always more than one. Examples are Facebook and Twitter. In social network theory, the existence of multi-relational schemas is very common especially for SM sites (Facebook), where we observe millions of users connecting with others forming millions of links that might represent actions such as messaging, (video) calling, sharing, browsing and so on (Fig. 2).
Characteristics:

- Object type = 1
- Relation type > 1

12.7.2 Bipartite Network

Another common schema observed in HINs is the bipartite network that models a relation or interaction in-between two (2) different types of objects such as multimedia-

Fig. 2 Multi-relation schema

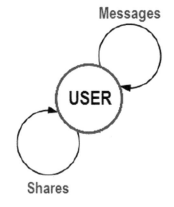

Fig. 3 Bipartite schema

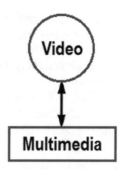

video, as depicted in Fig. 3. No special explanation is presented here, although there are extensions of bipartite networks modeling the existence of k-relations between objects forming links with other neighboring objects.
Characteristics:

a. Object type $= 2$
b. Relation type > 1
c. k-partite graph can be constructed.

Examples User-item, Document-word [5].

12.7.3   Star-Schema Network

This network schema is the most widely utilized in HINs. It represents the most common transformation of relational databases where an object that (might have attributes) generates a HIN acting as a hub where other objects can connect to. Common examples of such networks might be any typical relational database model like the DBLP database [39] which represents a bibliographic network with objects such as author, book, article etc. with links between them, or a movie database. Figure 4 depicts a possible star-schema network showing that the Movie node acts as the hub node.
Characteristics:

a. HIN that is using the target object as a hub node

Examples Bibliographic information network, Movie network, DBLP network

12.7.4   Multiple-hub Network

Multiple-hub networks come as an improvement of star-schemas with regards to information complexity. They are used to represent even more complex network structuring, as they consisted of many hubs. They are utilized when high specificity is required in data representation. Examples of implementation of such networks are common in the fields of bioinformatics. They could also be used in astrophysics or complex mathematic structures where huge fragmentation of network objects is required.

Figure 5 shows a non-realistic example of such a network structure. Although a multi-hub network for user-country-job etc. objects might stand realistically, depending on the task at hand.
Characteristics:

a. HIN that is using multiple hub nodes
b. High (probable) structural complexity

Examples: bioinformatics data, or any other network with high specialized information network.

Fig. 4 Star schema

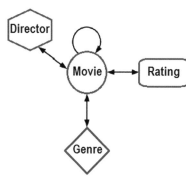

Fig. 5 Multiple-hub schema

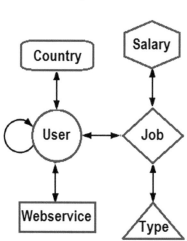

12.8 Task Categorization

During the past years, a great effort has been made by presenting methodologies, novel algorithms and implementations for these tasks. An appropriate categorization is the one proposed in a recent survey on HINs [5]. Next, an abstract presentation of the main mining tasks categorization follows:

Similarity One of the most researched and studied measures in DM. It is a way of finding out how similar objects are. It acts as the base for numerous other data mining techniques such as clustering, web searching etc. There is also a categorization of this measure found in literature: feature based approach and link-based approach.

Clustering is also a very well-known data mining process that groups data objects into sets or smaller groups (clusters) that have a high degree of similarity, although preserving dissimilarity with objects in neighboring clusters. Networked data are different from conventional data studied the past decades, where clustering was done relying on objects feature in the datasets [40]. Nowadays, there is a tendency to deal with networks as heterogeneous ones, which makes clustering a very useful task for discovering knowledge [41].

Classification comes to play when we need to calculate probable class labels, which is achievable through a generation of a classifier or a suitable new model [42]. In machine learning, classification is performed on identically structured objects, but require that links (associations) need also to be considered. Thus, a linked based object classification must take place where entities (nodes) relate to one another forming data graphs. Traditional methods are reused or extended to be able to handle this kind of connections [43].

Effective *link prediction* is an important issue when discussing link mining. It deals with algorithms we use that try to find out if a link between nodes exists, using as rules the following statements: (a) the observation of other nodes and (b) the attributes of these nodes. A great deal of works discuss prediction that takes into consideration only the structural properties of social networks using predictors [44]. Other methods use attribute information itself for link prediction [45].

Ranking functions are very important since they can measure an object's importance within a social network. For example, RankClus deals with bipartite networks creating groups (clusters) of objects preserving the equality of importance both on ranking and clustering concepts [9]. NetClus, is ideal for performing clustering when the network has the characteristics of a star type schema [46]. Many others exist, like HeProjI, OcdRank etc. [5].

Recommendation systems consist of a wide variety of algorithms originating from various fields of different sciences, such as information retrieval, machine learning, statistics, mathematics etc. [47]. The main aim is to make recommendations to users about every kind of objects or services that might be suitable for them. To do so, similarity measures must be utilized [48]. The past years, recommendation systems were naïve, just measuring user specific feedback information to recommend. Lately, ever smarter or updated techniques come into play, such as collaborative filtering [49], matrix factorization [50] or circle-based techniques [51] to boost to new levels the user recommendation experience.

Information fusion is one of the top priorities and tasks that highly characterize HINs. It implements powerful algorithms that fuse objects from the same or different networks whether they have completely relevant or irrelevant semantic meaning. SMN are full of with this type of information, creating the need for the research community to deal with this topic. The task is difficult, having to gather information and merge it to effectively mine it. Information fusion takes place also in bioinformatics [52] and in web semantics [53]. Fusing information of many different HINs and improve the elaboration and contemplation of the knowledge we receive is required.

This section also discusses *measures* we can use for HINs and SMN. Each measure gives useful feedback relevant with its specialization. An abstract presentation of the most fundamental measures follows:

Centrality: Generally, centrality is a measure that tries to calculate the central node in a graph. There are various versions of centrality developed to evaluate the importance of vertices within graphs.

Degree centrality, brings into play the degree value, calculating that the most central node is the one with the highest degree value [54].

Eigenvector centrality is a generalized implementation of the *degree centrality* that calculates the most important node (central) as the one that has the most connections with other important nodes. For this centrality measure the value is computed using an adjacency matrix for finding its eigenvector [55].

Every measure has its pros and cons; *Katz centrality* comes as an improvement for eigenvector centrality when dealing with directed graphs with the introduction of a bias term [56].

Google uses the *RageRank* centrality to rank webpages, and this implementation is considered a more normalized one compared with the *Katz* centrality [57].

Betweenness centrality, is utilized following the concept that the whole graph is formed of multiple hubs of nodes where the root of these hubs is always the central node [58].

Closeness centrality is a measure that considers central nodes to be marked as central because they are close to the rest of the nodes [60].

All these measures of centrality can be implemented in a more generic form (for a vaster measure implementation) where nodes are grouped together, and we can distinguish group degree centrality, group betweenness centrality and group closeness centrality [61].

Transitivity and reciprocity. One of the most common phenomena in SM instance handling is the friendship between nodes a.k.a. node linking. In the endeavor to understand the way nodes establish links between them, one can use the terms transitivity and reciprocity. Transitivity is "when a friend of my friend is my friend" and is considered for closed triads of edges, while reciprocity is a simpler version, "if you become my friend, I will be yours", considering only closed loops of length two (2) which can happen only in directed graphs.

To analyze this behavior, we use the clustering coefficient formulas [22], which can be distinguished to global clustering coefficient and local clustering coefficient. These formulas give a measure for analyzing transitivity of the whole network (global), as well as transitivity for individual nodes (local).

Balance and status: To decide whether relationships are formed in a consistent way in SM, we use some theories to perform this validation. For this work, a choice amongst various theories is made and we will present just the two (2) most common ones. The first is called "social balance" and the second is "social status" [22]. An abstract description of each of these theories follows:

Social balance, a.k.a. "structural balance theory" most often represents consistency in SM in terms of forming rules amongst entities. An informal paradigm of such rules could be:

The friend of my friend is my friend,
The friend of my enemy is my enemy,
The enemy of my enemy is my friend,
The enemy of my friend is my enemy.

Social status establishes another type of rule calculating how consistently entities assign status to the entities near them. Again, an informal paradigm of such rule is:

If X has a higher status than Y and Y has a higher status than Z, then X should have a higher status than Z.

Both theories are very important for measuring consistency between relationships of entities in SM mining. When used together they can be proved a very precise tool for forming, updating or recreating relationships to improve our social mining task.

Similarity: Since we deal with complex information networks and SM mining, similarity can be calculated by referring to structural equivalence. Loosely speaking, structural equivalence [62] is referring to the level of which two (2) nodes are considered similar when they have common neighboring nodes. This means that they share the same social environments, as well as they might have similar attitudes and behaviors (in the form of edges or vertex attributes). For measuring similarity, we use the concepts of *Cosine similarity* [63] and *Jaccard similarity* [64].

12.9 Machine Learning

Machine Learning (ML) [18] on Social networks is closely related to computational statistics, which also focuses on prediction-making through the use of computers. It has strong ties to mathematical optimization, which delivers methods, theory and application domains to the field. ML is sometimes contrasted to data mining, where the latter subfield focuses more on exploratory data analysis. ML can also be unsupervised and be used to learn and establish baseline behavioral profiles for various entities and then used to find meaningful anomalies. Within the field of data analytics, ML is a method used to devise complex models and algorithms that lend themselves to prediction; in commercial use, this is known as predictive analytics. These analytical models allow researchers, data scientists, engineers, and analysts to produce reliable, repeatable decisions and results and find insights through learning from historical relationships and trends in the data.

The algorithms in data mining are numerous and a categorization can be made, from very specific to a highly generic one. The categorization we will present is a rather simplistic one: Two (2) categories widely known, supervised learning and unsupervised learning [22].

In the first category (supervised learning) we have the class attribute and our aim is to make a prediction on what the class attribute value could be. In unsupervised learning, the dataset has no class attribute and the aim is to find entities/instances, that have a high level of similarity and try to group them. By effectively grouping them we create other opportunities on behalf of analysis tasks, such as pattern recognition.

12.9.1 Supervised Learning

Supervised learning can be distinguished to *classification* and *regression*. Classification refers to the discrete class labeling that occurs to feature values. Regression is the form of classification where these labels happen to be real numbers. Common classification algorithms are *decision tree learning, naïve Bayes, nearest neighbor* [65]. Yet, these are not suitable in their native version to fulfil our needs in HINs or SM mining every time.

Supervised learning algorithms split the labeled dataset into two (2) sets, the training and the testing set to perform a consistent evaluation. There are various algorithms that perform that task, for example *leave one out* and *k-fold cross validation* used on twitter data [66]. As mentioned earlier, common classification algorithms are used for objects that have the same structure, something that is not usable for real world datasets where links between entities exist.

To satisfy these needs, linked based object classification comes to play, supported by graph theory forming nodes linked with others via edges. A new wave of research has been conducted extending the capabilities of common classification methods [43]. In homogeneous networks our nodes and links are identical, on the other hand, in HINs other demands exist.

- The nodes do not have the same structural type allowing us to apply methods that classify many different types at the same time.
- Labels expand through links and spread into various dissimilar nodes.

For this type of networks, HINs can be considered when viewed macroscopically as a vast ocean of knowledge that is being transferred by waves of information. This specific incident triggers the opportunity to visualize and predict knowledge flow between dissimilar objects and links.

Many researchers experiment on rectifying the common classification methods to meet the needs of HINs. A list of some of these novel algorithms along with their functionality description follows:

- Transductive classification, is the prediction of labels of unlabeled data.
- Inductive classification is the generation of a decision function that decides based on data spread to the whole data space.

- Multi-label classification deals with simultaneous multiple labeling.
- Rank-base classification is a hybrid method for utilizing both ranking and classification at the same time.
- Information propagation with classification, another hybrid utilizing propagation with classification procedures.

Table 1 contains examples of the categorization of mining tasks [5].

Table 1 Supervised learning novel methods

Mining task	Method/research paper	Description
Transductive classification	GNetMine [67]	Models the link structure in information networks with arbitrary network schema and arbitrary number of object/link types
	HetPathMine [68]	Performs clustering with small labeled data on HIN through a novel meta path selection model
Inductive classification	IMBHN [69]	Establishes a classification model assigning weights to textual terms (in a bipartite HIN) representing textual document collections
Multi-label classification	Graffiti [70]	Performs multi-label graph-based classification for modeling the mutual influence between nodes as a random walk process
	Kong et al. [71]	Utilizes multiple types of relationships drawn from the linkage structure for multi-label classification
	Zhou et al. [72]	Performs edge-centric multi-label classification approach considering both the structure affinity and the label vicinity
Ranking-based classification	RankClass [9]	Utilizes a framework to perform more accurate analysis
	F-RankClass [73]	Utilizes classification framework that can be applied to binary or multi-class classification of unimodal or multimodal data

(continued)

Table 1 (continued)

Mining task	Method/research paper	Description
Information propagation-classification	Jendoubi et al. [74]	Classifies social messages based on self -spreading in the network and the theory of belief functions

12.9.2 Unsupervised Learning

Unsupervised learning deals with clustering which is the process we follow to create groups from our datasets. These groups are consisted of objects or entities that are characterized by the same or identical attributes but are adequative different from entities that belong to other clusters. For running a clustering algorithm, we need to specify the distance measure (e.g. Euclidean, Manhattan, Jaccard, Cosine distances) [75].

Other key points of common clustering methods are the process of object selection and the methods used for the evaluation of clustering results [40]. So, for evaluating results we can use quality measures like *cohesiveness* (measure for object-to-object distance), *separateness* (measure for cluster-to-cluster distance) and *silhouette index* (mix of cohesiveness and separateness) [22].

However, for studying clustering on networked data, things change considering that data are most commonly stored in homogeneous information networks, where community handling comes to play. Most of these methods are based on multiple subgraph generation and handling (e.g. normalized cuts [76]); much research is being conducted optimizing this procedure [77] and solving arising problems [78].

What is more, HINs with the ability to handle entities of different types as well as multi-typed links expand even more our ability to handle more packed data.

Despite the vast amount of information that can be stored in HINs and difficulties that arise, they allow to store much more information and develop considerably more methods for learning.

To give a glimpse of the tasks being developed lately (based on clustering), Table 2 depicts some of the novel models/algorithms for unsupervised tasks in HINs [5].

Table 2 Unsupervised learning novel methods

Task	Method/research paper	Clustering algorithm description/characteristic
Creation of balanced communities	Aggarwal et al. [79]	Introduction of local succinctness property
Modeling structure and content of social media networks using outlier links	Qi et al. [80]	Implementation based on heterogeneous random fields

(continued)

Table 2   (continued)

Task	Method/research paper	Clustering algorithm description/characteristic
Tackling of community detection problem	Cruz et al. [81]	Integration of structural and compositional dimensions for generating attributed graphs
Clusters detection based on connections and vertex attributes of the network	TCSC [82]	Density-based clustering model
Topic modeling of HIN in a unified way	Deng et al. [83]	Topic model with biased propagation
Cluster identifier	LSA-PTM [84]	Link-topic propagation between multi-typed entities for clustering
Unified topic model	Wang et al. [85]	Topic mining and multiple entities clustering
User guided cluster generation using path selection	Sun et al. [6]	Semi-supervised cluster algorithm
Cluster generator	RankClus [9]	Clustering of specific objects in a bipartite network based on qualities of clustering and ranking are mutually enhanced
Cluster generator	NetClus [46]	Handles clustering in a star-schema network
Clustering and ranking	ComClus [86]	Applies star schema network to merge the heterogeneous and homogeneous information
Rank-based clustering	HeProjI [87]	Arbitrary schema on HIN by projecting it into a sequence of subnetworks
Clustering and ranking	Chen et al. [88]	Probabilistic generative model on heterogeneous network with arbitrary schema
Clustering and ranking	Wang et al. [89]	Automated Construction of multi-typed topical hierarchies
Community detection and ranking in directed HIN	OcdRank [90]	Community detection and community-member ranking
Outlier detection	Gupta et al. [91]	Outlier-aware approach with joint non-negative matrix factorization for discovering community distribution patterns
Outlier detection	Zhuang et al. [92]	Uses queries and semantics for subnetwork outlier detection
Outlier measure	Kuck et al. [59]	Meta-path algorithm for mining outliers in HIN

12.10   Conclusions

SM have been attracting increasing research interest due to their wide expansion and user penetration. Researchers strive to come up with new techniques or improve existing ones that help exploit opportunities on this vast new source of multi-typed data. Any involvement with SM usually requires multidisciplinary knowledge.

This chapter aims to present the necessary theoretical concepts for grasping the basic steps of comprehending SM and performing SM data analytics. We discuss new waves of research regarding such complex information networks (like HINs) and we distinguish the types of analytics that can be performed on SM data. We refer to big data, data mining, social network structure and the basics of graph theory, which are essential theoretical background when engaging with SMA.

We explain the aims and nature of *descriptive*, *diagnostic*, *predictive* and *prescriptive analytics* along with examples of domains for actual implementation. We refer to *data*; structure, common types, preprocessing, modeling etc., before moving on with the presentation of methods for analytics. Any type of analysis involves the utilization of machine learning and data mining algorithms/techniques/frameworks, which utilize various tasks, depending on the goals at hand. We present novel algorithms for supervised and unsupervised learning using data from SM, implemented on HINs, which promise to handle enormous volumes of multi-typed data.

Future work of the discussed concepts constitutes the actual implementation of these ideas by producing a set of paradigms that can be easily reproduced and implemented for research, educational or even commercial purposes or act as a point of reference when engaging with any type of SMA. Another idea is to focus on application domains for analytics (e.g. industrial, educational, healthcare etc.) and record the statistics, thus producing a systematic literature review from this perspective. Furthermore, clustering and classification algorithms are widely used in SMA providing prospects for updating current literature regarding the types of SM [93]. A new taxonomy for SM types is our next step forward.

References

1. G. Bello-Orgaz, J.J. Jung, D. Camacho, Social big data: recent achievements and new challenges. Inf. Fusion **28**, 45–59 (2016)
2. M. Kivelä, A. Arenas, M. Barthelemy, J.P. Gleeson, Y. Moreno, M.A. Porter, Multilayer networks. J. Complex Netw. **2**(3), 203–271 (2014)
3. J. Han, in *International Conference on Discovery Science*. Mining Heterogeneous Information Networks by Exploring the Power of Links (Springer, Berlin, Heidelberg, Oct 2009), pp. 13–30
4. Y. Sun, J. Han, Mining heterogeneous information networks: a structural analysis approach. ACM SIGKDD Explor. Newsl. **14**(2), 20–28 (2013)
5. C. Shi, Y. Li, J. Zhang, Y. Sun, S.Y. Philip, A survey of heterogeneous information network analysis. IEEE Trans. Knowl. Data Eng. **29**(1), 17–37 (2017)
6. Y. Sun, B. Norick, J. Han, X. Yan, P.S. Yu, X. Yu, Pathselclus: integrating meta-path selection with user-guided object clustering in heterogeneous information networks. ACM Trans. knowl. Discov. Data (TKDD) **7**(3), 11 (2013)

7. X. Kong, P.S. Yu, Y. Ding, D.J. Wild, Meta path-based collective classification in heterogeneous information networks, in *Proceedings of the 21st ACM International Conference on Information and Knowledge Management*, (ACM, Oct 2012), pp. 1567–1571
8. C. Shi, X. Kong, P.S. Yu, S. Xie, B. Wu, Relevance search in heterogeneous networks, in *Proceedings of the 15th International Conference on Extending Database Technology* (ACM, Mar 2012), pp. 180–191
9. Y. Sun, J. Han, P. Zhao, Z. Yin, H. Cheng, T. Wu, Rankclus: integrating clustering with ranking for heterogeneous information network analysis, in *Proceedings of the 12th International Conference on Extending Database Technology: Advances in Database Technology* (ACM, Mar 2009), pp. 565–576
10. Y. Sun, J. Han, X. Yan, P.S. Yu, T. Wu, Pathsim: meta path-based top-k similarity search in heterogeneous information networks. Proc. VLDB Endow. **4**(11), 992–1003 (2011)
11. A. Banerjee, T. Bandyopadhyay, P. Acharya, Data analytics: hyped up aspirations or true potential? Vikalpa **38**(4), 1–12 (2013)
12. M. Minelli, M. Chambers, A. Dhiraj, *Big Data, Big Analytics: Emerging Business Intelligence and Analytic Trends for Today's Businesses* (Wiley, 2012)
13. T. Bayrak, A review of business analytics: a business enabler or another passing fad. Procedia—Soc. Behav. Sci. **195**, 230–239 (2015)
14. A. Abbasi, W. Li, V. Benjamin, S. Hu, H. Chen, Descriptive analytics: examining expert hackers in web forums, in *2014 IEEE Joint Intelligence and Security Informatics Conference (JISIC)* (IEEE, Sept 2014), pp. 56–63
15. G.F. Khan, Seven Layers of Social Media Analytics: Mining Business Insights from Social Media Text, Actions, Networks, Hyperlinks, Apps, Search Engine, and Location Data (2015)
16. M.A. Waller, S.E. Fawcett, Data science, predictive analytics, and big data: a revolution that will transform supply chain design and management. J. Bus. Logist. **34**(2), 77–84 (2013)
17. D. Bertsimas, N. Kallus, (2014). *From predictive to prescriptive analytics.* arXiv:1402.5481
18. T. Condie, P. Mineiro, N. Polyzotis, M. Weimer, Machine learning on big data, in *2013 IEEE 29th International Conference on Data Engineering (ICDE)* (IEEE, Apr 2013), pp. 1242–1244
19. G. George, M.R. Haas, A. Pentland, *Big Data and Management* (2014)
20. P. Gundecha, H. Liu, Mining social media: a brief introduction. *Tutorials in Operations Research* (2012), p. 1
21. D.M. Boyd, N.B. Ellison, Social network sites: definition, history, and scholarship. J. Comput.-Mediat. Commun. **13**(1), 210–230 (2007)
22. R. Zafarani, M.A. Abbasi, H. Liu, *Social Media Mining: an Introduction* (Cambridge University Press, 2014)
23. C.P. Chen, C.Y. Zhang, Data-intensive applications, challenges, techniques and technologies: a survey on big data. Inf. Sci. **275**, 314–347 (2014)
24. M.T. Thai, W. Wu, H. Xiong (eds.), *Big Data in Complex and Social Networks* (CRC Press, 2016)
25. G. Li, B.C. Ooi, J. Feng, J. Wang, L. Zhou, EASE: an effective 3-in-1 keyword search method for unstructured, semi-structured and structured data, in *Proceedings of the 2008 ACM SIGMOD International Conference on Management of Data* (ACM, Jun 2008), pp. 903–914
26. Digital image. Altova. Web. 2. https://www.altova.com/mapforce/data-sorting.html. Accessed 02 Nov 2016
27. R.Y. Wang, V.C. Storey, C.P. Firth, A framework for analysis of data quality research. IEEE Trans. Knowl. Data Eng. **4**, 623–640 (1995)
28. H. Xiong, G. Pandey, M. Steinbach, V. Kumar, Enhancing data analysis with noise removal. IEEE Trans. Knowl. Data Eng. **18**(3), 304–319 (2006)
29. X. Zhu, X. Wu (2004). Class noise versus attribute noise: a quantitative study. Artif. Intell. Rev. **22**(3), 177–210
30. M.I. Petrovskiy, Outlier detection algorithms in data mining systems. Program. Comput. Softw. **29**(4), 228–237 (2003)
31. S. Vijendra, P. Shivani, Robust *Outlier Detection Technique in Data Mining: A Univariate Approach* (2014). arXiv:1406.5074

32. J.W. Grzymala-Busse, J.W. Grzymala-Busse, Handling missing attribute values, in *Data Mining and Knowledge Discovery Handbook,* (Springer, Boston, MA, 2009), pp. 33–51
33. J.J. Tamilselvi, C.B. Gifta, Handling duplicate data in data warehouse for data mining. Int. J. Comput. Appl. (0975–8887) **15**(4), 1–9 (2011)
34. Y. Sun, J. Han, Mining heterogeneous information networks: principles and methodologies. Synth. Lect. Data Min. Knowl. Discov. **3**(2), 1–159 (2012)
35. R.W. Floyd, Algorithm 97: shortest path. Commun. ACM **5**(6), 345 (1962)
36. H. Lietz, Watts, Duncan J./Strogatz, Steven H. (1998). Collective dynamics of small-world networks. Nature **393**, S. 440–442. Schlüsselwerke der Netzwerkforschung (Springer VS, Wiesbaden), pp. 551–553
37. D.J. Watts, Networks, dynamics, and the small-world phenomenon. Am. J. Sociol. **105**(2), 493–527 (1999)
38. A.L. Barabási, R. Albert, Emergence of scaling in random networks. Science **286**(5439), 509–512 (1999)
39. Computer science bibliography (dblp). http://dblp.uni-trier.de/. Accessed 15 Oct 2016
40. A.K. Jain, Data clustering: 50 years beyond K-means. Pattern Recogn. Lett. **31**(8), 651–666 (2010)
41. Y. Kanellopoulos, P. Antonellis, C. Tjortjis, C. Makris, N. Tsirakis, k-Attractors: a partitional clustering algorithm for numeric data analysis. Appl. Artif. Intell. **25**(2), (2011), pp. 97–115
42. P. Tzirakis, C. Tjortjis, T3C: Improving a decision tree classification algorithm's interval splits on continuous attributes. Adv. Data Anal. Classif. **11**(2), 353–370 (2017)
43. J. Lafferty, A. McCallum, F.C. Pereira, *Conditional Random Fields: Probabilistic Models for Segmenting and Labeling Sequence Data* (2001)
44. D. Liben-Nowell, J. Kleinberg, The link-prediction problem for social networks. J. Am. Soc. Inform. Sci. Technol. **58**(7), 1019–1031 (2007)
45. A. Popescul, L.H. Ungar, Statistical relational learning for link prediction, in *IJCAI Workshop on Learning Statistical Models from Relational Data,* vol. 2003, (2003, August)
46. Y. Sun, Y. Yu, J. Han, Ranking-based clustering of heterogeneous information networks with star network schema, in *Proceedings of the 15th ACM SIGKDD International Conference on Knowledge Discovery and Data Mining* (ACM, Jun 2009), pp. 797–806
47. O. Nalmpantis, C. Tjortjis, The 50/50 recommender: a method incorporating personality into movie recommender systems, in *Proceedings of 8th International Conference on Engineering Applications of Neural Networks (EANN 17), Communications in Computer and Information Science (CCIS)* 744, (Springer, 2017), pp. 1–10
48. V.C. Gerogiannis, A. Karageorgos, L. Liu, C. Tjortjis, Personalised fuzzy recommendation for high involvement products in *IEEE International Conference on Systems, Man, and Cybernetics (SMC 2013),* (2013) pp. 4884–4890
49. C. Luo, W. Pang, Z. Wang, C. Lin, Hete-cf: Social-based collaborative filtering recommendation using heterogeneous relations, in *2014 IEEE International Conference on Data Mining (ICDM),* (IEEE Dec 2014), pp. 917–922
50. N. Srebro, T. Jaakkola, Weighted low-rank approximations, in *Proceedings of the 20th International Conference on Machine Learning (ICML-03)* (2003), pp. 720–727
51. X. Yang, H. Steck, Y. Liu, Circle-based recommendation in online social networks, in *Proceedings of the 18th ACM SIGKDD International Conference on Knowledge Discovery and Data Mining* (ACM, Aug 2012), pp. 1267–1275
52. Y.K. Shih, S. Parthasarathy, Scalable global alignment for multiple biological networks, in *BMC Bioinformatics* vol. 13, no. 3, (BioMed Central, Dec. 2012), p. S11
53. A. Doan, J. Madhavan, P. Domingos, A. Halevy, Ontology matching: a machine learning approach, in *Handbook on Ontologies* (Springer, Berlin, Heidelberg, 2004), pp. 385–403
54. L.C. Freeman, Centrality in social networks conceptual clarification. Soc. Netw. **1**(3), 215–239 (1978)
55. P.D. Straffin, Linear algebra in geography: eigenvectors of networks. Math. Mag. **53**(5), 269–276 (1980)

56. L. Katz, A new status index derived from sociometric analysis. Psychometrika **18**(1), 39–43 (1953)
57. L. Page, S. Brin, R. Motwani, T. Winograd, *The Pagerank Citation Ranking: Bringing Order to the Web* (Stanford InfoLab, 1999)
58. L.C. Freeman, A set of measures of centrality based on betweenness. Sociom 35–41 (1977)
59. J. Kuck, H. Zhuang, X. Yan, H. Cam, J. Han, Query-based outlier detection in heterogeneous information networks, in *Advances in database technology: proceedings*, in *International Conference on Extending Database Technology*, vol. 2015 (NIH Public Access, Mar 2015), p. 325
60. G. Sabidussi, The centrality index of a graph. Psychometrika **31**(4), 581–603 (1966)
61. C. Ni, C. Sugimoto, J. Jiang, Degree, closeness, and betweenness: application of group centrality measurements to explore macro-disciplinary evolution diachronically, in *Proceedings of ISSI* (2011), pp. 1–13
62. F. Lorrain, H.C. White, Structural equivalence of individuals in social networks. J. Math. Sociol. **1**(1), 49–80 (1971)
63. A. Rawashdeh, M. Rawashdeh, I. Díaz, A. Ralescu, Measures of semantic similarity of nodes in a social network, in *International Conference on Information Processing and Management of Uncertainty in Knowledge-Based Systems* (Springer, Cham, Jul 2014), pp. 76–85
64. P. Jaccard, Distribution de la flore alpine dans le bassin des Dranses et dans quelques régions voisines. Bull Soc. Vaudoise Sci. Nat. **37**, 241–272 (1901)
65. I.H. Witten, *Data mining with weka* (Department of Computer Science University of Waikato, New Zealand, Class Lesson, 2013)
66. T. Bodnar, M. Salathé, Validating models for disease detection using twitter, in *Proceedings of the 22nd International Conference on World Wide Web* (ACM, May 2013), pp. 699–702
67. M. Ji, Y. Sun, M. Danilevsky, J. Han, J. Gao, Graph regularized transductive classification on heterogeneous information networks, in *Joint European Conference on Machine Learning and Knowledge Discovery in Databases* (Springer, Berlin, Heidelberg, Sep 2010), pp. 570–586
68. C. Luo, R. Guan, Z. Wang, C. Lin, Hetpathmine: a novel transductive classification algorithm on heterogeneous information networks, in *European Conference on Information Retrieval* (Springer, Cham, Apr 2014), pp. 210–221
69. R.G. Rossi, T. de Paulo Faleiros, A. de Andrade Lopes, S.O. Rezende, Inductive model generation for text categorization using a bipartite heterogeneous network, in *2012 IEEE 12th International Conference on Data Mining (ICDM)* (IEEE, Dec 2012), pp. 1086–1091
70. R. Angelova, G. Kasneci, G. Weikum, Graffiti: graph-based classification in heterogeneous networks. World Wide Web **15**(2), 139–170 (2012)
71. X. Kong, B. Cao, P.S. Yu, (2013, August). Multi-label classification by mining label and instance correlations from heterogeneous information networks, in *Proceedings of the 19th ACM SIGKDD International Conference on Knowledge Discovery and Data Mining* (ACM, Aug 2013), pp. 614–622
72. Y. Zhou, L. Liu, Activity-edge centric multi-label classification for mining heterogeneous information networks, in *Proceedings of the 20th ACM SIGKDD International Conference on Knowledge Discovery and Data Mining,* (ACM, Aug 2014), pp. 1276–1285
73. S.D. Chen, Y.Y. Chen, J. Han, P. Moulin, A feature-enhanced ranking-based classifier for multimodal data and heterogeneous information networks, in 2013 *IEEE 13th International Conference on Data Mining (ICDM)* (IEEE, Dec 2013), pp. 997–1002
74. S. Jendoubi, A. Martin L. Liétard, B.B. Yaghlane, Classification of message spreading in a heterogeneous social network, in *International Conference on Information Processing and Management of Uncertainty in Knowledge-Based Systems* (Springer, Cham, July 2014), pp. 66–75
75. S.S. Choi, S.H. Cha, C.C. Tappert, A survey of binary similarity and distance measures. J. Syst., Cybern. Inform. **8**(1), 43–48 (2010)
76. J. Shi, J. Malik, Normalized cuts and image segmentation. IEEE Trans. Pattern Anal. Mach. Intell. **22**(8), 888–905 (2000)
77. Y. Zhou, H. Cheng, J.X. Yu, Graph clustering based on structural/attribute similarities. Proc. VLDB Endow. **2**(1), 718–729 (2009)

78. M. Sales-Pardo, R. Guimera, A.A. Moreira, L.A.N. Amaral, Extracting the hierarchical organization of complex systems. Proc. Natl. Acad. Sci. **104**(39), 15224–15229 (2007)
79. C.C. Aggarwal, Y. Xie, P.S. Yu, Towards community detection in locally heterogeneous networks, in *Proceedings of the 2011 SIAM International Conference on Data Mining* (Society for Industrial and Applied Mathematics, Apr 2011), (pp. 391–402)
80. G.J. Qi, C.C. Aggarwal, T.S. Huang, On clustering heterogeneous social media objects with outlier links, in *Proceedings of the Fifth ACM International Conference on Web Search and Data Mining* (ACM, Feb 2012), (pp. 553–562)
81. J.D. Cruz, C. Bothorel, Information integration for detecting communities in attributed graphs, in *2013 Fifth International Conference on Computational Aspects of Social Networks (CASoN)* (IEEE, Aug 2013), pp. 62–67
82. M.Z. Ratajczak, M. Kucia, M. Majka, R. Reca, J. Ratajczak, Heterogeneous populations of bone marrow stem cells–are we spotting on the same cells from the different angles? Folia Histochem. Cytobiol. **42**(3), 139–146 (2004)
83. H. Deng, J. Han, B. Zhao, Y. Yu, C.X. Lin, Probabilistic topic models with biased propagation on heterogeneous information networks, in *Proceedings of the 17th ACM SIGKDD international conference on Knowledge discovery and data mining* (ACM, Aug 2011), pp. 1271–1279
84. Q. Wang, Z. Peng, F. Jiang, Q. Li, LSA-PTM: a propagation-based topic model using latent semantic analysis on heterogeneous information networks, in *International Conference on Web-Age Information Management* (Springer, Berlin, Heidelberg, June 2013), (pp. 13–24)
85. X. Wang, C. Zhai, X. Hu, R. Sproat, Mining correlated bursty topic patterns from coordinated text streams, in *Proceedings of the 13th ACM SIGKDD International Conference on Knowledge Discovery and Data Mining* (ACM, Aug 2007), (pp. 784–793)
86. R. Wang, C. Shi, S. Y. Philip, B. Wu, Integrating clustering and ranking on hybrid heterogeneous information network, in *Pacific-Asia Conference on Knowledge Discovery and Data Mining* (Springer, Berlin, Heidelberg, Apr 2013), pp. 583–594
87. C. Shi, R. Wang, Y. Li, P.S. Yu, B. Wu, Ranking-based clustering on general heterogeneous information networks by network projection, in *Proceedings of the 23rd ACM International Conference on Information and Knowledge Management* (ACM, Nov 2014) (pp. 699–708)
88. J. Chen, W. Dai, Y. Sun, J. Dy, Clustering and ranking in heterogeneous information networks via gamma-poisson model, in *Proceedings of the 2015 SIAM International Conference on Data Mining* (Society for Industrial and Applied Mathematics, June 2015), (pp. 424–432)
89. C. Wang, J. Liu, N. Desai, M. Danilevsky, J. Han, Constructing topical hierarchies in heterogeneous information networks. Knowl. Inf. Syst. **44**(3), 529–558 (2015)
90. C. Qiu, W. Chen, T. Wang, K. Lei, Overlapping community detection in directed heterogeneous social network, in *International Conference on Web-Age Information Management* (Springer, Cham, June 2015), (pp. 490–493)
91. M. Gupta, J. Gao, J. Han, Community distribution outlier detection in heterogeneous information networks, in *Joint European Conference on Machine Learning and Knowledge Discovery in Databases* (Springer, Berlin, Heidelberg, Sept 2013), pp. 557–573
92. H. Zhuang, J. Zhang, G. Brova, J. Tang, H. Cam, X. Yan, J. Han, Mining query-based subnetwork outliers in heterogeneous information networks, in *2014 IEEE International Conference on Data Mining (ICDM)* (IEEE, Dec 2014), (pp. 1127–1132)
93. P. Gundecha, H. Liu, Mining social media: a brief introduction. In *New Directions in Informatics, Optimization, Logistics, and Production* (Informs, 2012) (pp. 1–17)

Paraskevas Koukaras holds a B.Sc. in Computer Science from A.T.E.I of Thessaloniki, a M.Sc. (with distinction) in ICT Systems from International Hellenic University, and he is a Ph.D. Student in Social Media Analytics at the same institution. He has been teaching programming in the private sector, sub-reviewing in various international conferences and acting as a lab assistant in database courses at the International Hellenic University. His current research focuses on influence detection, monitoring & intervention mechanisms and forecasting & prediction models in Social Media.

Christos Tjortjis holds a Deng (Hons) from Patras, Computer Eng. & Informatics, a B.Sc. (Hons) from Democritus Law School, Greece, an M.Phil. in Computation from UMIST, and a Ph.D. in Informatics from Manchester, U.K. He is Assistant Prof. in Decision Support & Knowledge Discovery and Director for the M.Sc. in Data Science at the School of Science and Technology, International Hellenic University. He was Lecturer at UMIST, Computation, and the Schools of Informatics and Computer Science, at the University of Manchester, adj. Senior Lecturer at Eng. Informatics & Telecoms, W. Macedonia, and at Computer Science, Ioannina. His research interests are in data mining and software engineering. He worked in 14 R&D projects, leading 3. He published over 50 papers in int'l referred journals and conf. and was PC member in over 80 int'l conf. He received some 630 citations (h-index 14).

Chapter 13
Machine Learning Methods for Opinion Mining In text: The Past and the Future

Athanasia Kolovou

Abstract Sentiment analysis, which is also referred as opinion mining, attracts continuous and increasing interest not only from the academic but also from the business domain. Countless text messages are exchanged on a daily basis within social media, capturing the interest of researchers, journalists, companies, and governments. In these messages people usually declare their opinions or express their feelings, their beliefs and speculations, i.e., their sentiments. The massive use of online social networks and the large amount of data collected through them, has raised the attention to analyze the rich information they contain. In this chapter we present a comprehensive overview of the various methods used for sentiment analysis and how they have evolved in the age of big data.

Keywords Sentiment analysis · Opinion mining · Emotion detection · Sentiment classification

13.1 Introduction

The study of sentiment analysis aims at exploring how emotions are conveyed in human communication. Most of the contents of human communication concern events from the external world. The importance of empathy in human interaction highlights the question of how emotions are represented mentally. For example memories, plans, judgments or even dreams make a person perceive and also express different meanings of a word.

Sentiment analysis is an ongoing field of research. It applies computational methods to extract opinions, sentiments and subjectivity from text. This survey presents a comprehensive overview of the last update in this field covering core topics of the research essential not only for learning but also for practical applications. Many

A. Kolovou (✉)
Department of Informatics and Telecommunications, National and Kapodistrian
University of Athens, Athens, Greece
e-mail: akolovou@di.uoa.gr

© Springer Nature Switzerland AG 2019   429
G. A. Tsihrintzis et al. (eds.), *Machine Learning Paradigms*,
Learning and Analytics in Intelligent Systems 1,
https://doi.org/10.1007/978-3-030-15628-2_13

recently proposed algorithms' enhancements and various applications are investigated and presented briefly. The main goal of this survey is to give nearly full image of the techniques and the related fields targeting mostly to a beginner at the field of sentiment analysis. This survey is covering published literature during 1994–2018, and it is organized on the basis of tasks to be performed, machine learning and natural language processing techniques used.

13.2   Terminology

The most often terms related to sentiment analysis are affect, feeling, emotion, sentiment, and opinion. Those terms are used interchangeably in research areas such as natural language processing, web and text mining, data mining and information retrieval. To spare any confusion about what concepts or features presented in this survey, we first present the definitions of the aforementioned terms in Table 13.1. The definitions are based on the Collins COBUILD (Collins Birmingham University International Language Database) dictionary [58].

13.3   Early Projects

Meanings of words are dynamic. They depend on the situation in which they are uttered and the words that they are combined with. The theory of semantics, developed in the context of cognitive linguistics, has played an import role in sentiment analysis. The semantic role of a sentence, or how the meanings of individual words

Table 13.1 Definitions of related terms from COBUILD dictionary

Term	Definition
Affect	The emotion associated with an idea or set of ideas; If something affects a person or thing, it influences them or causes them to change in some way
Feeling	A feeling is an emotion, such as anger or happiness
Emotion	Emotion is the part of a person's character that consists of their feelings, as opposed to their thoughts;any strong feeling, as of joy, sorrow, or fear
Sentiment	A sentiment that people have is an attitude which is based on their thoughts and feelings.
Opinion	A belief not based on absolute certainty or positive knowledge but on what seems true, valid, or probable to one's own mind; judgment
Sentiment analysis	It is the process of determining the emotional tone behind words, sentences or documents, used to gain an understanding of the the attitudes, opinions and emotions expressed
Subjective	Based on personal opinions and feelings rather than on facts
Sense	A word sense is one of the meanings of a word (some words have multiple meanings, some words have only one meaning)

are composed inspired a number of interesting novelties and formed the basis of sentiment analysis. Past work that involved the use of models inspired by cognitive linguistics [34] or approaches where sentences are regarded as subjective when they express the psychological point of view of a discourse character [109].

Previous research on sentiment-based classification has been at least partially knowledge-based. Some of that work focuses on classifying the semantic orientation of individual words or phrases, using linguistic features [32] or semi-manual construction of discriminant-word lexicons [18]. Other work, explicitly attempts to find features indicating that subjective language is being used, using a supervised learning algorithm that utilizes linguistic features of adjectives [33]. They explore the benefits that some lexical features of adjectives offer for the prediction of subjectivity at the sentence level. In particular they consider two such features: semantic orientation, which represents an evaluative characterization of a words deviation from the norm of its semantic group (e.g., beautiful is positively oriented, as opposed to ugly); and gradability, which characterizes a words ability to express a property in varying degrees (e.g. longer, taller, smartest). In addition adjective-adverb pairs [18, 32] are considered as "grading modifiers" that add emphasis to the adjectival phrase in which they are embedded instead of providing a grading effect (e.g., terribly wrong). Since subjectivity refers to aspects of language used to express opinions and evaluations, it can have a positive impact on sentiment classification tasks.

13.4 The Fascinating Opportunities that Sentiment Analysis Raises

Given the enormous amount of information on the Internet it is becoming more and more important to analyze the opinions and experiences of such information. Sentiment analysis and opinion mining have many applications ranging from e-commerce, marketing, to politics, health and other research areas. Organizations often do sentiment analysis to collect and analyze customer feedback about a product, a service, a marketing campaign or a brand.

In the year 2000 we find a sudden sprout of published papers on the subject. This is a result of the progress made in the field of natural language processing, the abundance of data available and the emergence of deep learning as a powerful machine learning technique.

13.5 Natural Language Processing for Sentiment Analysis

Natural Language Processing (NLP) refers to the application of computational techniques for the analysis and synthesis of natural language and speech. After all, "*it is astonishing what language can do*" [25]. Most tools aim at classifying text segments as positive, negative, or neutral. Sentiments and opinions are more complex than just having polarity. Utterances can convey that we are very angry, sad, elated, etc. In this context, it is often useful for applications to know the degree of intensity, that is the degree or amount of an emotion or degree of sentiment [69]. Sentiment extraction tasks differ according to the classification levels used (i.e., document, sentence, phrase, word or aspect level), the types of features considered, and the classification techniques used (supervised, unsupervised or semi-supervised). Various steps are needed to perform opinion extraction from given texts, since text is coming from several resources in diverse format. The first two mandatory steps in sentiment analysis are data collection and preprocessing.

As far as data collection is concerned, the majority of micro-blogging sites like Twitter, Facebook, Tripadvisor, Foursquare etc. made available their Application Programming Interface (API), allowing individuals to collect public data from their sites. But even though data collection is only a few programming lines away, is not only a time consuming task but has also introduced many challenges to the research community. The continuous increase in the volume and detail of data available online has brought up issues like data quality, noise and data heterogeneity, data collection capacity and even privacy, legal and regulatory issues. Those are only a few of considerations that one must not ignore. Words may have different meanings in different domains and irony and sarcasm require additional workout to derive the actual meaning of the content. Another challenge that systems have to cope with is data sparsity. For instance the maximum character limit in Twitter has been 140 characters until 2017, and it has been doubled ever since. Due to this limitation people may not express their opinion in clear manner. The challenges presented above are only a few to contemplate. Within a sentence we may encounter only one of the presented difficulties, but we may also encounter all of them simultaneously.

Pre-processing is an important step that deals with unstructured text data using traditional NLP methods, in order to utilize most of the information in a given text. The role of text pre-processing in sentiment analysis is discussed in depth in [51]. In particular, they explore the performance of various pre-processing options on sentiment analysis tasks, using well-known learning-based classification algorithms. Some popular steps are: tokenization, stop word removal, stemming, parts of speech (POS) tagging, spelling correction, identification of duplicate content etc. A text processing tool that performs text tokenization, word normalization, word segmentation and spell correction, geared towards Twitter, is presented in [7].

13.5.1 Affective Information for Sentiment Analysis

Over the years various models that represent emotional categories, and emotional dimensions became available for the English language (the creation of lexica for other languages is mostly accomplished from the automatic porting of data and techniques from existing English resources [4]). Those include psycholinguistic resources, sentiment and emotion lexicons. In general the lexica contain a list of words with an associated score, indicating their polarity. Such resources capture different nuances of affect and offer a rich representation for the sentiment conveyed in texts. In this section we present a comprehensive description of such resources and of their use in the context of sentiment analysis.

One of the most widely used lexicon has been the Subjectivity Lexicon [112] which is a list of subjectivity clues compiled from several sources, annotated both manually and automatically. Another popular resource is the Opinion Lexicon [37]. This list contains 2006 positive words and 4783 negative words that are compiled manually and are thus quite accurate. The useful properties of this resource are that it includes mis-spellings, morphological variants, slang, and social-media mark-up.

Many researchers claim that creating a flat list where words are annotated using a sharp division between a positive and negative sentiment value will not always account for the complexity of the semantic aspects of words and their polysemy. Working one step further in creating sentiment resources we find attempts assigning polarity values at the sense level rather than the word level. For example, the annotation scheme developed in MPQA Subjectivity Sense Annotations, a sense-aware lexicon, follows the work of [110] and relies on WordNet [43] as the sense inventory. This resource classifies a given word's sense as objective or subjective, without specifying, its polarity value. In addition, this lexicon includes part-of-speech information. While in this lexicon we encounter ambiguities, for example, a noun and also an adjective, could exhibit different polarity features, another resource, called SentiWordNet [2] addresses ambiguities both within and across parts of speech. SentiWordNet attaches positive, negative and neutral real-valued sentiment scores to WordNet synsets.[1] The sentiment value is graded over a scale in this lexicon as opposed to using a categorical value, thus addressing the limitations of the more simple resources highlighted above. Another sentiment-aware resource, called General Inquirer [92], is a lexicon attaching syntactic, semantic, and pragmatic information to part-of-speech tagged words. In this lexicon polarity is conceived as a binary value rather than a gradual concept.

Domain-dependent lexica are aiming at addressing word sense disambiguation, that is the case where a given word will exhibit only one sense in a specific context (for example the word "book" is indicative here, as in "book a flight" or "read a book"). There are two well known lexica that are built around specific domains, exploiting users reviews: the Yelp Lexicon, with a focus on restaurant reviews, automatically created from the very large Yelp challenge dataset, and the Amazon Lexicon, built on laptop reviews posted on Amazon [45]. Other resources that are domain-dependent and conceive polarity as a gradual rather than categorical value are, the AFINN lexicon [71], an affective lexicon of 2477 English words and phrases manually compiled and rated with an integer between −5 (very negative) and +5 (very positive) value, and the Sentiment140 lexicon [47]. Sentiment140 is composed of Sentiment140 Affirmative Context Lexicon and Sentiment140 Negated Context Lexicon, generated automatically from tweets with sentiment word hashtags and contain about 60,000 unigrams and 10 times more bigrams. This resource is part of a larger suite

---

[1]Nouns, verbs, adjectives and adverbs are grouped into sets of cognitive synonyms (synsets).

of lexica, automatically compiled by Said Mohammad and co-researchers.[2] Another more recent resource that is build automatically and exploits deep learning methods, is the Sentiment Treebank, which is a lexicon consisting of partial parse trees annotated with sentiment. This lexicon is introduced as part of a system which performs sentiment analysis on the sentence level [90]. The deep network system parses the sentence and identifies the sentiment of each node in the parse tree, starting at the leaves. The produced lexicon contains partial parse trees, each annotated with a sentiment score.

Handmade resources are always more difficult to create, they are smaller than automatically built ones but they are more accurate. Recently, a variety of affective lexica have been proposed. Affective lexica offer information about affect expressed in text according to dimensional approaches often studied at the field of emotion modeling. In particular they are not simply related to positive or negative sentiment polarity but to additional emotional categories such as joy, sadness, disgust, pleasantness and fear, providing values in finer levels of granularity. Moreover, various psycholinguistic resources are available that can give some additional measure about the emotional information disclosed in social media texts. The WordNet-Affect Lexicon [93] is a hand-curate collection of emotion-related words (nouns, verbs, adjectives, and adverbs), classified as "Positive", "Negative", "Neutral" or "Ambiguous" and categorized into 28 subcategories (Joy, Love, Fear, etc.). It was developed through the selection and labeling of the WordNet synsets which are representing affective concepts. The resource EmoLex [66], is a word-emotion association lexicon built manually through Amazons Mechanical Turk. It contains 14,182 words labeled according to eight primary emotions, joy, sadness, anger, fear, trust, surprise, disgust, and anticipation. It also contains annotations for negative and positive sentiments. Emolex has evolved, becoming available in over one hundred languages. This is performed by translating the English terms with the use of Google Translate. The underlying assumption is that, despite some cultural differences, it has been shown that a majority of affective norms are stable across languages.

Another psycholinguistic resource is The Dictionary of Affect in Language [108] which includes words rated on a three-point scale along three dimensions, namely activation, imagery, pleasantness. Furthermore, the Affective Norms for English Words [10] is a database developed by rating English words along three dimensions using the valence (positive/negative), arousal (active/passive), dominance (dominance/submissiveness) model [86]. Another related resource, is the Word Count (LIWC) dictionary [77] that assigns words to one or more emotional categories (such as sadness, anxiety, anger).

Handmade lexicon methods have two basic limitations. First they rely on static lists of words and secondly they do not consider that words' sentiments depend on their context. On the other hand, automatic methods are prone to noise and bias. Many ways for reducing noise in the automatic acquisition of resources have been explored by, for instance, the use of sentiment initializers such as emoticons [73] or, usually, bootstrapping models with manually annotated sets of seeds [60, 62].

---

[2]http://saifmohammad.com/WebPages/lexicons.html.

Particularly in [60], using the assumption that semantic similarity implies affective similarity, a fully automated algorithm is build that expands an affective lexicon with new entries. Starting from a set of manually annotated seed words, a linear model is trained using the least mean squares algorithm, followed by feature selection step. This lexicon-expansion method works very well, producing state-of-the-art results, and can be expanded to provide ratings for other expression-related dimensions [59].

In this section we have presented a partial list of sophisticated resources (many others exist). Those resources, publicly available, are allowing sentiment analysis systems to grasp the conceptual and affective information associated with natural language opinions.

13.5.2 Corpora Annotated for Sentiment Analysis Tasks

The increasing interest in the development of automatic systems for sentiment analysis has also prompted the production of sentiment-annotated corpora that could be exploited for the development of such systems in a machine learning fashion. The resources developed have assisted the creation of systems that provide meaningful sentiment analysis results. The list of resources we present in the following section is mostly limited to the English language and is not exhaustive, since the creation of such datasets is an ongoing project.

The resources developed within the SemEval competition (an annual event since 2013 that attracts a great number of participants), comprise tweet collections annotated for subjectivity, polarity, emotions, irony, and aspects/properties, and are not restricted to English [70, 84, 85]. In addition to the SemEval related data, other collections have been developed for the Twitter domain. The SMILE Twitter Emotion dataset [104], was created for the purpose of classifying emotions, expressed on Twitter towards arts and cultural experiences in museums. It contains 3,085 tweets, with 5 emotions namely anger, disgust, happiness, surprise and sadness. A small Twitter dataset[3] contains 5513 hand-classified tweets, where each tweet sentence is not annotated according to an aspect, but a specific topic is assigned to each tweet.

In the creation of sentiment datasets, there have been experiments with distant learning, where class labels are not assigned manually, but are rather derived from other information available. For example in [27] they are performing sentiment analysis on tweets, by training different classifiers using a dataset that was created using emoticons as noisy labels, and achieve an accuracy of about 80%. This dataset is available for research purposes and it is called Sentiment140 dataset. In another work, a new Twitter corpus was released [68], where a machine learning scheme is applied to automatically annotate a set of 2012 US presidential election tweets. It contains a set of tweets with annotations concerning different aspects of emotions, style (simple statement, sarcasm, hyperbole, understatement), and purpose (to point out a mistake, to support, to ridicule, etc.). The tweets were selected based on a set

---

[3]http://www.sananalytics.com/lab/twitter-sentiment/.

of hashtags related to the 2012 US presidential election and were annotated manually by reliance on crowdsourcing platforms. This dataset is the first tweets dataset annotated for all of these phenomena (related to irony, sentiment polarity and on emotions). In a similar context, a dataset that contains ironic vs non-ironic sentences (not tweets), is found in [103]. The Self-Annotated Reddit Corpus (SARC), is a large corpus utilized for training and evaluating systems build for sarcasm detection [42]. Another dataset related to sarcasm is The Sarcasm Corpus V2. It is a subset of the Internet Argument Corpus (IAC, also available for download), and contains data representing three categories of sarcasm: general sarcasm, hyperbole, and rhetorical questions [72]. This corpus has 1.3 million sarcastic statements.

In addition to the Twitter related data, other collections have been constructed, and a very good source of opinionated texts. For example, a widely used dataset is the Movie Review Data, which a collection of 1000 positive and 1000 negative movie reviews [75]. Another classic resource is the MPQA corpus [111], which contains news articles and other text documents manually annotated for opinions and other private states (i.e., beliefs, emotions, sentiments, speculations, etc.). This corpus has been expanded very recently with finer-grained information [20], and now includes entity and event target annotations (eTarget annotations). The eTarget annotations aim to specify the targets of subjectivities related to specific entities and events. The ISEAR dataset[4] contains student respondents, both psychologists and non-psychologists, that were asked to report situations in which they had experienced all of 7 major emotions (joy, fear, anger, sadness, disgust, shame, and guilt). In each case, the questions covered the way students had appraised the situation and how they reacted. The final data set contains reports on seven emotions each, by close to 3000 respondents in 37 countries on all 5 continents.

In line with the increasing interest in aspect-based sentiment analysis the work in [37], mines and summarizes reviews related to specific features of a product, on which customers have expressed their positive or negative opinion. Extraction of opinions on aspects (topics) related to a specific domain, is also the focus in [97], where the authors collected and processed 10,000 reviews from TripAdvisor.

In this section we presented a brief overview of datasets that can be used in sentiment analysis systems. The availability of such datasets opens the way to the possibility of building machine learning systems that predict emotion in unseen sentences.

13.5.3 Distributional Semantics and Sentiment Analysis

The context in which a word occurs is considered of great importance to the task of sentiment identification. The idea behind the distributional hypothesis of meaning, that is the assumption that words that occur in the same contexts tend to have similar meanings [31], has been exploited in the context of sentiment analysis. Traditional

---

[4]https://raw.githubusercontent.com/sinmaniphel/py_isear_dataset/master/isear.csv.

Distributional semantic models (DSMs) use vectors that keep track of the contexts (e.g., co-occurring words) in which target terms appear in a large corpus as proxies for meaning representations. Geometric techniques are applied to these vectors to measure the similarity in meaning of the corresponding words. Word embeddings or neural language models are the most recent and successful members of the distributional semantics family. Word embeddings are established as a highly effective tool in NLP tasks and have proven to substantially outperform traditional Distributional Semantic Models (DSMs) [6]. They are often used under the scope of an unsupervised approach for sentiment analysis [1]. A word embedding (or word representation), is a dense, low-dimensional and real-valued vector for a word [64, 65]. This technique, transforms words in a vocabulary to vectors of continuous real numbers (e.g., the word dog is represented by a vector: $(\ldots, 0.13, \ldots, 0.03, \ldots, 0.21, \ldots)$). Mikolov [64, 65] introduced continuous bag-of-words (CBOW) and continuous skip-gram, and released the popular and computational efficient word2vec toolkit. CBOW model predicts the current word based on the embeddings of its context words while Skip-gram model predicts surrounding words given the embeddings of current word. Another frequently used learning approach is Global Vector (GloVe [78]), which is trained on aggregated global word-word co-occurrence statistics from a large corpus.

Systems targeting at sentiment analysis do not only expend their efforts in feature engineering, which is important but labor intensive. As the list of candidate features grows, finding the best feature combination is not always feasible. It is clear that word embeddings played an important role in sentiment analysis models. Word embeddings can be used as features for neural but also for non-neural learning models. The most serious limitation in traditional word representation methods, is that they typically model the syntactic context of words but ignore the sentiment information of text. As a result, words with opposite polarity, such as good and bad, are mapped into close vectors. A first attempt at directly incorporating sentiment in learning word vectors is found in [57]. In this model words expressing similar sentiment end up having similar vector representations, following the probabilistic document model of [9], and assigning a sentiment predictor function to each word. The full objective of the model is to learn semantic vectors that are imbued with nuanced sentiment information.

The model developed in [95, 96] extends the existing word embedding learning algorithm [15] and trains a neural network by associating each n-gram with the polarity of a sentence. It is shown that sentiment-specific word embeddings (SSWE) effectively distinguish words with opposite sentiment polarity (the method is able to separate good and bad to opposite ends). A distant-supervised corpora without any manual annotations is used, leveraging massive tweets with emoticons. These automatically collected tweets contain noise so they cannot be directly used as gold training data to build sentiment classifiers, but they are effective enough to provide weakly supervised signals for training the sentiment specific embeddings. The model outperforms other models that use generally learned embeddings. Finally, the concept of paragraph vector is proposed in [53] to first learn fixed-length representation for variable-length pieces of texts, including sentences, paragraphs and documents. They experimented on both sentence and document-level sentiment classification tasks and achieved performance gains, which suggests that the proposed method is useful for capturing the semantics of the input text.

13.6 Traditional Models Based on Lexica and Feature Engineering

In this section we discuss the most tradition approach to sentiment analysis, which involves the development of a classification model, trained using pre-labeled dataset annotated with emotions (usually positive, negative or neutral). The papers described in this section are summarized in Table 13.2.

13.6.1 Lexicon Based

The Lexicon-based approach in general, relies on a sentiment lexicon, like the ones presented in Sect. 13.5.1. After the lexicon creation process, the lexicon based models incorporate statistical or semantic methods in order to find sentiment polarity of text segments.

The Semantic Orientation CALculator (SO-CAL) [94], uses dictionaries of words annotated with their semantic orientation (polarity and strength), and incorporates intensification and negation. The first step is the extraction of sentiment-bearing words (including adjectives, verbs, nouns, and adverbs), following their use in order to calculate semantic orientation, taking into account valence shifters (intensifiers, or negation). SO-CAL is applied to a polarity classification task, the process of assigning a positive or negative label to a text.

In another work, a dictionary-based approach is presented to identify sentiment sentences in contextual advertising [80]. They proposed an advertising strategy to improve ad relevance and user experience with the use of syntactic parsing and sentiment dictionary. A rule based approach is proposed that tackles topic word extraction and identifies consumers attitude related to advertising. Their results demonstrated the effectiveness of the proposed approach on advertising keyword extraction and ad selection.

Latent Semantic Analysis (LSA) is a statistical approach which is used to analyze the relationships between a set of documents and the terms mentioned in these documents [19]. In order to find the semantic characteristics from reviews researchers in [12] used LSA and examined the impact of the various features. The objective of their work is to understand the factors that determine the number of helpfulness votes, and why other reviews receive few or no votes at all. They worked on users feedback from CNET Download.com related to software programs. They showed that the semantic characteristics are more influential in affecting how many helpfulness votes a review receives.

Table 13.2 Summary of machine learning methods for sentiment classification

Refs.	Year	Preprocessing	Category	Methods	Aspect-based	Evaluation Dataset
[101]	2002	Minimal	Unsupervised ML	Semantic orientation of adjectives, adverbs using Pointwise Mutual and Information Retrieval	No	Reviews from Epinions
[54]	2009	Yes	Unsupervised ML	Extention of LDA (unsupervised hierarchical Bayesian model)	Yes	Movie reviews
[38]	2011	No	Unsupervised ML	Sentence-LDA and Aspect and Sentiment Unification Model	Yes	Amazon product reviews , Yelp restaurant reviews
[26]	2011	Yes	Unsupervised ML	Deep learning with Stacked denoising autoencoder	Yes	Amazon product reviews
[24]	2016	Yes	Unsupervised ML	Dependency parsing and sentiment features from lexicon	No	Cornell movie review, Obama-McCain debate and SemEval-2015
[81]	2009	Minimal	Unsupervised ML	Word similarity techniques across several domains	Yes	Gigaword corpus
[17]	2007	No	Supervised ML	Naive Bayes transfer classifier	Yes	20 Newsgroups, SRAA and reuters-21578
[76]	2011	Yes	Supervised ML	Naive Bayes and SVM	Yes	Twitter data, IMDB data, Blipr data, manually collected
[82]	2013	Yes	Supervised ML	Nave Bayes, and decision trees using discriminative features to detect irony	No	Manually collected Twitter corpus
[5]	2014	Yes	Supervised ML	Decision tree classifier using groups of features to represent irony, sarcasm	No	Manually collected Twitter corpus
[11]	2015	Yes	Enssemble	Ensemble of four different systems with various hand-crafted features and lexicons, using the average probability of the four classifiers	No	Semeval-2015

(continued)

Table 13.2 (continued)

Refs.	Year	Preprocessing	Category	Methods	Aspect-based	Evaluation Dataset
[74]	2016	Yes	Enssemble	Ensemble of different systems: hand-crafted and affective features, word embeddings, deep learning, topic modeling.	Yes	Semeval-2016
[99]	2016	Yes	Enssemble	Bagging, boosting, stacking, voting	No	Stanford Twitter Sentiment140, Health Care Reform (HCR) Twitter data. Obama-McCain Debate (OMD) Twitter data
[50]	2017	Yes	Enssemble	Ensemble of different systems: hand-crafted, irony based, aspect-based and affective features, word embeddings, deep learning, multy-step classification.	Yes	Semeval-2017
[67]	2013	Yes	Hybrid	IInear SVM classifier	No	SemEval-2013
[46]	2014	Yes	Hybrid	Linear SVM classifier	Yes	SemEval-2014
[61]	2014	Yes	Hybrid	NB tree classifier	No	SemEval-2014
[102]	2015	Yes	hybrid	Linear SVM classifier with sentiment-specific word embeddings	No	Twitter manually annotated dataset
[13]	2015	Yes	Hybrid	Rule-based and SVM (hand-crafted features and lexicon)	No	SemEval 2015
[41]	2016	Yes	Semi-supervised ML	Feature weight based on sentiment lexicon using SVM	No	Cornell movie review dataset, multi-domain sentiment dataset (MDSD)
[56]	2005	Yes	Rule-based	Hand-crafted features	Yes	Product reviews manually collected and annotated
[80]	2010	No	Lexicon based	Syntactic parsing to extract sentiment based topic words	No	Manually collected posts from Web forums

(continued)

Table 13.2 (continued)

Refs.	Year	Preprocessing	Category	Methods	Aspect-based	Evaluation Dataset
[12]	2011	Yes	Lexicon based	LSA text mining and logistic regression	No	Manually collected reviews from CNET Download.com related to software programs
[94]	2011	No	Lexicon based	Semantic orientation of individual words and contextual valence shifters	No	Epinions reviews, (MPQA) corpus, version 2.0, a collection of MySpace.com comments, 2007 SemEval headlines data
[39]	2015	No	Lexicon based	Custom sentiment analysis algorithm	No	Stanford Twitter Sentiment140, Movie reviews

More recently, [39] presents a new lexicon-based sentiment analysis algorithm that has been designed with the main focus on real time Twitter content analysis. The algorithm consists of two key components, namely sentiment normalization and an evidence-based combination function, which have been used in order to estimate the intensity of the sentiment rather than the traditional positive/negative label. In addition, a sentiment lexicon is build manually using SentiWordNet as a baseline. The application allows detailed visualization of the sentiment over time. In an effort to increase the accuracy of the algorithm, the evidence-based combination function is applied when positive and negative words co-occur in a message.

13.6.2 Machine Learning Based

Machine learning methods can be roughly divided into supervised and unsupervised techniques. The supervised methods make use of a large amount of labeled training data. The unsupervised methods are used when it is difficult to find such labeled training information. The first step in the sentiment classification problem is to extract and select text features. Those features (also called hand-crafted) include (but are not limited to):

- Syntactic attributes include part-of-speech (POS) tags, like for example finding adjectives, as we have already mentioned are important indicators of opinions. Other attributes are word n-grams or character n-grams.
- Punctuation marks, capitalized letters, character repetition, abbreviations, as they are considered sentiment intensifiers.
- Emoticons, to highlight the role played by facial expressions
- emojis, given the importance of visual icons for providing an additional layer of meaning to text.
- Negations,oppositions or contradictions, are usually expressions that are used to alter the sentiment orientation.
- Hashtags, for the social media domain, where the hashtag can be treated as a word or possibly a union of words, often indicating the semantic orientation of a sentence. (for example *Last minute "fixes" in the master branch are awesome! #not, More snow coming to Pittsburgh... #whooptydoo #sarcasm*).
- Intensifiers, are words such as very, quite, most, etc. These are words that change sentiment of the neighboring non-neutral terms.

The combination of two baseline approaches is explored in [13]. The application refers to classifying tweets as positive, negative or neutral. They use a rule-based classifier that first takes a decision about positive, negative, or unknown sentiment, using rules that are dependent on emoticons and lexicons (the Sentiment140 lexicon [47] and a lexicon from [56], which is a list of positive and negative words). A second classifier, based on support vector machines (SVM) [16], is trained on semantic, dependency, and sentiment lexicons and identifies positive, negative, and neutral messages. After developing the rule-based classifier and training the SVM, they combine them to refine the SVMs predictions.

The approach proposed by [82] incorporates features (like punctuation marks, emoticons, capitalized letters), to characterize irony. Additionally, features are often created in terms of elements related to sentiments [108]. Another related work by [5], considered the amount of positive and negative words from SentiWordNet [2] and introduced a novel set of features for sarcasm detection. Some of them are designed to detect imbalance and unexpectedness, others to detect common patterns in the structure of the sarcastic tweets (like type of punctuation, length, emoticons), and some others to recognize sentiments and intensity of the terms used. The model is tested by applying supervised machine learning methods to a Twitter corpus, manually collected for the task.

In [102] competitive results are achieved without the use of syntax, by extracting a rich set of automatic features. In particular, they split a tweet into a left context and a right context according to a given target, using distributed word representations and neural pooling functions to extract features. Both sentiment-driven and standard embeddings are used, and a rich set of neural pooling functions are explored. Sentiment lexicons are used as an additional source of information for feature extraction.

A semi-supervised approach based on sentiment lexicon construction was presented in [41]. The proposed framework is called SWIMS ("semisupervised subjective feature weighting and intelligent model selection"). A first step involves acquiring SentiWordNet as a labeled resource to extract adjectives, verbs, adverbs, and nouns that are subjective. The next step uses a feature-weighting mechanism, based on the well known pointwise mutual information criterion (PMI) [14], and trains a supervised SVM for sentiment classification.

Unsupervised models include lexicon-based or topic-modeling based approaches (topic modeling related literature is present in a next section). Despite the fact that they normally need a large volume of data to be trained accurately, unsupervised learning methods help us gain knowledge about the data without any act of annotation. As a first example in [101] the semantic orientation of a phrase is calculated as the mutual information between the given phrase and the word " excellent" minus the mutual information between the given phrase and the word "poor". Another unsupervised paradigm is [81], where distributional similarity is used to measure the similarity between words. The results indicate that word similarity techniques are suitable for applications that require sentiment classification across several domains. Linguistic information is exploited in [24], by studying the dependencies retrieved from a parsing analysis. The method is able to to predict sentiment in informal texts using unsupervised dependency parsing. In addition, they present their method for the creation of a sentiment lexicon, where the sentiment is adapted to a specific context. The starting point is a set of positive and negative words used as seeds. The polarities of the words are then expanded through a graph which is constructed using dependencies between them. Finally, in [50] task-dependent affective ratings are estimated, using a unsupervised domain adaptation technique. In particular two different corpora are utilized in this method. A web-harvested corpus that contains sentences, which is created by posing queries on a web search engine and aggregating the resulting snippets, and a Twitter corpus that contains tweets, collected using Twitter API. According to the proposed lexicon adaptation technique, they build a language model using domain relevant sentences, i.e., tweets. A sub-corpus is created, using a method called perplexity filtering. The sub-corpus is built by selecting only the top most relevant sentences from the web-harvested corpus, according to the Twitter corpus.

13.6.3 Hybrid

The hybrid approach combines both lexicon and machine learning methods with sentiment lexicons playing a dominant role. In an attempt to alleviate the aforementioned limitations of lexicon-based models, the affective lexicon expansion method in [61] is used to extract word-level features. Feature extraction is repeated for sub-strings and contrasting sub-string features are used to better capture complex phenomena like sarcasm. The resulting supervised system, using a Naive Bayes model, achieved high performance in classifying entire tweets.

The NRC-Canada sentiment system is designed to detect the sentiment of short textual messages. The system is based on a supervised text classification approach leveraging a variety of stylistic, semantic, and sentiment features generated from word and character n-grams, manually created and automatically generated sentiment lexicons, parts of speech, word clusters, hashtags and creatively spelled words and abbreviations (yummeee, lol, asap, fyi etc.). The sentiment features are primarily derived from the aforementioned tweet-specific sentiment lexicons. The system achieved the best results in SemEval-2013 shared task on Sentiment Analysis in Twitter [67], and again in 2014 when the shared task was repeated [46].

For a comparative analysis of five well-known classifiers, like for example Support Vector Machine (SVM), k-Nearest Neighbors (KNN), Logistic Regression (LR), etc., we refer the reader to [52]. This work which experimentally evaluates the most representative machine learning algorithms, can be used as a guideline to the decision of the implementation aspects for an effective and punctual algorithm for a sentiment analysis application.

Although the above-mentioned approaches are an important step toward the definition of robust systems, within the sentiment classification research field, there is no consensus regarding which method should be adopted. To overcome this limitation, many systems, combine different techniques following the ensemble learning paradigm. The idea behind ensemble mechanisms is to take advantage of several independent classifiers by combining them to achieve better performance. Various techniques can be used for the late fusion of classifiers, e.g. voting based methods [100], multi-classifiers fusion that use posteriors as features for training a new classifier [30, 48] or algebraic combinations [48]. Algebraic combinations are based on rules such as mean, median, product, max or min and combine the classifiers posteriors appropriately.

In [48] the authors showed that for various combinations schemes and for the cases of classifiers working in different feature spaces, the arithmetic mean is less sensitive to errors. The best performing system for SemEval 2015 Task 10 Subtask B, is an ensemble classifier that builds on well-performing classifiers of past SemEval competitions [11]. All four classifiers that constitute the ensemble system in this work employ polarity dictionaries in addition to hand crafted features. Other systems that combine different classifiers are presented at [50, 74]. Those systems try to leverage the diversity among different classifiers, ranging from traditional feature extraction-classification system, a lexicon-based approach, domain adaptation, topic modeling, neural networks and a pairwise classification system. In a most recent research in [99], well-known ensemble methods are evaluated for the Twitter domain. The methods are namely Bagging, Boosting, Stacking and Voting. This work proves that such methods can surprisingly surpass the traditional algorithms in performance and that the ensemble approach is a beneficial tool in the field of sentiment analysis.

13.7   Domain Adaptation and Topic Modeling

Domain adaptation is frequently used as a method to adjust a system to a specific task and has been extensively used in text classification and sentiment analysis. In this scheme, one aims to generalize a classifier that is trained on a source domain, for which plenty of training data is available, to a target domain, for which data is scarce. For example, a domain adaptation algorithm based on Nave Bayes and Expectation-Maximization (EM), that classifies text documents into several categories is proposed in [17]. In [76], a domain adaptation technique is performed in order to analyze the sentiment expressed in tweets that are related to movies. They utilized the information from other publicly available databases like IMDB and Blippr, and proposed two methods (based on Expectation Maximization (EM) and Support Vector Machines (SVM)), to identify instances that can improve the classifier. More recently, [26] applied stacked denoising auto-encoders with sparse rectifiers to domain adaptation in sentiment analysis (predicting whether a user liked a disliked a product based on a short review). The denoising autoencoders are stacked into deep learning architectures and the outputs of their intermediate layers are then used as input features for SVMs. The authors demonstrate that using the features learned from the autoencoders, in conjunction with linear SVM classifiers, outperforms the state-of-the-art methods for sentiment analysis tasks across different domains.

Topic modeling is a method for discovering "topics" that occur in collections of documents. Typically, a document contains multiple topics in various proportions, i.e., each document may be viewed as a mixture of various topics. The most well-known approach that was able to detect not only topics but also sentiments in reviews was introduced by [63]. In their approach they try to identify the topics in an article, associate each topic with sentiment polarities, and model each topic with its corresponding sentiments by using a topic-sentiment-mixture model. They are based on the assumption that a document can contain different topics and each topic consists of different sentiments using multinomial distributions. In the work of [54] a joint model of sentiment and topics (JST) is proposed. In this model there are distinct topic and sentiment labels. The model is based on Latent Dirichlet Allocation (LDA) [9] and is evaluated on the movie review dataset. With the help of Dirichlet distributions, two latent variables are integrated and the generative process is as follows: for each document choose a distribution with parameter 'γ', for each sentiment label choose a distribution with parameter 'α'. Then, for each word in the document, choose a sentiment label, a topic and a word from the distribution over words defined by the topic and the sentiment label. This model is completely unsupervised and does not require labeled data.

Another model proposed by [38] is an aspect-sentiment model. It analyzes the problem of how sentiments are expressed towards various aspects. Specifically, an LDA model is adapted to match the granularity of the discovered topics to the details of the reviews. In addition, they incorporate sentiment into a unified model (ASUM), that incorporates both aspect and sentiment. ASUM starts from a small set of sentiment priors and finds sentiment words related to specific aspects. The resulting language models represent the probability distributions over words for various pairs of aspect and sentiment. The proposed technique is applied over reviews of electronic devices and restaurants. In [113] they are able to combine topic models and sentiment analysis for the Twitter domain. First they build a baseline system that utilizes a rich set of features and lexica. The baseline is used to estimate the class probabilities for each unlabeled tweet (this step occurs only once). After this step, a subset of tweets with class probability higher than a confidence threshold, is selected and included in the labeled training set. Then they identify the topics for tweets using the aforementioned labeled set. The topic information is generated through topic modeling based on LDA. Once they topics are identified, they data is split into multiple subsets (clusters) based on topic distributions. For each cluster, a separate sentiment model is trained. The resulting sentiment mixture model is used to classify the unlabeled tweets. The process is iterative, and runs until a certain number of iterations has been reached or no more tweets have been promoted to the set of labeled tweets.

A recent study uses a deep learning system for topic-based sentiment analysis, with a context-aware attention mechanism utilizing the topic information [7]. This work uses a bidirectional Long short-term memory (LSTM) with shared weights to map the words of the tweet and the topic to the same vector space, in order to be able to make meaningful comparison between the two. The details behind neural network models and the attention mechanisms will be discussed in the next section.

13.8 Deep Learning for Sentiment Analysis

With the aid of Deep Learning algorithms, learning from large amounts of unsupervised data has become feasible [35]. Deep learning models have recently become very popular for sentiment analysis. We have to note that deep learning requires a mathematical and conceptual background covering relevant concepts in linear algebra, probability theory and information theory, numerical computation, and machine learning. Therefore this section, only presents existing, useful technology in the context of natural language processing. For more in-depth, general discussion of neural networks we refer the reader to [29] or [28]. We summarize the deep learning based techniques presented in this chapter, in Table 13.3.

13.8.1   Convolutional Neural Networks

We have already pointed out the importance of where in the sentence a word appears (for example consider the phrases not good and not bad). A successful computational model for sentiment analysis should be able to handle composition of concepts. The convolution-and-pooling (also called convolutional neural networks, or CNNs) architecture is an efficient and robust solution to this modeling problem. A convolutional neural network is designed to identify local sub-regions, which are called receptive fields, in a large structure, and combine them to produce a fixed size vector representation of the structure, capturing the aforementioned local aspects that are most informative for the prediction task at hand.

The models' initial step is to tokenize a sentence and create what is usually called in the literature, a *sentence matrix*, the rows of which are the word vector representations of each token. The word vectors are usually the outputs from training the Word2vec of Glove models which were presented in Sect. 13.5.3. We denote the dimensionality of the word vectors by d. The next step is to apply a non-linear (learned) function (also called "filter") using a k-word sliding window over the sentence. This function captures important properties of the words in the window. Then, a "pooling" [15] operation is used to combine the vectors resulting from the different windows into a single d-dimensional vector. Pooling takes the max or the average value observed in each of the d-channels over the different windows. The aim of the pooling operation is to allow the model to focus on the "strongest" information of the sentence, regardless of the individual tokens' location.

Numerous Convolutional network architectures have been introduced to the NLP community. For example, in [40], a Dynamic CNN (called DCNN) that consists of multiple convolutional layers is proposed. This model aims at representing sentences of varying length and uses as inputs word vectors, randomly initialized and induces a feature graph over the sentence that is capable of explicitly capturing short and long-range relations. This feature graph, resembling a syntactic parse tree, is able to capture word relations between noncontinuous phrases that are far apart in the input sentence. The proposed network is applied in various sentiment prediction tasks, including sentiment prediction in Twitter. A one-layer CNN architecture that employs word embeddings is proposed by [44] for sentence-level classification tasks. They experimented with several variants, namely CNN-rand (where word embeddings are randomly initialized), the CNN-static method (where the word embeddings are pre-trained to improve performance and fixed throughout the experiment),

Table 13.3 Summary of deep learning methods for sentiment classification (WE means word embeddings, CE is used to denote character embeddings)

Refs.	Year	Preprocessing	Semantic representation	Model	Attention	Evaluation dataset
[40]	2014	Minimal	WE	DCNN	No	Stanford Sentiment Treebank (SSTb), TREC questions dataset, Stanford Twitter Sentiment corpus (STS)
[44]	2014	Minimal	WE	CNN	No	Movie reviews, Stanford SST-1 and SST-2, Subjectivity dataset, TREC questions dataset, Customer reviews, MPQA
[87]	2014	Minimal	WE, CE	CNN	No	Stanford Sentiment Treebank (SSTb), Stanford Twitter Sentiment corpus (STS)
[89]	2015	Minimal	WE	CNN	No	Semeval-2015-Tasks A,B
[115]	2015	Minimal	One-hot encoding	char-CNN	No	Custom build large-scale datasets
[106]	2015	No	WE	LSTM	No	Stanford Twitter Sentiment corpus (STS), SemEval 2013
[105]	2016	No	WE	CNN-LSTM	No	Stanford Sentiment Treebank (SST), Chinese ValenceArousal Texts (CVAT)
[107]	2016	No	WE	LSTM	Yes	SemEval 2014 Task 4
[79]	2016	No	WE	BiLSTM	No	Movie Review, Stanford Sentiment Treebank (SST)
[114]	2017	Yes	WE	BiLSTM	Yes	Twitter
[55]	2017	No	WE	BiLSTM	Yes	MPQA corpus
[49]	2017	No	WE	Bi-TreeGRU	Yes	Stanford Sentiment Treebank (SST)
[7]	2017	Yes	WE	BiLSTM	Yes	SemEval-2017 Task 4

CNN-non-static (where word embeddings are pre-trained and fine-tuned) and the CNN-multichannel (where multiple variants of word embeddings are used).

Character-based networks are often useful for many tasks since a stream of characters builds up to formulate words. Those models offer more information about text structure without any knowledge on the syntactic or semantic structures of the language. Moreover, these approaches have the benefit of producing very small sized models, and being able to provide an embedding vector for every (unknown) word that may be encountered. The article in [115] explores character-level convolutional neural networks for text classification. They compare performance against traditional models such as bag of words, n-grams etc., on several large-scale datasets. A deep convolutional neural network that builds representations from characters up to sentences has also been reported in [87]. This network uses two layers that allow the manipulation of words and sentences of any size.

The recent advent of algorithms based on unsupervised pre-training [22, 23] have proven especially helpful in training deep architectures. For example in [89] the neural network is trained in three steps: (1) using unsupervised pre-training of word embeddings, (2) refinement of the embeddings using a weakly labeled corpus with emoticons, and (3) fine tuning the model with labeled data.

13.8.2 Recurrent Neural Networks (RNN) and Long Short-Term Memory Networks (LSTM)

Recurrent neural networks (RNNs), were first presented in 1990 [21] and try to address the problem of learning sequences (such as words (sequences of letters), sentences (sequences of words) and documents (sequenced of sentences)). They are a class of artificial neural networks that allow representing arbitrarily sized inputs in a fixed-size vector, while paying attention to the structured properties of the input. An RNN processes variable length input sequentially, in a way that resembles how humans do it. The connections between neurons form a directed cycle which allows information to be passed from one step of the network to the next. Their internal state, often referred as a "memory", allows them to process the sequence of inputs.

Researchers have developed many variations of the original RNN model, like for example the Bidirectional RNN [88] (based on the idea that the output at each time may not only depend on the previous elements in the sequence, but also on the next elements in the sequence) or Long Short-Term Memory network introduced in [36]. In theory, RNNs are absolutely capable of handling "long-term" dependencies. Sadly, in practice, RNNs don't seem to be able to do so. This problem was explored in depth in [8] where the fundamental aspects and reasons for this problem were analyzed. On the other hand, Long Short-Term Memory networks (LSTMs), which have been growing in popularity, are capable of learning the "long-term" dependencies.

As far as sentiment analysis is concerned, Wang [106] introduces an LSTM for Twitter sentiment classification which simulates the interactions of words during the compositional process. Multiplicative operations between word embeddings through gate structures are used to provide more flexibility and to produce better compositional results (compared to the additive ones of a simple recurrent neural network). The proposed architecture outperforms various classifiers and feature engineering approaches and the same time it is effective in dealing with negation.

In the context of affective text analysis, an interesting method is proposed in [105] which is a combination of a regional CNN model and an LSTM network, to predict the valence and arousal ratings of text. The proposed CNN, uses an individual sentence as a region, dividing an input text into several regions. This way the useful affective information in each region can be extracted and weighted according to its contribution to the valence-arousal prediction. Such information is sequentially integrated across the regions using the LSTM model. By combining the regional CNN and LSTM, both local (regional) information within sentences and long distance dependencies across sentences are considered in the prediction process. Though the variety of neural network models proposed have remarkably improved performance, they do not fully employ linguistic resources (e.g., sentiment lexicons, negation words, intensity words). In an attempt to model the linguistic role of a sentence, the work in [79], develops a simple sequence model with the aim to regularize the difference between the predicted sentiment distribution of the current position, and that of the previous or next positions (the current position in a sequence model encodes forward or backward contexts). The model is called linguistic regularized LSTM, and its performance is compared with many different neural network models. The results indicate that the model is able to capture the sentiment in sentences more accurately, by addressing it's linguistic role.

13.8.3 Attention Mechanism

The attention mechanism, a recent trend in deep learning models [83], allows a neural network to pay attention to a part of an input sequence, rather than encoding the full source sequence. The attention mechanism, initially introduced in [3] is inspired by the visual attention mechanism found in humans. Recent works have found that equipping an RNN or a CNN with an attention mechanism can achieve state-of-the-art results on a large number of NLP tasks.

A deep learning system for short-text sentiment analysis using an attention mechanism, in order to enforce the contribution of words that determine the sentiment of a message is presented in [7]. The main model is a bidirectional LSTM (BiLSTM) which has the ability to get word annotations that summarize the information from both directions of a sentence. Furthermore, the use of the attention mechanism aims to find the relative contribution (importance) of each word. The system ranked 1st in Subtask A of SemEval 2017 competition [85], and achieved very competitive results in the rest of the Subtasks.

Subsequent improvements of this line of research include the Structural Attention Neural Networks introduced by [49]. Their model expands the current recursive models by incorporating structural information around a node of a syntactic tree using both bottom-up and top-down information propagation. Basically it extracts informative nodes (structural attention) out of a syntactic tree and aggregates the representation of those nodes in order to form the sentence vector. This way the model is able to identify the most salient representations during the construction of the syntactic tree. The model uses word embeddings which are initialized using the Glove vectors and is evaluated on the task of sentiment classification of sentences sampled from movie reviews.

In the aspect-level sentiment classification field, the model in [107] also proposes an attention-based method. The model developed is an LSTM network which is able learn an embedding vector for each aspect (aspect embedding). The attention mechanism is designed in order to capture the key part of sentence in response to a given aspect. In the same manner, [114] proposed two attention-based bidirectional LSTMs to improve the classification performance while [55] extended the attention mechanisms by differentiating the attention obtained from the left context and the right context of a given target/aspect.

13.9   Conclusions

This contribution started by highlighting how much the interest for research in sentiment analysis has grown in the past few years. The complexity and challenges that exist were also pointed out. There's no denying that social media has a significant impact on the way we communicate nowadays. They expose our language to a constant state of alternation and regeneration. The daily challenges regarding sentiment analysis mainly focus on this constant evolution of the language in user-generated content. To survive in this type of environment, it has become important for sentiment analysis systems to perform deeper semantic processing. Natural language processing techniques can help in this direction with a drive towards more sophisticated and powerful algorithms.

In this chapter a survey of the different approaches motivations and resources (lexical and corpus based) were provided, highlighting how semantics are being modeled and used in this emerging and growing research field. This survey also discussed how sentiment analysis has expanded to include affect-related concepts, which are increasingly becoming more important in the processing and exploration of the meaning of words. Machine learning methods try to leverage rich set of features and apply sophisticated algorithms, while lexicon based approaches rely on the creation of polarity lexicons that represent rich and varied lexical knowledge of emotions hidden in text. On the other hand, hybrid approaches and the ensemble method are aiming at more impressive outcomes.

Deep learning has emerged as a powerful machine learning technique, capable of producing state-of-the-art prediction results. Deep learning methods use different techniques and algorithms than traditional machine learning methods. As we can see from Tables 13.2 and 13.3, the majority of traditional machine learning models incorporate a preprocessing step, while this is not the case for neural based models. It is also clear that word embeddings played an important role in deep learning based sentiment analysis models. But even without the use of deep learning, word embeddings are often used as features in non-neural machine learning models. Traditional approaches are based on a variety of manually extracted features and lexica, and this extraction process can yield strong baselines. The predictive capabilities of traditional models are also used in conjunction with the deep learning methods with the ability to surpass performance for the task at hand. In the future, we hope that neural models will also benefit from the exploitation of affect in language leading to even stronger data representations.

Even though significant progress has been made over the past few years there is still a lack of resources and researches concerning languages other than English, despite the fact that a great share of data is available. New models that leverage the power of word embeddings, Transfer learning [98] and Neural Networks can provide powerful end-to-end Sentiment Analysis pipelines which can also be language agnostic. Furthermore, one cannot ignore the fact that sentiment techniques are commonly based on written language. In fact, data comes in different forms including images, and videos which are also likely to express and convey peoples subtle feelings. Therefore, multimodal techniques are going to bring new opportunities for sentiment analysis research in the near future [91].

Sentiment analysis unifies different research fields such as psychology, sociology, natural language processing, and machine learning. It is evolving rapidly from a very simple (positive, negative, neutral) to more granular and deeper understanding of emotions conveyed in text. The wealth of information on the Web has lead to the growth of sentiment analysis as one of the most active research areas of the last 10 years. There is still room for much more to come.

References

1. R. Astudillo, S. Amir, W. Ling, B. Martins, M.J. Silva, I. Trancoso, INESC-ID: sentiment analysis without hand-coded features or linguistic resources using embedding subspaces, in *Proceedings of the 9th International Workshop on Semantic Evaluation (SemEval)* (2015), pp. 652–656
2. S. Baccianella, A. Esuli, F. Sebastiani, Sentiwordnet 3.0: an enhanced lexical resource for sentiment analysis and opinion mining, in *LREC*, vol. 10 (2010), pp. 2200–2204
3. D. Bahdanau, K. Cho, Y. Bengio, Neural machine translation by jointly learning to align and translate (2014). arXiv:1409.0473
4. C. Banea, R. Mihalcea, J. Wiebe, Porting multilingual subjectivity resources across languages. IEEE Trans. Affect. Comput. **4**(2), 211–225 (2013)
5. F. Barbieri, H. Saggion, F. Ronzano, Modelling sarcasm in twitter, a novel approach, in *Association for Computational Linguistics* (2014), pp. 50–58
6. M. Baroni, G. Dinu, G. Kruszewski, Don't count, predict! a systematic comparison of context-counting vs. context-predicting semantic vectors (2014)
7. C. Baziotis, N. Pelekis, C. Doulkeridis, Datastories at semeval-2017 task 4: deep LSTM with attention for message-level and topic-based sentiment analysis, in *Proceedings of the 11th International Workshop on Semantic Evaluation (SemEval-2017)*, pp. 747–754 (2017)

8. Y. Bengio, P. Simard, P. Frasconi, Learning long-term dependencies with gradient descent is difficult. IEEE Trans. Neural Netw. **5**(2), 157–166 (1994)
9. D.M. Blei, A.Y. Ng, M.I. Jordan, Latent Dirichlet allocation. J. Mach. Learn. Res. **3**(Jan), 993–1022 (2003)
10. M. Bradley, P. Lang, Affective norms for english words (ANEW): instruction manual and affective ratings. Technical report, (1999)
11. M.H.M.P.M. Büchner, B. Stein, Webis: an ensemble for twitter sentiment detection (2015)
12. Q. Cao, W. Duan, Q. Gan, Exploring determinants of voting for the helpfulness of online user reviews: a text mining approach. Decis. Support Syst. **50**(2), 511–521 (2011)
13. P. Chikersal, S. Poria, E. Cambria, Sentu: sentiment analysis of tweets by combining a rule-based classifier with supervised learning, in *Proceedings of the 9th International Workshop on Semantic Evaluation (SemEval 2015)* (2015), pp. 647–651
14. K.W. Church, P. Hanks, Word association norms, mutual information, and lexicography. Comput. Linguist. **16**(1), 22–29 (1990)
15. R. Collobert, J. Weston, L. Bottou, M. Karlen, K. Kavukcuoglu, P. Kuksa, Natural language processing (almost) from scratch. J. Mach. Learn. Res. **12**(Aug), 2493–2537 (2011)
16. C. Cortes, V. Vapnik, Support-vector networks. Mach. Learn. **20**(3), 273–297 (1995)
17. W. Dai, G.R. Xue, Q. Yang, Y. Yu, Transferring naive bayes classifiers for text classification (2007)
18. S.R. Das, M.Y. Chen, Yahoo! for Amazon: Sentiment extraction from small talk on the Web. Manag. Sci. **53**(9), 1375–1388 (2007)
19. S. Deerwester, S.T. Dumais, G.W. Furnas, T.K. Landauer, R. Harshman, Indexing by latent semantic analysis. J. Am. Soc. Inf. Sci. **41**(6), 391 (1990)
20. L. Deng, J. Wiebe, Mpqa 3.0: an entity/event-level sentiment corpus, in *Proceedings of the 2015 Conference of the North American Chapter of the Association for Computational Linguistics: Human Language Technologies* (2015), pp. 1323–1328
21. J.L. Elman, Finding structure in time. Cogn. Sci. **14**(2), 179–211 (1990)
22. D. Erhan, P.A. Manzagol, Y. Bengio, S. Bengio, P. Vincent, The difficulty of training deep architectures and the effect of unsupervised pre-training. AISTATS **5**, 153–160 (2009)
23. D. Erhan, Y. Bengio, A. Courville, P.A. Manzagol, P. Vincent, S. Bengio, Why does unsupervised pre-training help deep learning? J. Mach. Learn. Res. **11**(Feb), 625–660 (2010)
24. M. Fernández-Gavilanes, T. Álvarez-López, J. Juncal-Martínez, E. Costa-Montenegro, F.J. González-Castaño, Unsupervised method for sentiment analysis in online texts. Expert. Syst. Appl. **58**, 57–75 (2016)
25. G. Frege, Compound thoughts. Mind **72**(285), 1–17 (1963)
26. X. Glorot, A. Bordes, Y. Bengio, Domain adaptation for large-scale sentiment classification: a deep learning approach, in *Proceedings of the 28th International Conference on Machine Learning (ICML-11)* (2011), pp. 513–520
27. A. Go, R. Bhayani, L. Huang, Twitter sentiment classification using distant supervision. CS224N Project Report, Stanford **1**(12) (2009)
28. Y. Goldberg, A primer on neural network models for natural language processing. J. Artif. Intell. Res. (JAIR) **57**, 345–420 (2016)
29. I. Goodfellow, Y. Bengio, A. Courville, Y. Bengio, *Deep Learning*, vol. 1. MIT Press, Cambridge (2016)
30. P.A. Gutierrez, M. Perez-Ortiz, J. Sanchez-Monedero, F. Fernandez-Navarro, C. Hervas-Martinez, Ordinal regression methods: survey and experimental study. IEEE Trans. Knowl. Data Eng. **28**(1), 127–146 (2016)
31. Z.S. Harris, Distributional structure. Word **10**(2–3), 146–162 (1954)
32. V. Hatzivassiloglou, K.R. McKeown, Predicting the semantic orientation of adjectives, in *Proceedings of the 35th Annual Meeting of the Association for Computational Linguistics and Eighth Conference of the European Chapter of the Association for Computational Linguistics* (1997), pp. 174–181
33. V. Hatzivassiloglou, J.M. Wiebe, Effects of adjective orientation and gradability on sentence subjectivity, in *Proceedings of the 18th Conference on Computational Linguistics, Association for Computational Linguistics*, vol. 1 (2000), pp. 299–305

34. M.A. Hearst, Direction-based text interpretation as an information access refinement, in *Text-Based Intelligent Systems: Current Research and Practice in Information Extraction and Retrieval* (1992), pp. 257–274
35. G.E. Hinton, S. Osindero, Y.W. Teh, A fast learning algorithm for deep belief nets. Neural Comput. **18**(7), 1527–1554 (2006)
36. S. Hochreiter, J. Schmidhuber, Long short-term memory. Neural Comput. **9**(8), 1735–1780 (1997)
37. M. Hu, B. Liu, Mining and summarizing customer reviews, in *Proceedings of the Tenth ACM SIGKDD International Conference on Knowledge Discovery and Data Mining* (ACM, 2004), pp. 168–177
38. Y. Jo, A.H. Oh, Aspect and sentiment unification model for online review analysis, in *Proceedings of the fourth ACM International Conference on Web Search and Data Mining* (ACM, 2011), pp. 815–824
39. A. Jurek, M.D. Mulvenna, Y. Bi, Improved lexicon-based sentiment analysis for social media analytics. Secur. Inform. **4**(1), 9 (2015)
40. N. Kalchbrenner, E. Grefenstette, P. Blunsom, A convolutional neural network for modelling sentences (2014). arXiv:1404.2188
41. F.H. Khan, U. Qamar, S. Bashir, Swims: semi-supervised subjective feature weighting and intelligent model selection for sentiment analysis. Knowl.-Based Syst. **100**, 97–111 (2016)
42. M. Khodak, N. Saunshi, K. Vodrahalli, A large self-annotated corpus for sarcasm (2017). arXiv:1704.05579
43. A. Kilgarriff, *Wordnet: An Electronic Lexical Database* (2000)
44. Y. Kim, Convolutional neural networks for sentence classification (2014). arXiv:1408.5882
45. S. Kiritchenko, X. Zhu, C. Cherry, S. Mohammad, NRC-Canada-2014: detecting aspects and sentiment in customer reviews, in *Proceedings of the 8th International Workshop on Semantic Evaluation (SemEval 2014)* (2014), pp. 437–442
46. S. Kiritchenko, X. Zhu, C. Cherry, S. Mohammad: NRC-Canada-2014: detecting aspects and sentiment in customer reviews, in *Proceedings of the 8th International Workshop on Semantic Evaluation (SemEval 2014), Association for Computational Linguistics and Dublin City University, Dublin, Ireland* (2014), pp. 437–442, http://www.aclweb.org/anthology/S14-2076
47. S. Kiritchenko, X. Zhu, S.M. Mohammad, Sentiment analysis of short informal texts. J. Artif. Intell. Res. **50**, 723–762 (2014)
48. J. Kittler, M. Hatef, R.P. Duin, J. Matas, On combining classifiers. Pattern Anal. Mach. Intell. **20**(3), 226–239 (1998)
49. F. Kokkinos, A. Potamianos, Structural attention neural networks for improved sentiment analysis (2017). arXiv:1701.01811
50. A. Kolovou, F. Kokkinos, A. Fergadis, P. Papalampidi, E. Iosif, N. Malandrakis, E. Palogiannidi, H. Papageorgiou, S. Narayanan, A. Potamianos, Tweester at SemEval-2017 Task 4: Fusion of Semantic-Affective and pairwise classification models for sentiment analysis in twitter, in *Proceedings of the 11th International Workshop on Semantic Evaluation (SemEval-2017)* (2017), pp. 675–682
51. A. Krouska, C. Troussas, M. Virvou, The effect of preprocessing techniques on twitter sentiment analysis, in *2016 7th International Conference on Information, Intelligence, Systems & Applications (IISA)* (IEEE, 2016), pp. 1–5
52. A. Krouska, C. Troussas, M. Virvou, Comparative evaluation of algorithms for sentiment analysis over social networking services. J. Univers. Comput. Sci. **23**(8), 755–768 (2017)
53. Q. Le, T. Mikolov, Distributed representations of sentences and documents, in *International Conference on Machine Learning* (2014), pp. 1188–1196
54. C. Lin, Y. He, Joint sentiment/topic model for sentiment analysis, in *Proceedings of the 18th Conference on Information and Knowledge Management (CIKM)* (2009), pp. 375–384
55. J. Liu, Y. Zhang, Attention modeling for targeted sentiment, in *Proceedings of the 15th Conference of the European Chapter of the Association for Computational Linguistics: Volume 2, Short Papers*, vol. 2 (2017), pp. 572–577

56. B. Liu, M. Hu, J. Cheng, Opinion observer: analyzing and comparing opinions on the web, in *Proceedings of the 14th international conference on World Wide Web* (ACM, 2005), pp. 342–351
57. A.L. Maas, R.E. Daly, P.T. Pham, D. Huang, A.Y. Ng, C. Potts, Learning word vectors for sentiment analysis, in *Proceedings of the 49th Annual Meeting of the Association for Computational Linguistics: Human Language Technologies, Association for Computational Linguistics*, vol. 1 (2011), pp. 142–150
58. C. Mair, Collins cobuild english language dictionary (1988)
59. N. Malandrakis, S.S. Narayanan, Therapy language analysis using automatically generated psycholinguistic norms, in *Sixteenth Annual Conference of the International Speech Communication Association* (2015)
60. N. Malandrakis, A. Potamianos, E. Iosif, S. Narayanan, Emotiword: affective lexicon creation with application to interaction and multimedia data, in *Internatinal Workshop on computational Intelligence for Multimedia Understanding* (Springer, 2011), pp. 30–41
61. N. Malandrakis, M. Falcone, C. Vaz, J. Bisogni, A. Potamianos, S. Narayanan, SAIL: sentiment analysis using semantic similarity and contrast features, in *Proceedings of the 8th International Workshop on Semantic Evaluation (SemEval)* (2014), pp. 512–516
62. N. Malandrakis, A. Potamianos, K.J. Hsu, K.N. Babeva, M.C. Feng, G.C. Davison, S. Narayanan, Affective language model adaptation via corpus selection, in *2014 IEEE International Conference on Acoustics, Speech and Signal Processing (ICASSP)* (IEEE, 2014), pp. 4838–4842
63. Q. Mei, X. Ling, M. Wondra, H. Su, C. Zhai, Topic sentiment mixture: modeling facets and opinions in weblogs, in *Proceedings of the 16th International Conference on World Wide Web (ICWWW)* (2007), pp. 171–180
64. T. Mikolov, K. Chen, G. Corrado, J. Dean, Efficient estimation of word representations in vector space (2013)
65. T. Mikolov, I. Sutskever, K. Chen, G.S. Corrado, J. Dean, Distributed representations of words and phrases and their compositionality, in *Proceedings of Advances in Neural Information Processing systems (NIPS)* (2013), pp. 3111–3119
66. S.M. Mohammad, P.D. Turney, Crowdsourcing a word-emotion association lexicon. Comput. Intell. **29**(3), 436–465 (2013)
67. S.M. Mohammad, S. Kiritchenko, X. Zhu, NRC-Canada: building the state-of-the-art in sentiment analysis of tweets, in *Proceedings of the seventh international workshop on Semantic Evaluation Exercises (SemEval-2013). Atlanta, Georgia, USA* (2013)
68. S.M. Mohammad, X. Zhu, S. Kiritchenko, J. Martin, Sentiment, emotion, purpose, and style in electoral tweets. Inf. Process. Manag. **51**(4), 480–499 (2015)
69. S.M. Mohammad, F. Bravo-Marquez, M. Salameh, S. Kiritchenko, Semeval-2018 Task 1: affect in tweets, in *Proceedings of International Workshop on Semantic Evaluation (SemEval-2018). New Orleans, LA, USA* (2018)
70. P. Nakov, A. Nakov, S. Rosenthal, F. Sebastiani, V. Stoyanov, Semeval-2016 task 4: sentiment analysis in twitter, in *Proceedings of the 10th International Workshop on Semantic Evaluation (SemEval-2016)* (2016), pp. 1–18
71. F.Å. Nielsen, A new ANEW: evaluation of a word list for sentiment analysis in microblogs, in *Proceedings of the ESWC Workshop on Making Sense of Microposts* (2011), pp. 93–98
72. S. Oraby, V. Harrison, L. Reed, E. Hernandez, E. Riloff, M. Walker, Creating and characterizing a diverse corpus of sarcasm in dialogue (2017). arXiv:1709.05404
73. A. Pak, P. Paroubek, Twitter for sentiment analysis: when language resources are not available, in *2011 22nd International Workshop on Database and Expert Systems Applications (DEXA)* (IEEE, 2011), pp. 111–115
74. E. Palogiannidi, A. Kolovou, F. Christopoulou, E. Iosif, N. Malandrakis, H. Papageorgiou, S. Narayanan, A. Potamianos, Tweester at SemEval 2016: sentiment analysis in twitter using semantic-affective model adaptation, in *Proceedings of the 10th International Workshop on Semantic Evaluation (SemEval)* (2016)

75. B. Pang, L. Lee, A sentimental education: sentiment analysis using subjectivity summarization based on minimum cuts, in *Proceedings of the 42nd Annual Meeting on Association for Computational Linguistics* (2004), p. 271
76. V.M.K. Peddinti, P. Chintalapoodi, Domain adaptation in sentiment analysis of twitter, in *Proceedings of the Fifth AAAI Conference on Analyzing Microtext* (AAAI Press, 2011), pp. 44–49
77. J.W. Pennebaker, M.E. Francis, R.J. Booth, *Linguistic Inquiry and Word Count: LIWC 2001*, vol. 71 (Lawrence Erlbaum Associates, Mahway, 2001)
78. J. Pennington, R. Socher, C.D. Manning, Glove: global vectors for word representation, in *Empirical Methods in Natural Language Processing (EMNLP)* (2014), pp. 1532–1543. http://www.aclweb.org/anthology/D14-1162
79. Q. Qian, M. Huang, J. Lei, X. Zhu, Linguistically regularized LSTMs for sentiment classification (2016). arXiv:1611.03949 (2016)
80. G. Qiu, X. He, F. Zhang, Y. Shi, J. Bu, C. Chen, Dasa: dissatisfaction-oriented advertising based on sentiment analysis. Expert. Syst. Appl. **37**(9), 6182–6191 (2010)
81. J. Read, J. Carroll, Weakly supervised techniques for domain-independent sentiment classification, in *Proceedings of the 1st International CIKM Workshop on Topic-Sentiment Analysis for Mass Opinion* (ACM, 2009), pp. 45–52
82. A. Reyes, P. Rosso, T. Veale, A multidimensional approach for detecting irony in twitter. Lang. Resour. Eval. **47**(1), 239–268 (2013)
83. T. Rocktäschel, E. Grefenstette, K.M. Hermann, T. Kočiský, P. Blunsom, Reasoning about entailment with neural attention (2015). arXiv:1509.06664
84. S. Rosenthal, P. Nakov, S. Kiritchenko, S. Mohammad, A. Ritter, V. Stoyanov, Semeval-2015 task 10: sentiment analysis in twitter, in *Proceedings of the 9th International Workshop on Semantic Evaluation (SemEval 2015)* (2015), pp. 451–463
85. S. Rosenthal, N. Farra, P. Nakov, Semeval-2017 task 4: sentiment analysis in twitter, in *Proceedings of the 11th International Workshop on Semantic Evaluation (SemEval-2017)* (2017), pp. 502–518
86. J.A. Russell, A. Mehrabian, Evidence for a three-factor theory of emotions. J. Res. Pers. **11**(3), 273–294 (1977)
87. C.N. dos Santos, M. Gatti, Deep convolutional neural networks for sentiment analysis of short texts, in *COLING* (2014), pp. 69–78
88. M. Schuster, K.K. Paliwal, Bidirectional recurrent neural networks. IEEE Trans. Signal Process. **45**(11), 2673–2681 (1997)
89. A. Severyn, A. Moschitti, Unitn: training deep convolutional neural network for twitter sentiment classification, in *Proceedings of the 9th International Workshop on Semantic Evaluation (SemEval 2015)* (2015), pp. 464–469
90. R. Socher, A. Perelygin, J. Wu, J. Chuang, C.D. Manning, A. Ng, C. Potts, Recursive deep models for semantic compositionality over a sentiment treebank, in *Proceedings of the 2013 Conference on Empirical Methods in Natural Language Processing* (2013), pp. 1631–1642
91. M. Soleymani, D. Garcia, B. Jou, B. Schuller, S.F. Chang, M. Pantic, A survey of multimodal sentiment analysis. Image Vis. Comput. **65**, 3–14 (2017)
92. P.J. Stone, D.C. Dunphy, M.S. Smith, The general inquirer: a computer approach to content analysis (1966)
93. C. Strapparava, A. Valitutti et al., Wordnet affect: an affective extension of wordnet, in *LREC*, vol. 4 Citeseer (2004), pp. 1083–1086
94. M. Taboada, J. Brooke, M. Tofiloski, K. Voll, M. Stede, Lexicon-based methods for sentiment analysis. Comput. Linguist. **37**(2), 267–307 (2011)
95. D. Tang, F. Wei, N. Yang, M. Zhou, T. Liu, B. Qin, Learning sentiment-specific word embedding for twitter sentiment classification, in *Proceedings of the 52nd Annual Meeting of the Association for Computational Linguistics (Volume 1: Long Papers)*, vol. 1 (2014), pp. 1555–1565
96. D. Tang, F. Wei, B. Qin, N. Yang, T. Liu, M. Zhou, Sentiment embeddings with applications to sentiment analysis. IEEE Trans. Knowl. Data Eng. **28**(2), 496–509 (2016)

97. I. Titov, R. McDonald, A joint model of text and aspect ratings for sentiment summarization, in *Proceedings of ACL-08: HLT* (2008), pp. 308–316
98. L. Torrey, J. Shavlik, Transfer learning (2009)
99. C. Troussas, A. Krouska, M. Virvou, Evaluation of ensemble-based sentiment classifiers for twitter data, in *2016 7th International Conference on Information, Intelligence, Systems & Applications (IISA)* (IEEE, 2016), pp. 1–6
100. S. Tulyakov, S. Jaeger, V. Govindaraju, D. Doermann, Review of classifier combination methods, in *Machine Learning in Document Analysis and Recognition* (Springer, 2008), pp. 361–386
101. P.D. Turney, Thumbs up or thumbs down? Semantic orientation applied to unsupervised classification of reviews, in *Proceedings of the 40th Annual Meeting on Association for Computational Linguistics* (2002), pp. 417–424
102. D.T. Vo, Y. Zhang, Target-dependent twitter sentiment classification with rich automatic features
103. B.C. Wallace, Sociolinguistically informed natural language processing: Automating irony detection. BROWN UNIV PROVIDENCE RI, Technical report (2015)
104. B. Wang, M. Liakata, A. Zubiaga, R. Procter, E. Jensen, Smile: twitter emotion classification using domain adaptation, in *25th International Joint Conference on Artificial Intelligence* (2016), p. 15
105. J. Wang, L.C. Yu, K.R. Lai, X. Zhang, Dimensional sentiment analysis using a regional CNN-LSTM model, in *Proceedings of the 54th Annual Meeting of the Association for Computational Linguistics (Volume 2: Short Papers)*, vol. 2 (2016), pp. 225–230
106. X. Wang, Y. Liu, S. Chengjie, B. Wang, X. Wang, Predicting polarities of tweets by composing word embeddings with long short-term memory, in *Proceedings of the 53rd Annual Meeting of the Association for Computational Linguistics and the 7th International Joint Conference on Natural Language Processing (Volume 1: Long Papers)*, vol. 1 (2015), pp. 1343–1353
107. Y. Wang, M. Huang, L. Zhao et al., Attention-based LSTM for aspect-level sentiment classification, in *Proceedings of the 2016 Conference on Empirical Methods in Natural Language Processing* (2016), pp. 606–615
108. C. Whissell, Using the revised dictionary of affect in language to quantify the emotional undertones of samples of natural language. Psychol. Rep. **105**(2), 509–521 (2009)
109. J.M. Wiebe, Identifying subjective characters in narrative, in *Proceedings of the 13th Conference on Computational Linguistics, Association for Computational Linguistics*, vol. 2 (1990), pp. 401–406
110. J. Wiebe, R. Mihalcea, *Word sense and subjectivity*, in *Proceedings of the 21st International Conference on Computational Linguistics and the 44th Annual Meeting of the Association for Computational Linguistics* (2006), pp. 1065–1072
111. J. Wiebe, T. Wilson, C. Cardie, Annotating expressions of opinions and emotions in language. Lang. Resour. Eval. **39**(2–3), 165–210 (2005)
112. T. Wilson, J. Wiebe, P. Hoffmann, Recognizing contextual polarity in phrase-level sentiment analysis, in *Proceedings of the Conference on Human Language Technology and Empirical Methods in Natural Language Processing (HLT/EMNLP)* (2005), pp. 347–354
113. B. Xiang, L. Zhou, T. Reuters, Improving twitter sentiment analysis with topic-based mixture modeling and semi-supervised training. ACL **2**, 434–439 (2014)
114. M. Yang, W. Tu, J. Wang, F. Xu, X. Chen, Attention based LSTM for target dependent sentiment classification (2017)
115. X. Zhang, J. Zhao, Y. LeCun, Character-level convolutional networks for text classification, in *Advances in Neural Information Processing Systems* (2015), pp. 649–657

Part VI
Learning and Analytics in Intelligent Imagery and Video

Chapter 14
Ship Detection Using Machine Learning and Optical Imagery in the Maritime Environment

Jeffrey W. Tweedale

Abstract Machine Learning (ML) is increasingly being used to enable machines to aid our understanding of digital imagery. Convolution Neural Network (CNN) models have become a common method of implementing ML. In this paper, five award winning CNN models have been assessed for their ability to identify ship types in images from remote sensors. The CNNs are trained with a collection of known ship images using supervised learning techniques, and, due to the low number of images available for training, Transfer Learning (TL) is adopted to take advantage of pre-trained low-level features.

Keywords Convolution neural network · Computer vision · Deep learning · Imagenet large scale visual recognition competition · Machine learning · Maritime vision processing · Transfer learning

14.1 Introduction

Machine Learning (ML) employs Artificial Intelligence (AI) techniques to enable machines to mimic the type of adaptive associations made by humans [1]. Although production systems [2] are still used to pre-process information, ML and Computer Vision (CV) applications access data from the environment, and use this to reason and make decisions. Complex systems are often combined to simultaneously provide reasoning, learning, planning, object recognition, and natural language understanding.

In the 1980's, research into ML investigated the performance of machines conducting tasks such as data representation, acquisition, categorization and classification [3]. ML now includes techniques that generate symbolic representations, Motion Imagery (MI) and CV. These techniques enable scientists to design machines with the ability to recognise and assess in dynamic environments [4]. When Krizhevsky

J. W. Tweedale (✉)
Defence Science and Technology Group, Edinburgh, SA 5111, Australia
e-mail: Jeffrey.Tweedale@dst.defence.gov.au

© Springer Nature Switzerland AG 2019
G. A. Tsihrintzis et al. (eds.), *Machine Learning Paradigms*,
Learning and Analytics in Intelligent Systems 1,
https://doi.org/10.1007/978-3-030-15628-2_14

introduced Convolution Neural Networks (CNNs) architectures into ML models, his team increased the recognition rate by an order of magnitude [5]. Intense competition followed, with researchers creating models that exceeded 98% classification accuracies.

Pre-trained CNN models can be re-used in new models and classify new labels (such as ships). In the maritime domain, there are about 190,000 registered ships above 100 tonnes in 21,943 categories,[1] making the categorisation task a difficult one. To reduce the size of the task, the focus of this chapter is on a particular class of military ship (frigate) with an aim to identify ships by type within this class.

This chapter describes the re-use of existing ML models to reduce the new training required to classify new objects. The discussion shows that Transfer Learning (TL) enables researchers to re-train existing CNN architectures to achieve the same level of accuracy using less data. The focus for continued research is on extending the capability of existing models to recognise ships in a maritime environment. Given more training data, a new research plan would be aimed at generating machine readable labels from visual scenes (metadata), enabling CNNs to learn or adapt within a given context and reduce communications bandwidth.

14.2 Background

The concepts associated with ML are well defined and new models are being introduced that increase the accuracy of image recognition. To save time and effort, TL is also being used to exploit trained knowledge. Here TL relates to the cognitive science definition introduced by Thorndike and Woodworth early last century [6]. When applied in the ML domain, the term TL relates to storing knowledge gained while solving one problem prior to applying it to help solve a different problem that contains related issues [7]. For this research, TL is considered to be a process that is used to exploit the embedded knowledge of an existing CNN in order to create an application capable of predicting ships on water. However, Huang et al. [8] recently investigated TL obtained using the ImageNet dataset to train the reconstruction pathway of auto-encoders used for classification tasks of Synthetic Aperture Radar (SAR) images.

At Defence Science and Technology Group (DST Group) the Tactical Systems Integration (TSI) Branch has been asked to assist Navy with the integration challenges of fitting an Uninhabited Aerial System (UAS) onto military ships. These are aimed at enhancing its Intelligence, Surveillance and Reconnaissance (ISR) capability. A key aspect of this integration is in the exploitation of video imagery from lower angles of observation (30–45°). Unfortunately video imagery tasks currently rely on human operators to engage in persistent visual target detection, classification, identification and tracking. They need to constantly monitor the observed scene. Similarly, the

---

[1] See https://www.ihs.com/products/maritime-world-ship-register.html.

same operator must accumulate visual cues over time in order to achieve higher-order functions, such as maintaining situation awareness and discerning patterns-of-life.

CV techniques have been used to enable land based analysts to identify objects with approximately 60% accuracy. The literature demonstrates that due to their increased accuracy, ML tools have evolved as the tool of choice for object recognition. For instance ML techniques now achieve 98% accuracy when recognising salient objects. To verify these claims, five award winning CNN candidates were chosen to explore the use of TL for maritime image recognition.

14.3 Research Activities

This chapter describes the research presented at the Knowledge-Based Intelligent Information and Engineering Systems (KES) Smart Digital Futures international conference on Intelligent Decision Technology (IDT) [9], and discusses further research on the evolution of vision processing techniques and ML. The experiments conducted for that paper explored five state-of-the-art applications of TL and confirmed that they constitute a feasible methodology to classify ships in images. In this case, all models leverage off of existing knowledge learned from the same dataset.

Two datasets were considered (Pattern Analysis Statistical modelling and Computational Learning (PASCAL) Visual Object Classes (VOC) and ImageNet). The PASCAL VOC resource was established in Europe in 2005. The project contained annotated images that provided the vision processing algorithms used by the machine learning communities with this publicly available resource. At that time it contained 20,000 annotated images organised in twenty object classes. This dataset grew with the competition which ran between 2005–2012 [10]. ImageNet followed in 2010 with a set of over 60 million labelled high-resolution images organised in 22,000 categories. ImageNet was established by groups at Stanford and Princeton to support the ImageNet Large Scale Visual Recognition Competition (ILSVRC) (running from 2010–2018). Both datasets can be used to recognise the content of visual scenes and provide annotated labels of salient objects. This can include Neural Networks (NNs) that identify the object location within the image together with the locality of the scene presented. Researchers actively created algorithms to compete against each other to win challenges of the ILSVRC. Table 14.1 shows the statistics for the winning teams of the ILSVRC competition since 2012.

14.4 State of the Art Models

As stated above, there has been a steady improvement in the rate of recognition over the past decade. There are now a number of parallel research streams that support ML approaches to image processing (for instance, recognition, scene labelling and TL). In order to minimise training variability from different languages and tools, MatLab models for (AlexNet, VGG-16, VGG-19, GoogLeNet and SENet) were obtained.

Table 14.1 Winning models

Year	Model	Error rate [%]	Class accuracy [%]	Contributor(s)	Affiliation
2012	AlexNet [5]	15	85	Krizhevsky et al.	University of Toronto
2013	ZF Net [11]	11	89	Zeiler and Fergus	New York University
2014	VGG Net [12]	7	93	Simonyan and Zisserman	University of Oxford
2014	GoogLeNet [13]	7	93	Szegedy et al.	University of Oxford
2015	ResNet [14]	4	96	MRSA	Microsoft
2016	CUImage [15]	3	97	Trimps-Soushen	Hong Kong
2017	SENet [16]	2	98	Fu et al.	Momenta (Uni Oxford)

Other models were sought, but most of the promoted systems with closed Image Processing (IP) use commercially crowd-sourced activities to generate manually pre-processed data. For example, datasets are being provided via Amazon, Google, Facebook and Microsoft.

The award winning CNN models discussed in this chapter now employ Deep Learning (DL) techniques that can achieve results that surpass 98%. For instance, in 2012, AlexNet introduced the use of convolution layers to deliver a significant reduction in the classification error (as shown in Table 14.1). The error rate progressively reduced with Visual Geometry Group (VGG) and GoogLeNet in 2014, and ResNet in 2015. New entries were submitted in 2016–2017 with CUImage and SENet winning. More detail about each model is provided in Tweedale [9].

14.5   Transfer Learning Methodology

As indicated above, the term TL is borrowed from the cognitive sciences definition. For ML, the concept promotes the reuse of knowledge generated during a training session associated with another task. For instance, you need a large amount of labelled data is needed to create a classifier capable of isolating a specified salient object. At present, the process of TL employs supervised learning techniques using labelled data. For CV research, this predominantly relies on re-using successful CNN models, however Ganin and Lempitsky suggest that it is also possible to conduct unsupervised adaptation training with or without labels by employing back propagation techniques [17]. This approach assumes there is a large volume of data available for training. Unfortunately this is not the case for maritime images, especially when using sensors located at low angles of observation.

This chapter focuses on extending the capability of existing models to recognise ships in a maritime environment. For instance, using a pre-trained AlexNet CNN model to recognise ships. AlexNet was originally trained using ImageNet to classify 1,000 categories of object within a public scene. This involves training multiple convolution layers. The first layer is trained to recognise features, such as edges,

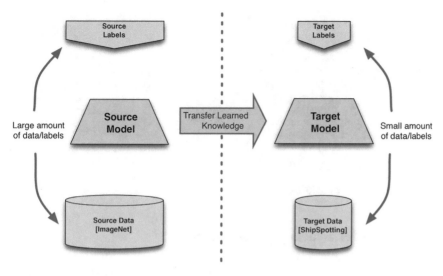

Fig. 14.1 Basic approach to transfer learning [19]

gradients, and solid colour areas. The second combines the data from the first layer to detect stripes, right angles, circles and other features. While the third and subsequent layers begin to recognise textures and groups of shapes until they are aggregated into people, animals, and other object classes. For ImageNet, the maritime image labels include: wreck, container ship, liner, fire boat, life boat, submarine, speed boat, aircraft carrier, pirate and paddle boat). By applying TL to AlexNet and using the existing *feature extraction* capabilities, it is possible to re-train and *fine-tune* the output layers to address new categories, such as frigate, destroyer and aircraft carrier. It would be easier to use divergent classes, however the experiments focused on differentiating only two classes (frigates and commercial vessels). The main objective was to distinguish between nine variants of ship within the frigate class.

This two step approach is shown in Fig. 14.1. Initially the network is trained with a large dataset to establish a set of filters and filter extraction networks. The classification layers are replaced and the network is again trained to establish links that recognise the class instances of the labels presented. This process transfers the existing knowledge available in the early layers of a pre-trained network [18, 19], like AlexNet or SENet, to augment the data used to create new classifiers that recognise ships. For instance, the training relies on the existing knowledge provided in the early layers (colours, shapes, textures and defining features) to create new links in the output layers that classify the labelled classes in the subsequent training data (in this case frigates).

Figure 14.2 shows the steps of TL when applied to a model's architecture. The final convolution and pooling layers of the original model are typically removed (shown on the left) to enable new layers to be added to the fully connected layer linking the previously trained feature layers (shown on the right). To ensure the TL process is robust and repeatable during experimentation, the same steps were applied to all models used.

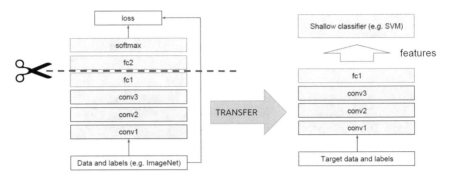

Fig. 14.2 Off the shelf approach transfer learning

14.6 Transfer Learning Experiments

The goal of these experiments was to determine if the output layers of one or more CNN models can be re-trained (using TL) to recognise ships. Experiments were conducted using MatLab to enable others access to repeat the process with their own dataset. Unfortunately ZF-Net, CUImage and SENet were not available in 'ConvNet' format for MatLab, therefore only AlexNet, GoogLeNet, VGG-16, VGG-19 and ResNet-50 were used. In each case, the original pre-trained model (architecture and weights that recognise 1,000 object classes) was loaded into memory and the first and last two neural network layers removed. These were replaced and were successfully re-trained using a collection of 6,365 ship images split into ten classes that were configured prior to employing TL across all five experiments (See Fig. 14.3). The class labels used and the number of images available included the Alvaro de Bazán (372), Anzac (428), De Zeven Provincien (393), F124 Sachsen (373), Halifax (777), La Fayette (320), Santa María (341), and UK Type 23 (1230). Commercial Ships (696) and a Destroyer (the Arleigh Burke with 1,435 images) as contrasting choices for the operators performing visual verification and validation. It is unclear how many images are required to provide robust recognition, however figures between 5,000 (with augmentation) and 20,000 images are stated in the literature [20]. The same hardware and software was used for all five experiments (one for each CNN model). Even with the low number of images and their raw state, the training results reflected those produced by the original CNN designs (Show in Table 14.1).

In the ML domain, TL was originally documented by Russakovsky and Fei-Fei [21]. In 2015, his team successfully demonstrated the re-use of a pre-trained CNN architectures to recognise new classification labels [22]. This proved that the number of parameters required to represent common constructs like colours, lines, shapes or related objects could be reused (at least given the same context).

The ImageNet dataset only contained images of 10 different classes of ship (wreck, container ship, liner, fire boat, life boat, submarine, speed boat, aircraft carrier, pirate and paddle boat). When recognising ships, few of these are relevant to navy and therefore a new dataset was created manually using images from the internet. Most

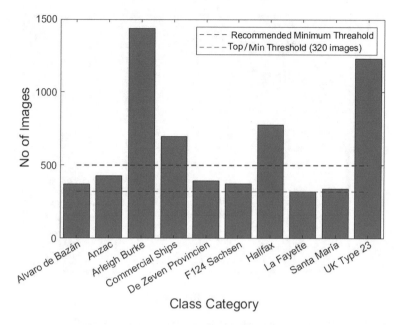

Fig. 14.3 Frequency distribution of training image dataset

of these were taken from a low angle to the horizontal and contained significant background noise (harbour or coasts). A total of 6,365 images were collected and divided into ten classes (eight Frigates, a Destroyer and one generalised background class of commercial vessels). Most of the existing datasets used for ML are pre-processed by humans prior to use. For example, backgrounds will be subtracted and images will be segmented to highlight the salient object. This process also ensures the ship becomes centred within the frame prior to being fed into the CNN. For our experiments, though, unprocessed ship images were used to train the models. With the release of new CNN models in 'ConvNet' format in MatLab in early 2018, two of the Inception models were also examined (Inceptionv2ResNet2 [23] and InceptionV3 [24]).

The results of a CNN are measured using either the percentage accuracy or loss value. For example, Fig. 14.4 displays the accuracy of each iteration plotted during the training run for each CNN. The results both Inception models have also been added for comparison [9]. The percentage accuracy value was plotted every 50 iterations. Only affected NN weights were updated each time an image was presented and the cycle repeated until the complete dataset was presented to the CNN. For training purposes, the complete cycle is labelled an epoch. In this case, 20 epochs are shown because two CNNs (both VGG models) required additional cycles to optimise the weighted values within the network. AlexNet, GoogLeNet and ResNet-50 only required 10 epochs to optimise their weighted values. These could have all been terminated at or prior to the fifth epoch. As suggested above, all of the training curves reported during the experiment closely matched to those generated by the original authors (rise-time, trend and characteristics).

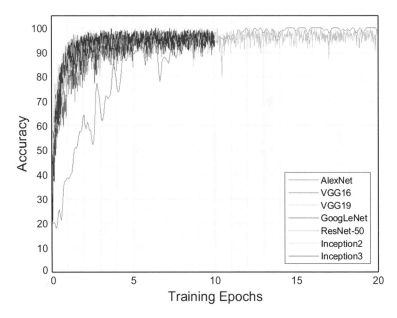

Fig. 14.4 Plot of training accuracy obtained using transfer learning (updated)

The training methodology used for each experiment was to isolate 5% of the 'Frigate Ships' dataset for validation and shuffle the image order within batches of 32 for the remaining images every epoch. The learning rate for all experiments was fixed at 0.001. This was reduced 'piecewise' each epoch, while L2 regularisation was set to 90% and momentum at $10\,e^{-4}$. Table 14.2 lists the results obtained during each experiment. The number of iterations for each experiment changed with respect to the weights in the model. The number of layers also varied, however only the convolution layers are trained as shown by the variation in network size in the last three columns. The classification results are encouraging and future experiments are planned to determine if synthetic sources can be used to enhance prediction or extend the scale. More research is required to enhance the classification accuracy.

Rather than repeating the predicted results for every test image in all five experiments, Fig. 14.5 shows a random selection of annotated test class images to demonstrate a two class prediction. To aid the explanation, black text on a yellow background is inserted at the top and bottom of each chip in the montage. The top message displays the original class label. The one at the bottom prepends a green tick (or red cross) to the NN confidence ranking. As shown, the predicted results produced in the two class experiment indicate a ranking that ranges between 96–100% (which is better than the training accuracy for the ResNet model used). A final validation was conducted by visually examining all of the predicted results for each experiment with a montage of the whole 'frigate ships' dataset. This process demonstrated the class with the most background clutter and the second least number of training images

Table 14.2 Experiment results (updated) [9]

Model name	Training iterations	Model layers	Conv layers	Accuracy [%]
AlexNet	1590	25	5	87.43
VGG-16	1273	41	16	75.78
VGG-19	1273	47	16	90.10
GoogLeNet	1297	144	53	91.35
ResNet	1790	177	54	91.04
Inception2	3580	825	207	92.60
Inception3	3580	316	84	98.84[a]

[a]This result was produced by removing the worst case class (the 'Santa Maria' with 38/51 predicted errors)

Fig. 14.5 Classification results following training using transfer learning [9]

produced the largest number of errors (where 38 of the 51 predicted errors were due to the presence of background clutter).

An alternative measure can be generated using a confusion matrix. Figure 14.6 displays a tabulated cross-reference of the results with individual classification accuracies shown with respect to all classes. This explains the classification issues with cross prediction between the Santa Maria and UK Type 23 classes. It is believed that more images from more angles with less noise and increased contrast are required to verify this concept.

Confusion Matrix

	Alvaro de Bazán	Anzac	Arleigh Burke	Commercial Ships	De Zeven Provincien	F124 Sachsen	Halifax	La Fayette	Santa María	UK Type 23	
Alvaro de Bazán	35 5.5%	0 0.0%	3 0.5%	0 0.0%	0 0.0%	0 0.0%	0 0.0%	1 0.2%	0 0.0%	3 0.5%	83.3% 16.7%
Anzac	0 0.0%	38 6.0%	1 0.2%	0 0.0%	0 0.0%	0 0.0%	0 0.0%	0 0.0%	1 0.2%	0 0.0%	95.0% 5.0%
Arleigh Burke	0 0.0%	0 0.0%	139 21.9%	0 0.0%	0 0.0%	0 0.0%	0 0.0%	0 0.0%	0 0.0%	0 0.0%	100% 0.0%
Commercial Ships	0 0.0%	0 0.0%	0 0.0%	70 11.0%	1 0.2%	0 0.0%	0 0.0%	0 0.0%	0 0.0%	0 0.0%	98.6% 1.4%
De Zeven Provincien	0 0.0%	2 0.3%	0 0.0%	0 0.0%	37 5.8%	0 0.0%	0 0.0%	0 0.0%	1 0.2%	1 0.2%	90.2% 9.8%
F124 Sachsen	1 0.2%	0 0.0%	0 0.0%	0 0.0%	0 0.0%	36 5.7%	1 0.2%	0 0.0%	2 0.3%	0 0.0%	90.0% 10.0%
Halifax	0 0.0%	1 0.2%	0 0.0%	0 0.0%	1 0.2%	1 0.2%	77 12.1%	0 0.0%	0 0.0%	2 0.3%	93.9% 6.1%
La Fayette	0 0.0%	0 0.0%	0 0.0%	0 0.0%	0 0.0%	0 0.0%	0 0.0%	31 4.9%	0 0.0%	0 0.0%	100% 0.0%
Santa María	0 0.0%	0 0.0%	0 0.0%	0 0.0%	0 0.0%	0 0.0%	0 0.0%	0 0.0%	6 0.9%	7 1.1%	46.2% 53.8%
UK Type 23	1 0.2%	2 0.3%	0 0.0%	0 0.0%	0 0.0%	0 0.0%	0 0.0%	0 0.0%	24 3.8%	110 17.3%	80.3% 19.7%
	94.6% 5.4%	88.4% 11.6%	97.2% 2.8%	100% 0.0%	94.9% 5.1%	97.3% 2.7%	98.7% 1.3%	96.9% 3.1%	17.6% 82.4%	89.4% 10.6%	91.0% 9.0%

Output Class (vertical axis) — Target Class (horizontal axis)

Fig. 14.6 Classification confusion matrix

The last form of verification is to examine the output layers of the CNN after they are trained. The process of visualising the output layers offers a pictorial view of robustness. As shown in Fig. 14.7, the definition of salient objects in the original output layers of the ImageNet trained CNN is far more pronounced than objects in the re-trained layer. For instance, Fig. 14.7a displays an aggregated set of features extracted in the output layer of the pre-trained network for the original 'Chicken' class (which is one of the 1,000 ImageNet labels). When visualised, the network reproduces what appears to represent a series of random 'Chicken' images in different locations and orientations. This clearly indicates that sufficient training has occurred. By contrast, Fig. 14.7b is one of the best ship object classes produced during the experiments. This shows the salient object is barely visible. Hence, by comparing the output layers of the original training effort with those of the re-trained networks, it is easy to see that more images are required.

During these experiments, each application of TL reflects the accuracy of the original training, however the definition of the output layers was less than recognisable.

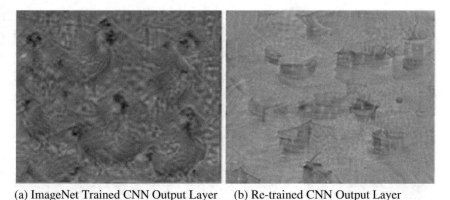

(a) ImageNet Trained CNN Output Layer   (b) Re-trained CNN Output Layer

Fig. 14.7 Visualisation example of one classification output layer class

Fig. 14.8 Montage10k0

It is believed that more images are required to enhance this definition. Augmentation techniques can be used to increase the dataset.

Most of the images are taken when ships are docked, capturing the hull with very little rotation in the Y axis. To accurately classify ships at sea within images taken from a UAS, images from all angles need to be used for training. That means images taken from 0–360° around the X axis and 0–90° around the Y axis (less than 45° may suffice). Finally changes in the Z axis could be considered if the sea state produces significant hull movement with respect to the horizon.

Brief experiments using synthetically generated images show that scenes with the ships oriented differently to those in the training set perform poorly. For instance, less than 30% produce accurate labels. As can be seen in Figs. 14.8 and 14.9, less than half display the hull and most of the images contain significant noise from waves and reflected light.

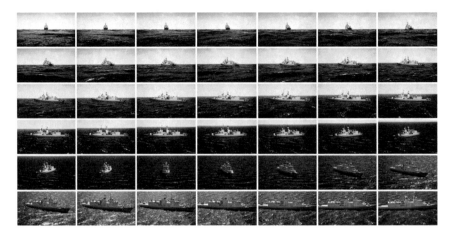

Fig. 14.9 Montage10k15

The results presented demonstrate there is room for improvement and further experiments are being conducted to evaluate enhanced performance with more recent models (like Inception, ResNet2 and NASNet). The TL experiments demonstrate that it is possible to transfer (retain) existing knowledge within CNN models and achieve similar results in the re-trained version. For instance the classification accuracies for each model measured during the experiments only varied 7.83% (or from −5.0% and +2.83% where 84.6% was the minimum and 98.84% maximum). This training verifies the original models and validates the ability to apply TL to recognise ships in the maritime domain.

14.7   Conclusions

This chapter highlights the successful use of TL when training a CNN to recognise military frigates (ships) in the maritime domain. Although existing research indicates that more labelled examples are required to successfully recognise objects in visual scenes, these experiments show encouraging classification results can be achieved using TL. The recognition rate indicates the need for more ship images to re-train the chosen network with an acceptable accuracy. For example, only two classes contained more than 1,000 images, while six contained less than 500 images. From the experiments even 1,435 images, the 'Arliegh Burke' class still reported errors. Each case is different, although suggestions range between 5,000–20,000 images per category. The evidence lies in the definition of the visualised output layers (as shown in Fig. 14.7). Noisy images can also increase prediction errors as demonstrated with the 'Santa Maria' class. Although the 'La Fayette' had fewer images, the 'Santa Maria' class had the worst visual definition and when removed the overall prediction accuracy rises from 91% to 96.86%.

Future effort should also be conducted to explore growing the dataset by augmenting the repository with modified versions of current images. This includes modifying the intensity, colour, cropping, shifting, flipping and rotating the original image. Other techniques can be used, however this author has already started investigating the ability to synthesise images using 3D computer generated renderings and taking pictures using calibrated camera runs from a variety of range and elevation settings in fixed steps around its centre.

References

1. M. Minsky, *Society of Mind* (Simon and Schuster, Pymble, Australia, 1985)
2. P.R. Thagard, *Computational Philosophy of Science* (MIT Press, 1993)
3. M. Tambe, D. Pynadath, C. Chauvat, C. Das, G. Kaminka, Adaptive agent architectures for heterogeneous team members, in *International Conference on Multi-Agents Systems (ICMAS2000), Boston, MA* (2000)
4. L. Galway, D. Charles, M. Black, Machine learning in digital games: a survey. Artif. Intell. Rev. **29**, 123–161 (2008).
5. A. Krizhevsky, I. Sutskever, G.E. Hinton, Imagenet classification with deep convolutional neural networks, in *Proceedings of the 25th International Conference on Neural Information Processing Systems. NIPS'12, USA* (Curran Associates Inc., 2012), pp. 1097–1105
6. E.L. Thorndike, R.S. Woodworth, The influence of improvement in one mental function upon the efficiency of other functions. Psychol. Rev. **8**(3), 247–261 (1901)
7. T. Dieterich, Special issue on inductive transfer. J. Mach. Learn. Kluwer Hingham, MA, USA **28**(1) (1997)
8. Z. Huang, Z. Pan, B. Lei, Transfer learning with deep convolutional neural network for SAR target classification with limited labeled data. Remote Sens. **9**(907), 1–21 (2017)
9. J.W. Tweedale, An application of transfer learning for maritime vision processing, in *KES Smart Innovation Systems and Technologies, Intelligent Decision Technologies* (Springer, 2018), pp. 87–97
10. M. Everingham, L. Van Gool, C.K.I. Williams, J. Winn, A. Zisserman, The pascal visual object classes (voc) challenge. Int. J. Comput. Vis. **88**(2), 303–338 (2010).
11. M.D. Zeiler, R. Fergus, Visualizing and understanding convolutional networks. CoRR (2013). arXiv:abs/1311.2901
12. K. Simonyan, A. Zisserman, Very deep convolutional networks for large-scale image recognition. CoRR **1409**(1556), 1–14 (2014)
13. C. Szegedy, W. Liu, Y. Jia, P. Sermanet, S.E. Reed, D. Anguelov, D. Erhan, V. Vanhoucke, A. Rabinovich, Going deeper with convolutions. CoRR (2014). arXiv:abs/1409.4842
14. K. He, X. Zhang, S. Ren, J. Sun, Deep residual learning for image recognition. CoRR (2015). arXiv:abs/1512.03385
15. X. Zeng, W. Ouyang, J. Yan, H. Li, T. Xiao, K. Wang, Y. Liu, Y. Zhou, B. Yang, Z. Wang, H. Zhou, X. Wang, Crafting GBD-net for object detection. IEEE Trans. Pattern Anal. Mach. Intell. **99**, 1 (2017)
16. J. Hu, L. Shen, G. Sun, Squeeze-and-excitation networks. CoRR (2017). arXiv:abs/1709.01507
17. Y. Ganin, V. Lempitsky, Unsupervised domain adaptation by backpropagation (2014), pp. 1–11
18. J. Yosinski, J. Clune, Y. Bengio, H. Lipson, How transferable are features in deep neural networks, in *Advances in Neural Information Processing Systems* (2014), p. 27
19. X. Giro-i-Nieto, E. Sayrol, A. Salvador, J. Torres, E. Mohedano, K. McGuinness, Transfer learning and domain adaptation. Deep Learning for Computer Vision, Summer Session, UniversityPolitecnica De Catalunya (2016), pp. 1–15

20. C. Sun, A. Shrivastava, S. Singh, A. Gupta, Revisiting unreasonable effectiveness of data in deep learning era. CoRR (2017), arXiv:1707.02968, pp. 1–13
21. O. Russakovsky, L. Fei-Fei, Attribute learning in large-scale datasets, in *Trends and Topics in Computer Vision*, ed. by K.N. Kutulakos. LNCS, vol. 6553 (Springer, Berlin, 2012), pp. 1–14
22. O. Russakovsky, J. Deng, H. Su, J. Krause, S. Satheesh, S. Ma, Z. Huang, A. Karpathy, A. Khosla, M. Bernstein, A.C. Berg, L. Fei-Fei, Imagenet large scale visual recognition challenge. Int. J. Comput. Vis. (IJCV) **115**(3), 211–252 (2015)
23. C. Szegedy, V. Vanhoucke, S. Ioffe, J. Shlens, Z. Wojna, Rethinking the inception architecture for computer vision. CoRR (2015). arXiv:1512.00567
24. C. Szegedy, S. Ioffe, V. Vanhoucke, Inception-v4, inception-resnet and the impact of residual connections on learning. Comput. Resour. Repos. Cornell Univ. **1602**(07261), 1–12 (2016)

Chapter 15
Video Analytics for Visual Surveillance and Applications: An Overview and Survey

Iyiola E. Olatunji and Chun-Hung Cheng

Abstract Owing to the massive amount of video data being generated as a result of high proliferation of surveillance cameras, the manpower to monitor such system is relatively expensive. Passively monitoring surveillance video however, incapacitates the usefulness of surveillance camera. Therefore, a drive to monitor events as they happen is expedient to fully harness the massive data generated by surveillance cameras. This is the main goal of video analytics. In this chapter, we extend the notion of surveillance. Surveillance refers not only to monitoring for security or safety purposes but encapsulates all aspects of monitoring to capture the dynamics of different application domains including retail, transportation, service industries and healthcare. This chapter presents a detailed survey of video analytics as well as its application. We present advances in video analytics research and emerging trends from subdomains such as behavior analysis, moving object classification, video summarization, object detection, object tracking, congestion analysis, abnormality detection and information fusion from multiple cameras. We also summarize recent development in video analytics and intelligent video systems (IVS). We evaluated the state-of-the-art approach to video analytics including deep learning approach and outlined research direction with emphasis on algorithm-based analytics and applications. Hardware-related issues are excluded from this chapter.

Keywords Video analytics · Intelligent video system · Video surveillance · Computer vision · Video survey and applications

I. E. Olatunji (✉) · C.-H. Cheng
Department of Systems Engineering and Engineering Management,
The Chinese University of Hong Kong, Sha Tin, Hong Kong
e-mail: iyiola@link.cuhk.edu.hk

C.-H. Cheng
e-mail: chcheng@se.cuhk.edu.hk

© Springer Nature Switzerland AG 2019   475
G. A. Tsihrintzis et al. (eds.), *Machine Learning Paradigms*,
Learning and Analytics in Intelligent Systems 1,
https://doi.org/10.1007/978-3-030-15628-2_15

15.1 Overview of Video Analytics

The widespread use of video cameras especially from mobile phones and low-cost high performance IP surveillance cameras are contributing substantially to the exponential growth of video data primarily used for surveillance. In China, over 170 million video surveillance cameras have been installed and expected to increase to over 600 million by 2020 [1].

Surveillance is usually synonymous to crime and intrusion detection due to the current increasing security issues in our society but we extend the ideology of surveillance to resource monitoring in this chapter.

Therefore, we define surveillance as the monitoring of people or object's activities and behavior for the purpose of intrusion detection, crime detection, scene monitoring and resource tracking.

In a traditional video surveillance system, an operator is assigned to actively watch videos captured by the cameras with the notion of tracking and detecting any suspicious persons or potentially dangerous abandoned object. However, this is quite unrealistic while considering the vast amount of cameras and several hours of recording currently available.

Subsequently, Caifeng et al. [2] has shown that the maximum attention span of any personnel monitoring a video task in 20 min.

Manually monitoring videos through personnel staring at several video screen for many hours is also relatively expensive and highly prone to errors causing several records of missed events. Therefore, a drive to automatically analyze these video data is of the essence. This is the goal of video analytics (VA); to ease the strenuous tasks of manually monitoring several hours of video, provide real-time alert when situation of interest occurs, and facilitate keyword search of enormous archives of video via automation of the task.

Video analytics or video content analysis refers to automatic processing and understanding of video content in order to determine or detect spatio-temporal events and extract information or knowledge about the observed scene. The generated video content can be from a single camera or multiple cameras. Video analytics is a constantly evolving field with novel techniques and algorithms continuously being developed in areas such as video semantic categorization, video retrieval from database, human action recognition, summarization and anomaly detection. Video analytics algorithms can either be implemented as a software package running in a centralized station where numerous servers are utilized for processing or as hardware on a dedicated video processing unit.

Sport, retail, automotive, transport, security, entertainment, traffic control including pedestrian crossing, digital real-time decision making devices, and healthcare are some of the several domains in which video analytics have been applied.

For example, the 2018 FIFA world cup took a new phase of monitoring football matches using videos to enhance referee's capability. The system is called Video Assistant Referee (VAR). VAR not only monitors the game but also analyzes player's performance in real-time. Another video system used for monitoring in the 2018 FIFA world cup is the Goal line technology (GLT). The Goal line technology is used to detect if the football passes the goal line which is usually difficult for linesman or assistant referee to see. These technologies driven by video surveillance on a football field provides the referee with sufficient information to make decision when dispute occur such as suspected penalty or a ferocious attack that deserves a yellow or red card.

For resource monitoring, a work by Cheng and Olatunji [3] showed that videos can be used in monitoring trolleys in an airport operation in real-time. This system provides an up-to-date inventory of the available resource (trolley) and significantly reduces replenishment time by about 70% due to its real-time alert system when there is shortage of resource.

For autonomous cars, obstacle detection and path planning are important for effective navigation. This is achieved by analyzing the data captured by Light Detection and Ranging (LiDAR) cameras or other 3D cameras installed in the car.

Cognitive factors such as attention span and time to react to an event can be investigated by analyzing the video content of an activity or event. Real-time situation awareness, target recognition, event prediction and prevention, and post-event data analysis are key goals of intelligent video system (IVS) fueled by video analytics.

Video analytics spans over broad application areas such as biometrics (face, tattoo, signature and iris recognition), detection and tracking (person, object, vehicle, abandoned object and logos), text recognition from video, search and retrieval of video content, geolocation and mapping, summarization and skimming, behavior analysis and event analysis.

Typically, videos are segmented into frames or set of still images. A single video camera can produce about 25–30 fps or more with 4 K and 3D video cameras which is equivalent to thousands or millions of frames depending on the sampling rate of the video. Similarly, it has been stated that video traffic will account for about 82% of the whole internet traffic [4]. Therefore, considering the volume and speed of data that can be generated by a single video data, algorithms to process the video must offer real-time or near real-time solutions.

Most of the existing IVS are based on centralized approach but there has been progressive research in the area of edge based architecture. Centralized approach involves routing videos to a base cloud or storage where all analytics take place whereas in edge-based architecture, video contents are analyzed near the source of data generation. Edge-based architecture provides better solution and avoids bandwidth and other cost associated with transporting data to the centralized station. However, this is beyond the scope of this chapter. Readers are referred to [5] for more information. Video analytics algorithm are parameter-oriented as they are key factors in determining the accuracy of the output. Such parameters include frame sampling, frame resolution and algorithmic parameters.

The most fundamental steps in automated video surveillance and monitoring (VSAM) is background image estimation and updating the background image to reflect changes in the background environment. However, since the emergence of deep learning and the breakthrough of convolutional neural network (CNN) in image, video and audio classification problems, there have been widespread adoption of

CNN especially in its application to large-scale video classification [6]. Due to high computational cost, graphics processing unit (GPU) are leveraged to run video-processing system's algorithms for effective performance gains by parallelizing a number of vision algorithms [7, 8]. However, computational details are excluded as they are not the focus of this chapter.

Video analytics systems based on deep learning models such as CNN forms the basis of state-of-the-art analytics systems applied in smart cities and real-time applications. The deep learning approach requires enormous amounts of data and training time to complete a task such as object segmentation, classification or detection.

Despite enormous effort in developing automated VSAM systems, current surveillance systems are not entirely capable of autonomously analyzing complex event from observed scene. To address this problem, independent work on several areas such as object tracking, behavior understanding, object classification, summarization and motion segmentation are combined to form a composite video analytic framework for video surveillance.

This paper presents a general overview of video analytics, current state-of-the-art method and integration of different aspect of video analytics algorithms to form an intelligent video system (IVS). The rest of the paper is described as follows: Sect. 2 discusses the theory of video analytics systems, more especially the deep learning approach which is the current state-of-the-art approach. Section 3 details survey of the algorithms and task involved in video analytics for surveillance. In Sect. 4, we discuss the application of analyzing video surveillance data in daily life operations. Section 5 presents research direction and concludes the paper.

15.2 Theory of Video Analytics

Video analytics have been a major area of research since the last few decades with several techniques been developed to overcome challenges such as accuracy and precision of analyzing video data. This section details the core concept of some early works and also recent works sprawled by deep learning. The core mathematical foundations discussed in this section has be applied to different application domain such as object tracking, segmentation and motion detection to name a few.

15.2.1 Background Initialization or Estimation

Background image estimation and updating the background image to reflect changes in the environment are the fundamental steps in automated VSAM. Several approach to background image estimation of moving object exist. They include pixel-based approach, block-based approach, neural network approach, and Gaussian Mixture Model (GMM). Readers are advised to read [9] for a comprehensive survey of scene background estimation.

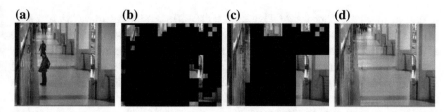

Fig. 1 Background estimation from a cluttered scene: **a** initial frame from video scene, **b–c** background initialization, **d** estimated background. Images adapted from [13]

Pixel-based approach extracts in-depth shape information of moving object by modelling each pixel separately. However, segmentation in pixel-based approach is highly susceptible to changing background. Block-based approach is less sensitive to changing background because the image is divided into blocks and the features extracted from these blocks are used for background modeling. However, it cannot obtain in-depth shape information of moving object like pixel-based method. To overcome the short coming of both methods, artificial neural network algorithm for background subtraction was proposed by Zhou et al. [10] to automatically extract background information and detect moving object based on the extracted background information. GMM requires large memory space and its convergence is slow. Ishii et al. [11] developed an online method for improving GMM where each Gaussian have a learning parameter that approach basic recursive learning after several observations by combining expectation maximization (EM) method with recursive filter observation. The drawback of the above method however is, their performance degrades when handing complex situations like different weather condition, camera instability and noise, and non-static background. Pham et al. [12] proposed Improved Adaptive Gaussian Mixture Model (IAGMM) to cater for some of the above problems. An example of background estimation is shown in Fig. 1.

Background subtraction is the fundamental method for dynamic image analysis. It models background for detecting moving objects by identifying statistically significant changes from a background model and separates it from foreground object.

However, when the camera is also in motion, it is computationally complex to model such backgrounds as consecutive frames cannot be compared. To solve this problem, Zhang et al. [14] used dense optical flow fields to estimate camera motion and segment the moving object over multiple frames. However, optical flow methods are prone to varying object boundaries. Gelgon and Bouthemy [15] combined color segmentation and motion-based regions to solve this problem but it is computationally expensive and not in real-time.

Angelov et al. [16] proposed a real-time and less computational expensive method for detecting moving object with moving camera. For Abandoned Object Detection (AOD), background subtraction and ground-truth homography were combined to detect objects and handle occlusion [17]. Several methods such as object tracking based method, stationary object detection algorithm, drop-off event detection algorithm, color and shape information of static foreground objects have been used either in part or fused together for AOD detection [18–20]. An interesting work by Guler [7] used two foreground images (one long-term and one short-term) to define AOD as a situation where an object's pixels are in the long-term foreground, but not in the short-term background. Ramirez-Alonso et al. [21] proposed a temporal weighted learning model for background estimation that can reinitialize and update parameters adaptively. Details of their proposed method is discussed below. There are 4 modules in the system with the first module handling scenes with static object. Module 2 deals with normal scenes that can contain dynamic object. In module 3, threshold is defined in order to separate background from foreground object and module 4 caters for dynamic scene changes using Speeded-Up Robust Features (SURF) algorithm to align information. Once the information about the background is aligned, the background model can be updated and background can be estimated.

Let $\mathbf{W}_{BS\_HLR}$ and $\mathbf{W}_{BS\_LLR}$ be the adaptive weight arrays updated by high and low parameters respectively, $\mathbf{L}_R$ be the learning rate values, $\mathbf{F}_{BE\_HLR}$ and $\mathbf{F}_{BE\_LLR}$ be the foreground models of the estimated background in the RGB color space. Classification of the foreground is performed by Eqs. (1) and (2):

$$\mathbf{F}_{BE\_HLR}(\mathbf{x}, t) = \begin{cases} 1 & \left\| \mathbf{W}_{BS\_HLR}(\mathbf{x}, t)^{RGB} - \mathbf{I}(\mathbf{x}, t)^{RGB} \right\|_2 > \varepsilon_1 \\ 0 & otherwise \end{cases} \tag{1}$$

$$\mathbf{F}_{BE\_LLR}(\mathbf{x}, t) = \begin{cases} 1 & \left\| \mathbf{W}_{BS\_LLR}(\mathbf{x}, t)^{RGB} - \mathbf{I}(\mathbf{x}, t)^{RGB} \right\|_2 > \varepsilon_2 \\ 0 & otherwise \end{cases} \tag{2}$$

where $\mathbf{x}$ is the pixel location $[x, y]$ of a M $\times$ N image size $(1 < x < M$ and $1 < y < N)$ and $t \in N$ is the time index. ε_1 and ε_2 are the threshold values. If the result of the Euclidean distance is more than the threshold values, the pixel is classified as foreground. A zero in the foreground indicates that the pixel is part of the background.

Weight are updated adaptively by Eq. (3)

$$\mathbf{W}_{BE}(\mathbf{x}, t + 1)^{RGB} = \mathbf{W}_{BE}(\mathbf{x}, t)^{RGB}(1 - \mathbf{L}_R(\mathbf{x}, t)) + \mathbf{L}_R(\mathbf{x}, t)\mathbf{I}(\mathbf{x}, t)^{RGB} \tag{3}$$

where $\mathbf{L}_R$ is the matrix holding the learning rate value for each weight defined by Eq. (4). The exponential factor in Eq. (4) generates fast learning in the initial frame. Therefore the value of t_a will be initialized to 0 and will increase by 1 if changes occur between consecutive input frames as given in Eq. (5).

$$\mathbf{L}_R(\mathbf{x}, t) = S_0^{\frac{-t_a}{T_f}} + A_0 A(\mathbf{x}, t) + L_0 \tag{4}$$

$$t_a = \begin{cases} t_a + 1 & \rho_v < 0.998 \\ t_a & otherwise \end{cases} \tag{5}$$

where ρ_v is the Pearson correlation coefficient between the input frame at time t and $t - 1$. L_0 is a classifier constraint whose value is < 0.1 and T_f is the number of frames. S_0 is the bootstrap learning constant and A_0 is the scaling factor for $A(\mathbf{x}, t)$. $A(\mathbf{x}, t)$ is the scene information matrix that defines the learning of each pixel and used for classifying pixels based on the values of $\mathbf{F}_{BE\_HLR}(\mathbf{x}, t)$ and $\mathbf{F}_{BE\_LLR}(\mathbf{x}, t)$. $A(\mathbf{x}, t)$ can be calculated as:

$$A(\mathbf{x}, t) = 1 - \frac{1}{2}\left[\mathbf{F}_{BE_{HLR}}(\mathbf{x}, t) + \mathbf{F}_{BE\_LLR}(\mathbf{x}, t)\right] \qquad (6)$$

when $A(\mathbf{x}, t) = 0$, pixels represent dynamic object or ghost detection. If $A(\mathbf{x}, t) = 0.5$, pixel is classified as a dynamic object only in one foreground model or ghost only in $\mathbf{F}_{BE\_LLR}$. If $A(\mathbf{x}, t) = 1$, then the pixel is classified as background in both foreground models.

15.2.2   State-of-the-Art Video Analysis Model

Handcrafted features such as Histogram Of Oriented Gradients (HOG) [22], Scale-invariant Feature Transform (SIFT) [23], Local Binary Pattern (LBP) [24], Local Ternary Patterns (LTP) [25], and Haar [26–28] have been used for video analysis and are usually considered as shallow network. These handcrafted features produce high dimensional feature vectors obtained by the aggregation of small local patches of subsequent video frames for motion and appearance information. However, these high dimensional feature vectors do not scale well for large scale video processing in uncontrolled environment.

Convolutional Neural Network (CNN) or ConvNet-based video analytics systems have shown superiority over shallow network based video analytic system in terms of precision and accuracy. Since the advent of CNN, several video classification methods [6, 29–31] based on CNN have been proposed to learn features from raw pixels but can only classify short video clips and still images. LeNet [32] was the first successful application of ConvNet for handwritten digit recognition which achieved an accuracy of 99.2% on the MNIST dataset [33] followed by deep CNN proposed by Alex Krizhevsky et al. [31] (AlexNet) for the ImageNet [34] dataset and achieved high accuracy. Several other methods have emerged since then. Dan Ciresan et al. [35] used a multi-column deep neural network to classify images on MNIST [33], CIFAR [36] and NORB [37] datasets. DeepFace was proposed by Yaniv Taigman et al. [38] for face verification from facial images on the Labeled Faces in the Wild (LFW) dataset [39]. However, these methods only perform well on images and not on videos. To solve this problems, Kang et al. [40] proposed temporal convolution network for extracting temporal information and fed it into CNN for object detection in video tubelets (sequence of detection bounding boxes).

For event detection, Kang et al. [42] proposed discriminative CNN. Karpathy et al. [6], Ng et al. [30] and Zha et al. [43] used CNN architectures to perform video classification by retaining the top layers of their neural network for performance optimization and reported better accuracy. High accuracy was reported by Simonyan et al. [44] for using CNNs for action recognition. Other variation of CNN such as GoogleNet [45], ResNet [46], ZFNet [47] and VGGNet [48] have emerged following

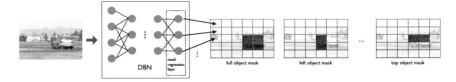

Fig. 2 Illustration of object detection using DNN proposed by Szegedy et al. [41]

the breakthrough of AlexNet [31]. Mathematical and architectural analysis of video analytics based on CNN model is what follows.

In a video analytics system, objects are detected and extracted from video and rescaled to a size say 150×150 pixels as shown in Fig. 2. Normalization of the object is performed for better accuracy by transforming pixel values from 0–255 to 0–1 before feeding them into the deep neural network.

Videos are firstly decoded into frames depending on the length of the input video data and other analysis are done on the generated video frames. For example, a 3 min (120 s) video can generate 4500 video frames at a rate of 25 fps (frame per second).

Let X be the training set and x_i be the decoded frames from video

$$X = x_1, x_2, x_3, \ldots, x_n \tag{7}$$

The number of channels from three (RGB) is reduced to one by converting the frame to gray scale and thus causing reduction in the processing time. Gray-scale conversion has no effect on algorithm accuracy. For the object detection phase, the converted gray scale frame is input into object detection algorithm and a bounding box is created around the region of interest. Haar cascade classifier [26] can be used for object detection.

Let (x, c) be a labeled frame where x is the frame data and c is the ground truth. The corresponding bounding box after an object is detected is given by:

$$R(x_0, y_0 \ldots x_n, y_n) \tag{8}$$

To extract desired object from video frame, cropping is performed around the detected area and the cropped area of the frame serves as input into the object classification phase. The extracted objects are rescaled to a size $w * h$, say $150 * 150$ pixels and normalized before feeding them into the deep neural network (DNN). Normalization of the object is performed for better accuracy by transforming pixel values from 0–255 to 0–1 before feeding them into the CNN as shown in Fig. 3.

CNN or DNN in general requires large amount of training data to give better performance and to avoid overfitting. When training data is scarce, data transformation (affine displacement field) such as contrast variation, skew, translation, rotation and flipping is performed on the input data set to generate additional training data to augment scarce data so as to increase the accuracy of the DNN. For details about

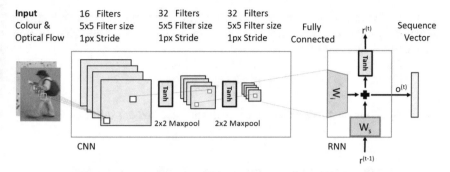

Fig. 3 Video-based person re-identification based on recurrent CNN architecture as proposed by [49]

how affine displacement can be used on video frames to generate additional data, Yaseen et al. [50] gives an intuitive explanation.

Based on the input training data and or transformed data, the CNN is trained to classify and discriminate the generated classes. A typical architecture consists of multiple alternating layers of convolutional and subsampling layers.

The convolutional and sub-sampling layers are denoted mathematically in Eqs. 9 and 10 respectively.

$$Conv_k, l = g(x_k, l * W_k, l + B_k, l) \tag{9}$$

$$Sub_k, l = g(x_k, l * w_k, l + b_k, l) \tag{10}$$

where $g(.)$ is the activation function, W and B represents the weight and biases of the system and the sub-sampling layer consist of downsampled inputs. * represents the convolution operation performed between the inputs and network weights.

Rectified Linear Units (ReLU) is the mostly used activation function for non-linearity. It models positive real numbers and helps in solving vanishing gradient problem with range from $[0, \infty]$. Other forms of activation functions include sigmoid, Leaky rectified linear unit (Leaky ReLU), Parametric rectified linear unit (PReLU), Randomized leaky rectified linear unit (RReLU), Exponential linear unit (ELU), Scaled exponential linear unit (SELU), S-shaped rectified linear activation unit (SReLU), Inverse square root linear unit (ISRLU), Adaptive piecewise linear (APL).

For the pooling layer, Max pooling is used. The purpose of the pooling layer is for dimensionality reduction, downsampling feature maps from convolutional layer and reducing the number of parameters so as to reduce computational cost. Yaseen et al. [50] used local response normalization as generalization technique. A typical video-based CNN architecture consists of two convolutional layer followed by two response normalization layers. Three max pooling layers are stacked underneath the response layer followed by the last convolutional layer. L2 regularization is used in the architecture in order to avoid overfitting by penalizing network weight. The output layer of the CNN network is the softmax layer used for optimizing negative log likelihood. It is given by:

$$l(i, x_i T) = M(e_i, f(x_i T)) \tag{11}$$

where $f(x_i T)$ is the function to calculate output value and e is the basis vector.

15.3 Algorithmic Domains and Task Involved in Video Analytics for Surveillance

15.3.1 Video Segmentation

Complex event is defined as a combination of several actions, objects and scenes [51]. Uncertainty in video content makes complex event detection an extremely challenging task. Moreover, semantic concept such as a single event class labels cannot explicitly capture the description of the event [52]. Therefore, it is necessary to combine multiple semantic concept to describe an event. For instance, the event of "playing soccer" can be easily associated with action concept of "running", "jumping" and "kicking", where the objects are "ball" and "players" and scene concept of "stadium". The single concept of "stadium" does not fully capture the description of the event as several activities can occur in the stadium such as running on the track, Super Bowl etc. Therefore, video segmentation is needed to extract informative segments about the event occurring in the video.

In video segmentation, it is important to consider temporal relations between key segments in a specific event for effective event detection. Considering all possible instances of event classes requires several training videos because intra-class variation exist in event videos. However, manual annotation of several videos is laborious. Song and Wu [53] proposed an intuitive approach of automatically extracting key segments for event detection in videos by learning from loosely labelled collection of web images and videos. Loosely labelled images were collected from Google and Flickr while videos were collected from YouTube. Their model is an adaptive latent structural Support Vector Machine (SVM) model where latent variables are the locations of key segments in the video. Set of semantic concepts were used for overall content description while single semantic concept were used for each local video segment. Temporal Relation Model (TRM) was proposed for the temporal relations between video segments and Segment-Event Interaction Model (SEIM) was used for evaluating correlation between key segments and events. The authors adapted labelled web images and videos from the web into their model and employed N adjacent point sample consensus (NAPSAC) [54] for noise elimination in videos and images.

Zhang et al. [55] created a knowledge base to reduce semantic gap between complex events by using tons of web images for learning noise-resistant classifiers to effectively model event-centric semantic concepts. Group incremental learning of target classifier was proposed by Wang et al. [56] where each concept group comprises of simple action videos and images querying from Web. Long et al. [57] and Duan et al. [58] proposed transfer kernel learning method and multiple source domain adaptation method respectively. In the multiple source domain adaptation method, relevant images sources are selected for annotating videos.

15.3.1.1   Extracting Video Segments for Action or Event Detection

Action recognition in videos involves both segmentation and classification. This problem can be addressed individually and sequentially with the use of sliding temporal window and aggregation, or perform both task simultaneously.

Song et al. [58] used segment-based approach by dividing video into a number of segments for feature extraction and classification. The extracted segments are used for event detection. Image sets were used by Song et al. [59] for detecting key segment from complex videos. Hidden Markov Model (HMM) [60] and global dynamic pooling structure [61] have been proposed for video segmentation. Habibian et al. [52] constructed a bag of 1346 concept detectors that were trained on the ImageNet [34] and TRECVID [62] dataset to generate a large vocabulary for event recognition.

Works in action recognition based on video segmentation can be categorized into three: Action segmentation, depth-based action recognition and deep learning based motion recognition. We briefly discuss this processes relative to video segmentation.

A.   *Similarity-based model*

Dynamic time warping (DTW) is the widely used method for action segmentation [63–65]. The first step is to obtain difference images. Two consecutive gray scale images are first subtracted to obtain difference images followed by partitioning of each difference image into 3×3 grid cells. The size of each cell is the average value of pixels within the cell while a motion feature flattens the difference image as a vector. The motion feature is extracted from both test and training video and calculated for each frame. It sums up to $9 \times (K - 1)$ matrix of motion features where K is the total number of frames. The two matrices represent two temporal sequences and the DTW distance between the two sequence is calculated by Viterbi algorithm to segment the actions.

Appearance based method of action segmentation from videos assumes similarity between the *start* and *end* frames of adjacent actions. Methods for identifying the *start* and *end* frames of actions are K-nearest neighbor (KNN) algorithm with HOG [66], and quantity of movement (QOM) [67]

B.   *Depth-based approach*

Besides appearance based approach, depth map based action recognition method have been proposed including combination of depth motion map (DMM) and HOG [68], using graphical model to encode temporal information [69], histogram of oriented 4D normals (HON4D) [70], capturing local motion and geometry information [71], and binary range-sample feature [72]. However, all the proposed methods are dataset-dependent and are based on hand-crafted features.

Deep learning methods have also been applied to depth based action recognition approach. A variant of DMM based on CNN was applied by [73, 74]. Wu. et al. [75] used 3D CNN as a feature extractor from depth data. Other techniques such as structured images [76, 77] technique has been proposed for depth based action recognition.

C. *Deep leaning-based motion recognition*

Deep learning approach for motion recognition can be partitioned into one of four categories:

Category 1 Video is viewed as a set of still images [29, 30]. In this category, each channel of the images is input onto one channel of a CNN. This approach is suboptimal but performs quite well.

Category 2 Video is represented as a volume and replaces the 2D filters of CNN with 3D filters thereby introducing temporal dimension [78, 79]. This approach doesn't work quite well probably due to lack of annotated training set.

Category 3 Video is regarded as a sequence of images and fed into Recurrent neural network (RNN) [80–82]. RNN allows for sequential parsing of video frames due to its sensitivity to both long term and short term patterns and its memory cell-like nature. It encodes frame-level information in the memory. It performs in similar magnitude to category 2.

Category 4 Video is represented as compact images and fed into pre-trained CNN architecture [83]. This approach achieved the state-of-the-art performance of action recognition due to the pretraining. Ochs et al. [84] proposed a method of segmenting moving objects using semi-dense point tracker based on optical flow to produce trajectories over several frames by long term analysis of the motion vector. They claim that intricate details can be extracted over a long period of analyzing videos by segmenting the meaningful or whole part of an object instead of a short time. Their method performs well than two-frame optical flow and color-based segmentation methods.

15.3.2 Moving Object Classification and Detection

Moving object detection from video is important due to its invaluable application to several application domains such as intelligent video surveillance, human behavior recognition, traffic control and action recognition.

15.3.2.1 Object Tracking

Object tracking in video surveillance involves the process of locating moving object(s) in video. Several applications of object tracking in video surveillance has been reported in the literature such as in resource tracking, customer queue analysis and transport [3].

The general step of object tracking in video is the extraction of foreground information to detect the object. Background subtraction algorithm such as IAGMM [7, 12] is then applied to capture background modelling of the scene. Shadow removal methods [85] can be subsequently applied to the foreground frame since shadows decreases tracking accuracy. Bounding box of an object is determined by connected component algorithm. Connected component algorithm scans object and groups its pixel into component while calculating the bounding box and area of the object. To ensure frame to frame matching of the detected object, method such as adaptive mean shift [86] can be used for comparison. The distance and size are some factors that defines object matching between frames. Subsequently, occlusion is detected and resolved using any of the occlusion method described later in 3.7 (Handling occlusion).

15.3.2.2 Motion Detection

Motion detection algorithm is similar to that of object tracking with the exception of consecutive frame matching with the detected object. Therefore, moving object can be detected via foreground images extracted by GMM background and connected component algorithm for noise removal. Upon applying connected component algorithm, the area of detection is well refined and produces the bounding box information of the moving objects [7]. Similarly, deep learning methods can be used. The goal of moving object detection is to capture video sequence from a fixed or moving camera. The output of the detection is a binary mask representing the moving object for each frame in a particular sequence. Shadows, variation in illumination and cloud movement makes object detection for moving object a difficult task [87]. Methods for moving object classification and detection can be categorized into moving camera with moving object and stationary camera with moving object.

A. **Moving object with stationary camera**

Analyzing moving objects using fixed cameras where background image pixels in the frames remain the same throughout a video sequence has been extensively studied in the literature [88–92]. The generic approach of handling the fixed camera problems is to model a stable background and apply background subtraction technique as described previously in Sect. 2 (theory of video analytics). Shantaiya et al. in [93] conducted an extensive review on object detection in video and grouped the methods into feature-based, motion-based, classifier-based and template-based models. Categorization of object tracking in videos into point tracking, kernel tracking and silhouette tracking as well as feature-based, region-based and contour-based was performed by [93, 94] respectively.

However, moving object detection with moving camera is relatively complicated than that of fixed camera due to camera motion and background modelling for generating foreground and background pixel fails [95].

B. Moving object with moving camera

Significant improvement has been made on cameras installed in drones and mobile phones with powerful imaging capabilities. These cameras are non-stationary and methods that can handle moving object detection for moving cameras are required. Although methods for analyzing moving object with fixed cameras is sufficient for some surveillance task since most cameras don't move. However, recent surveillance cameras have been equipped with pan-tilt-zoom (PTZ) functionalities or drones which may cause the cameras to move. Thus approach used in the fixed camera-moving object cannot be directly applicable where the background image pixel changes position throughout the video sequence.

Challenges faced in moving object detection includes defining the notion of moving object in terms of spatio-temporal relationship of pixels, variation in illumination or lighting condition, occlusion, changes in object appearance and reappearance, complex background such as moving cloud, sudden camera motion or other abrupt motion and shadows.

Readers are referred to [96] for an extensive survey on moving object detection in moving camera including the challenges of moving object detection in videos.

Solutions for moving object detection with moving cameras can be categorized into three:

1. Background modelling based methods [97, 98]: This approach aims at creating frame-by-frame background for each sequence using motion compensation method. Proposed algorithms includes Gaussian-based method [98], mixture of Gaussian (MoG) [99], adaptive MoG [100], double Gaussian model [101], kernel-based spatio-temporal model, Harris corner detector for feature point selection [102], multi-layer homography transform [103], complex homography [104], codebook modeling [105], thresholding [106], motion and appearance model [107], background keypoints and segmentation [108], and CNN-based method [109] for background modelling. The complexity of background modelling-based techniques is reasonable and thus well suited for real-time application.

2. Trajectory classification [110, 111]: Involves computing long trajectories for feature point and discriminating trajectories that belongs to different object from those background using clustering method. Proposed algorithms includes compensating long term motion based on flow optic technique [110, 111], bag-of-word classifier, and pre-trained CNN method for detecting moving object trajectories [112].

3. Extension of background subtraction method for static camera [113, 114]: low rank and sparse matrix decomposition method for static camera [115] are extended to moving camera. This tries to determine if there is coherency between

a set of image frame. If it exists, the low rank representation of the matrix created by these frame contains the coherency and the sparse matrix representation contains the outliers which represents the moving object in these frames. Low rank and sparse decomposition involves segmenting moving objects from the fixed background by applying principal component pursuit (PCP). It is a valuable technique in background modelling. Mathematical formula and optimization of this method can be found in [96].

15.3.3   Behavior Analysis

Interpreting behaviors from video footages is relatively fascinating as context needed for better understanding of the action can be extracted.

Human behavior can be understood through understanding of audience behavior by analyzing user interaction with digital display [116]. Typically, a camera is installed near the digital display or integrated into the display. Intel Anonymous Video Analytics (AIM) system [117] and Fraunhofer Avard [118] are commercial tools that have been deployed for understanding audience in terms of age, dwell time, gender, view time and distance from display through video processing. This can be combined with sales data to improve advertising campaigns or efforts [119]. Gillian et al. [120] developed a framework for analyzing user interaction with multiple displays i.e. across displays using depth-cameras. In surveillance, audience detection and tracking has been an interesting topic in crowd detection and estimation as well as single entity tracking within crowd. Crowd may refer to a group of people present at the same place with different (unstructured) or same (structured) reasons.

Crowd analysis involves studying crowd behavior and detecting abnormal behaviors in a static or dynamic (motion) scene. The analysis of crowd behavior becomes more challenging when the density of the crowd is high as shown in Fig. 4.

(a)   **(b)**

Fig. 4   **a** Crowd gathering at the train station during peak hours, **b** crowd gathering for religious activity. Image retrieved from alamy.com and www.thereformationroom.com respectively

The significance of crowd analysis has grown with the increase in the world population. For public safety, crowd management is very important in the construction of shopping malls, stadiums, and subway stations in order to avoid stampede or other disastrous outcomes. Therefore, using cameras for crowd analysis is important to detect or avoid terrorist attacks, bomb explosion, fire outbreak and other incidences that can cause havoc to public safety.

Crowd behavior analysis spans through several areas including pattern recognition, computer vision, mathematical modelling, artificial intelligence and data mining. The meaning of crowd differs based on the situation. For example, 10 persons gathering in a subway station can be regarded as crowd. Crowd analysis involves studying both group and individual behavior to determine abnormality. The definition abnormal behavior is quite ambiguous which makes crowd analysis quite an interesting area of research. Extensive survey of state-of-the-art deep learning method for crowd analysis can be found in [121].

The following attributes are used in analyzing crowd:

1. Counting and density estimation (congestion analysis)
2. Motion detection
3. Tracking
4. Behavior understanding.

Several factors must be considered when performing crowd analysis including terrain features, geometrical information as well as crowd flow.

15.3.3.1 Motion Feature Representation in Crowded Scenes

Motion features present an invaluable stance point for analysis of crowded scene. Existing work on motion feature analysis for crowded scenes can be divided into flow-based features, local spatio-temporal features and trajectory or tracklet [122]. These feature representations can be used for several task such as crowd behavior recognition, abnormality detection in crowd and motion pattern segmentation.

A. *Flow-based Features*

Tracking a person in a highly crowded environment is extremely difficult. However, in flow based features extraction, attention is only given to the occurrence not the actor (who is involved in what is happening). For example, singly looking at a person's action doesn't say much and may seem random but overall view of the crowd can be conclusive [123]. Flow based features are pixel level features. Several methods have been presented over the years [124–128]. Categorization of existing work is what follows:

1. *Optical Flow*:

Optical flow involves computing pixel-wise motion between consecutive frames. Optical flow handles multi-camera object motion and has been applied to detection crowd motion as well as crowd segmentation [129–132]. The drawback of Optical flow is that it cannot encapsulate spatio-temporal properties of the flow and does not capture long range dependencies.

2. *Particle Flow*:

Inspired by the Lagrangian framework of fluid dynamics [133], particle flow involves moving a grid of particles with the optical flow and providing trajectories that maps a particle's initial position to its future or current position. This method has shown dynamic application in crowd segmentation and detection of abnormal behavior in crowd [122]. However, there is time lag and cannot handle spatial changes.

3. *Streak Flow*:

Streakline was introduced by Mehran et al. [128] for analyzing crowd video by computing motion field. The proposed method is called Streak flow. Streak flow overcomes the challenges of particle flow. Although it captures motion information similar to particle flow, changes in the flow is faster and performs well in dynamic motion flow.

B. *Local Spatio-Temporal Features*:

Less structural (very crowded) scene have high variability and non-uniform movement. Motion in this type of scenes can be generated by any moving object and any of the optical flow features cannot provide useful information about the motion. Local spatio-temporal features are 2D patches or 3D cubes representation of the scene. They explore motion patterns and characterizes their spatio-temporal distributions on local 2D patches or 3D cubes. Spatio-temporal features are described below:

1. *Spatio-temporal Gradients*:

Kratz and Nishino [134] used spatio-temporal motion pattern model to capture steady-state motion behavior and their result shows that abnormal activities can be detected.

2. *Motion Histogram*:

Motion histograms considers motion information within the local region. Computing motion orientation on motion histogram takes considerable amount of time and it is highly susceptible to error. Thus, it is not suitable for crowd analysis. However, several improved methods based on motion histogram has been proposed by researchers.

Jodoin et al. [131] proposed orientation distribution function (ODF), a feature that does not have any information about the magnitude of flow but represents the probability density of a particular motion orientation. Multiscale histogram of optical flow (MHOF) was proposed by Cong et al. [135] as a feature descriptor that preserves both the spatio-contextual information and motion information.

C. *Trajectory/Tracklet*

Trajectory or tracklet represents motion by computing individual tracks. Motion features such as distance between object or motion energy can be extracted from trajectories of objects and can be used to analyze crowd activities.

However, object detection and tracking in a highly dense crowd is very difficult. Thus, the notion of tracklet emerges due to the inability to obtain complete trajectory in such setting.

Tracklet is a fragment or part of a trajectory obtained within a short period of time. When occlusion occurs, tracklet terminates. Traklets have been used in the area of human action recognition [136–138] by connecting them together to form a complete trajectory. Quite a number of tracklet-based approaches have been proposed for representing motion in crowded scenes [139–141]. The general ideology of tracklet is to extract them from dense region and enforce statial-temporal correlation between tracklet to detect patterns of behavior in crowded scenes.

In general, spatio-temporal feature have shown promising results in both motion understanding and crowd anomaly detection.

15.3.4 Anomaly Detection

Anomaly detection is an application of crowd behavior analysis. Detecting anomalies in videos is nontrivial due to variation in the definition of anomaly. i.e. an anomaly in one scene can be considered normal in another. This has attracted many researchers and several methods have been proposed. The general approach is to learn what is considered normal in a training video and use it to detect events that drifts from them (abnormality). Occlusion, distance between object and camera, and viewpoint may cause variation and thus contribute to anomaly in video. Existing methods in anomaly detection research can be categorized into trajectory-based method, global pattern-based method, and grid pattern based method [142]. Table 1 provides the summary of these methods.

A. **Trajectory-based method of anomaly detection**

Trajectory-based method segments scenes into different objects while the objects are tracked throughout the video sequence. The tracked object forms a trajectory which defines the behavior of the object [143]. String kernels clustering [144], single-class SVM [145], spatio-temporal path search [146], zone-based analysis [147], semantic tracking [148] and deep learning-based approach [149] have been used in evaluating abnormality in trajectory-based methods.

B. **Global pattern-based method of anomaly detection**

Global pattern-based method analyzes video sequence in entirety by extracting low or medium level features from video using spatio-temporal gradients or optical flow methods [150]. The advantage of this method is that it does not individually track each object in the video and thus suitable for crowd analysis. However, locating the position at which an anomaly occur is non-trivial. Gaussian mixture model (GMM)

Table 1 Summary of anomaly detection methods

Reference	Methods
Global pattern-based methods	
Popoola and Wang [150]	Optical flow methods
Yuan et al. [151]	Gaussian mixture model (GMM)
Xiong et al. [152]	Energy model
Wang et al. [153]	Stationary-map
Zhang et al. [154]	Social force model (SFM)
Cheng et al. [155]	Gaussian regression
Lee et al. [156]	Principal component analysis (PCA) model
Krausz et al. [157]	Global motion-map
Lee et al. [158]	Motion influence map
Chen et al. [159]	Salient motion map
Trajectory-based methods	
Brun et al. [144]	String kernels clustering
Piciarelli et al. [145]	Single-class SVM
Tran et al. [146]	Spatio-temporal path search
Cosar et al. [147]	Zone-based analysis
Song et al. [148]	Semantic tracking
Revathi and Kumar [149]	Deep learning-based method
Grid pattern based methods	
Xu et al. [161]	Sparse reconstruction of dynamic textures over an overcomplete basis set
Cong et al. [162]	Motion context descriptor
Thida et al. [163]	Spatio-temporal Laplacian Eigen map method
Yu et al. [164]	Hierarchical sparse coding
Li et al. [165]	Multiscale splitting of frames
Lu et al. [166]	Multiscale splitting of frames
Lu et al. [167]	Adaptive dictionary learning
Han et al. [168]	Online adaptive dictionary learning
Zhao et al. [169]	Sparse coding and sliding window
Xu et al. [142]	Stacked sparse coding (SSC) and SVM

[151], energy model [152], stationary-map [153], social force model (SFM) [154], Gaussian regression [155], principal component analysis (PCA) model [156], global motion-map [157], motion influence map [158], and salient motion map [159] are approaches used in the global pattern-based method.

C. Grid pattern-based method of anomaly detection

In contrast with the global pattern-based methods, the grid pattern-based methods do not consider frames as a whole but rather splits frames into blocks and individually analyze pattern on a block-level basis [160]. Grid pattern-based methods are more efficient due to the processing time reduction fueled by individual evaluation of pattern in the block level and ignoring inter-object connections. Spatio-temporal anomaly maps, local features probabilistic framework, joint sparsity model, mixtures of dynamic textures with GMM, low-rank and sparse decomposition (LSD), cell-based texture analysis, sparse coding (SC) and deep networks are used in evaluating grid pattern-based methods [142].

Xu et al. [161] used sparse reconstruction of dynamic textures over an overcomplete basis set to detect anomaly. Cong et al. [162] proposed the concept of searching for the best match in the training dataset using motion context descriptor. Thida et al. [163] extracted diverse crowd activities from videos using spatio-temporal Laplacian.

Eigen map method. All these methods are based on Sparse Coding (SC). The notion behind SC is that abnormal events in videos are characterized by sparse linear combinations of normal patterns with large reconstruction error while normal events are characterized by small reconstruction errors. Yu et al. [164] classified events as abnormal and normal using hierarchical sparse coding method. Li et al. [165] and Lu et al. [166] computed sparse representation in each scales by splitting frames in multiscale. To generate better representation of abnormal events, Lu et al. [167] and Han et al. [168] proposed adaptive dictionary learning and online adaptive dictionary learning respectively.

Sparse coding and sliding window was adopted by Zhao et al. [169] for detection of abnormal events in videos.

Xu et al. [142] proposed a method of detecting anomalies in video based on stacked sparse coding (SSC) with intra-frame classification. The video is first divided into blocks. The appearance and motion features for each block is described by the foreground interest point (FIP) descriptor and encoded by SSC. Support vector machine (SVM) is used to evaluate the intra-frame classification to determine abnormality in each block.

15.3.5 Video Summarization

Video summarization is the process of compressing a video by extracting only the important part of the video sequence. Video summarization is important due to current widespread of cameras leading to massive amount of data. It involves producing only significant or important highlights of a video that conveys the overall story. Definition of significant or important aspect of a video varies based on several criteria. The goal of video summarization is to generate a dense representation of a specific video. Determining informative or important sections of a video requires understanding of the video content. However, video content is diverse thereby making summarization a difficult task. Video summarization can also enhance video retrieval results.

Zhang et al. [170] defined a good summarization technique as one which is diverse, representative of videos of similar group and discriminative against videos in dissimilar groups. Domain-specific video summarization method was the early approach used for determining important segment of the video. For example, in sport, specific structure can facilitate the important segment according to the rules governing the sport. In movies, metadata such as movie script and captions can be used in generating video summaries.

Methods of dealing with video summarization can be categorized into unsupervised, supervised, query extractive and discriminative approach.

Supervised and unsupervised approaches have been developed to encapsulate domain knowledge for video summarization. Unsupervised summarization approach creates summary based on precise selection criteria. Supervised approach on the other hand, trains a summarization model using human-created summaries.

Potapov et al. [171] used classifier's confidence score to define important segment of a video. Methods in the supervised approach is difficult to generalize to other genres because it is highly dependent on domain knowledge but offers better performance.

The unsupervised approach is independent of domain knowledge and thus suitable for generic application. Yang et al. [172] used an auto-encoder to convert input video's features into a concise one and reconstruct the input using the decoder. Zhao et al. [173] proposed a method that reconstruct the rest of the original video based on a video summary.

Query extractive summarization methods [174, 175] are a variant of summarization methods that generates summary based on keyword input. This model assumes that a video can have multiple summaries. However, it may be unrealistic for real applications due to frame-level importance annotation for each keyword.

Panda et al. [176] introduced discriminative information by training a spatio-temporal CNN for classifying the category of each video and calculates the importance scores via gradient aggregation of the network's output. Kanehira et al. [170] proposed viewpoint-aware video summarization in which summary is built based on the aspect of the video that the viewer focuses on. To determine viewpoint, they leverage other videos in folders on the viewer's laptop or phone and performed semantic similarity and dissimilarity between the videos in the folder and the current video being watched to produce viewpoint-specific summaries. Otani et al. [177] proposed a method of improving video summarization techniques by using deep video features to encode various levels of content semantics such as actions, scenes and objects. The architecture used is a deep neural network for mapping videos and description to a semantic space. Clustering is applied to the segmented video content. Table 2 gives the taxonomy of video summarization methods.

Table 2 Taxonomy of video summarization methods

Reference	Year of publication	Summarization extraction method
Supervised summarization methods		
Gong et al. [178]	2014	Used human-created summaries to train a system to select informative and diverse subsets of video using sequential determinantal point process (seqDPP)
Gygli et al. [179]	2014	Video segmentation using set of consecutive frames where the beginning and end are aligned with positions of a video suitable for a cut (superframe segmentation)
Gygli et al. [180]	2015	Learnt important overall characteristics of a summary by jointly optimizing multiple objectives
Kulesza et al. [181]	2012	Built explanatory summaries by selecting diverse sentences using determinantal point processes (DPPs)
Lee et al. [182]	2012	Trained a regressor that predicts important regions of a video using egocentric features
Liu et al. [183]	2010	Elimination of irrelevant frames from video to generate informative summaries using window-level representation based on probabilistic graphical model
Plummer et al. [184]	2017	Used Semantically-aware video summarization technique by selecting a sequence of segments that best represent the content of input video
Potapov et al. [171]	2014	Proposed category-based video summarization method that transforms temporal segmentation into semantically-consistent segments and assigns scores to each segment
Sun et al. [185]	2014	Learning latent model for ranking domain-specific highlights and comparing raw video to edited video using latent linear ranking model
Zhang et al. [186]	2016	Used human-created summaries to automatically extract keyframe-based video summarization by transferring summary structures from annotated videos to unseen videos
Unsupervised summarization methods		
Chen et al. [187]	2011	Used combination of knowledge and individual narrative preference to generate summarized video content by segmenting video contents into local stories

(continued)

Table 2 (continued)

Reference	Year of publication	Summarization extraction method
Elhamifar and Kaluza [188]	2017	Proposed an incremental subset selection framework for generating summarized videos by updating set of representatives features based on previously selected set of representatives and new batch of data
Fleischman et al. [189]	2007	Proposed temporal feature induction method that extracts complex temporal information from video for classifying video highlights
Hong et al. [190]	2009	Used multi-video summarization technique to determine key shots as a combination of ranked list of web videos and user-defined skimming ratio
Khosla et al. [191]	2013	Used web-image based prior information to generate summarization obtained through crowdsourcing for poor quality videos
Kim et al. [192]	2014	Used storyline graph for creating structural video summaries that illustrates various events based on diversity ranking between images and video frames
Lu and Grauman [193]	2013	Used text analysis based method to determine random walk-based metric of influence between sub shots which captures event connectivity
Mahasseni et al. [194]	2017	Trained a system to learn a deep summarizer network based on autoencoder long short-term memory network (LSTM)
Song et al. [195]	2015	Proposed a video summarization framework called TVSum that detects important shots based on titles of the retrieved image
Query extractive summarization		
Sharghi et al. [174]	2016	Proposed a method based on Sequential and Hierarchical Determinantal Point Process (SH-DPP) to select key shot determined by the relevance of user query relative to the video context
Sharghi et al. [175]	2017	Used extracted semantic information for evaluating the performance of a video summarizer
Discriminative method of video summarization		
[170]	2018	Introduced a viewpoint approach to build a summary that depends on what the viewer focuses on using classification techniques that discriminates semantic similarity between different groups

15.3.6 Information Fusion from Multiple Camera

In this section, we review multi-camera surveillance systems for wide-area video monitoring. Multi-camera settings are typical in most real-world surveillance systems due to the numerous amount of cameras available and impracticability of one-camera to one-monitor methodology. Therefore, performing analytics with multiple camera is nontrivial since it requires modelling spatio-temporal relationship among objects, events, and sometimes careful configuration of camera views. Using multiple cameras for tracking can either be based on overlapping cameras or non-overlapping cameras. Multiple cameras with overlapping field of view is both economically and computationally expensive and requires well designed network to correlate the overlapping field of view. However, non-overlapping multiple camera field of views are quite realistic and applied in real-world application. In this section, we will focus on multiple cameras with non-overlapping field of view with the goal of tracking multiple targets across multiple cameras without losing track of any of the target as they move between and among the cameras.

Several methods such as Probabilistic Petri Net-based approach, Dominant sets clustering, Generalized maximum clique problem (GMCP), Generalized maximum multi clique problem (GMMCP), Multiple Instance Learning (MIL), Markov Random Fields (MRF), Multi-tracker ensemble framework, Structural Support Vector Machine (SSVM), Spatio-temporal context information, Top-push distance learning model (TDL), Recurrent neural network architecture and Constrained dominant sets clustering (CDSC) have been proposed to combine views from multiple cameras. Table 3 gives the details of these works.

Lu et al. [196] proposed a composite three-layer hierarchical framework using constrained dominant sets clustering (CDSC) technique for tracking object across multiple non-overlapping cameras. Within-camera tracking problem is solved in the first two layers of the framework while across-camera tracking is solved by concurrently combining tracks of the same person in all cameras in the third layer. The proposed CDSC method works by finding constrained dominant sets from a graph by generating cluster or clique that captures subset of the constraint set. This method can also be used for detection of person re-identification since the third layer can link broken tracks of the same person occurring during within-camera tracking.

15.3.7 Handling Occlusion

Occlusion is one of the major problem in video analytics. Effectively handling occlusion can greatly improve analytics accuracy. Occlusion can either be partial or total occlusion.

Objects can be occluded by other objects in the scene causing some parts of the object to be unseen (partial) or completely hidden from the observed video frame by other object (total). For instance, consider a mini cooper as the target object on a

Table 3 Recent works on object tracking using non-overlapping multiple camera

References	Method	Description
Wang et al. [197]	Probabilistic Petri Net-based approach	Tracks target based on appearance features of objects and spatio-temporal matching of target across camera views
Tesfaye et al. [198]	Dominant sets clustering	Pairwise relationship between detected objects in a temporal sliding window are considered and used as input into a fully connected edge-weighted graph
Zamir et al. [199]	Generalized maximum clique problem (GMCP)	Used motion and appearance feature to solve data association between object in multiple camera views based on generalized maximum clique problem (GMCP)
Deghan et al. [200]	Generalized maximum multi clique problem (GMMCP)	Formulated data association for tracking multiple object as a generalized maximum multi clique problem (GMMCP)
Kuo et al. [201]	Multiple instance learning (MIL)	Used on-line learned discriminative appearance affinity model to model association between tracked objects
Chen et al. [202]	Markov random Fields (MRF)	Used human part configurations to determine across-camera spatio-temporal constraints and pair-wise group activity constraints for multi-target tracking
Gao et al. [203]	Multi-tracker ensemble framework	Proposed a multi-tracker based approach that captures consistency between two successive frames and pair-wise correlation among different trackers
Zhang et al. [204]	Structural support vector machine (SSVM)	Expressed multi-target tracking as a network flow problem whose solution can be obtained by K-shortest paths algorithm
Cai et al. [205]	Spatio-temporal context information	Spatio-temporal context information is used for inter-camera tracking and for discriminating appearances of target

(continued)

Table 3 (continued)

References	Method	Description
Jinjie et al. [206]	Top-push distance learning model (TDL)	Top-push constrain is used for matching video features of persons instead of matching still images of a person across multiple camera. This approach provides high-level discriminative features and provides better matching for person re-identification in multiple non-overlapping camera views
Mclaughlin et al. [49]	Recurrent neural network architecture	CNN is used as a feature extractor for each frame combined with temporal pooling layer for person re-identification and detection across multiple cameras
Tesfaye et al. [196]	Constrained dominant sets clustering (CDSC)	Proposed a three-layer hierarchical framework using CDSC technique for tracking object across multiple non-overlapping cameras and detection of person re-identification

freeway. The mini cooper due to its size could be occluded by larger truck and vans which will in turn affect the detection of the mini cooper. Occlusion adversely affects object detection by changing the appearance model for a short time which can affect tracking of the object. Figure 5 shows an example of occlusion in real-life.

Several methods have been proposed for occlusion handling based on appearance model. Jepson et al. [207] used Estimation Maximization (EM) algorithm with appearance model based on filter responses from a steerable pyramid to deal with changing appearance of an object.

Spatio-temporal context information obtained from gradual analysis of the occlusion situation was proposed by Pan and Hu [208] to distinguish occluded object effectively. Contour-based approach was proposed by in [209]. In this approach, energy function evaluated in the contour is minimized which causes easy tracking of the object. Subsequently, maintaining appearance models of moving objects over time can effectively manage occlusion [210].

Method of handling occlusion for moving camera has been proposed by Hou et al. [211]. Their method used HOG and multiple kernel tracker based on mean shift to discriminate different condition of moving cameras thereby handling occlusion effectively.

(a)   **(b)**   **(c)**

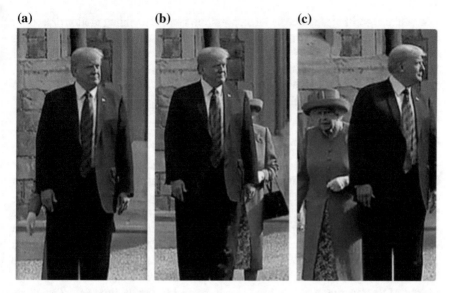

Fig. 5 President Trump occluding Queen Elizabeth II. **a** Full or total occlusion **b–c** Partial occlusion. Images obtained from twitter

15.4   Applications

Currently, video analytics drives many application domains from self-driving cars and drones to surveillance with majority of the models taking advantage of the deep neural networks paradigm.

Video analytics in surveillance applications can be grouped into three categories:

1. Predictive analytics: This involves making a prediction or forecast about the future. It identifies risk, opportunities and assesses the likelihood of a similar subject in different category exhibiting specific attribute. Simply put, it is the analysis of what will happen in the future. For example, predicting the next direction or move of a target based on the behavior of the target in the current frame.
2. Prescriptive analytics: Prescriptive analytics encompasses actions that should be taken in a given situation. For example; in resource tracking, when the availability of resource is low in a particular resource station, video analytics can proffer solution of using resource from a station with low resource usage to replenish the busy station requiring more resources.
3. Forensics: Categories of video analytics application in forensics involves using videos to analyze what has happened. In congestion analysis, when there is a situation of interest such as stampede, videos can be used to trace back the cause of the stampede.

General algorithms include facial recognition, vehicle tracking, motion detection, person (object) tracking and re-identification, loitering, vehicle license plate reading, slip and fall detection, abandoned object detection and crowd analysis. A full-fledge analytic system called VideoStorm [212] has been developed that tracks objects, classifies them and also allows querying of events from recorded video data. Some real-life application of predictive, prescriptive and forensics applications is what follows.

A. Application in Traffic and transport

According to Ananthanarayanan et al. [213], traffic-related accidents are one of the leading causes of death in the world. Proactive measures for risk identification and taking steps to prevent injuries on the road is required to mitigate the number of traffic-related deaths. Most of the time, there are early warning signs such as near-collision events at specific locations that can provide insight into why and when crashes can occur. Also monitoring traffic with videos and analyzing them can capture undocumented or unreported crashes. Traffic monitoring systems can use vision-based algorithms to detect and count the number of cars on the highways. Insight of the steering and braking behaviors can be equally provided. Therefore, analyzing video data generated by the large amount of traffic cameras installed in large cities is invaluable. VisonZero [214] is one successful real-world application of video analytics for eliminating traffic-related deaths by detecting situations in which something bad is about to happen and avoiding it (close calls) between road users (pedestrians, cars and bikers). For example, when there is an area where jaywalking is predominant, through the help of VisonZero, authorities can proffer solution of installing crosswalk or making adjustment to the traffic-light controllers.

Vehicle tracking is an important aspect of the application of video analytics in transport. Tracking of license plate, analysis of cause of collision, over speeding and vehicle size detection can be obtained by analyzing video data which provides in-depth information about the vehicle. A fully functional vehicle tracking system has been developed called Kestrel. Kestrel [215] is a video analytic system that uses information from heterogeneous camera networks of non-overlapping cameras to detect vehicle path. Detected path forms a large corpus that allows users to query based on location and event of interest in near real-time or after the fact fashion. The architecture of Kestrel is based on YOLO [216], a deep CNN for object detection that draws bounding boxes around detected object on a frame.

B. Application in self-driving cars and augmented reality (AR)

For self-driving cars, video analytics serves a great deal where the accuracy of algorithms are required to be at the optimum. Several onboard cameras are installed to make driving decisions and fusing all information from multiple video cameras is important with zero tolerance for anomalies. Several algorithms have been developed for self-driving cars including traffic light detection, human detection, pedestrian crossing detection, and automatic braking system. These algorithms are integrated together to enhance sensing and detection capabilities. Also dedicated short range communication (DSRC) can be used to communicate with other approaching self-driving car.

In AR, additional information is projected into user's view either by projecting holograms or by recording surrounding views and rendering objects directly on top of them when Virtual Reality (VR) headset are worn. AR systems uses combination of several algorithms such as object detection, face recognition and object count to analyze observed scenes.

C. Application in security

Body-worn cameras of law enforcement officers and government agencies have increased surveillance data. Therefore, the application of video analytics for surveillance video data is of uttermost importance. This has led to the application of video analytics to crime detection and suspicious vehicle license plates detection. Products such as Ring video doorbell [1] {https://ring.com/} and Nest video doorbell [2] {https://nest.com/} have brought a new light to cameras for home applications. They can identify people, motion and detect events such as stray animals, burglary and package delivery. They also allow home owner to communicate with the visitors or guest from mobile phones. Law enforcement and counter-terrorism agents can track public threats in real-time such as multiple coordinated attacks on public transport through video analytics. Drones can also be used for home and public surveillance.

D. Application to warfighters and unmannered aerial vehicles (UAV)

Warfighters operates in an uncertain world faced with the compulsion of making quick decisions. Video analytics from surveillance cameras can provide warfighters with decision-relevant information. To aid quick decision, moving objects can be detected and tracked in real-time. The tracked objects are classified and anomalies are detected [217]. Targeted suspect can be monitored via UAV also called drones.

Most drones are equipped with small low-cost high-resolution cameras and video footage taken from drones have contributed to growing data sources in the last few years [16]. Camera motion estimation and compensation task can be performed on moving objects from videos taken from a camera mounted on a UAV without prior knowledge of the object being detected thereby reducing information transmitted to the ground in real-time.

E. Application in customer service and queue analytics

Queuing is an important part of our daily activities either at the restaurant, bus station, train station, supermarket or coffee shop. Analysis of queue provides number of people waiting in line and the trends over a particular period of time. Queue length, body and face counting, although challenging can be tracked from camera views. This provides detailed insight of daily operations. Analyzing queues using existing surveillance cameras can enhance decision making of human resources scheduling for providing efficient services. It can also benefit the customers by displaying how long they will wait on the queue before being serviced.

Video analytics has been used by Seng [218] to quantify customer's emotion and satisfaction at the contact center. Emotion analysis can be used in identifying customer's perception about a product in terms of the presentation and interaction with company's representative. This can provide competitive advantage as well as improving customer's experience. Human emotion can also be recognized via these systems and translated to give customer satisfaction scores recovered from the continuous video data.

Customer's behavior can be monitored and understood which is a key factor for organization's success.

F. Application in resource tracking

Effective utilization of resource is of great importance in business operation considering that every organization is constrained in resources. Capacity management is a complex and difficult task. However, a link has been established between capacity management and resource management. It has been shown that managing resources can enhance capacity management especially in a capacity-constrained environment [3].

Olatunji and Cheng [1, 3] applied video analytics using existing surveillance cameras to track trolleys in an airport operation in real-time. Their algorithm is based on CNN and motif detection algorithm for sending timely alerts when replenishment of the resource is needed. The analytics algorithm running on CCTV footage can also detect abandoned bags and station's inventory of resource can be tracked.

G. Application in Healthcare and physical wellbeing

Healthcare is a crucial sector that requires special attention when applying video analytics. However, analyzing video data obtained from surveillance cameras can be very effective in understanding the behavior of the tracked person or object. For example, monitoring senior citizens or elderly homes can be quite challenging but video analytics on the surveillance footage obtained can be of great help if they run algorithms such as fall detection or detect any potential threat and triggers emergency notification to respective authorities like the emergency or ambulance unit. This also allows for remote monitoring via smartphone or tablet which provides more flexibility and alleviate the burden of physically being at the location of an elderly home. In gym and fitness centers, installed surveillance cameras can be analyzed to determine the most frequently used equipment as well as the duration or the amount of time a particular trainer spends on an equipment. This information can then be used for procurement of a type of equipment if frequently used or restructuring the fitness center. The fitness condition of gym members using an equipment can also be remotely monitored. For example, if a gym member falls off the treadmill, an alarm can be triggered and a staff attend to the person accordingly.

15.5   Conclusion and Research Direction

Information fusion between cameras and efficient analytic algorithm can greatly enhance the power of video analytics in several domains such as healthcare, transportation, security and resource tracking. The goal of video analytics is to convert observed scenes into quantitative and actionable intelligence that can provide real-time situational awareness of an event of interest as well as post-event evaluation. Combining other data such as data from geographical information system (GIS), location data and other metadata will form a composite suite of video analytics system and result in more efficient algorithms. Similarly, more research should be conducted on visualizing camera position and perspective including 3D projection, geospatial context and narrative time sequence as they affect the data quality and algorithm accuracy. Integration of social media data into video analytics systems and augmented reality (AR) views may increase algorithm robustness. Currently, application-specific systems exist such as burglary detection system, shoplifting detection system and fall detection system to mention a few. Moving towards a generic video analytic framework instead of current application-specific systems will be a good research direction.

In this chapter, we presented a concise survey of recent video analytics methods and techniques applied to surveillance camera data. We extended the definition of surveillance to resource management which is important for capacity-constrained industries. We discussed the theory behind current state-of-the-art approach of video analytics. It is envisaged that this chapter serves as a starting point for new researchers in the area of video analytics to have a broad knowledge of the field and also to streamline their focus on any of the modules that makes an automated video surveillance and monitoring (VSAM) system.

References

1. I.E. Olatunji, C.-H. Cheng, Dynamic threshold for resource tracking in observed scenes. in *IEEE International Conference on Information, Intelligence, Systems and Applications* (2018)
2. S. Caifeng, P. Fatih, X. Tao, G. Shaogang, *Video Analytics for Business Intelligence* (2012)
3. C.-H. Cheng, I. E. Olatunji, Harnessing constrained resources in service industries via video analytics. Arch. Ind. Eng. J. (2018)
4. C.V. Networking Index, *Forecast Methodol. 2016–2021 white Pap.*, vol. 1 (2016)
5. M. Ali, A. Anjum, M. U. Yaseen, A.R. Zamani, D. Balouek-Thomert, O. Rana, M. Parashar, Edge enhanced deep learning system for large-scale video stream analytics, in *2018 IEEE 2nd International Conference on Fog and Edge Computing (ICFEC)* (2018), pp. 1–10
6. A. Karpathy, G. Toderici, S. Shetty, T. Leung, R. Sukthankar, L. Fei-Fei, large-scale video classification with convolutional neural networks, in *Proceedings of the 2014 IEEE Conference on Computer Vision and Pattern Recognition* (2014), pp. 1725–1732
7. P. Guler, *Real-Time Multi-camera Video Analytics System on GPU*, no. (Mar 2013, 2015)
8. D.-S. Lee, Effective Gaussian mixture learning for video background subtraction. IEEE Trans. Pattern Anal. Mach. Intell. **27**(5), 827–832 (2005)
9. T. Bouwmans, L. Maddalena, A. Petrosino, Scene background initialization. *Pattern Recogn. Lett.* **96**, no. C, pp. 3–11, (2017)
10. Z. Zhou, D. Wu, X. Peng, Z. Zhu, C. Wu, J. Wu, Face, *Tracking Based on Particle Filter with Multi-feature Fusion* (2013)
11. I. Ishii, T. Ichida, Q. Gu, T. Takaki, 500-fps face tracking system. J. Real-Time Image Process. **8**(4), 379–388 (2013)

12. V. Pham, P. Vo, V.T. Hung, L.H. Bac, GPU implementation of extended gaussian mixture model for background subtraction, in *2010 IEEE RIVF International Conference on Computing & Communication Technologies, Research, Innovation, and Vision for the Future (RIVF)* (2010), pp. 1–4

13. V. Reddy, C. Sanderson, B.C. Lovell, A low-complexity algorithm for static background estimation from cluttered image sequences in surveillance contexts. *J. Image Video Process.* **2011**, 1:1–1:14 (2011)

14. G. Zhang, J. Jia, W. Xiong, T.-T. Wong, P.-A. Heng, H. Bao, Moving object extraction with a hand-held camera, *ICCV 2007*. in *IEEE 11th International Conference on Computer Visio* (2007), pp. 1–8

15. M. Gelgon, P. Bouthemy, A region-level motion-based graph representation and labeling for tracking a spatial image partition. Pattern Recognit. **33**(4), 725–740 (2000)

16. P. Angelov, P. Sadeghi-Tehran, C. Clarke, AURORA: autonomous real-time on-board video analytics. Neural Comput. Appl. **28**(5), 855–865 (2017)

17. E. Auvinet, E. Grossmann, C. Rougier, M. Dahmane, J. Meunier, Left-luggage detection using homographies and simple heuristics

18. D. Emeksiz, A. Temizel, A *Continuous Object Tracking System with Stationary and Moving Camera Modes*, vol. 854115, no. Oct 2012

19. P. Gil-Jiménez, R. López-Sastre, P. Siegmann, J. Acevedo-Rodríguez, S. Maldonado-Bascón, automatic control of video surveillance camera sabotage, in *Nature Inspired Problem-Solving Methods in Knowledge Engineering* (2007), pp. 222–231

20. A. Saglam, A. Temizel, Real-Time adaptive camera tamper detection for video surveillance, in *2009 Sixth IEEE International Conference on Advanced Video and Signal Based Surveillance* (2009), pp. 430–435

21. G. Ramirez-Alonso, J.A. Ramirez-Quintana, M.I. Chacon-Murguia, Temporal weighted learning model for background estimation with an automatic re-initialization stage and adaptive parameters update. Pattern Recognit. Lett. **96**, 34–44 (2017)

22. O. Déniz, G. Bueno, J. Salido, F. De la Torre, Face recognition using histograms of oriented gradients. Pattern Recognit. Lett. **32**(12), 1598–1603 (2011)

23. A.E. Abdel-Hakim, A.A. Farag, CSIFT: A SIFT descriptor with color invariant characteristics, in *2006 IEEE Computer Society Conference on Computer Vision and Pattern Recognition (CVPR'06)*, vol. 2 (2006), pp. 1978–1983

24. M.U. Yaseen, M.S. Zafar, A. Anjum, R. Hill, High performance video processing in cloud data centres. IEEE Symp. Serv.-Oriented Syst. Eng. (SOSE) **2016**, 152–161 (2016)

25. M.U. Yaseen, A. Anjum, N. Antonopoulos, Spatial frequency based video stream analysis for object classification and recognition in clouds, in *2016 IEEE/ACM 3rd International Conference on Big Data Computing Applications and Technologies (BDCAT)* (2016), pp. 18–26

26. M.U. Yaseen, A. Anjum, O. Rana, R. Hill, Cloud-based scalable object detection and classification in video streams. Futur. Gener. Comput. Syst. **80**, 286–298 (2018)

27. A.R. Zamani, M. Zou, J. Diaz-Montes, I. Petri, O. Rana, A. Anjum, M. Parashar, Deadline constrained video analysis via in-transit computational environments. IEEE Trans. Serv. Comput. 1 (2018)

28. A. Anjum, T. Abdullah, M. Tariq, Y. Baltaci, N. Antonopoulos, Video stream analysis in clouds: an object detection and classification framework for high performance video analytics. IEEE Trans. Cloud Comput. 1 (2018)

29. K. Simonyan, A. Zisserman, Two-stream convolutional networks for action recognition in videos, in *Proceedings of the 27th International Conference on Neural Information Processing Systems*, vol. 1 (2014), pp. 568–576

30. J.Y. Ng, M. Hausknecht, S. Vijayanarasimhan, O. Vinyals, R. Monga, G. Toderici, *Beyond Short Snippet : Deep Networks for Video Classification* (2014), p. 4842

31. A. Krizhevsky, I. Sutskever, G.E. Hinton, ImageNet classification with deep convolutional neural networks, in *Proceedings of the 25th International Conference on Neural Information Processing* Systems, vol. 1 (2012), pp. 1097–1105

32. Y. LeCun, L. Bottou, Y. Bengio, P. Haffner, Gradient-based learning applied to document recognition. Proc. IEEE **86**(11), 2278–2324 (1998)
33. L. Yann, C. Corinna, and J. C. B. Christopher, *MNIST Handwritten Digit Database* (2010)
34. J. Deng, W. Dong, R. Socher, L. Li, K. Li, L. Fei-Fei, ImageNet: a large-scale hierarchical image database, in *2009 IEEE Conference on Computer Vision and Pattern Recognition*, 2009, pp. 248–255
35. D. Ciregan, U. Meier, J. Schmidhuber, Multi-column deep neural networks for image classification, in *Proceedings of the 2012 IEEE Conference on Computer Vision and Pattern Recognition (CVPR)* (2012), pp. 3642–3649
36. K. Alex, N. Vinod, H. Geoffrey, *The CIFAR-10 dataset* (2014)
37. F.J. Huang, Y. LeCun, Large-scale learning with SVM and convolutional for generic object categorization, in *2006 IEEE Computer Society Conference on Computer Vision and Pattern Recognition (CVPR'06)*, vol. 1, (2006) pp. 284–291
38. Y. Taigman, M. Yang, M. Ranzato, L. Wolf, DeepFace: Closing the Gap to Human-Level Performance in Face Verification, in *2014 IEEE Conference on Computer Vision and Pattern Recognition* (2014), pp. 1701–1708
39. B.H. Gary, R. Manu, B. Tamara, L.-M. Erik, *Labeled Faces in the Wild: A Database for Studying Face Recognition in Unconstrained Environments* (2007)
40. K. Kang, X. Wang, Fully convolutional neural networks for crowd segmentation. *CoRR* (2014). abs/1411.4464
41. C. Szegedy, A. Toshev, D. Erhan, Deep neural networks for object detection, in *NIPS* (2013)
42. K. Kang, W. Ouyang, H. Li, X. Wang, Object detection from video tubelets with convolutional neural networks, in *Proceedings of the IEEE Conference on Computer Vision and Pattern Recognition* (2016), pp. 817–825
43. S. Zha, F. Luisier, W. Andrews, N. Srivastava, R. Salakhutdinov, Exploiting image-trained CNN architectures for unconstrained video classification, in *BMVC* (2015)
44. T. Pfister, K. Simonyan, J. Charles, A. Zisserman, Deep convolutional neural networks for efficient pose estimation in gesture video, *Asian Conf. Comput. Vis.* 538–552 (2014)
45. C. Szegedy, W. Liu, Y. Jia, P. Sermanet, S. Reed, D. Anguelov, D. Erhan, V. Vanhoucke, A. Rabinovich, Going deeper with convolutions, in *Proceedings of the IEEE conference on computer vision and pattern recognition* (2015), pp. 1–9
46. K. He, X. Zhang, S. Ren, J. Sun, Deep residual learning for image recognition, in *Proceedings of the IEEE Conference on Computer Vision and Pattern Recognition* (2016), pp. 770–778
47. M.D. Zeiler, R. Fergus, Visualizing and understanding convolutional networks, in *European Conference on Computer Vision* (2014), pp. 818–833
48. K. Simonyan, A. Zisserman, *Very Deep Convolutional Networks for Large-Scale Image Recognition* (2014). abs/1409.1556
49. N. McLaughlin, J.M.D. Rincon, P. Miller, Recurrent convolutional network for video-based person re-identification, in *2016 IEEE Conference on Computer Vision and Pattern Recognition (CVPR)* (2016), pp. 1325–1334
50. M.U. Yaseen, A. Anjum, N. Antonopoulos, Modeling and analysis of a deep learning pipeline for cloud based video analytics, in *Proceedings of the Fourth IEEE/ACM International Conference on Big Data Computing, Applications and Technologies (BDCAT 2017)*
51. S. Chen, N. Ram, DISCOVER: discovering important segments for classification of video events and recounting. IEEE Conf. Comput. Vis. Pattern Recognit. (2014)
52. A. Habibian, C.G.M. Snoek, Recommendations for recognizing video events by concept vocabularies. Comput. Vis. Image Underst. **124**, 110–122 (2014)
53. H. Song, X. Wu, *Extracting Key Segments of Videos for Event Detection by Learning From Web Sources*, vol. 20, no. 5 (2018), pp. 1088–1100
54. H. Wang, X. Wu, Y. Jia, Video annotation via image groups from the web. IEEE Trans. Multimed. **16**(5), 1282–1291 (2014)
55. X. Zhang, Y. Yang, Y. Zhang, H. Luan, J. Li, H. Zhang, T. Chua, Enhancing video event recognition using automatically constructed semantic-visual knowledge base. IEEE Trans. Multimed. **17**(9), 1562–1575 (2015)

56. H. Wang, H. Song, X. Wu, Y. Jia, Video annotation by incremental learning from grouped heterogeneous sources, in *Asian Conference on Computer Vision* (2014), pp. 493–507

57. M. Long, J. Wang, G. Ding, S.J. Pan, P.S. Yu, Adaptation regularization: a general framework for transfer learning. IEEE Trans. Knowl. Data Eng. **26**(5), 1076–1089 (2014)

58. L. Duan, D. Xu, S. Chang, Exploiting web images for event recognition in consumer videos: A multiple source domain adaptation approach, in *2012 IEEE Conference on Computer Vision and Pattern Recognition* (2012), pp. 1338–1345

59. H. Song, X. Wu, W. Liang, Y. Jia, Recognizing key segments of videos for video annotation by learning from web image sets. Multimed. Tools Appl. **76**(5), 6111–6126 (2017)

60. K. Tang, L. Fei-Fei, D. Koller, Learning latent temporal structure for complex event detection, in *2012 IEEE Conference on Computer Vision and Pattern Recognition* (2012), pp. 1250–1257

61. W. Li, Q. Yu, A. Divakaran, N. Vasconcelos, Dynamic pooling for complex event recognition, in *2013 IEEE International Conference on Computer Vision* (2013), pp. 2728–2735

62. P. Over, G. M. Awad, J. Fiscus, M. Michel, A. F. Smeaton, W. Kraaij, Trecvid 2009-goals tasks data evaluation mechanisms and metrics. *TRECVid Work, 2009* (2010)

63. H.J. Escalante, I. Guyon, V. Athitsos, P. Jangyodsuk, J. Wan, Principal motion components for one-shot gesture recognition. Pattern Anal. Appl. **20**(1), 167–182 (2017)

64. J. Wan, Q. Ruan, W. Li, S. Deng, One-shot learning gesture recognition from RGB-D data using bag of features. J. Mach. Learn. Res. **14**, 2549–2582 (2013)

65. J. Wan, V. Athitsos, P. Jangyodsuk, H.J. Escalante, Q. Ruan, I. Guyon, CSMMI: class-specific maximization of mutual information for action and gesture recognition. IEEE Trans. Image Process. **23**(7), 3152–3165 (2014)

66. D. Wu, F. Zhu, L. Shao, One shot learning gesture recognition from RGBD images, in *2012 IEEE Computer Society Conference on Computer Vision and Pattern Recognition Workshops* (2012), pp. 7–12

67. F. Jiang, S. Zhang, S. Wu, Y. Gao, D. Zhao, Multi-layered gesture recognition with kinect. J. Mach. Learn. Res. **16**, 227–254 (2015)

68. X. Yang, C. Zhang, and Y. Tian, Recognizing actions using depth motion maps-based histograms of oriented gradients, in *Proceedings of the 20th ACM International Conference on Multimedia* (2012), pp. 1057–1060

69. W. Li, Z. Zhang, Z. Liu, Action recognition based on a bag of 3D points, in *2010 IEEE Computer Society Conference on Computer Vision and Pattern Recognition—Workshops* (2010), pp. 9–14

70. O. Oreifej, Z. Liu, HON4D: histogram of oriented 4D normals for activity recognition from depth sequences, in *Proceedings of the 2013 IEEE Conference on Computer Vision and Pattern Recognition* (2013), pp. 716–723

71. X. Yang, Y. Tian, Super normal vector for activity recognition using depth sequences, in *2014 IEEE Conference on Computer Vision and Pattern Recognition* (2014), pp. 804–811

72. C. Lu, J. Jia, C. Tang, Range-Sample depth feature for action recognition, in *2014 IEEE Conference on Computer Vision and Pattern Recognition* (2014), pp. 772–779

73. P. Wang, W. Li, Z. Gao, C. Tang, J. Zhang, P. Ogunbona, ConvNets-Based action recognition from depth maps through virtual cameras and pseudocoloring, in *Proceedings of the 23rd ACM International Conference on Multimedia* (2015), pp. 1119–1122

74. P. Wang, W. Li, Z. Gao, J. Zhang, C. Tang, P.O. Ogunbona, Action recognition from depth maps using deep convolutional neural networks. IEEE Trans. Hum.-Mach. Syst. **46**(4), 498–509 (2016)

75. D. Wu, L. Pigou, P. Kindermans, N.D. Le, L. Shao, J. Dambre, J. Odobez, Deep dynamic neural networks for multimodal gesture segmentation and recognition. IEEE Trans. Pattern Anal. Mach. Intell. **38**(8), 1583–1597 (2016)

76. Y. Hou, S. Wang, P. Wang, Z. Gao, W. Li, Spatially and temporally structured global to local aggregation of dynamic depth information for action recognition. IEEE Access **6**, 2206–2219 (2018)

77. P. Wang, S. Wang, Z. Gao, Y. Hou, W. Li, Structured images for RGB-D action recognition, in *2017 IEEE International Conference on Computer Vision Workshops (ICCVW)* (2017), pp. 1005–1014

78. D. Tran, L. Bourdev, R. Fergus, L. Torresani, M. Paluri, Learning spatiotemporal features with 3D convolutional networks, in *Proceedings of the IEEE International Conference on Computer Vision* (2015), pp. 4489–4497
79. S. Ji, W. Xu, M. Yang, K. Yu, 3D convolutional neural networks for human action recognition. IEEE Trans. Pattern Anal. Mach. Intell. **35**(1), 221–231 (2013)
80. V. Veeriah, N. Zhuang, G.-J. Qi, Differential recurrent neural networks for action recognition, in *Proceedings of the IEEE international conference on computer vision* (2015), pp. 4041–4049
81. Y. Du, W. Wang, L. Wang, Hierarchical recurrent neural network for skeleton based action recognition, in *2015 IEEE Conference on Computer Vision and Pattern Recognition (CVPR)* (2015), pp. 1110–1118
82. J. Liu, A. Shahroudy, D. Xu, G. Wang, Spatio-temporal lstm with trust gates for 3d human action recognition, in *European Conference on Computer Vision* (2016), pp. 816–833
83. P. Wang, W. Li, S. Member, Z. Gao, C. Tang, P. O. Ogunbona, S. Member, *Depth Pooling Based Large-Scale 3-D Action Recognition With Convolutional Neural Networks*, vol. 20, no. 5 (2018), pp. 1051–1061
84. P. Ochs, J. Malik, T. Brox, *Segmentation of Moving Objects by Long Term Video Analysis*, vol. 36, no. 6 (2014), pp. 1187–1200
85. R. Cucchiara, C. Grana, M. Piccardi, A. Prati, Detecting objects, shadows and ghosts in video streams by exploiting color and motion information, in *Proceedings of 11th International Conference on Image Analysis and Processing, 2001*(2001), pp. 360–365
86. C. Beyan, A. Temizel, Adaptive mean-shift for automated multi object tracking. IET Comput. Vis. **6**(1), 1–12 (2012)
87. B. Risse, M. Mangan, B. Webb, L.D. Pero, Visual tracking of small animals in cluttered natural environments using a freely moving camera, in *2017 IEEE International Conference on Computer Vision Workshops (ICCVW)* (2017), pp. 2840–2849
88. A. Sobral, A. Vacavant, A comprehensive review of background subtraction algorithms evaluated with synthetic and real videos. Comput. Vis. Image Underst. **122**, 4–21 (2014)
89. T. Bouwmans, Recent advanced statistical background modeling for foreground detection—a systematic survey. Recent Patents Comput. Sci. **4**(3), 147–176 (2011)
90. V. Sharma, N. Nain, T. Badal, A survey on moving object detection methods in video surveillance. Int. Bull. Math. Res. **2**(1), 2019–2218 (2015)
91. A. Yilmaz, O. Javed, M. Shah, Object tracking. ACM Comput. Surv. **38**(4) (2006)
92. T. Bouwmans, Traditional and recent approaches in background modeling for foreground detection: an overview. Comput. Sci. Rev. **11–12**, 31–66 (2014)
93. S. Shantaiya, K. Verma, K. Mehta, A survey on approaches of object detection. Int. J. Comput. Appl. **65**(18), 14–20 (2013)
94. B. Deori, D.M. Thounaojam, A survey on moving object tracking in video. Int. J. Inf. Theory **3**(3), 31–46 (2014)
95. L. Leal-Taixé, A. Milan, K. Schindler, D. Cremers, I. Reid, S. Roth, Tracking the trackers: an analysis of the state of the art in multiple object tracking (2017). arXiv1704.02781
96. M. Yazdi, T. Bouwmans, New trends on moving object detection in video images captured by a moving camera: a survey. Comput. Sci. Rev. **28**, 157–177 (2018)
97. P. Delagnes, J. Benois, D. Barba, Active contours approach to object tracking in image sequences with complex background. Pattern Recognit. Lett. **16**(2), 171–178 (1995)
98. C.R. Wren, A. Azarbayejani, T. Darrell, A.P. Pentland, P finder: real-time tracking of the human body. IEEE Trans. Pattern Anal. Mach. Intell. **19**(7), 780–785 (1997)
99. Hayman and Eklundh, Statistical background subtraction for a mobile observer, in *Proceedings Ninth IEEE International Conference on Computer Vision*, vol. 1, (2003), pp. 67–74
100. Z. Zivkovic, F. Van Der Heijden, Efficient adaptive density estimation per image pixel for the task of background subtraction. Pattern Recognit. Lett. **27**(7), 773–780 (2006)
101. K.M. Yi, K. Yun, S.W. Kim, H.J. Chang, H. Jeong, J.Y. Choi, Detection of moving objects with non-stationary cameras in 5.8 ms: bringing motion detection to your mobile device, in *2013 IEEE Conference on Computer Vision and Pattern Recognition Workshops*, 2013, pp. 27–34

102. F.A. Setyawan, J.K. Tan, H. Kim, S. Ishikawa, Detection of moving objects in a video captured by a moving camera using error reduction, in *SICE Annual Conference, Sapporo, Japan, (Sept. 2014)* (2004), pp. 347–352
103. Y. Jin, L. Tao, H. Di, N. I. Rao, G. Xu, Background modeling from a free-moving camera by Multi-layer homography algorithm, in *2008 15th IEEE International Conference on Image Processing* (2008), pp. 1572–1575
104. P. Lenz, J. Ziegler, A. Geiger, M. Roser, Sparse scene flow segmentation for moving object detection in urban environments, in *Intelligent Vehicles Symposium (IV), 2011 IEEE* (2011), pp. 926–932
105. L. Gong, M. Yu, T. Gordon, Online codebook modeling based background subtraction with a moving camera," in *2017 3rd International Conference on Frontiers of Signal Processing (ICFSP)*, 2017, pp. 136–140
106. Y. Wu, X. He, T.Q. Nguyen, Moving Object Detection with a Freely Moving Camera via Background Motion Subtraction. IEEE Trans. Circuits Syst. Video Technol. **27**(2), 236–248 (2017)
107. Y. Zhu, A.M. Elgammal, A multilayer-based framework for online background subtraction with freely moving cameras, in *ICCV*, 2017, pp. 5142–5151
108. S. Minaeian, J. Liu, Y.-J. Son, Effective and Efficient Detection of Moving Targets from a UAV's Camera. IEEE Trans. Intell. Transp. Syst. **19**(2), 497–506 (2018)
109. M. Braham, M. Van Droogenbroeck, Deep background subtraction with scene-specific convolutional neural networks, in *IEEE International Conference on Systems, Signals and Image Processing (IWSSIP), Bratislava 23–25 May 2016* (2016), pp. 1–4
110. T. Brox, J. Malik, Object segmentation by long term analysis of point trajectories, in *European Conference on Computer Vision* (2010), pp. 282–295
111. X. Yin, B. Wang, W. Li, Y. Liu, M. Zhang, Background subtraction for moving cameras based on trajectory-controlled segmentation and label inference. KSII Trans. Internet Inf. Syst. **9**(10), 4092–4107 (2015)
112. S. Zhang, J.-B. Huang, J. Lim, Y. Gong, J. Wang, N. Ahuja, M.-H. Yang, *Tracking persons-of-interest via unsupervised representation adaptation* (2017). arXiv1710.02139
113. P. Rodríguez, B. Wohlberg, Translational and rotational jitter invariant incremental principal component pursuit for video background modeling, in *2015 IEEE International Conference on Image Processing (ICIP)* (2015), pp. 537–541
114. S.E. Ebadi, V.G. Ones, E. Izquierdo, Efficient background subtraction with low-rank and sparse matrix decomposition, in *2015 IEEE International Conference on Image Processing (ICIP)* (2015), pp. 4863–4867
115. T. Bouwmans, A. Sobral, S. Javed, S.K. Jung, E.-H. Zahzah, Decomposition into low-rank plus additive matrices for background/foreground separation: A review for a comparative evaluation with a large-scale dataset. Comput. Sci. Rev. **23**, 1–71 (2017)
116. I. Elhart, M. Mikusz, C.G. Mora, M. Langheinrich, N. Davies, F. *Informatics, Audience Monitor—an Open Source Tool for Tracking Audience Mobility in front of Pervasive Display*
117. Intel AIM Suite, *Intel Corporation*. https://aimsuite.intel.com/
118. Fraunhofer IIS, *Fraunhofer AVARD*. http://www.iis.fraunhofer.de/en/ff/bsy/tech/bildanalyse/avard.html
119. G. M. Farinella, G. Farioli, S. Battiato, S. Leonardi, G. Gallo, Face re-identification for digital signage applications, in *Video Analytics for Audience Measurement* (2014), pp. 40–52
120. N. Gillian, S. Pfenninger, S. Russell, and J. A. Paradiso, "Gestures Everywhere: A Multimodal Sensor Fusion and Analysis Framework for Pervasive Displays," in *Proceedings of The International Symposium on Pervasive Displays*, 2014, p. 98:98–98:103
121. G. Tripathi, K. Singh, D. Kumar, Convolutional neural networks for crowd behaviour analysis : a survey. Vis. Comput. (2018)
122. T. Li, H. Chang, M. Wang, B. Ni, R. Hong, *Crowded Scene Analysis : A Survey*, vol. 25, no. 3 (2015), pp. 367–386
123. R. Leggett, *Real-Time Crowd Simulation: A Review* (2004)

124. M. Hu, S. Ali, M. Shah, Detecting global motion patterns in complex videos, in *19th International Conference on Pattern Recognition, 2008. ICPR 2008* (2008), pp. 1–5
125. X. Wang, X. Yang, X. He, Q. Teng, M. Gao, A high accuracy flow segmentation method in crowded scenes based on streakline. Opt. J. Light Electron Opt. **125**(3), 924–929 (2014)
126. S. Wu, B.E. Moore, M. Shah, Chaotic invariants of Lagrangian particle trajectories for anomaly detection in crowded scenes, in *2010 IEEE Computer Society Conference on Computer Vision and Pattern Recognition* (2010), pp. 2054–2060
127. R. Mehran, A. Oyama, M. Shah, Abnormal crowd behavior detection using social force model, in *2009 IEEE Conference on Computer Vision and Pattern Recognition* (2009), pp. 935–942
128. R. Mehran, B.E. Moore, M. Shah, A streakline representation of flow in crowded scenes, in *Computer Vision—ECCV 2010* (2010), pp. 439–452
129. H. Su, H. Yang, S. Zheng, Y. Fan, S. Wei, The large-scale crowd behavior perception based on spatio-temporal viscous fluid field. IEEE Trans. Inf. Forensics Secur. **8**(10), 1575–1589 (2013)
130. M. Hu, S. Ali, M. Shah, Learning motion patterns in crowded scenes using motion flow field, in *2008 19th International Conference on Pattern Recognition* (2008), pp. 1–5
131. P. Jodoin, Y. Benezeth, Y. Wang, Meta-tracking for video scene understanding, in *2013 10th IEEE International Conference on Advanced Video and Signal Based Surveillance* (2013), pp. 1–6
132. Y. Benabbas, N. Ihaddadene, C. Djeraba, Motion pattern extraction and event detection for automatic visual surveillance. J. Image Video Process. **2011**, 7 (2011)
133. S.C. Shadden, F. Lekien, J.E. Marsden, Definition and properties of Lagrangian coherent structures from finite-time Lyapunov exponents in two-dimensional aperiodic flows. Phys. D Nonlinear Phenom. **212**(3–4), 271–304 (2005)
134. L. Kratz, K. Nishino, Tracking pedestrians using local spatio-temporal motion patterns in extremely crowded scenes. IEEE Trans. Pattern Anal. Mach. Intell. **34**(5), 987–1002 (2012)
135. Y. Cong, J. Yuan, J. Liu, Abnormal event detection in crowded scenes using sparse representation. Pattern Recognit. **46**(7), 1851–1864 (2013)
136. M. Lewandowski, D. Simonnet, D. Makris, S.A. Velastin, J. Orwell, Tracklet reidentification in crowded scenes using bag of spatio-temporal histograms of oriented gradients, in *Mexican Conference on Pattern Recognition* (2013), pp. 94–103
137. C. Kuo, C. Huang, R. Nevatia, Multi-target tracking by on-line learned discriminative appearance models, in *2010 IEEE Computer Society Conference on Computer Vision and Pattern Recognition* (2010), pp. 685–692
138. S. B\kak, D.-P. Chau, J. Badie, E. Corvee, F. Brémond, M. Thonnat, Multi-target tracking by discriminative analysis on Riemannian manifold, in *2012 19th IEEE International Conference on Image Processing (ICIP)* (2012), pp. 1605–1608
139. B. Zhou, X. Wang, X. Tang, Understanding collective crowd behaviors: learning a mixture model of dynamic pedestrian-agents, in *2012 IEEE Conference on Computer Vision and Pattern Recognition* (2012), pp. 2871–2878
140. W. Chongjing, Z. Xu, Z. Yi, L. Yuncai, Analyzing motion patterns in crowded scenes via automatic tracklets clustering. China Commun. **10**(4), 144–154 (2013)
141. B. Zhou, X. Wang, X. Tang, Random field topic model for semantic region analysis in crowded scenes from tracklets. CVPR **2011**, 3441–3448 (2011)
142. K. Xu, X. Jiang, T. Sun, *Anomaly Detection Based on Stacked Sparse Coding With Intraframe Classification Strategy*, vol. 20, no. 5 (2018), pp. 1062–1074
143. B.T. Morris, M.M. Trivedi, A survey of vision-based trajectory learning and analysis for surveillance. IEEE Trans. Circuits Syst. Video Technol. **18**(8), 1114–1127 (2008)
144. L. Brun, A. Saggese, M. Vento, Dynamic scene understanding for behavior analysis based on string Kernels. IEEE Trans. Circuits Syst. Video Technol. **24**(10), 1669–1681 (2014)
145. C. Piciarelli, C. Micheloni, G.L. Foresti, Trajectory-Based anomalous event detection. IEEE Trans. Circuits Syst. Video Technol. **18**(11), 1544–1554 (2008)
146. D. Tran, J. Yuan, D. Forsyth, Video event detection: from subvolume localization to spatiotemporal path search. IEEE Trans. Pattern Anal. Mach. Intell. **36**(2), 404–416 (2014)

147. S. Coşar, G. Donatiello, V. Bogorny, C. Garate, L.O. Alvares, F. Brémond, Toward abnormal trajectory and event detection in video surveillance. IEEE Trans. Circuits Syst. Video Technol. **27**(3), 683–695 (2017)
148. X. Song, X. Shao, Q. Zhang, R. Shibasaki, H. Zhao, J. Cui, H. Zha, A fully online and unsupervised system for large and high-density area surveillance: tracking, semantic scene learning and abnormality detection. ACM Trans. Intell. Syst. Technol. **4**(2), 35:1–35:21 (2013)
149. A.R. Revathi, D. Kumar, An efficient system for anomaly detection using deep learning classifier. Signal, Image Video Process. **11**(2), 291–299 (2017)
150. O.P. Popoola, K. Wang, Video-Based abnormal human behavior recognition—a review. IEEE Trans. Syst. Man Cybern. Part C (Applications Rev.) **42**(6), 865–878 (2012)
151. Y. Yuan, Y. Feng, X. Lu, Statistical hypothesis detector for abnormal event detection in crowded scenes. IEEE Trans. Cybern. **47**(11), 3597–3608 (2017)
152. G. Xiong, J. Cheng, X. Wu, Y.-L. Chen, Y. Ou, Y. Xu, An energy model approach to people counting for abnormal crowd behavior detection. Neurocomputing **83**, 121–135 (2012)
153. S. Yi, X. Wang, C. Lu, J. Jia, L0 regularized stationary time estimation for crowd group analysis, in *2014 IEEE Conference on Computer Vision and Pattern Recognition* (2014), pp. 2219–2226
154. Y. Zhang, L. Qin, R. Ji, H. Yao, Q. Huang, Social attribute-aware force model: exploiting richness of interaction for abnormal crowd detection. IEEE Trans. Circuits Syst. Video Technol. **25**(7), 1231–1245 (2015)
155. K. Cheng, Y. Chen, W. Fang, Video anomaly detection and localization using hierarchical feature representation and Gaussian process regression, in *2015 IEEE Conference on Computer Vision and Pattern Recognition (CVPR)* (2015), pp. 2909–2917
156. Y. Lee, Y. Yeh, Y.F. Wang, Anomaly detection via online oversampling principal component analysis. IEEE Trans. Knowl. Data Eng. **25**(7), 1460–1470 (2013)
157. B. Krausz, C. Bauckhage, Loveparade 2010: automatic video analysis of a crowd disaster. Comput. Vis. Image Underst. **116**(3), 307–319 (2012)
158. D. Lee, H. Suk, S. Park, S. Lee, Motion influence map for unusual human activity detection and localization in crowded scenes. IEEE Trans. Circuits Syst. Video Technol. **25**(10), 1612–1623 (2015)
159. C.C. Loy, T. Xiang, S. Gong, Salient motion detection in crowded scenes, in *2012 5th International Symposium on Communications, Control and Signal Processing* (2012), pp. 1–4
160. S. Vishwakarma, A. Agrawal, A survey on activity recognition and behavior understanding in video surveillance. Vis. Comput. **29**(10), 983–1009 (2013)
161. J. Xu, S. Denman, S. Sridharan, C. Fookes, R. Rana, Dynamic texture reconstruction from sparse codes for unusual event detection in crowded scenes, in *Proceedings of the 2011 Joint ACM Workshop on Modeling and Representing Events* (2011), pp. 25–30
162. Y. Cong, J. Yuan, Y. Tang, Video anomaly search in crowded scenes via spatio-temporal motion context. IEEE Trans. Inf. Forensics Secur. **8**(10), 1590–1599 (2013)
163. M. Thida, H. Eng, P. Remagnino, Laplacian eigenmap with temporal constraints for local abnormality detection in crowded scenes. IEEE Trans. Cybern. **43**(6), 2147–2156 (2013)
164. K. Yu, Y. Lin, J. Lafferty, Learning image representations from the pixel level via hierarchical sparse coding, in *2011 IEEE Conference on Computer Vision and Pattern Recognition (CVPR)* (2011), pp. 1713–1720
165. W. Li, V. Mahadevan, N. Vasconcelos, Anomaly detection and localization in crowded scenes. IEEE Trans. Pattern Anal. Mach. Intell. **36**(1), 18–32 (2014)
166. C. Lu, J. Shi, J. Jia, Abnormal event detection at 150 FPS in MATLAB, in *2013 IEEE International Conference on Computer Vision* (2013), pp. 2720–2727
167. C. Lu, J. Shi, J. Jia, Scale adaptive dictionary learning. IEEE Trans. Image Process. **23**(2), 837–847 (2014)
168. S. Han, R. Fu, S. Wang, X. Wu, Online adaptive dictionary learning and weighted sparse coding for abnormality detection, in *2013 IEEE International Conference on Image Processing* (2013), pp. 151–155

169. B. Zhao, L. Fei-Fei, E.P. Xing, Online detection of unusual events in videos via dynamic sparse coding. CVPR **2011**, 3313–3320 (2011)
170. A. Kanehira, L. Van Gool, Y. Ushiku, T. Harada, *Viewpoint-aware Video Summarization*
171. D. Potapov, M. Douze, Z. Harchaoui, C. Schmid, Category-Specific Video Summarization, in *Computer Vision—ECCV 2014* (2014), pp. 540–555
172. H. Yang, B. Wang, S. Lin, D.P. Wipf, M. Guo, B. Guo, Unsupervised extraction of video highlights via robust recurrent auto-encoders, in *2015 IEEE International Conference on Computer Vision* (2015), pp. 4633–4641
173. B. Zhao, E.P. Xing, Quasi real-time summarization for consumer videos, in *2014 IEEE Conference on Computer Vision and Pattern Recognition* (2014), pp. 2513–2520
174. A. Sharghi, B. Gong, M. Shah, Query-focused extractive video summarization, in *European Conference on Computer Vision* (2016), pp. 3–19
175. A. Sharghi, J.S. Laurel, B. Gong, Query-focused video summarization: dataset, evaluation, and a memory network based approach, in *The IEEE Conference on Computer Vision and Pattern Recognition (CVPR)*, (2017), pp. 2127–2136
176. R. Panda, A. Das, Z. Wu, J. Ernst, A.K. Roy-Chowdhury, Weakly supervised summarization of web videos, in *2017 IEEE International Conference on Computer Vision (ICCV)* (2017), pp. 3677–3686
177. M. Otani, Y. Nakashima, E. Rahtu, N. Yokoya, *Video Summarization using Deep Semantic Features*, pp. 1–16
178. B. Gong, W.-L. Chao, K. Grauman, F. Sha, Diverse sequential subset selection for supervised video summarization, in *Advances in Neural Information Processing Systems 27*, ed. by Z. Ghahramani, M. Welling, C. Cortes, N.D. Lawrence, K.Q. Weinberger (Curran Associates, Inc., 2014), pp. 2069–2077
179. M. Gygli, H. Grabner, H. Riemenschneider, L. Van Gool, Creating Summaries from User Videos, in *Computer Vision—ECCV 2014* (2014), pp. 505–520
180. M. Gygli, H. Grabner, L. Van Gool, Video summarization by learning submodular mixtures of objectives, in *2015 IEEE Conference on Computer Vision and Pattern Recognition (CVPR)* (2015), pp. 3090–3098
181. A. Kulesza, B. Taskar, others, Determinantal point processes for machine learning. *Found. Trends®in Mach. Learn.* 5(2–3), 123–286 (2012)
182. Y. J. Lee, J. Ghosh, K. Grauman, Discovering important people and objects for egocentric video summarization, in *2012 IEEE Conference on Computer Vision and Pattern Recognition* (2012), pp. 1346–1353
183. D. Liu, G. Hua, T. Chen, A hierarchical visual model for video object summarization. IEEE Trans. Pattern Anal. Mach. Intell. **32**(12), 2178–2190 (2010)
184. B.A. Plummer, M. Brown, S. Lazebnik, Enhancing video summarization via vision-language embedding, in *Computer Vision and Pattern Recognition*, vol. 2 (2017)
185. M. Sun, A. Farhadi, T. Chen, S. Seitz, ranking highlights in personal videos by analyzing edited videos. IEEE Trans. Image Process. **25**(11), 5145–5157 (2016)
186. K. Zhang, W.-L. Chao, F. Sha, K. Grauman, Summary transfer: exemplar-based subset selection for video summarization, in *Proceedings of the IEEE conference on computer vision and pattern recognition* (2016), pp. 1059–1067
187. F. Chen, C. De Vleeschouwer, Formulating team-sport video summarization as a resource allocation problem. IEEE Trans. Circuits Syst. Video Technol. **21**(2), 193–205 (2011)
188. E. Elhamifar, M.C.D.P. Kaluza, Online summarization via submodular and convex optimization, in *CVPR* (2017), pp. 1818–1826
189. M. Fleischman, B. Roy, D. Roy, Temporal feature induction for baseball highlight classification, in *Proceedings of the 15th ACM international conference on Multimedia* (2007), pp. 333–336
190. R. Hong, J. Tang, H.-K. Tan, S. Yan, C. Ngo, T.-S. Chua, Event driven summarization for web videos, in *Proceedings of the First SIGMM Workshop on Social Media* (2009), pp. 43–48
191. A. Khosla, R. Hamid, C. Lin, N. Sundaresan, Large-Scale video summarization using Web-Image priors, in *2013 IEEE Conference on Computer Vision and Pattern Recognition* (2013), pp. 2698–2705

192. G. Kim, L. Sigal, E.P. Xing, Joint summarization of large-scale collections of web images and videos for storyline reconstruction, in *2014 IEEE Conference on Computer Vision and Pattern Recognition* (2014), pp. 4225–4232
193. Z. Lu, K. Grauman, Story-Driven summarization for egocentric video, in *2013 IEEE Conference on Computer Vision and Pattern Recognition* (2013), pp. 2714–2721
194. B. Mahasseni, M. Lam, S. Todorovic, Unsupervised video summarization with adversarial lstm networks, in *The IEEE Conference on Computer Vision and Pattern Recognition (CVPR)*, vol. 1 (2017)
195. Y. Song, J. Vallmitjana, A. Stent, A. Jaimes, TVSum: summarizing web videos using titles, in *2015 IEEE Conference on Computer Vision and Pattern Recognition (CVPR)* (2015), pp. 5179–5187
196. Y.T. Tesfaye, S. Member, E. Zemene, S. Member, *Multi-target Tracking in Multiple Non-overlapping Cameras using Constrained Dominant Sets*, pp. 1–15
197. Y. Wang, S. Velipasalar, M.C. Gursoy, Distributed wide-area multi-object tracking with non-overlapping camera views. Multimed. Tools Appl. **73**(1), 7–39 (2014)
198. Y.T. Tesfaye, E. Zemene, M. Pelillo, A. Prati, Multi-object tracking using dominant sets. IET Comput. Vis. **10**(4), 289–297 (2016)
199. A. Roshan Zamir, A. Dehghan, and M. Shah, GMCP-Tracker: global multi-object tracking using generalized minimum clique graphs, in *Computer Vision—ECCV 2012* (2012), pp. 343–356
200. A. Dehghan, S.M. Assari, M. Shah, GMMCP tracker: globally optimal generalized maximum multi clique problem for multiple object tracking, in *2015 IEEE Conference on Computer Vision and Pattern Recognition (CVPR)* (2015), pp. 4091–4099
201. C.-H. Kuo, C. Huang, R. Nevatia, Inter-Camera association of multi-target tracks by on-line learned appearance affinity models, in *Computer Vision—ECCV 2010* (2010), pp. 383–396
202. D. Cheng, Y. Gong, J. Wang, Q. Hou, N. Zheng, Part-Aware trajectories association across non-overlapping uncalibrated cameras. Neurocomputing **230**, 30–39 (2017)
203. Y. Gao, R. Ji, L. Zhang, A. Hauptmann, symbiotic tracker ensemble toward a unified tracking framework. IEEE Trans. Circuits Syst. Video Technol. **24**(7), 1122–1131 (2014)
204. S. Zhang, Y. Zhu, A. Roy-Chowdhury, Tracking multiple interacting targets in a camera network. Comput. Vis. Image Underst. **134**, 64–73 (2015)
205. Y. Cai, G. Medioni, Exploring context information for inter-camera multiple target tracking, in *IEEE Winter Conference on Applications of Computer Vision* (2014), pp. 761–768
206. J. You, A. Wu, X. Li, W.-S. Zheng, Top-Push video-based person re-identification, in *2016 IEEE Conference on Computer Vision and Pattern Recognition* (2016), pp. 1345–1353
207. A.D. Jepson, D.J. Fleet, T.F. El-Maraghi, Robust online appearance models for visual tracking. IEEE Trans. Pattern Anal. Mach. Intell. **25**(10), 1296–1311 (2003)
208. J. Pan, B. Hu, Robust occlusion handling in object tracking, in *2007 IEEE Conference on Computer Vision and Pattern Recognition* (2007), pp. 1–8
209. A. Yilmaz, X. Li, M. Shah, Contour-based object tracking with occlusion handling in video acquired using mobile cameras. IEEE Trans. Pattern Anal. Mach. Intell. **26**(11), 1531–1536 (2004)
210. A. Senior, A. Hampapur, Y.-L. Tian, L. Brown, S. Pankanti, R. Bolle, Appearance models for occlusion handling. Image Vis. Comput. **24**(11), 1233–1243 (2006)
211. L. Hou, W. Wan, K.-H. Lee, J.-N. Hwang, G. Okopal, J. Pitton, Robust human tracking based on DPM constrained multiple-kernel from a moving camera. J. Signal Process. Syst. **86**(1), 27–39 (2017)
212. H. Zhang, G. Ananthanarayanan, P. Bodik, M. Philipose, P. Bahl, M.J. *Freedman, Live Video Analytics at Scale with Approximation and Delay-Tolerance*
213. G. Ananthanarayanan, P. Bahl, P. Bodík, K. Chintalapudi, M. Philipose, L. Ravindranath, S. Sinha, Real-Time video analytics: the killer app for edge computing. *Computer (Long. Beach. Calif)* **50**(10), 58–67 (2017)
214. F. Loewenherz, V. Bahl, Y. Wang, Video analytics towards vision zero. ITE **87**, 25–28 (2017)

215. H. Qiu, X. Liu, S. Rallapalli, A.J. Bency, K. Chan, *Kestrel: Video Analytics for Augmented Multi-camera Vehicle Tracking* (2018), pp. 48–59
216. J. Redmon, S. Divvala, R. Girshick, A. Farhadi, You only look once: unified, real-time object detection, in *Proceedings of the IEEE conference on computer vision and pattern recognition* (2016), pp. 779–788
217. E. K. Bowman, M. Turek, P. Tunison, S. Thomas, E.K. Bowman, M. Turek, P. Tunison, R. Porter, V. Gintautas, P. Shargo, J. Lin, Q. Li, X. Li, R. Mittu, C.P. Rosé, K. Maki, Advanced text and video analytics for proactive decision making, no. May 2017 (2018)
218. K.P. Seng, *Video Analytics for Customer Emotion and Satisfaction at Contact Centers*, vol. 48, no. 3 (2018), pp. 266–278

Part VII
Learning and Analytics in Integrated Circuits

Chapter 16
Machine Learning in Alternate Testing of Integrated Circuits

John Liaperdos, Angela Arapoyanni and Yiorgos Tsiatouhas

Abstract Integrated circuit (IC) fabrication involves sophisticated and sensitive equipment, while design complexity and scale of integration are increasing in order to obtain the largest functionality within the smallest possible chip area. As ICs scale down, increasing uncertainties in the manufacturing processes lead to increased numbers of malfunctioning chips that have to be detected and withdrawn early in the production stage in order to assure product quality. Testing an IC for compliance with its specifications might require a considerable amount of time and resources, since a wide range of different tests—each tailored for a specific performance characteristic—should be performed, resulting in the collection of extremely large data sets that correspond to test data. In order to reduce test time and cost, several machine learning techniques have been applied for the identification of patterns and non-linear correlations among test data and performance characteristics. Furthermore, data analytics have achieved promising results in areas such as test quality prediction and fault diagnosis. In this chapter, the Alternate Test paradigm is presented since it is probably the most well-established machine learning application in the field of IC testing. Alternate Test is performed by the exploitation of a set of common test observables—produced using simple stimuli and cost-effective equipment of low complexity—that are sufficiently correlated to all performance characteristics of interest. Following the supervised learning approach, non-linear regression models are constructed for each performance characteristic, that provide accurate performance predictions based on the values of the measured test observ-

J. Liaperdos (✉)
Department of Computer Engineering, Technological Educational Institute of Peloponnese, Kladas, Valiotis Building, 23100 Sparta, Greece
e-mail: i.liaperdos@teipel.gr

A. Arapoyanni
Department of Informatics and Telecommunications, National and Kapodistrian University of Athens, Panepistimiopolis, Ilissia, 15784 Athens, Greece
e-mail: arapoyanni@di.uoa.gr

Y. Tsiatouhas
Department of Computer Science and Engineering, University of Ioannina, Panepistimioupolis, 45110 Ioannina, Greece
e-mail: tsiatouhas@cse.uoi.gr

© Springer Nature Switzerland AG 2019   519
G. A. Tsihrintzis et al. (eds.), *Machine Learning Paradigms*,
Learning and Analytics in Intelligent Systems 1,
https://doi.org/10.1007/978-3-030-15628-2_16

ables. Using statistical feature selection techniques, a low dimensionality can be obtained for the set of test observables, by which a further reduction in test time and cost can be achieved.

16.1 Introduction

The application of machine learning principles and techniques to IC design should come as no surprise, especially in the semiconductor industry that manufactures billions of devices per year. IC manufacturing involves increasingly complex design and production cycles, resulting in an exponential growth of the amount of related data. Ensuring that all devices that reach their target market perform as specified is an interesting challenge for both chip designers and quality-control engineers, which today exploit data analysis to ensure quality in IC mass production [18, 47].

Several problems that can be addressed using machine learning methods are common in semiconductor manufacturing, such as *function learning* (to determine the optimum circuit that, given a specific input, produces a certain output), *prediction* (to indirectly determine chip performance), *classification* (to classify a device as functional or faulty, given a set of test results and previous training data), *recommendation* (to recommend a new combination of design choices based on prior designers' experience), etc. These problems appear in several procedures that span the whole chip design and production cycle, namely *physical design* [19], *routing* [35], *design optimization* [33, 34, 43], *fault diagnosis* [15, 28, 29, 39], *performance representation* [5], *testing* [3, 7, 11, 24, 31, 44], *calibration* [7, 12–14, 24], *health estimation* [45] and *security* [2, 36], to mention only the most representative.

Among existing machine learning techniques that are successfully applied to IC design and production, the *Alternate Test* [3, 44] paradigm is presented in this chapter since it is probably the most well-established technique in the field of IC testing, leading to significant test time and cost reduction when applied to the analog blocks of System-on-Chip (SoC) and System-in-Package (SiP) integrated circuits.

The following sections provide an overview of the Alternate Test paradigm as a supervised statistical learning technique, as well as a detailed presentation of a typical Alternate Test workflow to test and calibrate analog/radio-frequency (RF) blocks in SoC or SiP integrated circuits. Specifically, Sect. 16.2 provides a description of Alternate Test, including its advantages and limitations, as well as typical applications of this paradigm to IC production. In Sect. 16.3 a typical workflow for testing and calibration of analog/RF ICs is presented, that is based on the Alternate Test methodology. Finally, Sect. 16.4 briefly describes techniques that have been proposed in order to improve the efficiency of Alternate Test.

16.2   Alternate Test

By contributing to a large portion of the total production cost of integrated circuits, testing becomes a major concern especially in the case of high frequency analog ICs. The standard industry practice for production testing of analog circuits is specification testing, wherein all performance characteristics are measured and, subsequently, compared to their specification limits to reach a pass or fail decision regarding chip functionality.

A main contributor to the raise of conventional specification testing cost is the expensive automatic test equipment (ATE) used to measure the performance characteristics of the circuit under test (CUT). Although these measurements are simple, they require a variety of test resources which, together with the long test application times, increase the total manufacturing cost. Limitations on conventional testing are also posed by the difficulty of the ATE to directly access all, or even part of the internal nodes of an IC especially in SoC or SiP designs. Although some internal signals can be made available to the external tester, frequency limitations due to lower speed of the input/output (I/O) interface may not permit their direct observation. Thus testing cost is increasing and tends to be comparable to the rest manufacturing cost.

To overcome the cost and inabilities of functional testing, the concept of Alternate Test was proposed [44], according to which all circuit performance characteristics are indirectly determined (predicted) from the circuit's response to a suitable test stimulus. Test response collection is significantly simplified by using a common test configuration (exploiting a low cost ATE) and by applying the same stimulus in order to test multiple specifications concurrently.

16.2.1   Description

With Alternate Test, a whole series of conventional specification tests is replaced by a single common test procedure. The principle of Alternate Test is presented in Fig. 16.1 and can be synopsized as follows:

An optimally selected stimulus is applied to the CUT and circuit response is measured at specific test points (test observables). The resulting test observables are processed in order to concurrently predict all circuit performance characteristics of interest [44]. The procedure is concluded with the comparison of the predicted performances to their specified counterparts, by which a decision can be made regarding the circuit's compliance with its specifications.

Performance prediction is possible due to the correlation between process variations, performance characteristics and test response, as shown in Fig. 16.2, since performance variation is apparently a result of process variation. Due to the large number of the pertinent device parameters and the similarly large number of performance characteristics that have to be determined, it is impossible to find a closed-form mathematical relationship to describe the correlation between performance

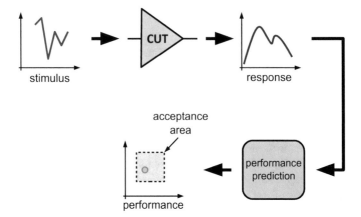

Fig. 16.1 Alternate test principle

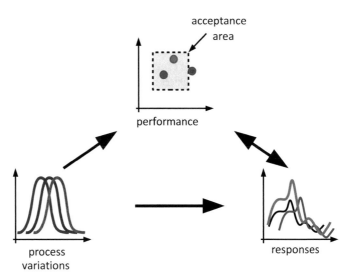

Fig. 16.2 Correlation between process variations, performance and response

and process variations. Moreover, the measurement of all circuit device parameters is impractical. Alternate Test exploits the idea that, with a proper stimulus selection, the corresponding response will be affected by the same variation of circuit device parameters that also affect performance. In such a case, statistical analysis of a large sample of circuits will be able to reveal the correlation between response and performance, providing a prediction mechanism for the latter avoiding direct measurement.

The aforementioned prediction procedure follows a nonlinear multivariate statistical analysis, that allows the construction of a set of functions (one for each performance characteristic) which map circuit response to the value of the corresponding performance characteristic.

Critical factors that determine prediction accuracy are the following:

- Proper stimulus selection, to maximize correlation between response and performance
- Proper selection of test observables, to allow prediction for multiple performance characteristics
- Exploitation of a suitable supervised learning procedure, that is applied to a training set of circuits, for which performance values are predetermined using conventional methods.

The first Alternate Test implementations for low-frequency analog circuits were using time-domain oversampling for response acquisition. However, the application of similar techniques for high frequencies imposes serious difficulties. To address these problems, frequency conversion has been proposed for both the stimulus and the test response using mixing techniques [46]. Alternatively, to avoid sampling issues, the use of sensors has been proposed in order to convert high-frequency responses to DC voltages [6].

For analog blocks embedded in SoCs and SiPs, built-in Alternate Test implementations have been proposed, to overcome accessibility and observability problems that are related to circuit stimulation and response measurement [3].

16.2.2   Advantages and Limitations

Compared to conventional test methods, Alternate Test exhibits significant advantages, the most important of which are the following:

- Allows prediction of multiple performances, based on a single circuit response.
- Uses a common and rather simple test configuration for all performance characteristics and also a common stimulus, leading to o significant reduction in test time and cost.
- Uses less complex and less expensive equipment for both circuit stimulation and response measurements.

Compromised prediction accuracy is probably the main disadvantage of Alternate Test, in the case of not well established test procedures. Proper selection of the test observables is crucial in order to obtain accurate prediction models [11]. Furthermore, high prediction reliability is obtained by using defect filters [17, 40] that detect and exclude outlier circuits for which performance prediction is expected to be inaccurate.

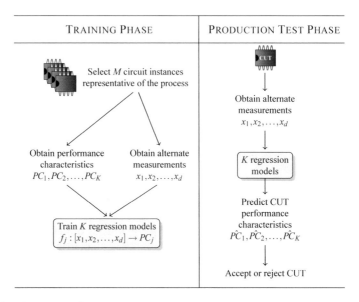

Fig. 16.3 Alternate Test flow

16.2.3 Methodology

The Alternate Test flow is illustrated in Fig. 16.3. Let $PC_1, PC_2, \ldots, PC_K$ denote K CUT performance characteristics that need to be determined. Let also $x = [x_1, x_2, \ldots, x_d]$ denote a set of d low-cost alternate measurements. The mapping between x and performance characteristic PC_j, denoted by f_j,

$$f_j : x \to PC_j, \ j = 1, 2, \ldots, K \tag{16.1}$$

is constructed through regression modeling in a preliminary training phase that employs M circuits collected from different lots, wafers, and sites on the same wafer, such that they are as representative as possible of the fabrication process.

In actual production testing alternate measurements are taken from each CUT, that allow the prediction of its performance characteristics $\hat{PC}_1, \hat{PC}_2, \ldots, \hat{PC}_K$. Based on the values of the latter, a binary pass/fail decision regarding acceptance or rejection of the CUT is taken.

Meaningful application of the Alternate Test approach should be assured by the exclusion of defective circuits from the training set, since the statistical distribution of defective circuits does not match the distribution of the fabrication process. For this reason, it is assumed that defective circuits are filtered-out a priori, by the application of defect filtering techniques proposed in the literature [40].

16.2.4 Typical Applications

Apart from being a self-sufficient IC testing method, Alternate Test can be also used in combination with other testing techniques, in order to achieve increased test efficiency. Furthermore, it can be exploited in order to perform chip performance restoration, as described in the following.

16.2.4.1 Testing

As shown in Fig. 16.3, the final outcome of the production test phase of Alternate Test consists of the predicted CUT performances, from which a binary pass/fail decision can be made regarding CUT functionality. If at least one performance characteristic violates the corresponding specification, the CUT is discarded with no further action.

Overall testing accuracy can be significantly improved when Alternate Test is combined with defect-oriented testing (DOT) techniques, as shown in Sect. 16.3. Defect-oriented testing, or structural testing, follows the assumption that most or all defect mechanisms manifest themselves in more fundamental observables than the specifications, thus simplifying the test procedure and also reducing cost. DOT efficiency is primarily determined by the defect detection capability obtained by the selected observables and the cost for their stimulation and measurement [38].

16.2.4.2 Calibration

Built-in circuitry that provides IC post-fabrication adjustment capabilities may allow restoration of failing performance characteristics to the acceptable limits, as described below. This circuit *calibration* procedure addresses the problem of increased yield loss, that is a result of parametric variations, adequate control of which is extremely difficult with the continuous reduction of IC dimensions. In the case of analog/RF circuits, this results in large portions of the manufactured ICs exhibiting performance outside their prescribed nominal limits, with a direct impact on the cost of the final product.

After fabrication, calibration of circuits found to be non-compliant with the specifications is achieved by shifting them to a compliant state of operation either by the direct adjustment of a circuit element (e.g. by changing the effective length of an integrated variable inductor [8], or by switching on or off transistors in a digitally controllable current source [37]). Modification of the value of an external circuit parameter (e.g. the supply voltage [13], or a specific bias voltage [9]) has also been proposed to perform calibration. A critical issue to be addressed in the calibration procedure is the appropriate selection of the specific circuit's state at which performance is restored to acceptable levels, or, equivalently, the accurate correlation of the adjustable elements' values to the corresponding circuit's performance.

For non-adjustable circuits, common Alternate Test practice implies that the test observables are measured at the circuit's single state in order to derive regression models for the prediction of the circuit's performance characteristics. In the case of an adjustable circuit, however, where more than one states are available, this approach, where a single state is utilized, may be characterized by reduced test efficiency since the adjustment mechanism is not extensively utilized. Consequently, test observables must be measured in various states or even all states aiming to increase the test efficiency.

The fundamental principle of Alternate Test-based calibration methods consists of two phases: *the measurement phase*, during which the performance characteristics of the circuit are determined, and the *adjustment phase* where performances are restored to their acceptable ranges, if necessary and if possible.

- **Measurement phase**: In order to determine if calibration is necessary, all performance characteristics of interest are determined during the measurement phase. This is achieved either through conventional specification-based tests—where performance characteristics are measured directly—or using the Alternate Test approach.
- **Adjustment phase**: The performance characteristic values that have been determined during the measurement phase are exploited, in order to find the optimal circuit adjustment that ensures restoration of performance.

16.3 A Typical Workflow for Testing and Calibration of Analog/RF ICs

In this section we present a representative strategy for testing and calibration of adjustable analog/RF circuits [24]. The testing part combines the defect-oriented and the alternate test approaches, exploiting the advantages of both. Test observables in multiple states are considered in order to develop regression models of high accuracy for the prediction of the circuit's performance characteristics in each state of operation. The set of test observables is also measured in multiple states during the test phase of a fabricated circuit. This multi-state measurement strategy drastically improves the prediction of the circuit's performance characteristics and provides high test coverage, which is maximized by extending the test observables used for defect detection. Performance prediction also permits the accurate characterization of a defect-free circuit as compliant with the specifications or not, in every state of operation using the predicted performance characteristics in each state. In case that the circuit is compliant in at least one state, effective 'one-step' calibration [13, 14] is performed by selecting the state that best fits to the circuit's specifications and tuning accordingly the existing adjustment mechanism.

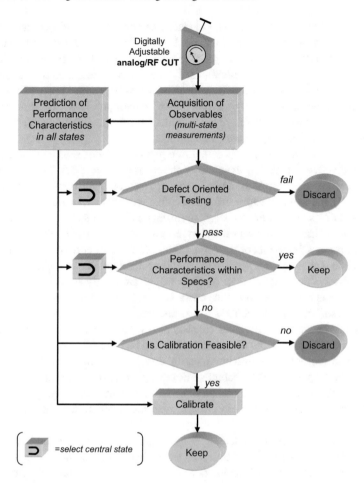

Fig. 16.4 Testing and calibration flow

16.3.1 Overview

Digitally adjustable circuits are considered, that are designed such as to be able to operate in several discrete states providing the capability to vary their performance characteristics around their post-fabrication values. This functionality is obtained by the use of at least one adjustable element, the value of which is related to the circuit's performance characteristics under consideration.

The described test and calibration methodology is illustrated in Fig. 16.4 and can be synopsized as follows. The adjustable analog/RF CUT is measured in various states to obtain a specific set of test observables according to the alternate test approach. Then its performance characteristics are predicted for all states of operation using pre-developed regression models. The set of measured observables and the predicted performance characteristics are used for defect detection. The predicted performance characteristics in a single state (the central state) of operation can be sufficient for defect detection, as it will be shown in Sect. 16.3.2. Defect-free circuits are examined to determine if their predicted performance characteristics in the central state comply with the specifications. For each circuit found to be incompliant, the predicted performance characteristics in the remaining states are used to explore the ability to calibrate it. Finally, circuits for which their predicted performance characteristics in at least one state are compliant with the specifications are calibrated exploiting the existing adjustable element.

Let us assume that the circuit's calibration procedure focuses on a set of K performance characteristics (PCs), denoted by the vector $PC = (\text{PC1}, \text{PC2}, \ldots, \text{PCK})$. If, furthermore, the circuit is designed to operate in N discrete states S_i, ($i = 1, 2, \ldots, N$), the value of each performance characteristic will differ in each state of operation. For a specific circuit design, performance is also expected to differ among manufactured circuit instances subject to process variations and device mismatches.

An optimally selected set of test observables (voltages or currents) is exploited for performance characteristics' prediction and defect detection. The optimal selection of these observables implies that: (a) their values are highly sensitive to the occurrence of defects and (b) performance characteristics' predictions based on these values provide sufficient coefficients of determination, allowing an adequately effective calibration. The set of observables consists of q values per circuit's state of operation; however a subset of the N circuit's states can be used. The set of observables is denoted by a vector O with maximum cardinality $N \times q$. This vector is used to predict the circuit's performance characteristics using pre-developed regression models, to accomplish defect detection and finally—in a defect-free case—to perform a calibration procedure so that the circuit will meet the specifications, as illustrated in Fig. 16.5.

In order to accurately predict all K performance characteristics in all N states of operation for a given circuit instance, proper predictive models are required that are capable to map the obtained vector O to these performance characteristics PC, as analyzed in the following subsection.

16.3.1.1 Predictive Models' Training Phase

Following the alternate testing approach, test observable measurements on a sufficiently large number (M) of actual circuit instances, corresponding to common process perturbations, are used to build $N \times K$ predictive models (training phase), where N and K stand for the number of the circuit's states and the number of performance characteristics under consideration, respectively. These are nonlinear models, which are constructed using statistical regression methods (e.g. multivariate adaptive regression splines (MARS) [10]), and map the vector O of the observables to all K performance characteristic values for all N states of the circuit's operation.

An alternative option, in case that it is not feasible to perform measurements on actual circuits or the existing sample of circuits is small, is to obtain the set of observables O and the set of performance characteristics PC for the M circuit

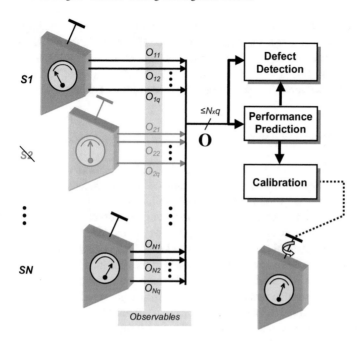

Fig. 16.5 Circuit measurements, defect detection and calibration (overview)

instances by Monte Carlo simulations, exploiting the statistical models of the used technology, in order to develop the required predictive models.

16.3.1.2   Variation Bands Estimation Phase

The first step towards defect detection is the derivation of the expected range of values, due to process variations and device mismatches, for each individual observable as well as for each predicted performance characteristic. The derivation of these ranges, for which we will adopt the term "*variation bands*" [24], assumes that the circuit is set to operate in its "central" state (denoted by S_c). Note that in this state and under typical process conditions the circuit exhibits performance characteristic values which correspond to their nominal counterparts. In other words, a fabricated circuit instance with device parameters that correspond to typical process parameters is identical to the so-called "golden circuit" while operating in its central state of operation.

To maximize defect detection efficiency, we may consider to extend the initial set of observables (O) to a superset of extended observables $E = (E1, E2, \ldots, Ep)$, with $p \geq$ cardinality(O), where Ei ($1 \leq i \leq p$) represents either an initial observable or a simple linear combination of these observables. The latter can be determined by specifying correlations between elements of O, either through empirical observation or via principal component analysis (PCA).

The derivation of the corresponding variation bands is performed by statistical analysis either on the M actual instances or on the M Monte-Carlo simulated instances considered in the training phase, using the same set of observables (O) as in the previous subsection, since the required input data are identical. Figure 16.6a summarizes the overall procedure followed for the derivation of the variation bands.

16.3.1.3 Defect Detection Phase

During the actual testing of a fabricated circuit the test observables are measured in multiple states of operation and performance characteristics are predicted for all states using the regression models already available.

Defect detection is carried out after the extended observables are calculated, as illustrated in Fig. 16.6b. In addition, the predicted performance characteristics for the circuit's central state of operation are considered. Defect detection is accomplished according to the following rule: If at least one of the extended observables (in a DOT sense) or at least one of the PCs (in an alternate test sense) fails to fall within its corresponding variation band, the CUT is classified as defective and discarded, otherwise it is considered to be free of defects. In the latter case, a calibration procedure is initiated in case that the predicted PCs in the central state do not comply with the specifications, in order to reduce the parametric yield loss, as described in the next subsection. We should mention that the specifications define an operation band for the PCs that is a sub-space of the variation band. Also, note that the above procedure actually implements a p-input hyper-rectangular defect filter [17, 40] of $K + p$ dimensions which incorporates a conventional p-dimensional filter augmented by the K predicted PCs, which are introduced in order to reveal possible correlations between the elements of E and, thus, improve the filter's selectivity.

16.3.1.4 Calibration Phase

The calibration procedure determines the circuit's state of operation for which all predicted performance characteristics (PCs) comply with their specifications. This is done using the regression models available from the training phase for all K PCs and all N states of the circuit's operation, according to the principle described in Fig. 16.7.

The calibration procedure attempts to find the state(s) of operation (S_i) for which all PCs are compliant with the specifications. In case the number of such states is greater or equal to one, then the specific circuit instance is correctable, else it is uncorrectable. In Fig. 16.7, an example of a correctable circuit is shown, since for the state $S2$ all predicted PCs meet the specifications simultaneously. On the contrary, for the uncorrectable circuit instance shown in the same figure, no state exists for which all predicted PCs fall inside their acceptable ranges. In such a case the calibration procedure either classifies the circuit as unacceptable, and the circuit is discarded,

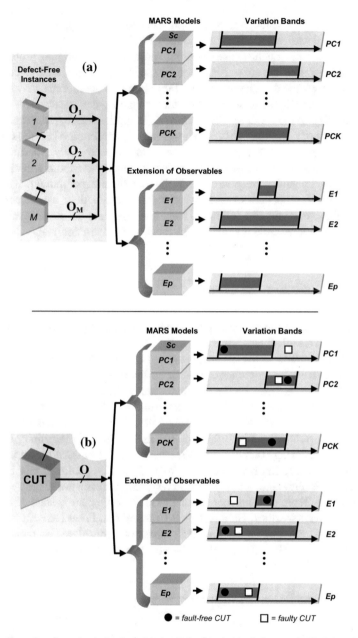

Fig. 16.6 Procedure for **a** the derivation of the variation bands (shaded areas) and **b** defect detection

or as a second-class instance to be used in less competitive applications, operating in the state that sets its PCs closer to the specifications.

Mainly depending on the density of states, there is a non-negligible probability for the existence of multiple states for which all specifications are met in a correctable circuit; therefore, the designer may tailor the calibration procedure such as to define a priority list for PCs or to set other custom rules in order to select the preferred state, according to the needs and the nature of the application (e.g. the state with the lowest power consumption [16]). When this state is chosen, the circuit is forced to operate in the specific state, and calibration terminates successfully. Alternatively, the proposed calibration procedure can be applied to all analog/RF circuits which form a system's full transmit or receive chain. If the specifications of individual stages are not well defined, relaxation of various individual specifications can be considered, based on the preceding and following stages, to maintain an overall performance within specifications, even if some blocks function far from nominal.

16.3.2 RF Mixer Case Study

The effectiveness of the previously presented methodology has been evaluated by simulations on a typical receiver RF mixer.

The RF mixer under consideration, presented in Fig. 16.8, is designed in a 0.18μm Mixed-Signal/RF CMOS technology with an intermediate frequency (IF) of 150 MHz. Its topology is based on the double balanced Gilbert cell Mixer [30].

The mixer consists of an RF stage (transistors M1, M2) and a differential local oscillator (LO) stage (transistors M3–M6). The RF transistors transform the input voltage into current, which is further commutated to the complementary IF outputs each LO period through output load resistances R1, R2. Degeneration inductors L1, L2 are introduced to improve linearity. The main device parameters of the design are presented in Fig. 16.8, while the mixer's typical performance characteristics are summarized in Table 16.1 and correspond to the mixer's central state of operation. From these characteristics, Gain (G), 1dB Compression Point (1dB CP) and Input Referred 3rd Order Intercept Point (IP3) have been considered as targets for the calibration process ($K = 3$); therefore the vector of performance characteristics is $PC = $ (G, 1dB CP, IP3).

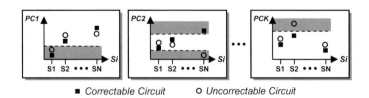

Fig. 16.7 Calibration principle. (Shaded areas indicate non-compliant performance characteristic ranges, while S_i ($i = 1, 2, \ldots, N$) correspond to the circuit's states of operation)

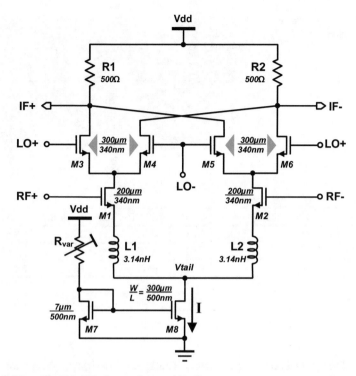

Fig. 16.8 The adjustable RF mixer under consideration

Table 16.1 RF mixer performance characteristics

Parameter	Typical value
f_{RF}	1.9 GHz
f_{IF}	150 MHz
LO power	5.0 dBm
Gain (G)[a]	4.4 dB
NF SSB	12.5 dB
Input IP3[a]	9.7 dBm
Input 1dB CP[a]	−0.2 dBm
Supply voltage	3.3 V
Power consumption	15.4 mW

[a]Calibration target performance characteristics

A digitally controllable resistor R_{var} in the mixer's bias circuitry has been used as the adjustable element, by which the mixer's current is controlled and states of the circuit's operation are selected. The implementation of this resistor is presented in Fig. 16.9, where transistor switches m0–m3 controlled by bits r0–r3 determine the value of R_{var} and, consequently, the mixer's tail current (I), as summarized in

Fig. 16.9 Variable resistor implementation

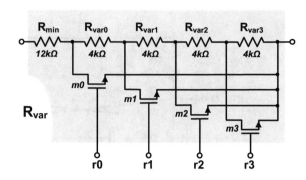

Table 16.2 RF Mixer states of operation

State ID (Si)	Control bits ($r0 \rightarrow r3$)	G (dB)	1dB CP (dBm)	IP3 (dBm)	I (mA)
S1	0000	3.68	−1.77	7.52	3.63
S2	0001	4.03	−0.92	8.60	4.08
S3[a]	*0011*	*4.39*	*−0.19*	*9.73*	*4.67*
S4	0111	4.72	0.62	10.88	5.46
S5	1111	4.85	0.85	10.72	6.58

[a]Central state ($S_c = S3$)

Table 16.2 for typical process parameters. This specific design provides a total number of five states ($N = 5$. Since deviated performance characteristics are expected to lie around their typical values, state *S3* is selected as the central state of operation ($S_c = $ S3) to allow both positive and negative calibration.

16.3.2.1 Voltage Observables' Selection

The use of the LO signal is adopted as the test stimulus at the RF inputs of the mixer [11, 13, 20, 21, 42]. The self-mixing of the LO signal forces the mixer to operate as a homodyne (zero IF) mixer, generating DC voltage levels at its "IF" outputs, accompanied by the higher order mixing products which can be easily removed by low-pass filtering. The aforementioned DC levels (IF_+, IF_-) are used as the main observables, together with the DC voltage component of V_{tail} as shown in Fig. 16.8 ($q = 3$).

Prediction accuracy is expected to improve analogously if the voltage observables are obtained for more than one of the available states provided by the mixer's adjustable element. Moreover, according to the results presented in Sect. 16.3.2.2, only two states are enough to provide very high prediction accuracy. The states adopted in the simulations are the central state ($Sc = S3$) and the maximum tail current state ($S5$). Both voltage triads obtained from these states form the vector of observables **O**, as summarized in Table 16.3.

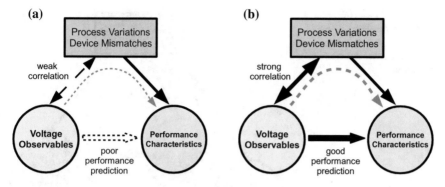

Fig. 16.10 Impact of voltage observable selection on performance prediction efficiency

Theoretically, performance characteristics could be precisely derived if the values of all circuit parameters affected by process variations and device mismatches were known. This is apparently impossible in practice, therefore nonlinear regression models are used to indirectly estimate these parameters through voltage observables and to map the latter to the performance characteristics of interest, as shown in Fig. 16.10. As a result, prediction efficiency is expected to be determined by the strength of the correlation of the selected voltage observables to process variations and device mismatches, according to Fig. 16.10. If voltage observables are weakly correlated to either process variations or device mismatches (Fig. 16.10a), poor performance prediction is expected. On the contrary, a stronger correlation would lead to a more effective prediction (Fig. 16.10b).

While process variations globally affect circuit parameters in a systematic manner, device mismatches are randomly distributed across the circuit's topology. Therefore, the exploitation of voltage observables which are obtained from different circuit nodes (e.g. IF_+, IF_-, V_{tail}) would potentially improve the ability of the regression models to take into account the influence of mismatches, leading to a more effective performance prediction.

A further improvement in performance prediction can be achieved if the input of the regression models is augmented in order to include voltage observable values

Table 16.3 List of voltage observable vector (O) elements

Element ID	DC voltage	Mixer state
O1	IF_+	$S3$[a]
O2	IF_-	$S3$[a]
O3	V_{tail}	$S3$[a]
O4	IF_+	S5
O5	IF_-	S5
O6	V_{tail}	S5

[a]Central state ($S_c = S3$)

Fig. 16.11 Impact of
voltage observables obtained
from different circuit states
on performance prediction
efficiency

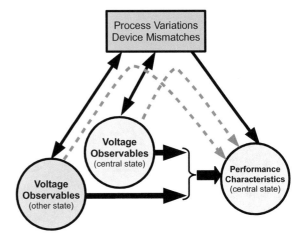

obtained from different circuit states. In such a state, both performance characteristics and voltage observable values differ from their central-state counterparts, while process variations and device mismatches remain common. In Fig. 16.11, the case of an additional state used to enhance the performance prediction accuracy is illustrated.

16.3.2.2 Training—Prediction Models' Development

In the training phase the models required to provide performance prediction in all states of operation are obtained using MARS. Since we consider three performance characteristics (G, 1dB CP, IP3) and five mixer states, there are fifteen prediction models to be derived during the training procedure described in Sect. 16.3.1.1.

The effectiveness of the selected elements in O is justified by the prediction results regarding Gain (G), 1dB Compression Point (1dB CP) and Input Referred 3rd Order Intercept Point (IP3) performance characteristics shown in Table 16.4. In

Table 16.4 Prediction accuracy metrics (Mixer's Central State—S_c) for different voltage observable vectors (O)

O[a]	G		1dB CP		IP3	
	R^2	$RRMSE$	R^2	$RRMSE$	R^2	$RRMSE$
1–2	0.5259	0.6885	0.2077	0.8901	0.2168	0.8850
(1,2)	0.8787	0.3483	0.3100	0.8307	0.3081	0.8318
(1,2,3)	0.9688	0.1766	0.5960	0.6356	0.5693	0.6563
(1,2,4,5)	0.9702	0.1727	0.8891	0.3330	0.8727	0.3567
(1,2, ..., 6)	0.9731	0.1641	0.9464	0.2314	0.9438	0.2371

[a]O elements represented by their ID indices (see Table 16.3)

this table, R^2 and *RRMSE* stand for the prediction's coefficient of determination and the Relative Root Mean Squared Error in predicted values, respectively [4]. *RRMSE* is defined according to the following equation:

$$RRMSE \triangleq \sqrt{\frac{\sum_{i=1}^{N_{ev}} (\widehat{PC_i} - PC_i)^2}{\sum_{i=1}^{N_{ev}} (PC_i - \overline{PC})^2}}, \tag{16.2}$$

where N_{ev} is the number of observations used for model evaluation, PC_i are the actual performance characteristic values, $\widehat{PC_i}$ are the predicted values computed by the model and $\overline{PC}$ represents the sample mean of actual values.

This metric assures that actual values PC_i which are close to the sample mean $\overline{PC}$ lead to a decreased *RRMSE* only if their predicted counterparts $\widehat{PC_i}$ are correspondingly close. On the contrary, actual values that lay far from the mean value lead to an increased *RRMSE* only if their predicted counterparts are correspondingly distant. In other words, prediction error evaluation through *RRMSE* is conducted in a weighted fashion, where error in observations close to their expected value is heavier penalized, compared to observations at the tails of their distribution.

The results shown in Table 16.4 were obtained by using a training set of $M = 500$ defect-free mixer instances, following the procedure described in Sect. 16.3.1.1, while the prediction accuracy metrics were calculated using another set of 100 instances for which performance characteristics were obtained by simulation and then compared to their predicted counterparts. In both sets of instances Monte Carlo simulations were performed, exploiting the statistical models of the used technol-

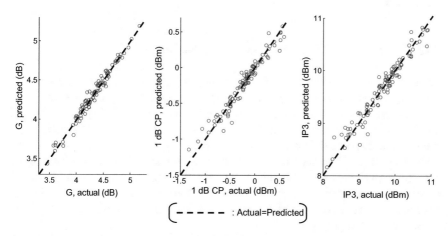

Fig. 16.12 Graphical representation of performance prediction accuracy (Central state of operation)

Table 16.5 Prediction accuracy metrics for different mixer states

State	G		1dB CP		IP3	
	R^2	*RRMSE*	R^2	*RRMSE*	R^2	*RRMSE*
S1	0.9769	0.1521	0.8861	0.3375	0.8941	0.3254
S2	0.9761	0.1545	0.9597	0.2008	0.9597	0.2008
S3	0.9731	0.1641	0.9464	0.2314	0.9438	0.2371
S4	0.9651	0.1868	0.9248	0.2742	0.8869	0.3363
S5	0.8518	0.3849	0.9454	0.2336	0.9165	0.2889

ogy. These models allow statistical variations of device parameters based on real IC manufacturing process variations and device mismatches.

In Table 16.4 we observe that while IF_+ and IF_- alone ($O = (O1,O2)$) provide a satisfactory accuracy in Gain predictions, they fail to achieve similarly high confidence levels for 1dB CP and IP3. In [13] the ΔIF ($= IF_+ - IF_-$) observable has been proposed for performance prediction. However, according to Table 16.4 the ($IF_+ - IF_-$) observable alone ($O = O1$–$O2$), fails to provide acceptable confidence levels for the specific mixer design. Furthermore, although the introduction of V_{tail} ($O = (O1,O2,O3)$) increases the prediction scores for all three target performance characteristics, only when more than one voltage triads are jointly used (i.e. by measurements from more than one states of operation), high determination coefficients are achieved, as shown in Table 16.4 and graphically represented in Fig. 16.12 for two voltage triads ($O = (O1,O2,O3,O4,O5,O6)$), following the multi-state measurement approach which we have proposed in the previous section. Moreover, the predictive models obtained by the proposed multi-state approach maintain a similarly high accuracy for predictions regarding all states of the mixer's operation, as indicated in Table 16.5.

16.3.2.3 Extended Observables and Variation Bands

As already proven, the mixer's differential output ($\Delta IF = IF_+ - IF_-$) is not adequate for the prediction of PCs if used alone as a test observable; however earlier research has shown that it provides effective defect detection [13, 21]. Indeed, ΔIF, IF_+ and IF_- are dependent parameters since $\Delta IF = IF_+ - IF_-$. This correlation allows us to improve defect detection by introducing ΔIF as an extended observable. Figure 16.13 illustrates this improvement as a transition from a coarse 2D defect filter (using IF_+ and IF_- as inputs) to its finer 3D counterpart (where ΔIF is used as an additional input). In conclusion, instead of the vector of voltage observables O we use its extended counterpart E which includes the mixer's differential output, as shown in Table 16.6.

The eight voltage elements shown in Table 16.6 together with the three predicted performance characteristics for the central state S_c result in a total of eleven variation

Table 16.6 List of extended observable vector (E) elements

Element ID	DC voltage	Mixer state
E1	IF_+	$S3^a$
E2	IF_-	$S3^a$
E3	V_{tail}	$S3^a$
E4	IF_+	S5
E5	IF_-	S5
E6	V_{tail}	S5
E7	$IF_+ - IF_-$	$S3^a$
E8	$IF_+ - IF_-$	S5

[a] Central state ($S_c = S3$)

bands to be constructed for defect detection, according to the procedure described in Sect. 16.3.1.2. These variation bands are obtained exploiting the data from the 500 3σ Monte Carlo simulations used in Sect. 16.3.2.2 for the defect-free mixer case.

16.3.2.4 Defect Detection

The defects under consideration are hard (a) opens in circuit branches, (b) shorts between device terminals and (c) bridgings between adjacent circuit nodes. While shorts and bridgings have been simulated using ideal resistive shorts ($1m\Omega$), the simulation of opens (broken lines) has followed the capacitive open fault model, assuming a capacitance value equal to 0.1 fF [1]. All possible defects (38 opens, 43 shorts, 13 bridgings) have been simulated in the presence of process variations and

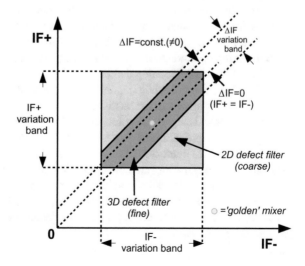

Fig. 16.13 Defect detection improvement due to the extension of observables

Table 16.7 Defect detection probability results

Type of defect	Defect detection probability (%)		
	Defect-oriented (E)	Alternate (PC)	Combined (E, PC)
Shorts	78.58	99.14	100
Opens	100	89.26	100
Bridgings	100	70.15	100
Overall	*90.20*	*91.14*	*100*

device mismatches (100, 3σ Monte Carlo iterations per defect), setting the mixer in both selected states, namely the central state ($S_c = S3$) and the maximum tail current state ($S5$). Then, each defect has been characterized as detectable or not according to the procedure described in Sect. 16.3.1.2.

Defective mixer instances with values for a particular element of E or PC inside the corresponding variation band are considered undetectable when this particular element is examined. However, the final decision regarding the detectability of a specific defect is derived after checking its behavior with respect to the variation bands of all elements of E and PC (eleven variation bands).

A graphical representation of this procedure is shown in Fig. 16.14, where all defects are depicted one over another in the form of a stack, where the horizontally distributed traces (horizontal clouds of points) consist of the 100 Monte Carlo variants obtained for each defect case. Since it would be meaningless to depict all variation bands together, these have been unified to a common one by normalizing all minimum and maximum band boundary values to -1 and 1, respectively and by scaling simulation data obtained for all defect traces, accordingly. Furthermore, among the eleven points that correspond to the elements of E and PC for a specific simulation run regarding a specific defect, only the point with the maximum absolute distance from the center of the common variation band has been depicted, for the sake of simplicity. The choice of this point is further justified as it is most likely to fall outside the common variation band and, thus, it is the best candidate to provide detection of the corresponding defect.

Defect detection probabilities, defined as the percentage of trace points located outside the common variation band, have been calculated and the results are summarized in Table 16.7, where columns labeled "Defect Oriented" and "Alternate" correspond to the probabilities obtained by the extended observables (E) and the predicted performance characteristics (PC), respectively, while "Combined" indicates the result obtained by using all eleven variation bands. According to these results, all defects, according to the defect model under consideration, can be detected successfully since a detection probability of 100% is provided. This ensures that all mixer instances entering the succeeding calibration phase are free of defects and, hence, candidate for calibration.

Fig. 16.14 Defect stacks
(Normalized bands, all
defects under consideration)

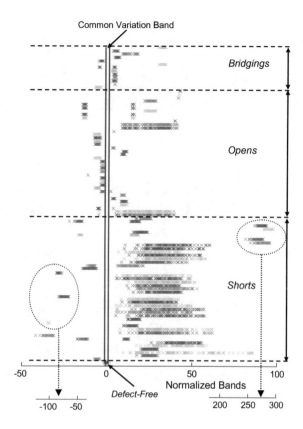

16.3.2.5   Calibration

To evaluate the efficiency of the calibration procedure, a set of 100 defect-free mixer
instances were generated by Monte Carlo 6σ simulations. As a result of the selection
of 6σ parameter distributions, some of the generated instances will present extreme
perturbations of process parameters and device mismatches. For this reason it is
expected that these instances will be filtered-out during the defect detection phase.
Indeed, from the specific set of 100 instances, 30 are excluded from the calibration
procedure beforehand, since they are classified as outliers. The calibration procedure
is designed following the principle described in Sect. 16.3.1.4 and shown in Fig. 16.7,
in order to shift the circuit's performance characteristics as close as possible to the
center of their corresponding specification ranges, yet designers might choose other
approaches, according to their needs. When the procedure concludes to a state (S_i) for
which successful calibration is possible, the corresponding control bit values shown
in Table 16.2 are applied to the variable resistor inputs, in the bias circuitry, and the
circuit operates in that specific state ever since. If such a state does not exist, the
procedure reports a calibration failure and the circuit under test should be considered
as not usable.

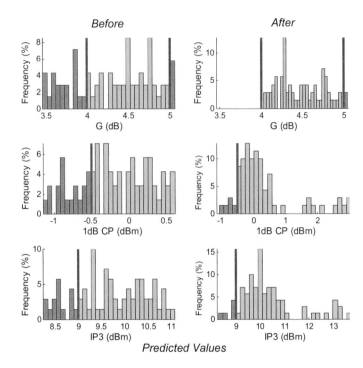

Fig. 16.15 Distributions of the predicted performance characteristics, before and after the calibration phase

Let us assume that specifications for the mixer under consideration (see Table 16.1 for typical values) require: 4dB ≤ G ≤ 5dB, 1dB CP ≥ −0.5dBm and IP3 ≥ 9dBm. The calibration procedure provides the results based on predicted PC values, as shown in Fig. 16.15, where the prediction-based performance characteristic distributions are presented. Bold vertical lines indicate the edges of the PC values' space as these are set by the specifications. One should note that, according to the discussion in Sect. 16.3.1.4, there is a small number of instances for which the calibration procedure declares inability to find any state of operation at which all specifications are met simultaneously, and thus remain out of specs.

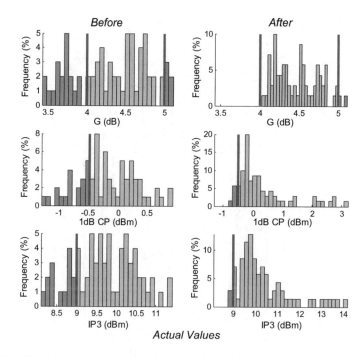

Fig. 16.16 Distributions for the actual performance characteristics, before and after the calibration phase

Figure 16.16, which illustrates the actual efficiency of the calibration procedure, corresponds to Fig. 16.15 where actual PC data for the same circuit's instances are used, instead. Due to the high accuracy of the regression models used, estimated calibration results based on performance prediction are found to closely match the actual results achieved, as observed by the comparison of the aforementioned figures.

For the mixer's specifications under consideration, 48.57% of the instances involved in the calibration procedure are found to comply with the specifications before calibration. After calibration, the amount of compliant mixer instances corresponds to 88.57%, which indicates a +82.35% relative yield improvement. Note that this relative yield improvement maintains the same value either if it is calculated over the set of instances that successfully pass the defect-filtering procedure or over the original set of 100 instances under consideration.

By modifying the edge values set by the specifications the impact of calibration on parametric yield may increase or decrease, as we can conclude from the results shown in Table 16.8, from which the calibration efficiency may be evaluated for various sets of specifications. As the latter become less strict, relative yield improvement degrades since a large amount of instances already complies with the specifications, before calibration. Calibration for the remaining instances is more difficult to achieve, since their PCs present severe deviations from their nominal values and, furthermore, their calibration might become contradicting, since the convergence of one PC towards its

Table 16.8 Impact of calibration on 6σ parametric yield for various specifications (Instances undergoing calibration: 70. Yield values in *italics* correspond to the total yield, calculated over the original set of 100 Monte Carlo instances)

Case ID	Specifications									Yield		
	G		1dB CP		IP3					Before	After	Improvement
	(dB)		(dBm)		(dBm)					[1]	[2]	$\left\{\frac{[2]-[1]}{[1]}\right\}$
	min	max	min	max		min	max			(%)	(%)	(%)
1	4	5	−0.5	0.5		9	10			25.71 *18.00*	48.57 *34.00*	+88.89
2	4	5	−0.5	+∞		9	+∞			48.57 *34.00*	88.57 *62.00*	+82.35
3	4	+∞	−0.5	+∞		9	+∞			54.29 *38.00*	90.00 *63.00*	+65.79
4	4	+∞	−∞	+∞		−∞	+∞			71.43 *50.00*	100 *70.00*	+40.00
5	−∞	+∞	−0.5	+∞		−∞	+∞			74.29 *52.00*	90.00 *63.00*	+21.15
6	−∞	+∞	−∞	+∞		9	+∞			80.00 *56.00*	94.29 *66.00*	+17.86

nominal range might shift another PC outside its specifications. However, this issue may be addressed by increasing the density of the mixer's states of operation in order to obtain a finer coverage for the ranges of interest.

16.4 Improvements

Alternate Test predictive models are constructed during the preliminary training phase, before the main testing procedure is carried out. Training relies on a relatively small set of fully-characterized circuits, the information richness of which determines the predictive strength of the constructed models. Such a training set should be, ideally, large enough and representative of the manufacturing process (i.e., consisting of real circuits corresponding to different lots, obtained over a long period of time) [41], which is usually not the case at the time of test development. More specifically, the statistical properties of the training set are expected to differ from those of the much larger set of circuits that are subject to test, especially those that correspond to extreme process perturbations (i.e., process corners).

In order to address the problem described above, several solutions have already been proposed. In [41], synthetic data having the statistical properties of real data are used to enrich the predictive model training sets and, consequently, their predictive strength. In [32], performance prediction is exploited in order to calibrate circuits that violate their specifications. Furthermore, several techniques have been developed that leverage prior knowledge from simulation data and fuse this information with data from a few real circuits to construct accurate regression models across the whole design space [23, 25]. To the same direction, statistically treating the training set in a weighted fashion, in order to derive more accurate models at the most sensitive regions of a circuit's performance, namely the regions that are in proximity to the specification limits, allows significant prediction accuracy improvement [26].

Finally, feature selection algorithms have been applied to determine the optimum Alternate Test observables from the test response, that maintain both high prediction accuracy and low test complexity [22, 27].

References

1. E. Acar, S. Ozev, Defect-based RF testing using a new catastrophic fault model, in *Proceedings of the IEEE International Test Conference (ITC)* (2005), pp. 429–437
2. A. Ahmadi, M.M. Bidmeshki, A. Nahar, B. Orr, M. Pas, Y. Makris, A machine learning approach to fab-of-origin attestation, in *2016 IEEE/ACM International Conference on Computer-Aided Design (ICCAD)* (2016), pp. 1–6. https://doi.org/10.1145/2966986.2966992
3. S.S. Akbay, A. Halder, A. Chatterjee, D. Keezer, Low-cost test of embedded RF/analog/mixed-signal circuits in SOPs. IEEE Trans. Comput.-Aided Des. Integr. Circuits Syst. **27**(2), 352–363 (2004)
4. M. Allen, *Understanding Regression Analysis* (Springer, US, 1997)

5. F.D. Bernardinis, M.I. Jordan, A. SangiovanniVincentelli, Support vector machines for analog circuit performance representation, in *Proceedings 2003. Design Automation Conference* (2003), pp. 964–969. https://doi.org/10.1145/775832.776074
6. S. Bhattacharya, A. Chatterjee, Use of embedded sensors for built-in-test RF circuits., in *Proceedings of International Test Conference (ITC)* (2004), pp. 801–809
7. S. Bou-Sleiman, M. Ismail, Built-in-self-test and digital self-calibration for RF SoCs, in *SpringerBriefs in Electrical and Computer Engineering* (Springer, New York, 2011)
8. T. Das, A. Gopalan, C. Washburn, P. Mukund, Self-calibration of input-match in RF front-end circuitry. IEEE Trans. Circuits Syst. II **52**(12), 821–825 (2005)
9. K. Dufrene, R. Weigel, A novel IP2 calibration method for low-voltage downconversion mixers, in *Proceedings of the IEEE International Symposium on Radio Frequency Integrated Circuits (RFIC), San Jose, CA, USA* (2006), pp. 292–295
10. J.H. Friedman, Multivariate adaptive regression splines. Ann. Stat. **19**, 1–141 (1991)
11. E. Garcia-Moreno, K. Suenaga, R. Picos, S. Bota, M. Roca, E. Isern, Predictive test strategy for CMOS RF mixers. Integr. VLSI J. **42**, 95–102 (2009)
12. A. Goyal, M. Swaminathan, A. Chatterjee, A novel self-healing methodology for RF amplifier circuits based on oscillation principles, in *Proceedings of the IEEE Design Automation & Test in Europe (DATE)* (2009), pp. 1656–1661
13. A. Goyal, M. Swaminathan, A. Chatterjee, Self-calibrating embedded RF down-conversion mixers, in *Proceedings of the IEEE Asian Test Symposium (ATS), Taichung, Taiwan* (2009), pp. 249–254
14. D. Han, B.S. Kim, A. Chatterjee, DSP-driven self-tuning of RF circuits for process-induced performance variability. IEEE Trans. VLSI Syst. **18**(2), 305–314 (2010)
15. K. Huang, H.G. Stratigopoulos, S. Mir, Fault diagnosis of analog circuits based on machine learning, in *2010 Design, Automation Test in Europe Conference Exhibition (DATE 2010)* (2010), pp. 1761–1766. https://doi.org/10.1109/DATE.2010.5457099
16. R. Khereddine, L. Abdallah, E. Simeu, S. Mir, F. Cenni, Adaptive logical control of RF LNA performances for efficient energy consumption, in *Proceedings of the IFIP/IEEE International Conference on Very Large Scale Integration (VLSI-SoC)* (2010), pp. 518–525
17. N. Kupp, P. Drineas, M. Slamani, Y. Makris, Confidence estimation in non-RF to RF correlation-based specification test compaction, in *Proceedings of the 13th European Test Symposium (ETS)* (2008), pp. 35–40
18. N. Kupp, Y. Makris, Integrated optimization of semiconductor manufacturing: a machine learning approach, in *2012 IEEE International Test Conference* (2012), pp. 1–10. https://doi.org/10.1109/TEST.2012.6401531
19. B. Li, P.D. Franzon, Machine learning in physical design, in *2016 IEEE 25th Conference on Electrical Performance of Electronic Packaging and Systems (EPEPS)* (2016), pp. 147–150. https://doi.org/10.1109/EPEPS.2016.7835438
20. I. Liaperdos, L. Dermentzoglou, A. Arapoyanni, Y. Tsiatouhas, Fault detection in RF mixers combining defect-oriented and alternate test strategies, in *Conference on Design of Circuits and Integrated Systems (DCIS)* (2011)
21. I. Liaperdos, L. Dermentzoglou, A. Arapoyanni, Y. Tsiatouhas, A test technique and a BIST circuit to detect catastrophic faults in RF mixers, in *Conference on Design and Technology of Integrated Systems in the Nanoscale Era (DTIS)* (2011)
22. J. Liaperdos, A. Arapoyanni, Y. Tsiatouhas, Adjustable RF mixers' alternate test efficiency optimization by the reduction of test observables. IEEE Trans. Comput.-Aided Des. Integr. Circuits Syst. **32**(9), 1383–1394 (2013). https://doi.org/10.1109/TCAD.2013.2255128
23. J. Liaperdos, A. Arapoyanni, Y. Tsiatouhas, A method to adjust the accuracy of analog/RF alternate tests, in *Proceedings 28th Conference on Design of Circuits and Integrated Systems (DCIS)* (2013)
24. J. Liaperdos, A. Arapoyanni, Y. Tsiatouhas, A test and calibration strategy for adjustable RF circuits. Analog. Integr. Circuits Signal Process. **74**(1), 175–192 (2013). https://doi.org/10.1007/s10470-012-9981-x

25. J. Liaperdos, H.G. Stratigopoulos, L. Abdallah, Y. Tsiatouhas, A. Arapoyanni, X. Li, Fast deployment of alternate analog test using Bayesian model fusion, in Proceedings of Conference on Design, Automation and Test in Europe (DATE), pp. 1030–1035 (2015)
26. J. Liaperdos, A. Arapoyanni, Y. Tsiatouhas, Improved alternate test accuracy using weighted training sets, in *2016 Conference on Design of Circuits and Integrated Systems (DCIS)* (2016), pp. 1–6. https://doi.org/10.1109/DCIS.2016.7845263
27. J. Liaperdos, A. Arapoyanni, Y. Tsiatouhas, State reduction for efficient digital calibration of analog/RF integrated circuits. Analog. Integr. Circuits Signal Process. **90**(1), 65–79 (2017). https://doi.org/10.1007/s10470-016-0880-4
28. C. Lim, Y. Xue, X. Li, R.D. Blanton, M.E. Amyeen, Diagnostic resolution improvement through learning-guided physical failure analysis, in *2016 IEEE International Test Conference (ITC)* (2016), pp. 1–10. https://doi.org/10.1109/TEST.2016.7805824
29. Q. Ma, Y. He, F. Zhou, A new decision tree approach of support vector machine for analog circuit fault diagnosis. Analog. Integr. Circuits Signal Process. **88**(3), 455–463 (2016). https://doi.org/10.1007/s10470-016-0775-4
30. S.A. Maas, *Microwave Mixers* (Artech House Publishers, 1993)
31. D. Maliuk, H. Stratigopoulos, Y. Makris, Machine learning-based BIST in analog/RFICs, in *Mixed-Signal Circuits, Devices, Circuits, and Systems*, ed. by T. Noulis (CRC Press, 2016), pp. 266–290
32. V. Natarajan, S. Sen, S.K. Devarakond, A. Chatterjee, A holistic approach to accurate tuning of RF systems for large and small multiparameter perturbations, in *Proceedings of the 28th VLSI Test Symposium (VTS)* (2010), pp. 331–336
33. S.J. Park, H. Yu, M. Swaminathan, Preliminary application of machine-learning techniques for thermal-electrical parameter optimization in 3-d ic, in *2016 IEEE International Symposium on Electromagnetic Compatibility (EMC)* (2016), pp. 402–405. https://doi.org/10.1109/ISEMC.2016.7571681
34. S.J. Park, B. Bae, J. Kim, M. Swaminathan, Application of machine learning for optimization of 3-D integrated circuits and systems. IEEE Trans. Very Large Scale Integr. (VLSI) Syst. **25**(6), 1856–1865 (2017). https://doi.org/10.1109/TVLSI.2017.2656843
35. C.W. Pui, G. Chen, Y. Ma, E.F.Y. Young, B. Yu, Clock-aware ultrascale FPGA placement with machine learning routability prediction: (invited paper), in *2017 IEEE/ACM International Conference on Computer-Aided Design (ICCAD)* (2017), pp. 929–936. https://doi.org/10.1109/ICCAD.2017.8203880
36. X. Ren, V.G. Tavares, R.D.S. Blanton, Detection of illegitimate access to JTAG via statistical learning in chip, in *2015 Design, Automation Test in Europe Conference Exhibition (DATE)* (2015), pp. 109–114
37. S. Rodriguez, A. Rusu, L.R. Zheng, M. Ismail, Digital calibration of gain and linearity in a CMOS RF mixer, in *Proceedings of the IEEE International Symposium on Circuits and Systems (ISCAS), Seattle, USA* (2008), pp. 1288–1291
38. M. Sachdev, J. de Gyvez, Defect-oriented testing for nano-metric CMOS VLSI circuits, in *Frontiers in Electronic Testing* (Springer, 2007)
39. P. Song, Y. He, W. Cui, Statistical property feature extraction based on FRFT for fault diagnosis of analog circuits. Analog. Integr. Circuits Signal Process. **87**(3), 427–436 (2016). https://doi.org/10.1007/s10470-016-0721-5
40. H.G. Stratigopoulos, S. Mir, E. Acar, S. Ozev, Defect filter for alternate RF test, in *Proceedings of the IEEE European Test Symposium* (2009), pp. 161–166
41. H.G. Stratigopoulos, S. Mir, Y. Makris, Enrichment of limited training sets in machine-learning-based analog/RF test, in *Conference on Design, Automation and Test in Europe (DATE)* (2009), pp. 1668–1673
42. K. Suenaga, R. Picos, S. Bota, M. Roca, E. Isern, E. Garcia-Moreno, Built-in test strategy for CMOS RF mixers, in *Conference on Design of Circuits and Integrated Systems (DCIS)* (2005)
43. H.M. Torun, M. Swaminathan, A.K. Davis, M.L.F. Bellaredj, A global Bayesian optimization algorithm and its application to integrated system design. IEEE Trans. Very Large Scale Integr. (VLSI) Syst. **26**(4), 792–802 (2018). https://doi.org/10.1109/TVLSI.2017.2784783

44. P. Variyam, S. Cherubal, A. Chatterjee, Prediction of analog performance parameters using fast transient testing. IEEE Trans. Comput.-Aided Des. Integr. Circuits Syst. **21**(3), 349–361 (2002)
45. A.S.S. Vasan, M.G. Pecht, Electronic circuit health estimation through kernel learning. IEEE Trans. Ind. Electron. **65**(2), 1585–1594 (2018). https://doi.org/10.1109/TIE.2017.2733419
46. R. Voorakaranam, S. Cherubal, A. Chatterjee, A signature test framework for rapid production testing of RF circuits, in *Proceedings of Design, Automation and Test in Europe Conference and Exhibition (DATE)* (2002), pp. 186–191
47. Y. Zhu, J. Xiong, Modern big data analytics for "old-fashioned" semiconductor industry applications, in *2015 IEEE/ACM International Conference on Computer-Aided Design (ICCAD)* (2015), pp. 776–780. https://doi.org/10.1109/ICCAD.2015.7372649

Printed in the United States
By Bookmasters